Applications of Symmetry in Discrete Mathematics

Applications of Symmetry in Discrete Mathematics

Editor: Isla Harvey

MURPHY & MOORE
www.murphy-moorepublishing.com

www.murphy-moorepublishing.com

MURPHY & MOORE

Cataloging-in-Publication Data

Applications of symmetry in discrete mathematics / edited by Isla Harvey.
p. cm.
Includes bibliographical references and index.
ISBN 978-1-63987-746-1
1. Symmetry (Mathematics). 2. Discrete mathematics. 3. Group theory.
4. Numerical analysis. I. Harvey, Isla.
QA174.7.S96 A67 2023
516.1--dc23

Murphy & Moore Publishing
1 Rockefeller Plaza,
New York City,
NY 10020, USA

ISBN 978-1-63987-746-1

Contents

Preface

Discrete mathematics refers to the study of mathematical structures that are countable or otherwise separable or unique. Logical statements, combinations and graphs are examples of discrete structures. Symmetry is a proportionate and balanced resemblance between two halves of an object. It implies that one half is the mirror image of the other. Discrete symmetry in mathematics is a symmetry that explains non-continuous changes in a system. Symmetry is frequently used in discrete mathematics to show relationships between two or more elements of a set, where the first element relates to the second element and vice-versa. This book explores all the important aspects of discrete mathematics and symmetry in the present day scenario. It presents researches and studies performed by experts across the globe. This book, with its detailed analyses and data, will prove immensely beneficial to professionals and students involved in this area of mathematics at various levels.

This book is a result of research of several months to collate the most relevant data in the field.

When I was approached with the idea of this book and the proposal to edit it, I was overwhelmed. It gave me an opportunity to reach out to all those who share a common interest with me in this field. I had 3 main parameters for editing this text:

1. Accuracy – The data and information provided in this book should be up-to-date and valuable to the readers.

2. Structure – The data must be presented in a structured format for easy understanding and better grasping of the readers.

3. Universal Approach – This book not only targets students but also experts and innovators in the field, thus my aim was to present topics which are of use to all.

Thus, it took me a couple of months to finish the editing of this book.

I would like to make a special mention of my publisher who considered me worthy of this opportunity and also supported me throughout the editing process. I would also like to thank the editing team at the back-end who extended their help whenever required.

Editor

1

The Characteristic Polynomials of Symmetric Graphs

Nafaa Chbili [1,*], Shamma Al Dhaheri [1], Mei Y. Tahnon [2] and Amna A. E. Abunamous [1]

1 Department of Mathematical Sciences, United Arab Emirates University, Al Ain 15551, UAE; 201403947@uaeu.ac.ae (S.A.D.); 201250600@uaeu.ac.ae (A.A.E.A.)
2 College of Engineering, United Arab Emirates University, Al Ain 15551, UAE; 201606519@uaeu.ac.ae
* Correspondence: nafaachbili@uaeu.ac.ae

Abstract: In this paper, we study the way the symmetries of a given graph are reflected in its characteristic polynomials. Our aim is not only to find obstructions for graph symmetries in terms of its polynomials but also to measure how faithful these algebraic invariants are with respect to symmetry. Let p be an odd prime and Γ be a finite graph whose automorphism group contains an element h of order p. Assume that the finite cyclic group generated by h acts semi-freely on the set of vertices of Γ with fixed set F. We prove that the characteristic polynomial of Γ, with coefficients in the finite field of p elements, is the product of the characteristic polynomial of the induced subgraph $\Gamma[F]$ by one of $\Gamma \setminus F$. A similar congruence holds for the characteristic polynomial of the Laplacian matrix of Γ.

Keywords: automorphism group; adjacency matrix; characteristic polynomial

---

1. Introduction

The purpose of this paper is to address the question of how the symmetries of a graph are reflected in its algebraic invariants. In other words, we study the interaction between the algebraic invariants of a graph and its automorphism group. Indeed, our study will be limited to discussing how the information about graph periodicity is carried by its characteristic polynomials. This question is primarily motivated by the results about knots and spatial graph periodicity obtained in terms of their quantum invariants. We shall start by fixing some notation. Let n be a positive integer and $V = \{v_1, \ldots, v_n\}$ be a finite set. A weight on V is a symmetric function w: $V \times V \rightarrow \mathbb{Z}_+ \cup \{0\}$. The pair $\Gamma = (V, w)$ is called a weighted graph. Indeed, a weighted graph can be seen as a finite graph possibly with multiple edges and loops. In particular, if the graph Γ is simple, then the corresponding symmetric function w is defined by $w(u, v) = 1$ if u and v are adjacent and 0 otherwise. Throughout this paper, wherever no confusion may arise, we will simplify notation and use the term graph instead of weighted graph. Given a graph Γ with vertex set $V = \{v_1, \ldots, v_n\}$, the adjacency matrix of Γ is the n-square matrix $A(\Gamma) = [w_{ij}]$ where $w_{ij} = w(v_i, v_j)$. Since the weight function w is symmetric, so is the matrix $A(\Gamma)$. It is worth mentioning here that while the matrix $A(\Gamma)$ depends on the ordering of the vertices, its characteristic polynomial, defined as $det(xI_n - A(\Gamma))$, is known to be independent of that order. It is denoted hereafter by $\varphi_\Gamma(x)$. On the other hand, if $D(\Gamma) = [d_{ij}]$ is the degree matrix of Γ, which is the diagonal n-square matrix defined by $d_{ii} = deg(v_i)$, then the Laplacian matrix of Γ is defined by $L(\Gamma) = D(\Gamma) - A(\Gamma)$. The characteristic polynomial of $L(\Gamma)$ is also independent of the ordering of the vertices and is denoted hereafter by $\psi_\Gamma(x)$. Both polynomials $\varphi_\Gamma(x)$ and $\psi_\Gamma(x)$ are well-known graph invariants which carry important information about the graph.

An automorphism of a weighted graph Γ is a permutation σ of the set of vertices V such that for any pair of vertices u and v, we have $w(\sigma(u), \sigma(v)) = w(u, v)$. The set of all automorphisms of Γ forms a group denoted here by $Aut(\Gamma)$.

Let $p \geq 2$ be an integer. A graph Γ is said to be *p-periodic* if its automorphism group $Aut(\Gamma)$ contains an element h such that $h^p = Id$ and for any vertex v, we have $h^i(v) \neq v$ for all $1 \leq i \leq p-1$. In other words, the finite cyclic group $\mathbb{Z}_p$ acts freely on the set of vertices of the graph. A graph Γ is said to be *semi-free p-periodic* if $Aut(\Gamma)$ contains an element h such that $h^p = Id$ and the set of fixed vertices by h is non empty. This set of fixed vertices will be denoted hereafter by F. It is worth mentioning here that for simplicity we will sometimes abuse notation and denote by F the induced subgraph $\Gamma[F]$ as well. Notice that if Γ is semi-free p-periodic then $\Gamma \setminus F$ is p-periodic.

Let r and s be two positive integers. The complete 4-partite graph $K_{r,s,p,p}$ is semi-free p-periodic with fixed subgraph F the complete bipartite graph $K_{r,s}$. The graph $\Gamma \setminus F$ is the complete bipartite graph $K_{p,p}$.

Given a p-periodic weighted graph Γ, we define its quotient graph $\overline{\Gamma}$ as the weighted graph $(\overline{V}, \overline{w})$, where $\overline{V}$ is the quotient set of V under the action of h and $\overline{w}$ is the weight function on $\overline{V}$ defined by $\overline{w}(\overline{u}, \overline{v}) = \sum_{i=0}^{p-1} w(u, h^i(v))$, see [1].

The natural question of how the information about symmetries of graphs are carried by their algebraic invariants has been subject to relatively extensive study. In particular, several results about the characteristic polynomial of periodic graphs and graph coverings in general have been obtained, see [2–4] for instance. In [1], the first author suggested a more elementary approach by considering these polynomials with coefficients in the finite field of p elements $\mathbb{F}_p$. Actually, he studied the characteristic polynomial of p-periodic graphs and proved that $\varphi_\Gamma(x)$ is determined by the characteristic polynomial of the quotient graph $\overline{\Gamma}$. More precisely, $\varphi_\Gamma(x) \equiv \varphi_{\overline{\Gamma}}(x^p)$ in $\mathbb{F}_p[x]$. Furthermore, the polynomial $\psi_\Gamma(x)$ is divisible by x^p in $\mathbb{F}_p[x]$. Other necessary conditions for a graph to be p-periodic have been also obtained in terms of the Tutte polynomial and its generalizations [1,5]. In this paper, we discuss the case of semi-free actions and show that some similar congruences hold.

Theorem 1. *Let p be an odd prime and Γ be a semi-free p-periodic graph with a fixed subgraph F. Then the following congruences hold in $\mathbb{F}_p[x]$.*

$$\varphi_\Gamma(x) \equiv \varphi_F(x)\varphi_{\Gamma\setminus F}(x) \equiv \varphi_F(x)\varphi_{\overline{\Gamma\setminus F}}(x^p).$$

Example 1. *The complete 4-partite graph $K_{2,2,3,3}$ admits a semi-free action of $\mathbb{Z}_3$, with the 4 vertices of degree 8 fixed. Hence $F = K_{2,2}$ and $\Gamma \setminus F = K_{3,3}$. Computations of the characteristic polynomials show that $\varphi_{K_{2,2,3,3}}(x) = -108x^6 - 120x^7 - 37x^8 + x^{10}$, $\varphi_{K_{2,2}}(x) = -4x^2 + x^4$ and $\varphi_{K_{3,3}}(x) = -9x^4 + x^6$. Thus, the congruences given in Theorem 1 hold modulo 3.*

Theorem 2. *Let p be an odd prime and Γ be a semi-free p-periodic graph with a fixed subgraph F. Then $\psi_\Gamma(x)$ is divisible by $\psi_F(x)$ with quotient in $\mathbb{F}_p[x^p]$.*

Example 2. *For the complete 4-partite graph $K_{2,2,3,3}$, we have modulo 3: $\psi_{K_{2,2,3,3}}(x) \equiv 2x + 2x^2 + x^3 + 2x^5 + x^6 + 2x^8 + x^9 + x^{10}$; $\psi_{K_{2,2}}(x) \equiv 2x + 2x^2 + x^3 + x^4$ and $\psi_{K_{3,3}}(x) \equiv x^6$. It can be easily checked that $\psi_{K_{2,2,3,3}}(x) \equiv \psi_{K_{2,2}}(x)\psi_{K_{3,3}}(x-1)$ modulo 3. Hence, the condition given by Theorem 2 is satisfied.*

2. Block Circulant Matrices

In this section we will introduce block circulant matrices and briefly explain how to compute their characteristic polynomials. More details can be found in [6]. Let p and s be two positive integers.

For $0 \le i \le p-1$, let A_i be a s-square matrix. The block circulant matrix $C = C(A_0, \dots, A_{p-1})$ is the ps-square matrix of the form:

$$\begin{bmatrix} A_0 & A_1 & A_2 & \dots & A_{p-1} \\ A_{p-1} & A_0 & A_1 & \dots & A_{p-2} \\ \vdots & \vdots & \vdots & \vdots & \vdots \\ A_1 & A_2 & A_3 & \dots & A_0 \end{bmatrix}.$$

The characteristic polynomial of the circulant matrix C, with coefficients in the finite filed $\mathbb{F}_p$, satisfies the following property.

Lemma 1. *Let p be an odd prime and $C = C(A_0, \dots, A_{p-1})$ be a block circulant matrix. Then, $\varphi_C(x) \in \mathbb{F}_p[x^p]$.*

Proof. Let ζ be a pth primitive root of unity. For $0 \le k \le p-1$, we define the s-square matrix $T_k = \sum_{j=0}^{p-1} \zeta^{kj} A_j$. Let $\varphi_{T_k}(x)$ denotes the characteristic polynomial of the matrix T_k. The characteristic polynomial of the circulant matrix C is $\varphi_C(x) = \prod_{k=0}^{p-1} \varphi_{T_k}(x)$, see [6]. Since p is prime, the following homomorphism is well defined:

$$\begin{array}{rccc} f_p: & \mathbb{Z}[\zeta] & \longrightarrow & \mathbb{Z}_p \\ & \sum n_i \zeta^i & \longmapsto & \overline{\sum n_i}. \end{array}$$

Furthermore, it extends to a homomorphism of rings $f_p : \mathbb{Z}[\zeta][x] \longrightarrow \mathbb{Z}_p[x]$. Notice that for all $0 \le k \le p-1$ we have $f_p(\varphi_{T_k}(x)) = f_p(\varphi_{T_0}(x))$. Thus $f_p(\varphi_C(x)) = f_p(\varphi_{T_0}(x))^p$. In conclusion, $\varphi_C(x)$ is congruent to $\varphi_{T_0}(x^p)$ modulo p. This completes the proof of the lemma. $\square$

Let Γ be a semi-free p-periodic graph with fixed set of vertices F. Assume that F is made up of r vertices $u_1, u_2, \dots, u_r$. Since p is prime and $\Gamma \setminus F$ is p-periodic, the finite cyclic group $\mathbb{Z}_p = \langle h \rangle$ acts freely on the set of vertices of $\Gamma \setminus F$. This set splits into a partition made up of s orbits, where s is the number of vertices of the quotient graph $\overline{\Gamma \setminus F}$. Since the action is free and p is prime, each orbit is made up of exactly p elements. Let us label the vertices of $\overline{\Gamma \setminus F}$ as $v_1, \dots, v_s$. Let π be the canonical surjection from $V \setminus F$ to $\overline{V \setminus F}$. We label the elements of $\pi^{-1}(v_i)$ as $v_i^0, \dots, v_i^{p-1}$, so that $h(v_i^k) = v_i^{k+1}$ if $0 \le k < p-1$ and $h(v_i^{p-1}) = v_i^0$. The order of the vertices of $\overline{\Gamma \setminus F}$ extends to a natural order of the vertices of $\Gamma \setminus F$, namely, $v_1^0, v_2^0, \dots, v_s^0, v_1^1, v_2^1, \dots, v_s^1, v_1^2, \dots, v_s^{p-1}$. Notice that for all $1 \le i \le r$, $1 \le k, k' \le s$ and $0 \le j \le p-1$, we have :

1. $w(u_i, h^m(v_k^j)) = w(u_i, v_k^j)$,
2. $w(h^m(v_{k'}^0), h^m(v_k^j)) = w(v_{k'}^0, v_k^j)$.

Consequently, if we consider the order of the vertices of Γ obtained by placing the vertices of F first then those of $\Gamma \setminus F$ in the order described above, then the adjacency matrix of Γ is of the following form:

$$A(\Gamma) = \left[\begin{array}{c|ccccc} A(F) & R & R & \dots & R \\ \hline R^t & A_0 & A_1 & \dots & A_{p-1} \\ R^t & A_{p-1} & A_0 & \dots & A_{p-2} \\ \vdots & \vdots & \vdots & \vdots & \vdots \\ R^t & A_1 & A_2 & \dots & A_0 \end{array}\right],$$

where $A(F)$ denotes the adjacency matrix of the subgraph F, R is an $r \times s$-matrix and for $0 \le i \le p-1$, A_i is a s-square matrix. Notice that the block circulant sub-matrix $(A_0, A_1, \dots, A_{p-1})$ is indeed the adjacency matrix of the periodic graph $\Gamma \setminus F$. For instance the adjacency matrix of the semi-free 3-periodic graph displayed in Figure 1 is of the form.

$$\left[\begin{array}{c|cc|cc|cc}
0 & 1 & 0 & 1 & 0 & 1 & 0 \\ \hline
1 & 0 & 2 & 0 & 0 & 0 & 1 \\
0 & 2 & 0 & 1 & 1 & 0 & 1 \\ \hline
1 & 0 & 1 & 0 & 2 & 0 & 0 \\
0 & 0 & 1 & 2 & 0 & 1 & 1 \\ \hline
1 & 0 & 0 & 0 & 1 & 0 & 2 \\
0 & 1 & 1 & 0 & 1 & 2 & 0
\end{array}\right].$$

The adjacency matrix of the semi-free 3-periodic graph $K_{2,2,3,3}$, see Figure 2, can be written in the form:

$$\left[\begin{array}{cccc|cc|cc|cc}
0 & 0 & 1 & 1 & 1 & 1 & 1 & 1 & 1 & 1 \\
0 & 0 & 1 & 1 & 1 & 1 & 1 & 1 & 1 & 1 \\
1 & 1 & 0 & 0 & 1 & 1 & 1 & 1 & 1 & 1 \\
1 & 1 & 0 & 0 & 1 & 1 & 1 & 1 & 1 & 1 \\ \hline
1 & 1 & 1 & 1 & 0 & 1 & 0 & 1 & 0 & 1 \\
1 & 1 & 1 & 1 & 1 & 0 & 1 & 0 & 1 & 0 \\ \hline
1 & 1 & 1 & 1 & 0 & 1 & 0 & 1 & 0 & 1 \\
1 & 1 & 1 & 1 & 1 & 0 & 1 & 0 & 1 & 0 \\ \hline
1 & 1 & 1 & 1 & 0 & 1 & 0 & 1 & 0 & 1 \\
1 & 1 & 1 & 1 & 1 & 0 & 1 & 0 & 1 & 0
\end{array}\right].$$

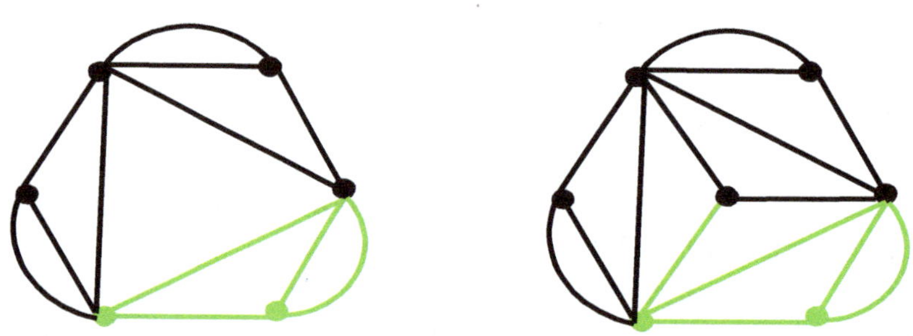

Figure 1. A 3-periodic graph (**left**) and a semi-free 3-periodic graph with a fixed vertex (**right**).

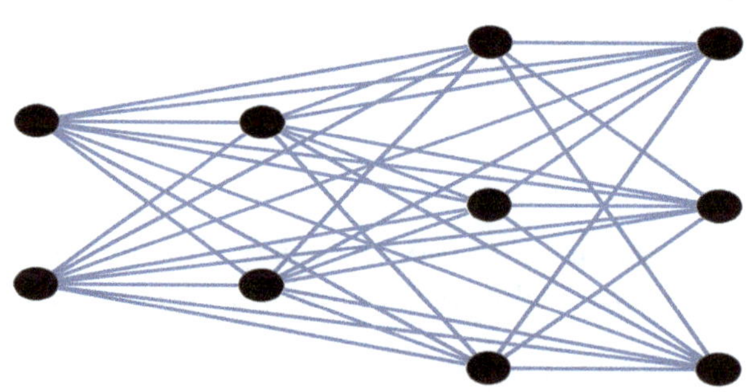

Figure 2. The complete 4-partite graph $K_{2,2,3,3}$.

3. Proofs

We will first prove Theorem 1. The proof of Theorem 2 will be based on similar arguments. Let us compute the characteristic polynomial of Γ, $det(A(\Gamma) - xI_n)$ by expanding with respect to the first row. The basic observation here is that since $w(u_1, h^m(v_k^j)) = w(u_1, v_k^j)$ for $1 \leq m \leq p$, the co-factors corresponding to each of these p entries are equal. Hence their contribution to the value of $det(A(\Gamma) - xI_n)$ will add to zero modulo p. We will prove that for the co-factors which correspond to deleting the first row of $A(\Gamma)$ and the columns $r+1, r+s+1, \ldots, r+(p-1)s+1$ of $A(\Gamma)$. The sub-matrices obtained by such operations will be denoted hereafter by $\Delta_{r+1}, \Delta_{r+s+1}, \ldots, \Delta_{r+(p-1)s+1}$, respectively. These sub-matrices are displayed below.

$$\Delta_{r+1} = \left[\begin{array}{c|cccc} A(F) & S' & S & \ldots & S \\ \hline R^t & A_0' & A_1 & \ldots & A_{p-1} \\ R^t & A'_{p-1} & A_0 & \ldots & A_{p-2} \\ \vdots & \vdots & \vdots & \vdots & \vdots \\ R^t & A_1' & A_2 & \ldots & A_0 \end{array}\right], \Delta_{r+s+1} = \left[\begin{array}{c|cccc} A(F) & S & S' & \ldots & S \\ \hline R^t & A_0 & A_1' & \ldots & A_{p-1} \\ R^t & A_{p-1} & A_0' & \ldots & A_{p-2} \\ \vdots & \vdots & \vdots & \vdots & \vdots \\ R^t & A_1 & A_2' & \ldots & A_0 \end{array}\right],$$

$$\ldots, \Delta_{r+(p-1)s+1} = \left[\begin{array}{c|cccc} A(F) & S & S & \ldots & S' \\ \hline R^t & A_0 & A_1 & \ldots & A'_{p-1} \\ R^t & A_{p-1} & A_0 & \ldots & A'_{p-2} \\ \vdots & \vdots & \vdots & \vdots & \vdots \\ R^t & A_1 & A_2 & \ldots & A_0' \end{array}\right]$$

where, S is the matrix obtained from R by deletion of the first row. Moreover, if M is a matrix, then we denote by M' the matrix obtained from M by deletion of the first column.

Let us explain how to transform the first matrix Δ_{r+1} into Δ_{r+s+1} using elementary operations on rows and columns. These operations are known to keep the determinant unchanged up to a sign. First, we interchange column $r+i$ and $r+s+i$ for all $1 \leq i \leq s-1$, then we apply $s-1$ column interchanges to move column $r+2s-1$ to column $r+s$, our matrix is then transformed into:

$$\left[\begin{array}{c|cccc} A(F) & S & S' & \ldots & S \\ \hline R^t & A_1 & A_0' & \ldots & A_{p-1} \\ R^t & A_0 & A'_{p-1} & \ldots & A_{p-2} \\ \vdots & \vdots & \vdots & \vdots & \vdots \\ R^t & A_2 & A_1' & \ldots & A_0 \end{array}\right].$$

Now, we shall rearrange rows using elementary row operations. Indeed, using $(p-1)s$ interchanges of rows we will be able to transform our matrix into:

$$\left[\begin{array}{c|cccc} A(F) & S & S' & \ldots & S \\ \hline R^t & A_2 & A_1' & \ldots & A_0 \\ R^t & A_1 & A_0' & \ldots & A_{p-1} \\ R^t & A_0 & A'_{p-1} & \ldots & A_{p-2} \\ \vdots & \vdots & \vdots & \vdots & \vdots \\ \Gamma & & & & \end{array}\right].$$

By interchanging column $r+i$ and column $r+p(s-1)+i-1$, for $1 \leq i \leq s$, we obtain:

$$\left[\begin{array}{c|cccc} A(F) & S & S' & \ldots & S \\ \hline R^t & A_0 & A_1' & \ldots & A_2 \\ R^t & A_{p-1} & A_0' & \ldots & A_1 \\ R^t & A_{p-2} & A'_{p-1} & \ldots & A_0 \\ \vdots & \vdots & \vdots & \vdots & \vdots \end{array}\right]$$

Finally, we can perform $(p-3)s$ column operations to rearrange the matrix into the desired form:

$$\left[\begin{array}{c|cccc} A(F) & S & S' & \dots & S \\ \hline R^t & A_0 & A'_1 & \dots & A_{p-1} \\ R^t & A_{p-1} & A'_0 & \dots & A_{p-2} \\ \vdots & \vdots & \vdots & \vdots & \vdots \\ R^t & A_1 & A'_2 & \dots & A_0 \end{array}\right].$$

Notice that the total number of rows and columns interchanges performed to transform Δ_{r+1} into Δ_{r+s+1} is $2(s-1)+(p-1)s+s+(p-3)s = 2(s-1)+(2p-3)s$. Thus $|\Delta_{r+1}| = (-1)^s|\Delta_{r+s+1}|$. Similarly, one can show that $|\Delta_{r+1}| = (-1)^{ms}|\Delta_{r+ms+1}|$, for $1 \leq m \leq p-1$. It can be easily checked that

$$(-1)^{r+1}w_{1r+1}|\Delta_{r+1}| = (-1)^{r+1+s}w_{1r+s+1}|\Delta_{r+s+1}| = \dots = (-1)^{r+(p-1)s+1}w_{1r+s+1}|\Delta_{r+(p-1)s+1}|.$$

Hence, the contributions of these p sub-matrices to the value of the determinant add to zero modulo p. Consequently, to compute the $\varphi_\Gamma(x)$ modulo p, one may consider only the co-factors corresponding to entries in the first row of F. With an elementary induction argument on the size of F, we can prove that $\varphi_\Gamma(x) \equiv \varphi_F(x)\varphi_{\Gamma\setminus F}(x)$ modulo p. Notice that since $\Gamma \setminus F$ is a p-periodic graph, we have $\varphi_{\Gamma\setminus F}(x) \equiv \varphi_{\overline{\Gamma\setminus F}}(x^p)$ modulo p, as proven in [1]. This completes the proof of Theorem 1.

The proof of Theorem 2 is also based on the study of block circulant matrices. Using the order of the vertices described above, the Laplacian matrix of Γ is of the form:

$$L(\Gamma) = \left[\begin{array}{c|cccc} l(F) & M & M & \dots & M \\ \hline M^t & B_0 & B_1 & \dots & B_{p-1} \\ M^t & B_{p-1} & B_0 & \dots & B_{p-2} \\ \vdots & \vdots & \vdots & \vdots & \vdots \\ M^t & B_1 & B_2 & \dots & B_0 \end{array}\right],$$

where $l(F)$ is an r-square matrix, M is an $r \times s$-matrix and for $0 \leq i \leq p-1$, B_i is a s-square matrix. In a similar way, we can prove that $det(L(\Gamma) - xI_n)$ is the product of $det(l(F) - xI_n)$ by $det(C(B_0, \dots, B_{p-1}) - xI_n)$. On the other hand, it can be easily seen that $l(F)$ and $L(F)$ coincide out of the diagonal. Furthermore, for any i, the diagonal entries are congruent modulo p, $d_{ii}(l(F)) \equiv d_{ii}(L(F))$ modulo p. Thus, $det(l(F) - xI_n) \equiv det(L(F) - xI_n)$ modulo p. Consequently, $\psi_F(x)$ divides $\psi_\Gamma(x)$ in $\mathbb{F}_p[x]$.

Finally, by Lemma 1 $det(C(B_0, \dots, B_{p-1}) - xI_n)$ is a polynomial on x^p. This ends the proof of Theorem 2.

4. Conclusions

By using elementary properties of block circulant matrices, we established two obstruction criteria for a graph to be symmetric. More precisely, we proved that if a given graph Γ is semi-free p-periodic with set of fixed vertices F, then its characteristic polynomial, with coefficients reduced modulo p, is the product of the characteristic polynomial of the induced subgraph $\Gamma[F]$ by the one of $\Gamma \setminus F$. We also proved that the characteristic polynomial of the Laplacian matrix of a semi-free p-periodic graph satisfies a similar necessary condition.

Author Contributions: Conceptualization, N.C.; Investigation, S.A.D., M.Y.T. and A.A.E.A.; Software, S.A.D., M.Y.T. and A.A.E.A.; Supervision, N.C.; Writing-review and editing, N.C.

References

1. Chbili, N. Graph polynomials and symmetries. *J. Algebra Appl.* **2018**. [CrossRef]
2. Feng, R.; Kwak, J.H.; Lee, J. Characteristic polynomials of graph coverings. *Bull. Austral. Math. Soc.* **2004**, *69*, 133–136. [CrossRef]
3. Lee, J.; Kim, H.K. Characteristic polynomials of graphs having a semi-free action. *Linear Algebra Appl.* **2000**, *307*, 35–46. [CrossRef]

4. Wang, K. Characteristic polynomials of symmetric graphs. *Linear Algebra Appl.* **1983**, *51*, 121–125. [CrossRef]
5. Chbili, N. A note on the Tutte polynomial and automorphism group of a graph. *Asian Eur. J. Math.* **2014**, *7*, 1450001. [CrossRef]
6. Friedman, B. Eigenvalues of composite matrices. *Math. Proc. Camb. Philos. Soc.* **1961**, *57*, 37–49. [CrossRef]

2

Duality in Geometric Graphs: Vector Graphs, Kirchhoff Graphs and Maxwell Reciprocal Figures

Tyler Reese, Randy Paffenroth and Joseph D. Fehribach *

Department of Mathematical Sciences, Worcester Polytechnic Institute, 100 Institute Road, Worcester, MA 01609-2280, USA; tmreese@wpi.edu (T.R.); rcpaffenroth@wpi.edu (R.P.)
* Correspondence: bach@wpi.edu

Abstract: We compare two mathematical theories that address duality between cycles and vertex-cuts of graphs in geometric settings. First, we propose a rigorous definition of a new type of graph, vector graphs. The special case of $\boldsymbol{R}^2$-vector graphs matches the intuitive notion of drawing graphs with edges taken as vectors. This leads to a discussion of Kirchhoff graphs, as originally presented by Fehribach, which can be defined independent of any matrix relations. In particular, we present simple cases in which vector graphs are guaranteed to be Kirchhoff or non-Kirchhoff. Next, we review Maxwell's method of drawing reciprocal figures as he presented in 1864, using modern mathematical language. We then demonstrate cases in which $\boldsymbol{R}^2$-vector graphs defined from Maxwell reciprocals are "dual" Kirchhoff graphs. Given an example in which Maxwell's theories are not sufficient to define vector graphs, we begin to explore other methods of developing dual Kirchhoff graphs.

Keywords: Kirchhoff graphs; geometric graphs; duality; Maxwell reciprocals

---

1. Introduction

This paper will discuss a newly-defined type of graph: vector graphs. Graphs in which edges are identified as vectors arise in the theory of Kirchhoff graphs developed by Fehribach [1,2]. Kirchhoff graphs comprise a special class of vector graphs, which reflect the orthogonal complementarity of the row and null space of integer matrices. The original development of Kirchhoff graphs was motivated by the study of chemical reaction networks. Given the stoichiometric matrix of a chemical reaction network, any Kirchhoff graph for the transpose of that matrix represents a circuit diagram for that network. Kirchhoff graphs were so-named because they satisfy both the Kirchhoff potential law and the Kirchhoff current law. The use of Kirchhoff graphs as reaction network diagrams was discussed by Fehribach [1], as well as by Fishtik, Datta *et al.* [3–6], who discuss a related graph, the "reaction route graph". A number of other graphs have been used in the study of reaction networks; a summary can be found in [7,8]. Within the context of reaction networks, the vertices of a Kirchhoff graph represent combinations of component chemical species, whereas the vector edges represent the reaction steps required to move between these combinations. Therein lies the motivation for developing graphs with "vector" edges: within a chemical reaction network, the same reaction process can occur given a number of different initial states of the chemical system. In a graphical representation, taking the edges corresponding to this reaction to be identical in the vector sense allows us to graphically identify this similarity. To the authors' knowledge, graphs with vector edges, where edges identified as the same vector are considered the same edge, have not been previously well studied.

The primary focus of this paper will be the relationships between Kirchhoff graphs and Maxwell's theory of reciprocal figures [9,10]. Both theories deal with geometric graphs in which the length and direction of each edge are taken into consideration. In particular, this leads to a discussion of a new type of geometric duality: Kirchhoff dual graphs. Section 2 presents a rigorous definition of vector graphs, including explicit examples in the two-dimensional Euclidean plane. After defining the edge, cycle and cut spaces on this new type of graph, in Section 3, we define and discuss Kirchhoff graphs. Here, we present some cases in which we are guaranteed that simple vector graphs are either Kirchhoff or non-Kirchhoff. In order to ease the comparison of the two theories, Section 4 reviews the beginnings of Maxwell's reciprocal methods, using modern mathematical language. Section 5 then provides connections between these two theories and demonstrates that in general, Maxwell reciprocals can lead to pairs of Kirchhoff graphs (in R^2), which are dual to each other. Given examples in which these methods no longer agree, Section 6 moves beyond Maxwell's reciprocals and begins to explore other methods of identifying Kirchhoff duals.

2. Vector Graphs

Kirchhoff graphs are distinctive among all graphs in that their edges are identified as vectors. In this section, we first give an abstract definition of this class of graphs, vector graphs, then present concrete examples which are both useful in applications and have motivated these ideas. These definitions may seem opaque upon first reading; subsequent examples will be used to illustrate the intuition behind these abstract notions. Speaking generally, vector graphs are graphs with edges identified as vectors, which satisfy the vector space axioms. The following discussion makes this idea exact.

Let $D = (V(D), E(D))$ be a finite directed multi-graph with vertex set $V(D) = \{v_1, \ldots, v_p\}$ and edge set $E(D) = \{e_1, \ldots, e_n\}$. A cycle L of D is an alternating sequence of vertices and edges $L = v_{L_1}(e_{L_1})v_{L_2}(e_{L_2}) \ldots v_{L_k}(e_{L_k})v_{L_{k+1}}$, where $v_{L_{k+1}} = v_{L_1}$ and edge e_{L_j} is directed between vertices v_{L_j} and $v_{L_{j+1}}$. (We allow that cycle L traverses any edge or vertex of D more than once, and edges may be traversed in either direction, independent of orientation. Some authors would instead refer to this as a "closed walk" in graph D.) We say edge e_{L_j} is oriented as L when $v_{L_j}v_{L_{j+1}} = e_{L_j}$; otherwise, when $v_{L_{j+1}}v_{L_j} = e_{L_j}$, e_{L_j} is not oriented as L [11]. L can also be written as $L = (\pm e_{L_1})(\pm e_{L_2}) \cdots (\pm e_{L_k})$ (where $+$, respectively $-$, is chosen if e_{L_k} is oriented, respectively not oriented, as L). Let F be a vector space with zero element $\bar{0}$ and $S = \{s_1, \ldots, s_m\}$ be a set of m non-zero vectors in F.

Definition 1. Given a multi-digraph D and some set $S \subset F$ as above, a (D, S)**-vector assignment** is a surjective map, $\varphi : E(D) \to S$, which assigns to each edge $e_i \in E(D)$ some element $s_j \in S$. We will notate $\varphi : e_i \mapsto s_j$ as $\varphi(e_i) = s_j$. A (D, S)-vector assignment is **consistent** if for any cycle $L = (\pm e_{L_1})(\pm e_{L_2}) \cdots (\pm e_{L_k})$ in D, then $\pm\varphi(e_{L_1}) \pm \varphi(e_{L_2}) \pm \cdots \pm \varphi(e_{L_k}) = \bar{0}$ in F.

Definition 2. A **vector graph** $\overline{D} = (D, \varphi)$ consists of a multi-digraph D, together with a (D, S)-vector assignment φ that is consistent. A vector graph has a natural set of vertices $V(\overline{D}) = \{\overline{v_1}, \ldots, \overline{v_p}\}$ and edges $\{\overline{e}_1, \ldots, \overline{e}_n\}$ (where $v_i \in V(D) \mapsto \overline{v_i} \in V(\overline{D})$; e_i in $D \mapsto \overline{e}_i$ in $\overline{D}$). Two edges $\overline{e}_i, \overline{e}_j$ of $\overline{D}$ are **identified** if $\varphi(e_i) = \varphi(e_j)$; otherwise, they are **distinct**. $E(\overline{D})$ is now used to denote the set of distinct edges in $\overline{D}$, meaning $|E(\overline{D})| = |S|$ (φ is surjective).

Remark 1. The underlying vector space F and set of vectors S are pivotal in the definition of a vector graph, though not included in the notation (D, φ), as they are implicit by the inclusion of φ. We will sometimes refer to a vector graph as an F**-vector graph** to make clear the ambient space F of vectors; for example, R^2-vector graphs will be detailed below in Section 2.1.

Remark 2. Under this definition, a vector graph must maintain the vector space axioms of F. For example, let $\overline{D} = (D, \varphi)$ be a vector graph where $(e_1)(e_2)(e_3)(e_4)$ is a cycle in D and $\varphi(e_i) = s_i$ for $1 \leq i \leq 4$. Then, for example, if the vectors of S satisfy $s_1 = -s_3$ in F, it must be the case that $s_2 = -s_4$ in F, as well.

The two-dimensional Euclidean plane is a natural space in which we visualize both graphs and vectors; therefore, the next section is dedicated to an explicit description of vector graphs in the plane.

2.1. Vector Graphs in the Plane

We may choose vector space $F = R^2$, so that each $s_i \in S$ is a two-space vector. Then, given multi-digraph D and (D, S)-vector assignment φ, if $\overline{D} = (D, \varphi)$ is a vector graph, $\overline{D}$ can be drawn in the plane where each edge $\overline{e}_i$ *is drawn as* its assigned vector s_j. This is guaranteed by the consistency of vector assignment φ. In such a drawing, identified edges in $\overline{D}$ are represented by identical vectors in the plane. Thus, R^2-vector graphs may be considered as geometric graphs constructed on vector edges. Conversely, any drawing of a digraph in R^2 in which edges are taken as vectors, those with the same length and direction are considered identical, *defines* an R^2-vector graph.

Remark 3. Herein lies the intuition behind requiring φ to be *consistent*; any graph drawn on vector edges gives a well-defined vector graph. For instance, let D be a graph *drawn* on vector edges with cycle $(e_1)(e_2)(-e_3)(-e_4)$. If e_1 and e_3 are identical vectors, edges e_2 and e_4 are then *forced* to be identical, as well. The definition of vector graphs maintains such forced identifications through the consistency of φ.

All vector graphs illustrated in this paper will be shown as R^2-vector graphs: points connected by vectors in two-space. For simplicity, we will omit Euclidean axes, but label all vector edges with labels s_i to make clear which edges are identified. Although vector graphs will always be presented in the plane, these graphs are not inherently two-dimensional. In fact, for applications, it can be advantageous to consider R^n-vector graphs, where two-space representations can arise as plane projections of these higher dimensional objects.

Example 1. Consider a simple digraph, $D = (V(D), E(D))$, where:

$$V(D) = \{v_1, v_2, v_3, v_4\} \qquad E(D) = \{e_1 = v_1v_4, e_2 = v_4v_3, e_3 = v_2v_3, e_4 = v_1v_2, e_5 = v_1v_3, e_6 = v_2v_4\}$$

Take $F = R^2$ and $S = \{s_1 = [0,2], s_2 = [2,0], s_3 = [2,2], s_4 = [-2,2]\}$. Let φ be the following (D, S)-vector assignment:

$$e_1 \mapsto s_1 \qquad e_2 \mapsto s_2 \qquad e_3 \mapsto s_1 \qquad e_4 \mapsto s_2 \qquad e_5 \mapsto s_3 \qquad e_6 \mapsto s_4$$

One may verify that φ is consistent. For example, $(e_1)(-e_6)(-e_4)$ is a cycle in D and:

$$\varphi(e_1) - \varphi(e_6) - \varphi(e_4) = s_1 - s_4 - s_2 = [0,2] - [-2,2] - [2,0] = [(0+2-2), (2-2+0)] = [0,0] = \overline{0} \in R^2$$

Therefore, $\overline{D} = (D, \varphi)$ is a vector graph, and $\overline{D}$ can be drawn in the plane where each edge is drawn as (and labeled by) its assigned vector s_j, as shown in Figure 1.

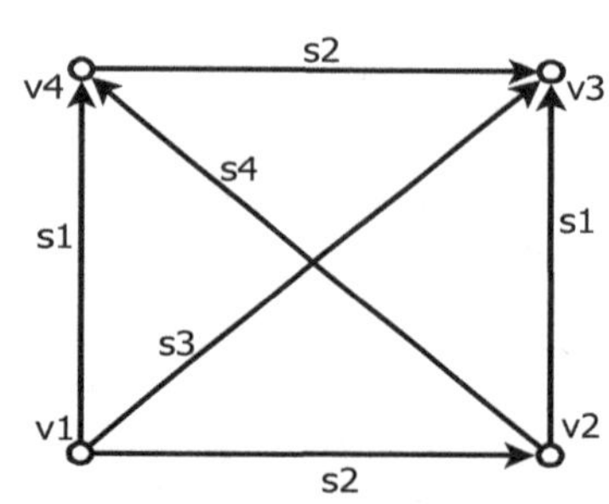

Figure 1. Vector graph $\overline{D} = (D, \varphi)$.

Example 2. Figure 2 displays two vector graphs, $\overline{A} = (D_0, \varphi_1)$ and $\overline{B} = (D_0, \varphi_2)$, where φ_i is a (D_0, S_i)-vector assignment and $S_1 \neq S_2$. Thus, two different (consistent) (D_0, S_i)-vector assignments on the same multi-digraph D_0 produce visibly different R^2-vector graphs. $\overline{A}$ is a vector graph on five vertices with eight distinct edges, whereas $\overline{B}$ is a vector graph on five vertices with four distinct edges. By definition, vector graphs are allowed to have multiple edges connecting two vertices. In an R^2-vector graph, all such multiples must be identified with each other; Roman numerals will be used to indicate the multiples of each edge when drawing vector graphs.

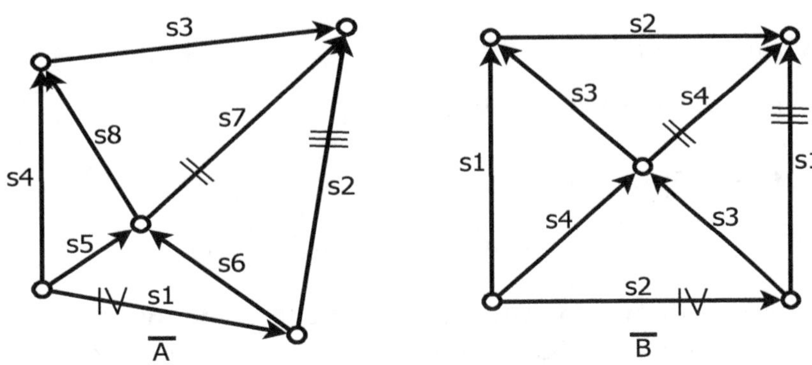

Figure 2. Two vector graphs generated by different vector assignments on the same multi-digraph.

2.2. The Edge Space of Vector Graphs

In classical graph theory, the edge space of a directed graph D is defined to be the vector space of functions from $E(D)$ into any scalar field (for example, C as in [11]) with distinct elements $-1, 0$ and 1. The standard basis of the edge space is the functions $\{f_1, \ldots, f_n\}$ where $f_i(e_j) = \delta_{i,j}$ (the Kronecker delta). Therefore, any a in the edge space of D may be written as $a = \sum_{i=1}^{n} \alpha_i f_i = [\alpha_i]_{i=1}^{n}$. In particular, any element of the edge space of D can be represented by a vector in C^n. When considering vector graphs, which are both directed and allow multiple (identified) edges, we require a scalar field F that contains Z as a sub-ring (we choose C). Let $\overline{D} = (D, \varphi)$ be a vector graph as above: in particular, $|V(D)| = |V(\overline{D})| = p$, $|E(D)| = n$ and $|S| = |E(\overline{D})| = m$.

Definition 3. The **edge space** of vector graph $\overline{D}$, denoted by $C_e(\overline{D})$, is the complex vector space of all functions from $E(\overline{D})$ into C; that is, from the set of distinct edges of $\overline{D}$ into C. For any subset $S_1 = \{s_{(1,1)}, \ldots, s_{(1,k)}\}$ of distinct edges in $E(\overline{D})$ and any $\mathcal{G} \in C_e(\overline{D})$, take $\mathcal{G}(S_1) = \sum_{i=1}^{k} \mathcal{G}(s_{(1,i)})$.

Lemma 4. *Given $\overline{D}$ as in Definition 3,* $\dim(C_e(\overline{D})) = m$.

Proof. For each $1 \leq i \leq m$, let $\mathcal{E}_i \in C_e(\overline{D})$ be the function defined by $\mathcal{E}_i(s_j) = \delta_{i,j}$. In $C_e(\overline{D})$, the zero element, $\bar{0}$, is the function that assigns value zero to each distinct $s_i \in E(\overline{D})$. If there exist scalars $\{\gamma_1, \ldots, \gamma_m\}$, such that $\gamma_1 \mathcal{E}_1 + \ldots \gamma_m \mathcal{E}_m = \bar{0}$, then, since $\mathcal{E}_i(s_j) = \delta_{i,j}$, for each i,

$$0 = \bar{0}(s_i) = (\gamma_1 \mathcal{E}_1 + \ldots \gamma_m \mathcal{E}_m)(s_i) = \gamma_i \mathcal{E}_i(s_i) = \gamma_i$$

Therefore $\mathcal{E}_1 \ldots \mathcal{E}_m$ are linearly independent in $C_e(\overline{D})$. Furthermore, let $\mathcal{G} \in C_e(\overline{D})$ be arbitrary, such that $\mathcal{G} : s_i \mapsto y_i \in \mathbb{C}$. Then, $\mathcal{G} = \sum_{i=1}^{m} y_i \mathcal{E}_i$, meaning $\mathcal{E}_1 \ldots \mathcal{E}_m$ span $C_e(\overline{D})$, as well. Therefore, $\mathcal{E}_1 \ldots \mathcal{E}_m$ are a basis of $C_e(\overline{D})$ and $\dim(C_e(\overline{D})) = m$. □

We call $\{\mathcal{E}_1 \ldots \mathcal{E}_m\}$ the standard basis of $C_e(\overline{D})$ and introduce the inner product under which this basis is orthonormal. If $\mathcal{G}_1 = \sum_{i=1}^{m} \beta_i \mathcal{E}_i$ and $\mathcal{G}_2 = \sum_{i=1}^{m} b_i \mathcal{E}_i$ in $C_e(\overline{D})$, then define $\left\langle \mathcal{G}_1, \mathcal{G}_2 \right\rangle_{C_e(\overline{D})} := \sum_{i=1}^{m} \beta_i \overline{b_i}$. We can represent each $\mathcal{G} \in C_e(\overline{D})$ as a vector in C^m. In particular, given $\mathcal{G} = \sum_{i=1}^{m} \beta_i \mathcal{E}_i$ represent $\mathcal{G}$ by $[\beta_i]_1^m = [\beta_1, \ldots, \beta_m] \in C^m$, then for any $\mathcal{G}_1$, $\mathcal{G}_2$ as above, $\left\langle \mathcal{G}_1, \mathcal{G}_2 \right\rangle_{C_e(\overline{D})} = \sum_{i=1}^{m} \beta_i \bar{b_i} = \left\langle [\beta_i]_1^m, [b_i]_1^m \right\rangle_{C^m}$. Next, we consider two special subspaces of $C_e(\overline{D})$.

2.2.1. The Cycle Space

Let $L = \overline{v}_{L_1}(\overline{e}_{L_1})\overline{v}_{L_2} \ldots \overline{v}_{L_k}(\overline{e}_{L_k})\overline{v}_{L_{k+1}}$ be any cycle in vector graph $\overline{D}$. L can be identified with an element $z_L \in C_e(\overline{D})$. As $z_L = \sum_{i=1}^{m} z_L(s_i)\mathcal{E}_i$, it suffices to define $z_L(s_i)$ for each $s_i \in E(\overline{D})$. Namely, for each distinct edge $s_i \in E(\overline{D})$, define:

$$\mathcal{O}_L^{(+)}(s_i) := |\{j : 1 \leq j \leq k \text{ and } \overline{v}_{L_j}\overline{v}_{L_{j+1}} = \overline{e}_{L_j} \text{ and } \varphi(e_{L_j}) = s_i\}|$$
$$\mathcal{O}_L^{(-)}(s_i) := |\{j : 1 \leq j \leq k \text{ and } \overline{v}_{L_{j+1}}\overline{v}_{L_j} = \overline{e}_{L_j} \text{ and } \varphi(e_{L_j}) = s_i\}|$$

For each $1 \leq i \leq m$, define:

$$z_L(s_i) := \mathcal{O}_L^{(+)}(s_i) - \mathcal{O}_L^{(-)}(s_i)$$

Observe that $z_L(s_i)$ is the net number of times edges identified as s_i are traversed in L. This number is positive when more edges identified as s_i are traversed oriented as L and negative if more occurrences of s_i are traversed that are not oriented as L.

Definition 5. Let $Z(\overline{D})$ denote the subspace of $C_e(\overline{D})$ spanned by the elements z_L for all cycles L in $\overline{D}$. $Z(\overline{D})$ is the **cycle space** of $\overline{D}$.

Any $z_L \in C_e(\overline{D})$ can be represented as a vector in C^m. We call this vector the *characteristic vector* of L, denoted by $\chi(L)$. This vector is also sometimes called the *cycle vector* corresponding to L. The cycle space of $\overline{D}$ is the subspace of C^m spanned by the cycle vectors corresponding to all cycles L in $\overline{D}$.

2.2.2. The Cut Space

Let $P = (V_1, V_2)$ be a partition of $V(\overline{D})$, such that $V_1 \cap V_2 = \emptyset$ and $V_1 \cup V_2 = V(\overline{D})$. Let $E(V_1, V_2)$ denote the set of all directed edges in $\overline{D}$ between V_1 and V_2. More specifically,

$$E(V_1, V_2) = \{\overline{e} \text{ in } \overline{D} : \overline{e} = \overline{v}_i\overline{v}_j \text{ where } \overline{v}_i \in V_1 \text{ and } \overline{v}_j \in V_2 \text{ or } \overline{v}_i \in V_2 \text{ and } \overline{v}_j \in V_1\}$$

Definition 6. A partition $P = (V_1, V_2)$ of $V(\overline{D})$ as described above is called a **cut** in $\overline{D}$, and $E(V_1, V_2)$ denotes the set of **cut edges**. (Some authors refer to $E(V_1, V_2)$ as the cut: the edges one would need to "cut" to separate the vertex sets V_1 and V_2 in D.)

Any cut P of $\overline{D}$ is identified with an element $u_P \in C_e(\overline{D})$. Once again, it suffices to define $u_P(s_i)$ for each $s_i \in E(\overline{D})$. Namely, define:

$$\mathcal{P}^{(+)}(s_i) := |\{j : 1 \leq j \leq n \text{ and } \overline{e}_j \text{ is directed from } V_1 \text{ to } V_2 \text{ and } \varphi(e_j) = s_i\}|$$
$$\mathcal{P}^{(-)}(s_i) := |\{j : 1 \leq j \leq n \text{ and } \overline{e}_j \text{ is directed from } V_2 \text{ to } V_1 \text{ and } \varphi(e_j) = s_i\}|$$

For each $1 \leq i \leq m$, define:

$$u_P(s_i) := \mathcal{P}^{(+)}(s_i) - \mathcal{P}^{(-)}(s_i)$$

Therefore, $u_P(s_i)$ is the net number of edges identified as s_i directed from V_1 to V_2. This number is positive when more edges identified as s_i are directed from V_1 to V_2 and negative if more are directed from V_2 to V_1.

Definition 7. Let $U(\overline{D})$ denote the subspace of $C_e(\overline{D})$ spanned by the elements u_P for all cuts P of $\overline{D}$. $U(\overline{D})$ is the **cut space** of $\overline{D}$.

As $u_P \in C_e(\overline{D})$, it can be represented as a vector $\Lambda(P) \in C^m$. We will refer to this vector as the *cut vector* corresponding to P. The cut space of $\overline{D}$ is the subspace of C^m spanned by the cut vectors corresponding to all cuts P of $\overline{D}$.

2.3. Computing the Cycle Space and Cut Space

The cycle space and cut space of multi-digraph D are defined in a similar way. For example, any cycle $l = v_{l_1}(e_{l_1})v_{l_2}\dots v_{l_k}(e_{l_k})v_{l_{k+1}}$ has cycle vector $\chi(l) = [\lambda_i]_{i=1}^n \in C^n$ where λ_i is the net number of times cycle l traverses edge e_i. The dimension of the cycle space of D is the dimension of the subspace of C^n spanned by the elements $\chi(l)$ for all cycles l of D.

Observe that if $l = v_{l_1}(e_{l_1})v_{l_2}\dots v_{l_k}(e_{l_k})v_{l_{k+1}}$ is a cycle in D, then there is a corresponding cycle $L = \overline{v}_{l_1}(\overline{e}_{l_1})\overline{v}_{l_2}\dots\overline{v}_{l_k}(\overline{e}_{l_k})\overline{v}_{l_{k+1}}$ in $\overline{D}$. Moreover, the cycle vectors $\chi(l) \in C^n$ and $\chi(L) \in C^m$ are related by the (D,S)-vector assignment φ. In particular, we can represent the action of this vector assignment in matrix form. Define:

$$T = \left[T_{i,j}\right] \in Z^{n\times m} \text{ where } T_{i,j} = \begin{cases} 1 & \text{if } \varphi(e_i) = s_j \\ 0 & \text{Otherwise} \end{cases}$$

Let A be any matrix with n columns. Observe that the j-th column of the product AT is the sum of all columns of A with index i, such that $\varphi(e_i) = s_j$. For example, given any cycle L of $\overline{D}$ and its corresponding cycle l in D, one may verify that (using row vectors) $\chi(L) = \chi(l)T$. Moreover, this reduces finding the dimension of $Z(\overline{D})$ to an exercise in linear algebra. Let $Z(D)$ represent the cycle space of multi-digraph D, and let M be any matrix with n columns whose rows span $Z(D)$ (for example, take the rows of M to be the cycle vectors of some set of fundamental cycles). Then dim($Z(D)$) = Rank(M) and, more importantly,

$$\dim(Z(\overline{D})) = \text{Rank}(MT)$$

In particular, it follows that $\dim(Z(\overline{D})) \leq \dim(Z(D))$. This inequality can be strict when nontrivial cycles in D are reduced to null cycles with respect to edge identification or distinct cycles in D become identified or dependent under φ. This will be better illustrated in an example below.

If $p = (\{v_{1,1}, v_{1,2}, \dots v_{1,k}\}, \{v_{2,1}, \dots v_{2,p-k}\})$ and $P = (\{\overline{v}_{1,1}, \overline{v}_{1,2}, \dots \overline{v}_{1,k}\}, \{\overline{v}_{2,1}, \dots \overline{v}_{2,p-k}\})$ are corresponding cuts of D and $\overline{D}$, respectively, one may verify that $\Lambda(P) = \Lambda(p)T$. Let $U(D)$ represent the cut space of D and N be any matrix with n columns whose rows span $U(D)$. Then dim($U(D)$) = Rank(N) and, similarly,

$$\dim(U(\overline{D})) = \text{Rank}(NT)$$

Once again, it follows that $\dim(U(\overline{D})) \leq \dim(U(D))$. This is best illustrated with an example.

Example 3. Recall the following vector graph $\overline{D} = (D,\varphi)$ from Example 1.

$$V(D) = \{v_1, v_2, v_3, v_4\} \qquad E(D) = \{e_1 = v_1v_4, e_2 = v_4v_3, e_3 = v_2v_3, e_4 = v_1v_2, e_5 = v_1v_3, e_6 = v_2v_4\}$$

$$\varphi : e_1 \mapsto s_1 = [0,2] \quad e_2 \mapsto s_2 = [2,0] \quad e_3 \mapsto s_1 \quad e_4 \mapsto s_2 \quad e_5 \mapsto s_3 = [2,2] \quad e_6 \mapsto s_4 = [-2,2]$$

Then φ can be represented by matrix:

$$T = \begin{bmatrix} 1 & 0 & 0 & 0 \\ 0 & 1 & 0 & 0 \\ 1 & 0 & 0 & 0 \\ 0 & 1 & 0 & 0 \\ 0 & 0 & 1 & 0 \\ 0 & 0 & 0 & 1 \end{bmatrix}$$

One set of fundamental cycles in D is $\{l_1 = v_1v_4v_3v_1, l_2 = v_1v_2v_3v_1, l_3 = v_1v_4v_2v_1\}$, so $\dim(Z(D)) = 3$. (Note that D is simple, and we can notate cycles using only the order of vertices.) Given corresponding cycles $\{L_1, L_2, L_3\}$ of $\overline{D}$, take:

$$M_0 = \begin{bmatrix} \chi(l_1) \\ \chi(l_2) \\ \chi(l_3) \end{bmatrix} = \begin{bmatrix} 1 & 1 & 0 & 0 & -1 & 0 \\ 0 & 0 & 1 & 1 & -1 & 0 \\ 1 & 0 & 0 & -1 & 0 & -1 \end{bmatrix} \text{; where } M_0T = \begin{bmatrix} 1 & 1 & -1 & 0 \\ 1 & 1 & -1 & 0 \\ 1 & -1 & 0 & -1 \end{bmatrix} = \begin{bmatrix} \chi(L_1) \\ \chi(L_2) \\ \chi(L_3) \end{bmatrix}$$

Therefore, $\dim(Z(\overline{D})) = \text{Rank}(M_0T) = 2$. Observe that the cycle $l_4 = v_1v_4v_3v_2v_1$ with $\chi(l_4) = [1, 1, -1, -1, 0, 0]$ is nontrivial in D. However, the corresponding cycle $L_4 = \overline{v}_1\overline{v}_2\overline{v}_3\overline{v}_4\overline{v}_1$ satisfies $\chi(L_4) = \chi(l_4)T = [0, 0, 0, 0, 0, 0]$, meaning that L_4 is a null cycle in $\overline{D}$.

A set of fundamental cuts for D is $\{p_1 = (v_4, \{v_1, v_2, v_3\}), p_2 = (v_3, \{v_1, v_2, v_4\}), p_3 = (v_2, \{v_1, v_3, v_4\})\}$, so $\dim(U(D)) = 3$. Given corresponding cuts $\{P_1, P_2, P_3\}$ of $\overline{D}$, take:

$$N_0 = \begin{bmatrix} \Lambda(p_1) \\ \Lambda(p_2) \\ \Lambda(p_3) \end{bmatrix} = \begin{bmatrix} -1 & 1 & 0 & 0 & 0 & -1 \\ 0 & -1 & -1 & 0 & -1 & 0 \\ 0 & 0 & 1 & -1 & 0 & 1 \end{bmatrix} \text{; where } N_0T = \begin{bmatrix} -1 & 1 & 0 & -1 \\ -1 & -1 & -1 & 0 \\ 1 & -1 & 0 & 1 \end{bmatrix} = \begin{bmatrix} \Lambda(P_1) \\ \Lambda(P_2) \\ \Lambda(P_3) \end{bmatrix}$$

Therefore, $\dim(U(\overline{D})) = \text{Rank}(N_0T) = 2$. Though this vector graph does not contain a cut P in $\overline{D}$ satisfying $\Lambda(P) = [0, 0, 0, 0]$, examples of such cuts (in the form of null vertices) will be shown in later examples; for instance, the center vertex of vector graph $\overline{C_1}$ in Figure 4 has cut vector $[0, 0, 0, 0]$.

Remark 4. Observe that these bases were determined by choosing the spanning tree $\mathcal{T}$ of D where $V(\mathcal{T}) = V(D)$ and $E(\mathcal{T}) = \{e_1, e_4, e_5\}$.

3. Kirchhoff Graphs

For any vertex $\overline{v}_i \in V(\overline{D})$, we can choose a cut in $\overline{D}$ of the form $(\overline{v}_i, V(\overline{D}) - \{\overline{v}_i\})$. In the special case of such a cut, call the cut vector in C^m the *incidence vector* of vertex $\overline{v}_i$, and denote it by $\lambda(\overline{v}_i)$. (As we associate each edge with a vector in F and each vertex with a vector in C^m, the authors recognize the possibility that Kirchhoff graphs may be studied using sheaves on graphs [12]. Such connections are not explored any further here.) For example, given $\overline{D}$ as in Figure 3, consider the cut $P = (\overline{v}_1, V(\overline{D}) - \overline{v}_1)$. Then, $\lambda(\overline{v}_1) = [1, 1, 1, 0]$.

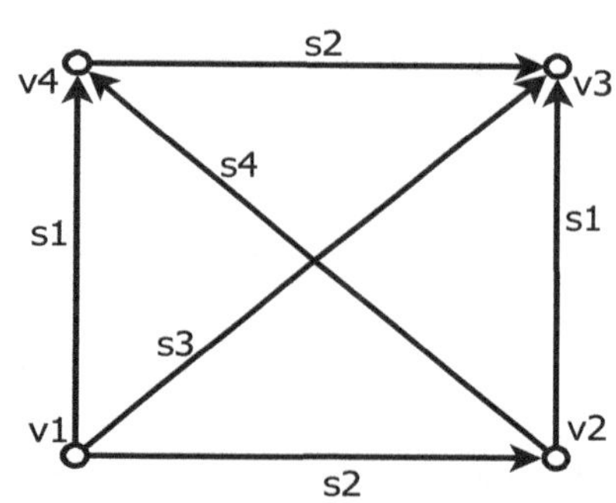

Figure 3. Vector graph $\overline{D} = (D, \varphi)$.

Definition 8. A vector graph $\overline{D}$ is a **Kirchhoff graph** if and only if for all vertices $\overline{v} \in V(\overline{D})$ and all cycles L of $\overline{D}$,

$$\left\langle \chi(L), \lambda(\overline{v}) \right\rangle_{\mathbb{C}^{|E(\overline{D})|}} = 0$$

If $\overline{D}$ is an F-vector graph that is also Kirchhoff, we will sometimes call $\overline{D}$ an ***F*-Kirchhoff graph**.

Example 4. Consider the two R^2-vector graphs in Figure 4, $\overline{C}_1$ and $\overline{C}_2$. We will demonstrate that one of these vector graphs is a Kirchhoff graph, while the other is not.

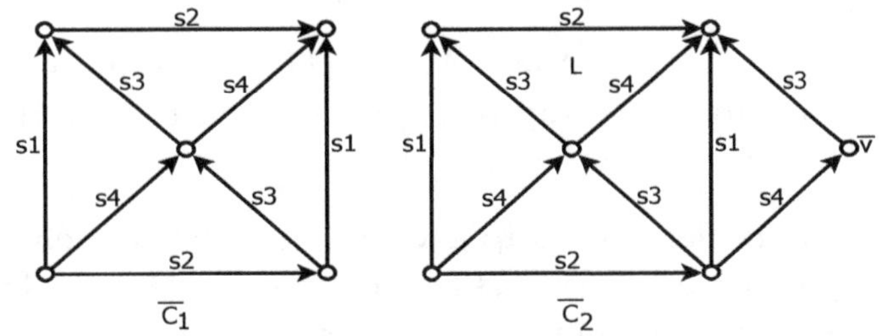

Figure 4. Two vector graphs, $\overline{C}_1$ and $\overline{C}_2$.

It is easy to check that:

$$Z(\overline{C}_1) = Z(\overline{C}_2) = \text{span}\{ \begin{bmatrix} 1 & 0 & -1 & -1 \end{bmatrix}, \begin{bmatrix} 0 & 1 & 1 & -1 \end{bmatrix} \}$$

For each vertex $\overline{v}_i$ in vector graph $\overline{C}_1$, $\lambda(\overline{v}_i) \in \text{span}\{[1,1,0,1], [-1,1,-1,0]\}$. Therefore, $\overline{C}_1$ is a Kirchhoff graph. On the other hand, consider the vertex labeled $\overline{v}$ in vector graph $\overline{C}_2$ satisfying $\lambda(\overline{v}) = [0,0,1,-1]$. The cycle labeled L in $\overline{C}_2$ has cycle vector $\chi(L) = [0,1,1,-1]$. Then $\overline{C}_2$ is not a Kirchhoff graph, because:

$$\left\langle \lambda(\overline{v}), \chi(L) \right\rangle_{C_e(\overline{C}_2)} = \left\langle \begin{bmatrix} 0 & 0 & 1 & -1 \end{bmatrix}, \begin{bmatrix} 0 & 1 & 1 & -1 \end{bmatrix} \right\rangle_{\mathbb{C}^4} = 2 \neq 0$$

Next, we demonstrate broad cases in which we are guaranteed that a vector graph is either a Kirchhoff graph or non-Kirchhoff. These results hold independent of the dimension or choice of vector space F.

3.1. Simple Vector Graphs

Definition 9. A vector graph $\overline{D} = (D, \varphi)$ is **simple** if D is a simple digraph (*i.e.*, for all $i \neq j$, there is at most one edge directed from v_i to v_j).

In this section, we will present two general cases of simple vector graphs: one in which we are guaranteed that the vector graph is Kirchhoff and one in which we are guaranteed the graph is not Kirchhoff.

Definition 10. Let $D = (V(D), E(D))$ be a multi-digraph. A vector graph $\overline{D} = (D, \varphi)$ is **general** if for all pairs of edges $\overline{e}_i = \overline{v}_{i_1}\overline{v}_{i_2}$ and $\overline{e}_j = \overline{v}_{j_1}\overline{v}_{j_2}$ in $\overline{D}$,

$$\varphi(e_i) = \varphi(e_j) \text{ only if } \overline{v}_{i_1} = \overline{v}_{j_1} \text{ and } \overline{v}_{i_2} = \overline{v}_{j_2}$$

That is, with the exception of multiple edges between the same two vertices, every pair of edges in a general vector graph is distinct.

Theorem 11. *Any simple, general vector graph is a Kirchhoff graph.*

Proof. Given that D has no multiple edges and vector graph $\overline{D}$ is general, *every* edge occurring in vector graph $\overline{D}$ is distinct. Therefore, $|E(D)| = |S|$, and φ is bijective: its matrix representation T is some $n \times n$ permutation matrix $T = \mathcal{P}$. For any cycle L in $\overline{D}$ with corresponding cycle l in D, $\chi(L) = \chi(l)\mathcal{P} \in C^n$. Similarly, for every vertex $\overline{v}_i$ of $\overline{D}$ with corresponding vertex v_i in D, $\lambda(\overline{v}_i) = \lambda(v_i)\mathcal{P} \in C^n$. Therefore, for any vertex $\overline{v}_i$ and cycle L in vector graph $\overline{D}$, because permutation matrix $\mathcal{P}$ is unitary,

$$\left\langle \lambda(\overline{v}_i), \chi(L) \right\rangle = \left\langle \lambda(v_i)\mathcal{P}, \chi(l)\mathcal{P} \right\rangle = \left\langle \lambda(v_i), \chi(l) \right\rangle = 0$$

Therefore, $\overline{D}$ is a Kirchhoff graph. The fact that $\left\langle \lambda(v_i), \chi(l) \right\rangle = 0$ follows from the orthogonality of the cut space and cycle space of simple digraphs. A proof of this standard graph-theoretic result can be found in [11]. □

Next, we present a general class of simple vector graphs, which are not Kirchhoff graphs. Let $\overline{D}_2 = (D, \varphi_2)$ be a simple vector graph (as before, take $V(\overline{D}_2) = \{\overline{v}_1, \ldots, \overline{v}_p\}$ and $E(\overline{D}_2) = \{s_1, \ldots, s_m\}$) in which all edges are distinct except for one pair of identified edges. Without loss of generality, we may re-label the vectors in set S, so that exactly two edges of $\overline{D}_2$ are identified as s_1, and for all $2 \leq i \leq m$, φ_2 assigns vector s_i to exactly one edge of $\overline{D}_2$.

Theorem 12. *$\overline{D}_2$ is a Kirchhoff graph if and only if the two edges identified as s_1 are either both bridges in D or form a directed (oriented) cycle, which is also a set of cut-edges in D.*

Proof. As defined above, $\overline{D}_2$ is a vector graph on p vertices and a total of $m+1$ edges. We may re-label the edges of D so that, without loss of generality, $\varphi_2(e_1) = \varphi_2(e_{m+1}) = s_1$: in $\overline{D}_2$ edges $\overline{e}_1$ and $\overline{e}_{m+1}$ are the one pair of identified edges. Since D is a simple digraph, for any vertex v_k and any cycle l of D,

$$\left\langle \lambda(v_k), \chi(l) \right\rangle = \sum_{i=1}^{m+1} \lambda(v_k)_i \chi(l)_i = 0 \tag{1}$$

Considering vector graph $\overline{D}_2$, we can observe the effects of the pair of identified edges under φ_2. For vertex $\overline{v}_k$ in $\overline{D}_2$ corresponding to v_k,

$$\lambda(\overline{v}_k)_i = \lambda(v_k)_i \text{ for } 2 \leq i \leq m \qquad \text{and} \qquad \lambda(\overline{v}_k)_1 = \lambda(v_k)_1 + \lambda(v_k)_{m+1} \tag{2}$$

Similarly, for any cycle $L = \overline{v}_{i_1}\overline{v}_{i_2}\ldots\overline{v}_{i_1}$ in $\overline{D}$ with corresponding cycle l in D,

$$\chi(L)_i = \chi(l)_i \text{ for } 2 \leq i \leq m \qquad \text{and} \qquad \chi(L)_1 = \chi(l)_1 + \chi(l)_{m+1} \tag{3}$$

Therefore, for any vertex $\overline{v}_k$ and any cycle L of $\overline{D}_2$:

$$\begin{aligned}\left\langle \lambda(\overline{v}_k), \chi(L) \right\rangle &= \sum_{i=1}^{m} \lambda(\overline{v}_k)_i \chi(L)_i = \lambda(\overline{v}_k)_1 \chi(L)_1 + \sum_{i=2}^{m} \lambda(\overline{v}_k)_i \chi(L)_i \\ &= \left(\lambda(v_k)_1 + \lambda(v_k)_{m+1}\right)\left(\chi(l)_1 + \chi(l)_{m+1}\right) + \sum_{i=2}^{m} \lambda(v_k)_i \chi(l)_i \text{ by Equations (2) and (3)}\end{aligned}$$

Therefore,

$$\begin{aligned}\left\langle \lambda(\overline{v}_k), \chi(L) \right\rangle &= \left(\lambda(v_k)_1 \chi(l)_{m+1} + \lambda(v_k)_{m+1} \chi(l)_1\right) + \sum_{i=1}^{m+1} \lambda(v_k)_i \chi(l)_i \\ &= \left(\lambda(v_k)_1 \chi(l)_{m+1} + \lambda(v_k)_{m+1} \chi(l)_1\right) + 0 \text{ by Equation (1)}\end{aligned}$$

That is,

$$\left\langle \lambda(\overline{v}_k), \chi(L) \right\rangle = \lambda(v_k)_1 \chi(l)_{m+1} + \lambda(v_k)_{m+1} \chi(l)_1 \tag{4}$$

First, assume that $e_1 e_{m+1}$ is a directed cycle and set of cut edges in D. Then, every vertex of D is either incident with neither or both of e_1 and e_{m+1}. If v_k is incident with neither, $\lambda(v_k)_1 = \lambda(v_k)_{m+1} = 0$ (and notably, $\lambda(v_k)_1 = -\lambda(v_k)_{m+1}$); if incident with both, $\lambda(v_k)_1 = -\lambda(v_k)_{m+1}$. Moreover, as directed cycle $e_1 e_{m+1}$ is a set of cut edges, any cycle l in D must traverse e_1 and e_{m+1} the same net number of times. In particular, $\chi(l)_1 = \chi(l)_{m+1}$. Therefore, for all v_k and l,

$$\lambda(v_k)_1 \chi(l)_{m+1} + \lambda(v_k)_{m+1} \chi(l)_1 = \lambda(v_k)_1 \chi(l)_1 - \lambda(v_k)_1 \chi(l)_1 = 0$$

By Equation (4), $\left\langle \lambda(\overline{v}_k), \chi(L) \right\rangle = 0$, and $\overline{D}$ is a Kirchhoff graph. On the other hand, assume that both edges identified as s_1, e_1 and e_{m+1}, are bridges in D. Then, given *any* cycle l in D,

$$\chi(l)_1 = \chi(l)_{m+1} = 0 \tag{5}$$

Thus, for any vertex $\overline{v}_k$ and any cycle L in $\overline{D}_2$, by Equations (4) and (5):

$$\begin{aligned} \left\langle \lambda(\overline{v}_k), \chi(L) \right\rangle &= \lambda(v_k)_1 \chi(l)_{m+1} + \lambda(v_k)_{m+1} \chi(l)_1 \\ &= \lambda(v_k)_1 (0) + \lambda(v_k)_{m+1} (0) = 0 \end{aligned}$$

Therefore, $\overline{D}_2$ is a Kirchhoff graph.

Conversely, assume that at least one of e_1 and e_{m+1} is not a bridge in D, and either $e_1 e_{m+1}$ is not a directed cycle in D or *is* a directed cycle, but not a cut-set. Assume, without loss of generality, e_1 is not a bridge in D. Then, there exists some cycle l_0 in D, such that $\chi(l_0)_1 \neq 0$.

Case 1: $e_1 e_{m+1}$ is not a directed cycle in D. As D was a simple digraph, e_1 and e_{m+1} cannot have the same initial and terminal vertices. Moreover, because $e_1 e_{m+1}$ is not a directed cycle in D, there must exist a vertex v_j that is incident to edge e_{m+1}, but *not* edge e_1. In particular,

$$\lambda(v_j)_{m+1} \neq 0 \text{ and } \lambda(v_j)_1 = 0$$

Then, taking the corresponding vertex $\overline{v}_j$ and cycle L_0 in $\overline{D}_2$, by Equation (4),

$$\begin{aligned} \left\langle \lambda(\overline{v}_j), \chi(L_0) \right\rangle &= \lambda(v_j)_1 \chi(l_0)_{m+1} + \lambda(v_j)_{m+1} \chi(l_0)_1 = 0 + \lambda(v_j)_{m+1} \chi(l_0)_1 \\ &= \lambda(v_j)_{m+1} \chi(l_0)_1 \neq 0 \end{aligned}$$

Therefore, $\overline{D}_2$ is not a Kirchhoff graph.

Case 2. $e_1 e_{m+1}$ is a directed cycle, but not a cut-set in D. Let v_i and v_j denote the end-vertices of e_1 and e_{m+1}, so that $v_i e_1 v_j e_{m+1} v_i$ is the directed cycle in D. In particular,

$$\lambda(v_i)_1 = 1 \qquad \text{and} \qquad \lambda(v_i)_{m+1} = -1$$

As $\{e_1, e_m\}$ is not a cut-set of D, there exists a $v_j - v_i$ path P, which does not traverse e_1 or e_{m+1}. Now, consider the cycle:

$$l_P = v_i e_1 v_j P v_i$$

Then, in particular,

$$\chi(l_P)_1 = 1 \qquad \text{and} \qquad \chi(l_P)_{m+1} = 0$$

Therefore, by Equation (4):

$$\left\langle \lambda(\overline{v}_i), \chi(L_P) \right\rangle = \lambda(v_i)_1 \chi(l_P)_{m+1} + \lambda(v_i)_{m+1} \chi(l_P)_1 = 0 + \lambda(v_i)_{m+1} \chi(l_P)_1$$
$$= (-1)(1) = -1 \neq 0$$

Therefore, $\overline{D}_2$ is not a Kirchhoff graph. This completes the proof of the theorem. □

Theorems 11 and 12 illustrate that Kirchhoff graphs are a very special class of vector graphs. While a general simple vector graph is guaranteed to be Kirchhoff, if we change vector assignment φ slightly, and identify only one pair of identical edges, the Kirchhoff property is easily lost.

3.2. Kirchhoff Graphs and Matrices

Of particular interest in applications [1,2] is the discussion of Kirchhoff graphs in relation to integer-valued matrices. We define (this definition is identical in mathematical content, though presented in a different form, to that given in [2]) the following:

Definition 13. For a matrix $A \in Z^{p \times m}$, a vector graph $\overline{D}$ is a **Kirchhoff graph for** A if and only if:

(i) $|S| = |E(\overline{D})| = m$

(ii) dim(Z($\overline{D}$)) = dim(Null(A))

(iii) For any cycle L of $\overline{D}$, $\chi(L) \in \text{Null}(A)$

(iv) For any $\overline{v} \in V(\overline{D})$, $\lambda(\overline{v}) \in \text{Row}(A)$

In the case that (i)–(iv) are satisfied, we also say that $\overline{D}$ is a Kirchhoff graph **generated** by A.

Under this definition, if $\overline{D}$ is a Kirchhoff graph generated by A, $\overline{D}$ is also a Kirchhoff graph for any matrix B that is row-equivalent to A.

Definition 14. Let $A \in Z^{p \times m}$ be an integer matrix, and suppose Null(A) is the column span of B^T for $B \in Z^{k \times m}$. A vector graph $\overline{D}$ is a Kirchhoff graph **generated by Null (A)** if $\overline{D}$ is a Kirchhoff graph generated by matrix B.

Definition 15. Two Kirchhoff graphs $\overline{D}$ and $\overline{D'}$ are **dual Kirchhoff graphs** if there exists some matrix A, such that $\overline{D}$ is a Kirchhoff graph generated by A, and $\overline{D'}$ is a Kirchhoff graph generated by Null(A). We say each vector graph is a **Kirchhoff dual** of the other.

One question we will begin to explore in this paper is:

Question 1. *Given a Kirchhoff graph $\overline{D}$ generated by matrix A, can we use vector graph $\overline{D}$ to construct a vector graph $\overline{D'}$ that is a Kirchhoff graph generated by Null(A), that is, a Kirchhoff dual of $\overline{D}$?*

Answering this question as "yes" would be an important step towards proving the existence of a Kirchhoff graph generated by any (arbitrary) integer matrix, one of the primary open problems in the theory of Kirchhoff graphs [1,2]. This would mean that given any integer matrix A, if one can construct a Kirchhoff graph for either A or Null(A), then there exists a Kirchhoff graph generated by A.

A natural place to begin exploring this idea of dual Kirchhoff graphs, and in particular dual R^2-Kirchhoff graphs, is the well-developed theory of reciprocal figures of James Clerk Maxwell. Given a geometric graph satisfying certain properties, Maxwell's work provides a concrete and *universal* method of constructing a reciprocal diagram, where the vertex cuts and cycles in one diagram correspond exactly with the cycles and vertex cuts in the other. We will begin by presenting a re-description of Maxwell's original work in [9] using modern mathematical terms. Given these foundations, we will describe commonalities between Maxwell figures and Kirchhoff graphs. Cases in which Maxwell's theory is no longer applicable lead us to consider other methods of finding R^2-Kirchhoff duals.

4. Maxwell Reciprocal Figures

James Clerk Maxwell introduced his idea of reciprocal figures in the mid 1860s [9]. He describes a type of geometric reciprocity, which has applications to mechanical problems. A "frame" is a geometric system of points connected by straight lines. In order to study equilibrium within these frames, he suggests we think of the points as mutually acting on each other, with forces in the direction of the lines adjoining them. Each line is thus endowed with one of two types of forces: if the force acting between its endpoints draws the points closer together, we call the force a *tension*, and if the force tends to separate the points, we call the force a *pressure*. Maxwell outlines a method for drawing a diagram of force corresponding to this frame, where every line of the force diagram represents, in magnitude and direction, the force acting on a line in the frame. Thus, the frame and its forces are presented in two separate diagrams, henceforth referred to as the *frame diagram* and the *force diagram*. In representing the forces in a separate diagram, Maxwell recognizes the loss of some interpretability: it is not immediately clear which forces act on which points in the frame. On the other hand, this method reduces finding the equilibrium of forces to examining whether the force diagram has closed polygons.

We draw a force diagram, such that each line in the force diagram is parallel to the line on which it acts in the frame. The lengths of these lines are proportional to the forces acting on the frame. For a frame that is in equilibrium, when any number of lines meet at a point in the frame, the corresponding lines in the force diagram form a closed polygon. In certain cases, Maxwell observes, for any lines that meet in the force diagram, the corresponding lines in the frame form a closed polygon. This is the type of geometric reciprocity Maxwell desires: two figures are considered "'reciprocal" if either diagram can be taken as a frame diagram, and the other represents the system of forces that keeps that frame in equilibrium. More precisely, in [9], he gives:

Definition 16. Two plane rectilinear figures are **reciprocal** when they consist of an equal number of straight lines, so that the corresponding lines that meet in a point in the one figure form a closed polygon in the other.

The two diagrams in Figure 5 are the original example of reciprocal figures as presented by Maxwell [9]. Note that the same letters are used to label corresponding (parallel) lines in the frame diagram and the force diagram.

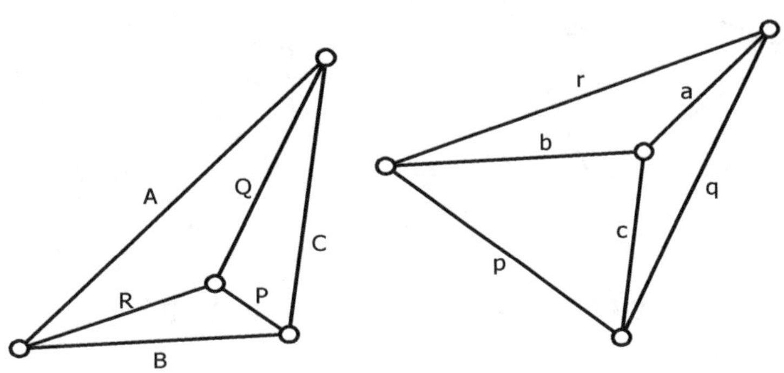

Figure 5. Maxwell's reciprocal figures as presented in [9]. Corresponding parallel edges are labeled with the same letter; capitalized in the frame diagram and lower-case in the force diagram.

Given that the frame is in equilibrium, we may draw the force diagram as outlined below. For convenience, refer to points of the frame by the labels of the lines incident to it. For example, the center point in the frame diagram is $\boldsymbol{PQR}$.

(1) Begin by drawing line $\boldsymbol{p}$ in the force diagram, parallel to line $\boldsymbol{P}$ in the frame. Choose a length for this line that represents the magnitude of the force acting along $\boldsymbol{P}$. Although the length of this first edge is an arbitrary choice, once the length of line $\boldsymbol{p}$ has been fixed, the lengths of all other lines in the force diagram are determined.
(2) The forces acting along lines $\boldsymbol{P}$, $\boldsymbol{Q}$ and $\boldsymbol{R}$ in the frame are in equilibrium. Therefore, in the force diagram, draw from one endpoint of $\boldsymbol{p}$ a line parallel to $\boldsymbol{Q}$ and from the other endpoint of $\boldsymbol{p}$ a line parallel to $\boldsymbol{R}$. Thus, form a (closed) triangle $\boldsymbol{pqr}$ in the force diagram, representing the equilibrium of forces acting on point $\boldsymbol{PQR}$. Steps (1) and (2) are illustrated in Figure 6. Note that once we fixed the length of edge $\boldsymbol{p}$, the lengths of $\boldsymbol{q}$ and $\boldsymbol{r}$ are therefore predetermined and represent the magnitude of the forces acting on $\boldsymbol{Q}$ and $\boldsymbol{R}$.

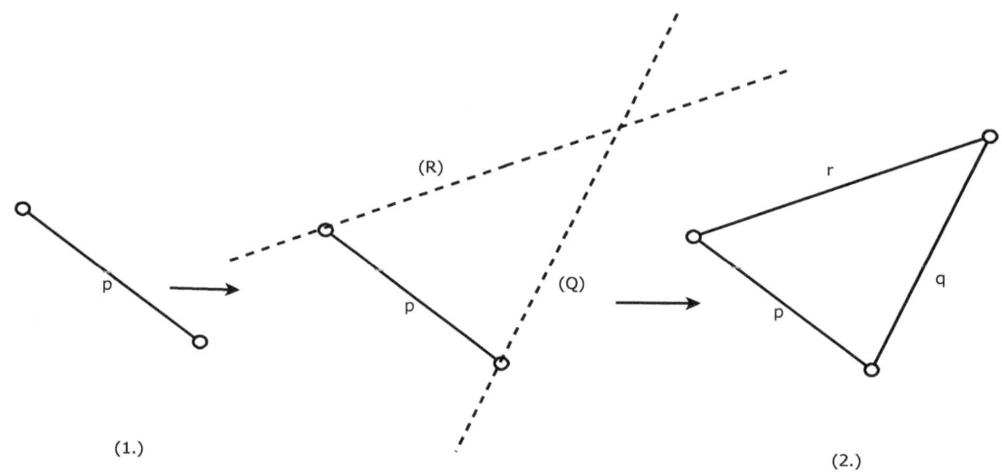

Figure 6. Steps (1) and (2).

(3) The other end of line $\boldsymbol{P}$ in the frame diagram meets lines $\boldsymbol{B}$ and $\boldsymbol{C}$ at a point. As the forces at point $\boldsymbol{PBC}$ are in equilibrium, draw a triangle in the force diagram with $\boldsymbol{p}$ as one side and lines parallel to $\boldsymbol{B}$ and $\boldsymbol{C}$ as the others. This can be done in one of two ways, illustrated in Figure 7.

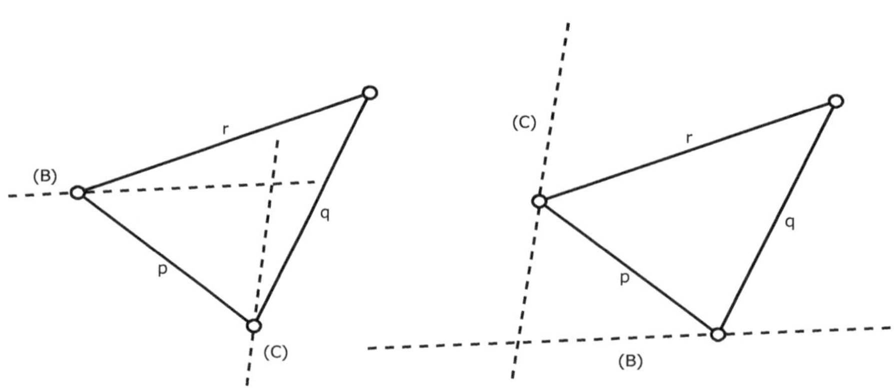

Figure 7. Two possible choices for Step (3).

Only one of these triangles belongs in the force diagram. As the force diagram is the reciprocal figure of the frame, look to the frame diagram to determine which of these two choices is correct. The endpoints of line $\boldsymbol{p}$ in the force diagram correspond to closed polygons in the frame diagram: namely, those polygons that contain $\boldsymbol{P}$ as an edge. That is, the endpoints of $\boldsymbol{p}$ in the force diagram correspond to polygons $\boldsymbol{PRB}$ and $\boldsymbol{PQC}$. As $\boldsymbol{PRB}$ is a closed polygon in the frame diagram, line $\boldsymbol{b}$ in the force diagram (parallel to $\boldsymbol{B}$) must meet lines $\boldsymbol{r}$ and $\boldsymbol{p}$ in a point. Similarly, since $\boldsymbol{PQC}$ is a closed polygon in

the frame diagram, line c in the force diagram (parallel to C) must meet lines q and c in a point. This corresponds to choosing the first case shown in Figure 7.
(4) Consider the equilibrium of forces at point QCA in the frame diagram. Two of the corresponding lines of force, q and c, have already been determined in the force diagram. Therefore, the only choice for line a (parallel to A) in the force diagram must complete triangle qca. Step (4) is illustrated in Figure 8.

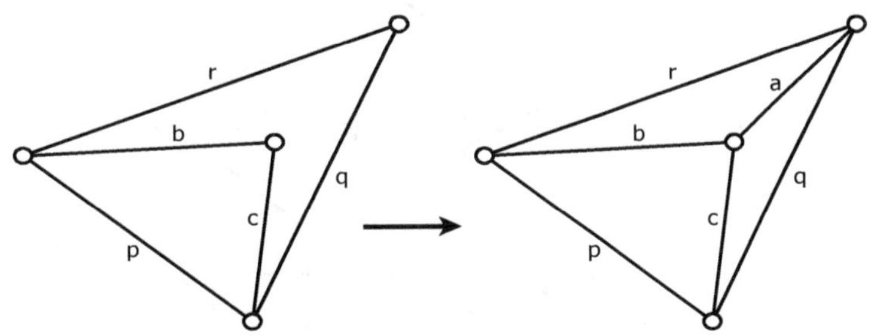

Figure 8. Step (4).

Remark 5. We can use an alternate notation to label the frame and force diagrams, known as *Bow's notation* or *interval notation*. In this method, we label regions of the frame diagram and the corresponding vertices in the force diagram with the same label. A full description of this method can be found in [10].

Maxwell's reciprocal figures are, in fact, geometric graphs that are dual to one another. Therefore of particular interest to us is Maxwell's description of diagrams, which have reciprocal figures: namely, plane projections of (closed) polyhedra. Consider a closed polyhedron $\mathcal{P}$ as the region bounded by some (finite) number of intersecting planes, $\{M_i\}$.

(1) Let $N_1, \ldots, N_n$ be the normal vectors to planes $M_1, \ldots M_n$. Choose a plane of projection, M, that satisfies:

(a) M does not intersect the closed polyhedron $\mathcal{P}$.

(b) For all $1 \le i \le n$, a line through normal vector N_i of plane M_i intersects plane M.

The standard plane projection of polyhedron $\mathcal{P}$ onto plane M gives one of the two figures. Now, construct a *second* polyhedron $\mathcal{P}'$, which, with respect to some paraboloid of revolution, is a geometric reciprocal to the first polyhedron. The projection of $\mathcal{P}'$ onto plane of projection M is then a reciprocal figure of the projection of $\mathcal{P}$ onto M. These two figures are reciprocal in the sense that corresponding lines are perpendicular to each other. One may obtain reciprocal figures with parallel orientation simply by rotating one figure by 90 degrees in the plane. Construct $\mathcal{P}'$ as follows:

(2) Let z_0 be a fixed-point in three-space that does not lie in M, which we will call the *origin*. The line perpendicular to plane of projection M that passes through point z_0 will be called the *axis*, denoted by z. Lastly, let z_M be the point of intersection of line z with plane M.

(3) For each $1 \le k \le n$, draw a line that is normal to plane face M_k of $\mathcal{P}$ and passes through point z_0. This line will intersect (projection) plane M at a unique point, call it m_k. Each point m_k may be thought of as the "projection" of face M_k onto plane M.

(4) For each $1 \le k \le n$, let z_k be the point of intersection of axis z with face M_k. This intersection always exists, since we chose M which is not parallel to any of the normal vectors N_k (and z is a normal vector of M). Let d_k be the (three-space) vector $z_M z_k$. That is, d_k is the vector between the intersection points of z with M and M_k. Ultimately, let:

$$p_k = m_k - d_k$$

We will call this point in space p_k the "point corresponding to face M_k of the polyhedron $\mathcal{P}$". Steps 2–4 of finding a point p_k corresponding to a particular face M_k are illustrated in Figure 9.

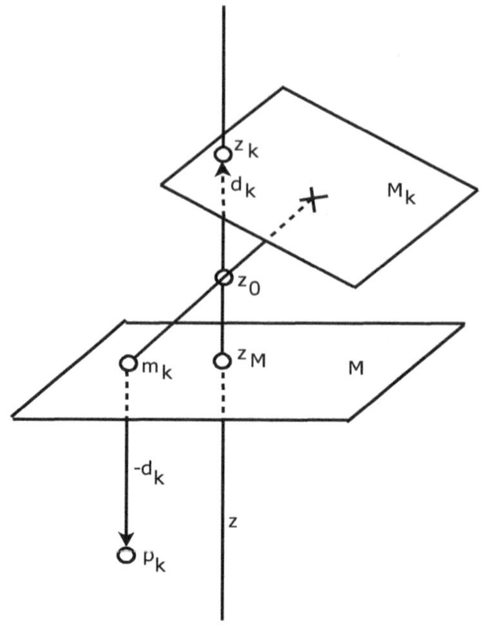

Figure 9. Finding a point p_k corresponding to a particular face M_k.

In this way, we determine n points $p_1, \cdots, p_n$ corresponding to the faces $M_1, \cdots M_n$ of $\mathcal{P}$. These points form the vertices of polyhedron $\mathcal{P}'$.

(5) For any two faces M_i and M_j that meet in an edge in polyhedron $\mathcal{P}$, draw a line between the corresponding points p_i and p_j. These new lines give us the edges of polyhedron $\mathcal{P}'$.

If we let M_iM_j denote the edge at the intersection of planes M_i and M_j and p_ip_j denote the line joining the corresponding points, one may easily check geometrically that the projection of line M_iM_j onto M is perpendicular to the projection of line p_ip_j onto M. In this way, the projection of $\mathcal{P}'$ onto M produces a figure, every line of which is perpendicular to the projection of the corresponding line of $\mathcal{P}$ onto M. What is more, for any lines that meet at a point in one projection, the corresponding lines form a closed polygon in the other projection. That is, the projection onto M of $\mathcal{P}$ and $\mathcal{P}'$ forms a pair of reciprocal figures in the sense of Maxwell.

5. Maxwell Reciprocal Figures and Kirchhoff Graphs

Through his theory of reciprocal figures, Maxwell deals with a pair of geometric graphs in which the vertex-cuts of one graph correspond to the cycles in the other, and *vice versa*. In this section, we discuss the relationship between Maxwell's figures, R^2-Kirchhoff graphs and, particularly, R^2-Kirchhoff duals. To make a connection between these Maxwell figures and Kirchhoff graphs, we define an R^2 vector graph corresponding to a given Maxwell diagram. First, we define a pair of embedded digraphs corresponding to Maxwell figures and then derive vector graphs accordingly.

Label the edges of the frame diagram as e_j and the corresponding (parallel) edges of the force diagram as f_j. We will adopt notation that indexes both vertices and cycles by these edge labels. Let $\{e_i, e_j, e_k\}$ denote the vertex incident to edges e_i, e_j and e_k, and use standard cycle notation for simple graphs, $f_if_jf_k$. Observe that if $f_if_jf_k$ is a cycle in the force diagram, there is a vertex of the form $\{e_i, e_j, e_k\}$ in the frame diagram, and *vice versa*. By arbitrarily assigning directions to each edge in the frame diagram, construct a corresponding plane-embedded digraph, H. We can then construct a plane-embedded digraph corresponding to the force diagram, $\boldsymbol{H'}$, by carefully assigning directions to each edge f_i. In particular, assign directions, such that for each cycle $f_i f_j f_k \ldots$ in the force diagram,

$$\chi(f_i f_j f_k \ldots) = \lambda(\{e_i, e_j, e_k, \ldots\}) \tag{6}$$

Finally, derive a pair of vector graphs, $\overline{\boldsymbol{D}}$ and $\overline{\boldsymbol{D'}}$, from $\boldsymbol{H}$ and $\boldsymbol{H'}$ by identifying pairs of geometric edges with the same length and direction (*i.e.*, the same two-space vector). Re-label all edges (with labels $\{s_i\}$ in $\overline{\boldsymbol{D}}$ and $\{t_i\}$ in $\overline{\boldsymbol{D'}}$) to reflect this edge-identification. One may verify that the result is two $\boldsymbol{R}^2$-vector graphs derived from the original pair of Maxwell figures.

Definition 17. Let $\overline{\boldsymbol{D}}$ and $\overline{\boldsymbol{D'}}$ be the pair of $\boldsymbol{R}^2$-vector graphs defined above. We say that $\overline{\boldsymbol{D}}$ and $\overline{\boldsymbol{D'}}$ are a pair of **corresponding $\boldsymbol{R}^2$-vector graphs** to the Maxwell figures. Note that we say "a pair", as there are many such sets of corresponding vector graphs, depending on the arbitrary assignment of directions to edges in the frame diagram.

Example 5. Figure 10 displays the Maxwell reciprocal figures as in Figure 5, now re-labeled as described above.

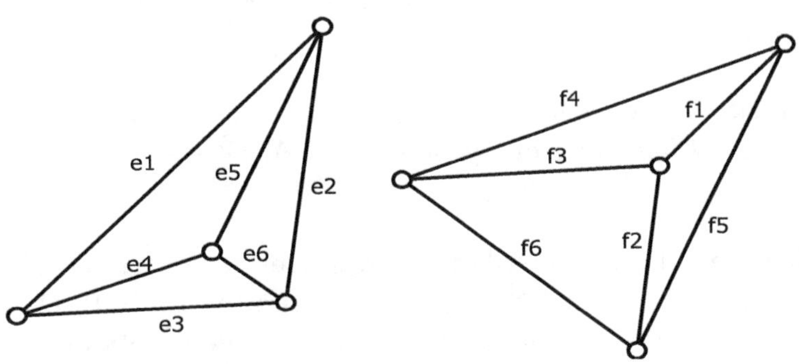

Figure 10. Maxwell reciprocal figures.

Figure 11 illustrates one pair of $\boldsymbol{R}^2$-vector graphs corresponding to these Maxwell reciprocals. In this example, the re-labeling of vector edges seems to be trivial. An example in which this process is nontrivial appears at the end of this section. Taking:

$$A = \begin{bmatrix} 1 & 0 & 1 & 1 & 0 & 0 \\ 0 & 1 & 1 & 0 & 0 & 1 \\ 0 & 0 & 0 & -1 & 1 & 1 \end{bmatrix} \text{ and } B = \begin{bmatrix} 1 & 0 & 0 & -1 & -1 & 0 \\ 0 & 1 & 0 & 0 & 1 & -1 \\ 0 & 0 & 1 & -1 & 0 & -1 \end{bmatrix}$$

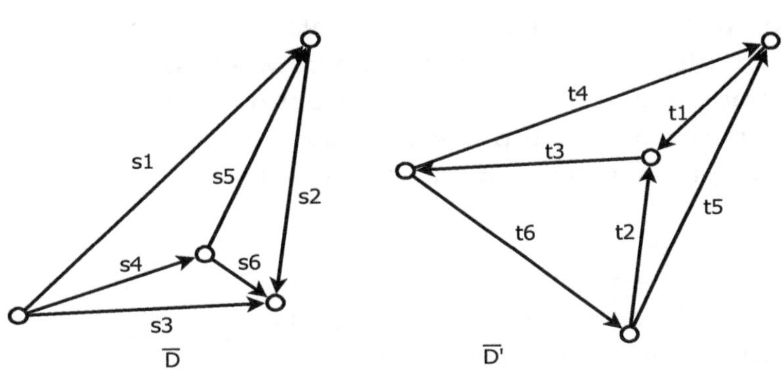

Figure 11. A pair of corresponding $\boldsymbol{R}^2$-vector graphs.

It is easy to verify that $\overline{D}$ is a Kirchhoff graph generated by A. Moreover, Null(A) is the column span of B^T, and most importantly, $\overline{D'}$ is a Kirchhoff graph generated by B. Beginning with Maxwell reciprocal diagrams, we have arbitrarily assigned directions to one diagram, then consistently assigned directions to the other diagram. The result is two R^2-vector graphs that are both Kirchhoff and indeed are each others' Kirchhoff dual.

The observation above is actually one example of more general results. Let G be a geometric (simple) graph, which has a reciprocal figure in the sense of Maxwell.

Lemma 18. *If no pair of edges of G is parallel in the plane, then any vector graph corresponding to G is a Kirchhoff graph.*

Proof. Let $\overline{D}$ be any vector graph derived by assigning directions to each edge in G, and let $\theta(s_i)$ be the angle vector s_i makes with the positive x-axis. Then, because no edges of G were parallel in the plane, every pair of vectors s_i, s_j for $\overline{D}$ satisfies $\theta(s_i) \neq \theta(s_j)$. That is, each edge e_i is assigned a distinct two-space vector, and $\overline{D}$ is a general vector graph. Therefore as $\overline{D}$ is also simple, $\overline{D}$ is Kirchhoff by Theorem 11. □

Now, let G and G' be a pair of Maxwell reciprocal figures and $\overline{D}$ and $\overline{D'}$ be any pair of R^2-vector graphs corresponding to G and G'.

Theorem 19. *If G has no parallel edges in the plane, both $\overline{D}$ and $\overline{D'}$ are Kirchhoff graphs. Moreover, for any matrix A, such that $\overline{D}$ is a Kirchhoff graph generated by A, $\overline{D'}$ is generated by Null(A): $\overline{D}$ and $\overline{D'}$ are Kirchhoff duals.*

Proof. Since G has no parallel edges in the plane, its Maxwell reciprocal G' also has no pair of parallel edges. Therefore, both $\overline{D}$ and $\overline{D'}$ are Kirchhoff graphs by Lemma 18. Both $\overline{D}$ and $\overline{D'}$ are *general* vector graphs; therefore, in Step (4) above, we use the following simple re-labeling:

$$e_i \longmapsto s_i \text{ and } f_i \longmapsto t_i \text{ for all } i$$

For any matrix A, such that $\overline{D}$ is generated by A, the fact that $\overline{D'}$ is generated by Null(A) follows from Equation (6). □

Remark 6. Lemma 18 and Theorem 19 above demonstrate that any Maxwell frame diagram without parallel edges leads to a number of pairs of dual R^2-Kirchhoff graphs. This means that *most* Maxwell diagrams will lead directly to pairs of Kirchhoff duals: given a randomly-chosen closed polyhedron, a randomly-chosen plane projection will lead to a frame diagram with no parallel edges.

Though Remark 6 demonstrates that there is significant overlap between Maxwell reciprocals and simple R^2-Kirchhoff duals, arguably, the cases that make vector graphs interesting are those in which vector edges are identified. The connection between Maxwell figures and R^2-Kirchhoff graphs becomes considerably more delicate when considering geometric graphs with parallel edges. The care with which the polyhedron and projection must be constructed in the following example, however, demonstrates that, in some sense, these situations arise as particular special cases of general Maxwell diagrams.

Example 6. In this example, we explicitly construct a graph G with the following properties:
(1) When considered as a geometric graph, G has a Maxwell reciprocal, G'.
(2) There exists an R^2-vector graph $\overline{G}$ corresponding to G that is not Kirchhoff.
(3) Any R^2-vector graph $\overline{G'}$ corresponding to G' *is* Kirchhoff.

Consider the following system of five planes in $\boldsymbol{R}^3$:

$$-x+z=0 \qquad -x+3y-2z=0 \qquad y+z=0 \qquad 5x-y+4z=0 \qquad z=0 \tag{7}$$

The three-dimensional region bounded by these five planes forms a closed, bounded polyhedron in $\boldsymbol{R}^3$, as illustrated in Figure 12.

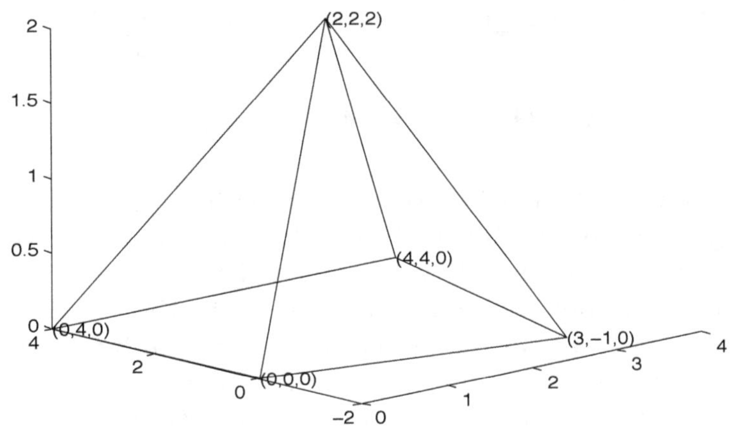

Figure 12. Closed polyhedron defined by Equation (7).

The projection of this figure onto the xy-plane has a reciprocal figure in the sense of Maxwell. The projection of this polyhedron is shown in Figure 13.

Remark 7. This was a very specifically-chosen embedding of this polyhedron, which forces the projection of two edges to be parallel with the same length in the plane. As these two edges were not parallel in three-space, a randomly-chosen plane projection of this polyhedron would, in general, result in a geometric graph having no pair of parallel edges.

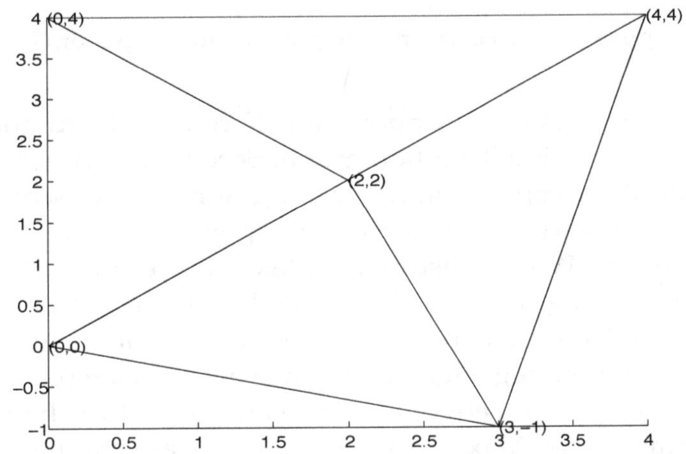

Figure 13. Projection on the xy-plane of the Polyhedron in Figure 12.

Now, consider the projection in Figure 13 as a frame diagram and, being the plane projection of a convex polyhedron, as shown in Figure 14, construct the associated force diagram.

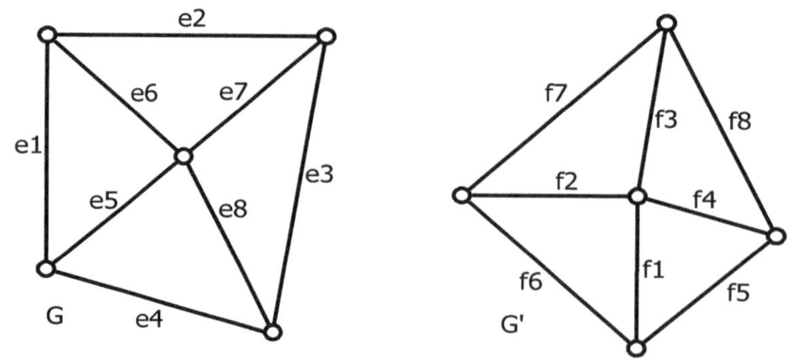

Figure 14. G and G', Maxwell reciprocals.

The vector graph $\overline{G}$ shown in Figure 15 is one corresponding vector graph for the frame diagram G in Figure 14. Every edge in vector graph $\overline{G}$ has multiplicity one, and each vector edge occurs exactly once, except for edge s_1, which occurs exactly twice. Given that both occurrences of s_1 are not bridges (nor a directed cycle), it follows from Theorem 12 that $\overline{G}$ is *not* a Kirchhoff graph. On the other hand, consider the force diagram G'. It is clear that given *any* vector graph $\overline{G'}$ corresponding to G', every pair of vector edges t_i, t_j satisfies *either* $|t_i| \neq |t_j|$ or $\theta(t_i) \neq \theta(t_j)$. That is, each edge is assigned a distinct vector in R^2. Therefore, every R^2-vector graph corresponding to G' is general and, thus, Kirchhoff by Theorem 11.

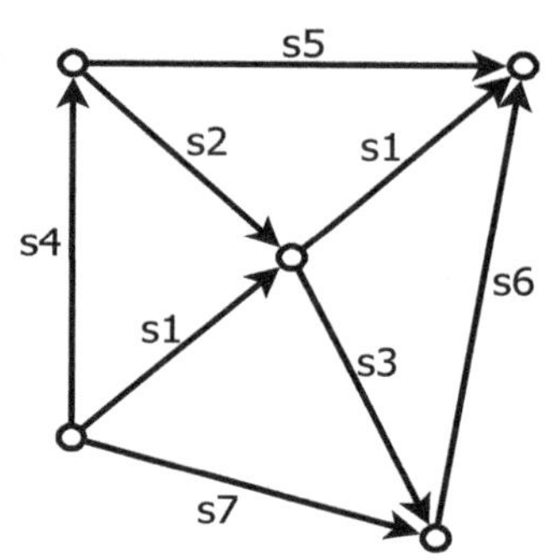

Figure 15. $\overline{G}$, a corresponding R^2-vector graph for G.

Remark 8. This example illustrates a few important differences. First, not every R^2-vector graph corresponding to a Maxwell figure is a Kirchhoff graph. Second, it is possible that a pair of Maxwell reciprocals generates a pair of corresponding R^2-vector graphs, one of which is Kirchhoff, while the other is not. These discrepancies arise because the corresponding vector graphs may have differing numbers of distinct vector edges. When constructing Maxwell's reciprocal figures, we are guaranteed that every pair of corresponding edges is *parallel* to each other. The *length* of each edge in the reciprocal figure is prescribed in Maxwell's construction. However, if two parallel edges in the frame diagram have the same length, the corresponding edges in the force diagram are parallel, but need not have the same length. For example, in Figure 14, edges e_5 and e_7 are parallel *and* have the same length. In the reciprocal diagram, as guaranteed by Maxwell, edges f_5 and f_7 are parallel. However, f_5 and f_7 have different lengths.

This is not to suggest that the *only* Maxwell figures that correspond to dual Kirchhoff graphs are those with no parallel edges. Symmetry in the frame diagram can lead to symmetry in the force diagram, which results in a number of identified edges in the corresponding vector graphs.

Example 7. Consider the pair of Maxwell reciprocal figures in Figure 16 and pair of corresponding $\boldsymbol{R}^2$-vector graphs in Figure 17.

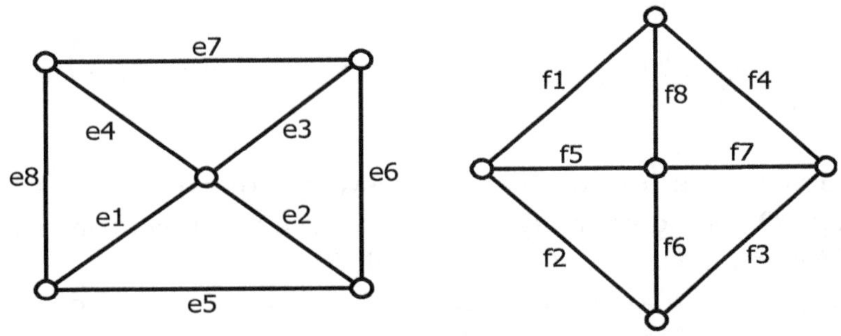

Figure 16. A pair of Maxwell reciprocal figures.

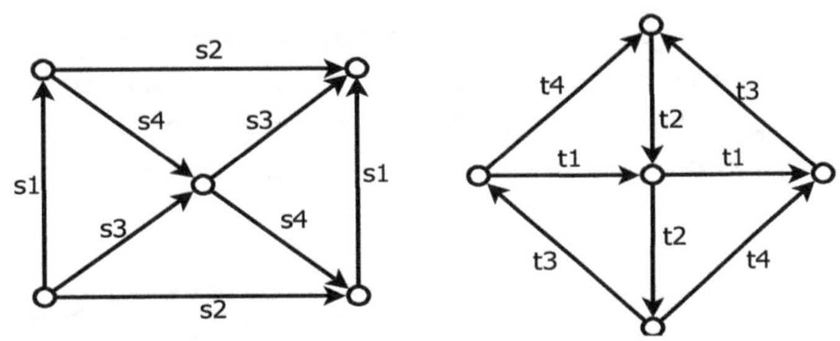

Figure 17. A pair of corresponding vector graphs.

One may verify that these two $\boldsymbol{R}^2$-vector graphs are both Kirchhoff graphs, generated by matrices $\boldsymbol{A}$ and $\boldsymbol{B}$, respectively, where:

$$\boldsymbol{A} = \begin{bmatrix} \mathbf{1} & \mathbf{1} & \mathbf{1} & \mathbf{0} \\ \mathbf{-1} & \mathbf{1} & \mathbf{0} & \mathbf{1} \end{bmatrix} \quad \text{and} \quad \boldsymbol{B} = \begin{bmatrix} \mathbf{1} & \mathbf{0} & \mathbf{-1} & \mathbf{1} \\ \mathbf{0} & \mathbf{-1} & \mathbf{1} & \mathbf{1} \end{bmatrix}$$

Moreover, Null($\boldsymbol{A}$) = colspan($\boldsymbol{B}^T$); these Kirchhoff graphs are, in fact, a pair of Kirchhoff duals.

6. Non-Simple, Non-General and Non-Planar $\boldsymbol{R}^2$-Vector Graphs

Remark 6 and Example 7 illustrate that in many cases, Maxwell reciprocal figures can lead to pairs of $\boldsymbol{R}^2$-Kirchhoff duals. Example 6, however, demonstrates that these two theories do not always agree: $\boldsymbol{R}^2$ vector graphs corresponding to a Maxwell figure are not necessarily Kirchhoff when edges are identified as vectors. Even if the vector graph corresponding to a Maxwell figure *is* Kirchhoff, the vector graph corresponding to the Maxwell reciprocal *need not* be Kirchhoff. Moreover, there are a number of classes of vector graphs that will never correspond to a Maxwell figure; for instance, vector graphs with multiple edges and vector graphs that are not the projection of a polyhedron. This leads us to begin exploring other methods of constructing dual vector graphs, in particular, dual $\boldsymbol{R}^2$-vector graphs.

6.1. *Planar $\mathbf{R}^2$-Kirchhoff Graphs*

Before considering non-simple Kirchhoff graphs, we first illustrate an alternative method of constructing the pair of $\boldsymbol{R}^2$-Kirchhoff duals given in Figure 17, based on planarity. Rather than Maxwell reciprocals, we consider very specific embeddings of digraphs, which are dual in the standard sense.

Example 8. Begin with the first vector graph in Figure 17, $\overline{\boldsymbol{D}}$. Next, construct a dual $\boldsymbol{R}^2$-vector graph, $\overline{\boldsymbol{D}'}$, for $\overline{\boldsymbol{D}}$, as follows. First, view $\overline{\boldsymbol{D}}$ as a standard (planar) directed graph, and form the standard dual graph (we will use the convention of assigning directions to edges in $\overline{D'}$ by traversing cycles in $\overline{D}$ in a *clockwise* orientation). For each edge labeled s_i in $\overline{D}$, we draw a corresponding edge in $\overline{D'}$, which we label as t_i. This process is illustrated in Figure 18, and leads to the following:

Question 2. *Can we position/embed the vertices of* $\overline{D'}$ *in* $\mathbf{R}^2$*, such that for each* j*, all copies of every edge with label* t_j *are identical vectors in* $\mathbf{R}^2$*?*

We see that in the case of vector graph $\overline{D}$, the answer to this question is "yes". In particular, we can arrange vector graph $\overline{D'}$ as shown in Figures 19. Observe that this is precisely the $\mathbf{R}^2$-vector graph $\overline{D'}$ corresponding to Maxwell reciprocal G' in Example 7.

Therefore, in the example above, much like the standard case of plane-embedded digraphs, one can interchange the roles of faces and vertices to find Kirchhoff duals. Moreover, this method can address Kirchhoff graphs with multiple edges: multiple vector edges are replaced by consecutive (identical) vector edges in the dual $\mathbf{R}^2$-vector graph, and *vice versa*.

Remark 9. This relation between vector edges corresponds to the desired result in reference to matrices. Non-unit entries in an incidence vector indicate multiple edges, whereas non-unit entries in a cycle vector indicate consecutive edges. Therefore, replacement of multiple edges by consecutive edges (and *vice versa*) causes the cut space and cycle space to interchange roles.

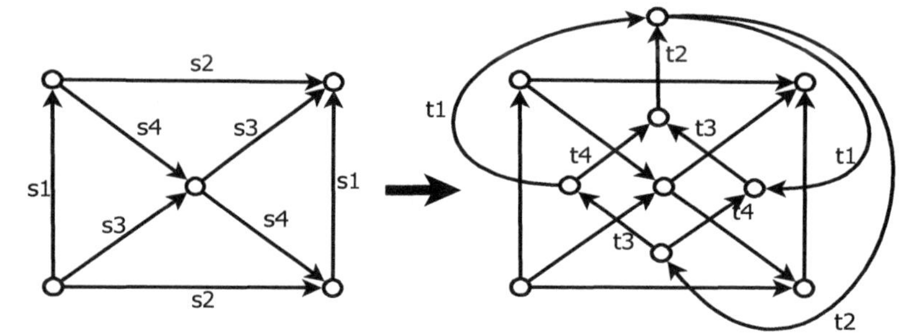

Figure 18. The standard dual of $\mathbf{D}$.

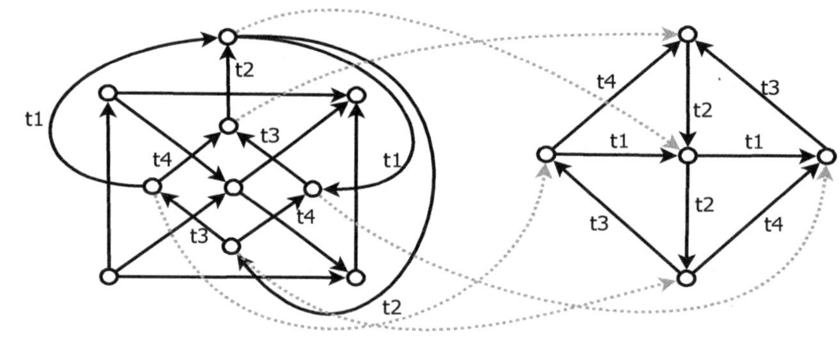

Figure 19. Arranging vector graph $\overline{D'}$.

Example 9. Consider the matrix:

$$A = \begin{bmatrix} 2 & 0 & -1 \\ 0 & 2 & 1 \end{bmatrix} \text{ where } \mathrm{Null}(A) = \mathrm{span}\left\{ \begin{bmatrix} 1 & -1 & 2 \end{bmatrix}^T \right\}$$

The first $\mathbf{R}^2$-vector graph in Figure 20 is a Kirchhoff graph $\overline{D}$ generated by A. We also re-draw this vector graph in order to better illustrate the method of addressing multiple edges.

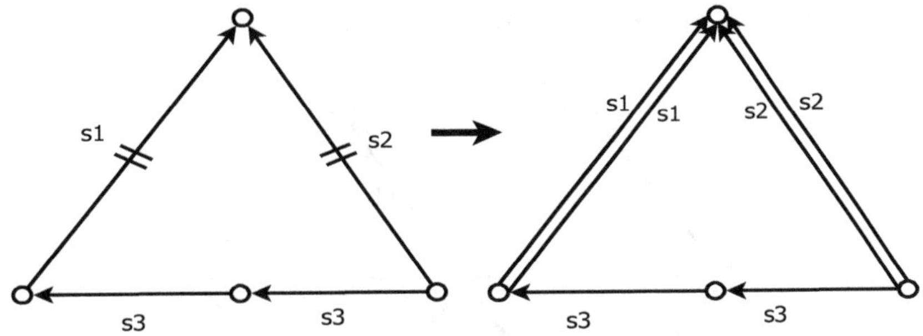

Figure 20. Kirchhoff graph $\overline{D}$ generated by matrix A.

As before, construct a dual directed graph and arrange the vertices to give a dual R^2 vector graph, shown in Figure 21. Observe that multiple edges in the R^2-vector graph result in consecutive edges in the dual R^2-vector graph, and *vice versa*. One can easily verify that this is an R^2-Kirchhoff graph generated by Null(A).

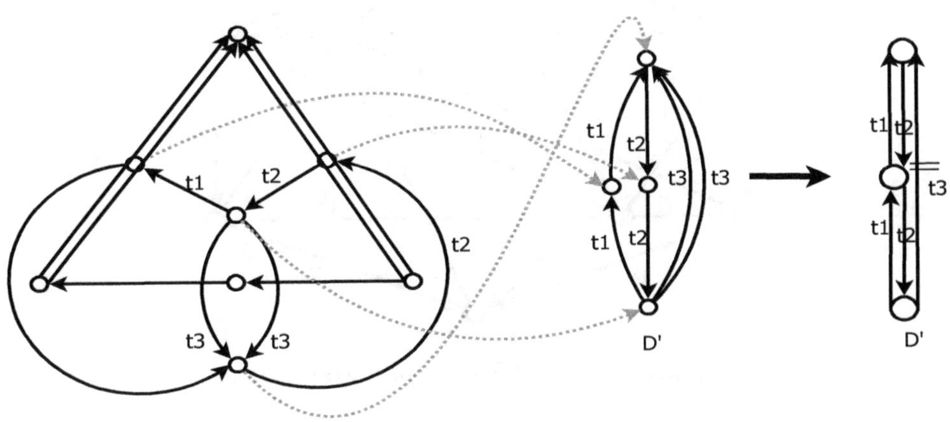

Figure 21. Arranging vector graph $\overline{D'}$.

One advantage of using dual methods is graph construction. In some cases, given a matrix A, it may be relatively straightforward to construct an R^2-Kirchhoff graph for Null(A). Using a dual vector graph method may then lead to an R^2-Kirchhoff graph for A without constructing one directly. In the example above it may not be intuitively clear how to construct an R^2-Kirchhoff graph for the matrix $[1, -1, 2]$. On the other hand, it is a fairly easy task to construct an R^2-Kirchhoff graph with *cycle space* spanned by $[1, -1, 2]$, and an R^2-Kirchhoff graph for $[1, -1, 2]$ was then obtained via a Kirchhoff dual.

6.2. Non-Planar R^2-Kirchhoff Graphs

The planar R^2-Kirchhoff graphs given above are not representative of all R^2-Kirchhoff graphs, which need neither be planar nor have a geometric dual. One advantage of Definition 15 is the opportunity to construct dual R^2-Kirchhoff graphs in cases where no dual graph exists under standard notions. In the following example, we consider $\overline{H} = (K_{3,3}, \varphi)$, an R^2-vector graph representation of $K_{3,3}$, and present a Kirchhoff dual, $\overline{H'}$. This Kirchhoff dual leads to a few interesting observations: $\overline{H'}$ is an embedding of a digraph, which *has* a geometric dual, and embedding this geometric dual to create an R^2-vector graph $\overline{H''}$ results in a Kirchhoff graph that is, nearly, the original $\overline{H}$.

Example 10. Consider the vector graphs $\overline{H}$ shown in Figure 22, and $\overline{H'}$ in Figure 23.

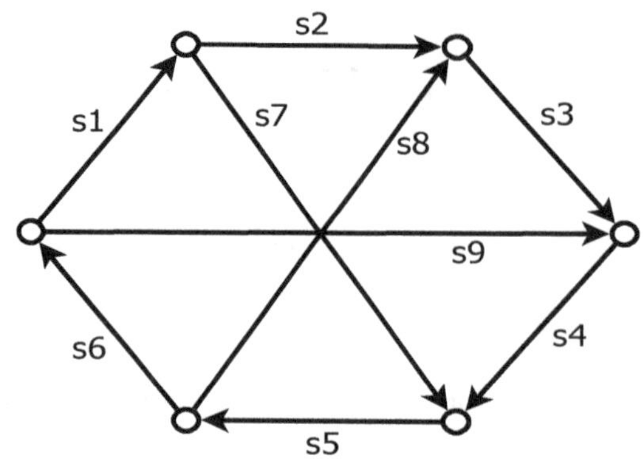

Figure 22. An R^2-Kirchhoff graph, $\overline{H} = (K_{3,3}, \varphi)$.

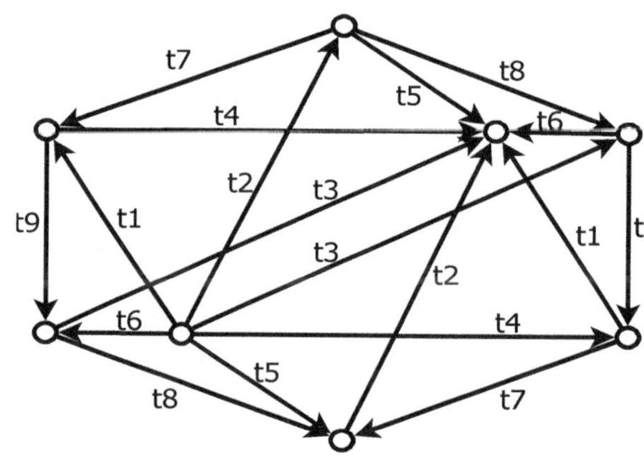

Figure 23. Dual R^2-Kirchhoff graph, $\overline{H'}$.

$\overline{H}$ is both simple and general. Therefore, by Theorem 11, $\overline{H}$ is an R^2-Kirchhoff graph. $\overline{H}$ is an embedding of $K_{3,3}$, which is non-planar and has no geometric dual. However, $\overline{H}$ is an R^2-Kirchhoff graph generated by matrix A, where $\text{Null}(A) = B^T$. Moreover, vector graph $\overline{H'}$ is an R^2-Kirchhoff graph generated by B. That is, $\overline{H'}$ is a Kirchhoff dual of $\overline{H}$.

$$A = \begin{bmatrix} -1 & 1 & 0 & 0 & 0 & 0 & 1 & 0 & 0 \\ 0 & -1 & 1 & 0 & 0 & 0 & 0 & -1 & 0 \\ 0 & 0 & -1 & 1 & 0 & 0 & 0 & 0 & -1 \\ 0 & 0 & 0 & -1 & 1 & 0 & -1 & 0 & 0 \\ 0 & 0 & 0 & 0 & -1 & 1 & 0 & 1 & 0 \end{bmatrix}$$

$$B = \begin{bmatrix} 1 & 1 & 1 & 0 & 0 & 0 & 0 & 0 & -1 \\ 0 & 1 & 1 & 1 & 0 & 0 & -1 & 0 & 0 \\ 0 & 0 & 1 & 1 & 1 & 0 & 0 & 1 & 0 \\ 0 & 0 & 0 & 1 & 1 & 1 & 0 & 0 & 1 \end{bmatrix}$$

Therefore, although $\overline{H}$ does not have a dual graph under any standard notions, when taken as an R^2-vector graph, it has a dual in the Kirchhoff sense. Moreover, this Kirchhoff dual $\overline{H'}$ arises as a plane projection of a polyhedron in R^3, meaning $\overline{H'}$ has a geometric dual. This polyhedron and its geometric dual are illustrated in Figure 24 (for each edge labeled t_j in $\overline{H'}$, we label the corresponding edge as s_j in the geometric dual)

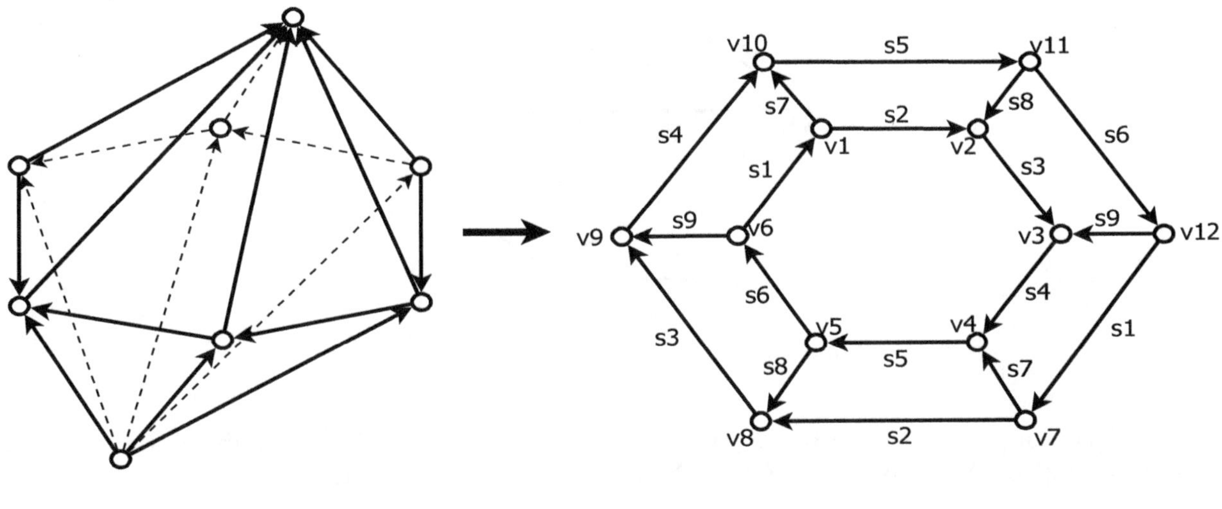

Figure 24. The polyhedron of $\overline{H'}$ and its geometric dual.

In order to construct this geometric dual as an R^2-vector graph, $\overline{H''}$, we must embed the vertices in the plane so that all vector edges with label s_i are identical. This can be accomplished by, for each $1 \leq j \leq 6$, mapping the pair of vertices v_j and v_{j+6} to the same point in the plane. This embedding is illustrated in Figure 25.

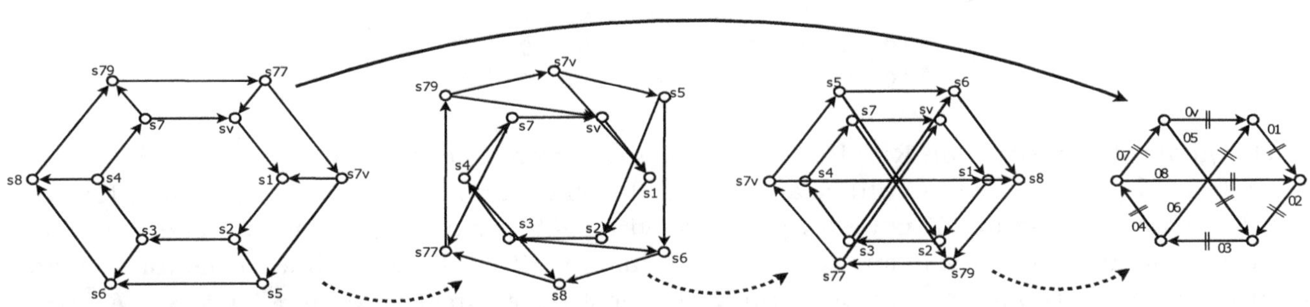

Figure 25. Constructing R^2-vector graph $\overline{H''}$.

This R^2-vector graph is vector graph $\overline{H}$, with every edge doubled. As a result, $\overline{H''}$ is both an R^2-Kirchhoff graph generated by A and a Kirchhoff dual for $\overline{H'}$.

By following essentially the reverse process as that outlined in Example 10, we can illustrate how to directly construct an R^2-Kirchhoff dual for K_5, the complete graph on five vertices, which is also non-planar.

Example 11. Consider the R^2-vector graph in Figure 26, $\overline{K} = (K_5, \varphi)$, which is a vector graph constructed from an R^2-vector assignment of K_5:

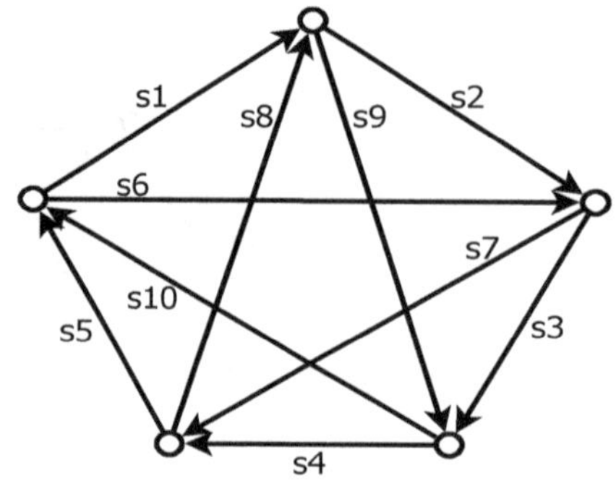

Figure 26. R^2-vector graph $\overline{K} = (K_5, \varphi)$.

As an R^2-vector graph, $\overline{K}$ is both simple and general. Therefore, $\overline{K}$ is Kirchhoff and generated by matrix A, given below. Moreover, Null(A) is the column span of B^T where:

$$A = \begin{bmatrix} -1 & 1 & 0 & 0 & 0 & 0 & 0 & -1 & 1 & 0 \\ 0 & -1 & 1 & 0 & 0 & -1 & 1 & 0 & 0 & 0 \\ 0 & 0 & -1 & 1 & 0 & 0 & 0 & 0 & -1 & 1 \\ 0 & 0 & 0 & -1 & 1 & 0 & -1 & 1 & 0 & 0 \end{bmatrix}$$

$$B = \begin{bmatrix} 1 & 1 & 0 & 0 & 0 & -1 & 0 & 0 & 0 & 0 \\ 0 & 0 & 1 & 0 & 0 & 1 & 0 & 0 & 0 & 1 \\ 0 & 0 & 0 & 1 & 1 & 0 & 0 & 0 & 0 & -1 \\ 0 & 0 & 0 & 1 & 0 & 0 & 0 & 1 & 1 & 0 \\ 0 & 1 & 1 & 0 & 0 & 0 & 0 & 0 & -1 & 0 \\ 0 & 0 & 1 & 1 & 0 & 0 & -1 & 0 & 0 & 0 \end{bmatrix}$$

Our goal is to construct an R^2-Kirchhoff graph generated by B. Rather than attempting to do so directly from the matrix, we can utilize any R^2-Kirchhoff graph generated by A. Clearly, R^2-vector graph $\overline{K''}$, shown in Figure 27, is vector graph $\overline{K}$ with each vector edge doubled. Therefore, $\overline{K''}$ is also a Kirchhoff graph for A, and any R^2-Kirchhoff dual for $\overline{K''}$ is also a Kirchhoff dual for $\overline{K}$. We will demonstrate that we can use R^2-vector graph $\overline{K''}$ to construct a Kirchhoff dual, $\overline{K'}$. Clearly, $\overline{K''}$ results from a plane embedding of a digraph with 20 edges. For a moment, however, suppose that this digraph had more than five vertices; some of these vertices were then embedded at the same point when forming R^2-vector graph $\overline{K''}$. In particular, consider digraph D as in Figure 28, on **10** vertices and 20 edges.

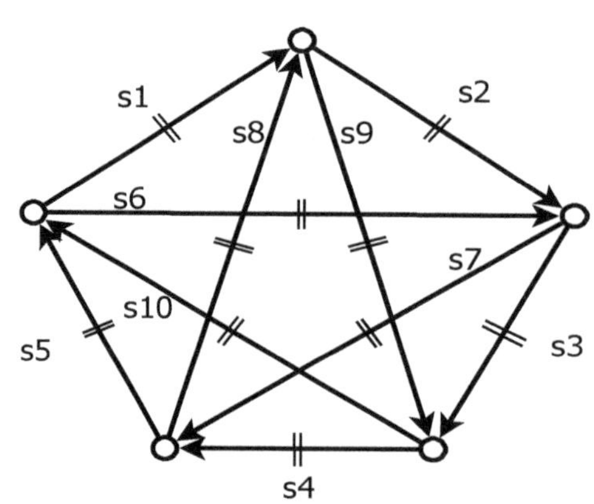

Figure 27. R^2-vector graph $\overline{K''}$.

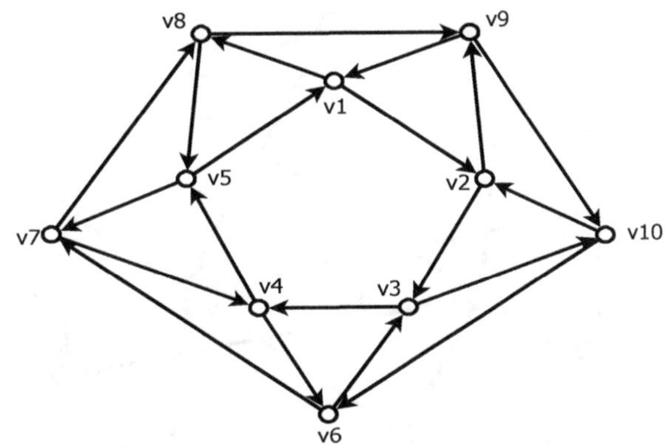

Figure 28. Digraph $\boldsymbol{D}$ with 10 vertices and 20 edges.

An embedding of $\boldsymbol{D}$, which, for each $\mathbf{1} \leq j \leq \mathbf{5}$, maps the pair of vertices v_j and v_{j+5} to the same point in the plane will produce $\mathbb{R}^2$-vector graph $\overline{K''}$. We alternately render digraph $\boldsymbol{D}$ in Figure 29, labeling edges to indicate which become identical vectors under this embedding. This digraph $\boldsymbol{D}$ is clearly planar, so we may construct its standard dual, $\boldsymbol{D'}$. For each edge labeled s_j in $\boldsymbol{D}$, we label the corresponding edge as t_j in $\boldsymbol{D'}$. Given dual picture $\boldsymbol{D'}$, shown in Figure 30, we have returned to Question 2: can we embed the vertices of $\boldsymbol{D'}$ in $\boldsymbol{R}^2$, such that all directed edges with label t_j are identical vectors? One such embedding of $\boldsymbol{D'}$ is illustrated in Figure 31, forming an $\boldsymbol{R}^2$-vector graph $\overline{K'}$. More importantly, one may verify that this vector graph $\overline{K'}$ is a Kirchhoff graph generated by $\boldsymbol{B}$. That is, $\overline{K'}$ is a Kirchhoff dual of $\overline{K} = (K_5, \varphi)$, and we have constructed $\overline{K'}$ directly from one of its $\boldsymbol{R}^2$-Kirchhoff duals, $\overline{K''}$.

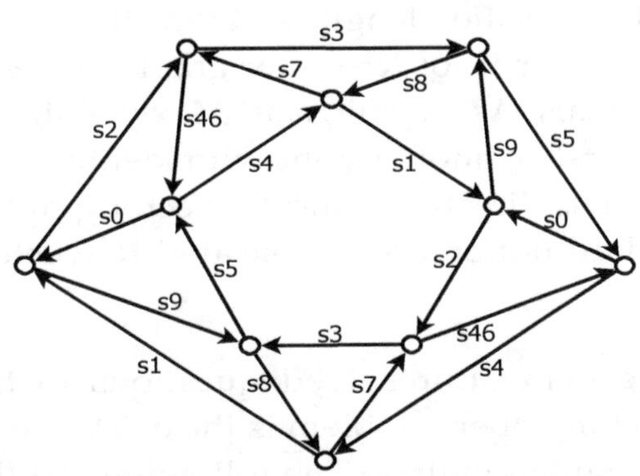

Figure 29. Identified edges in digraph $\boldsymbol{D}$.

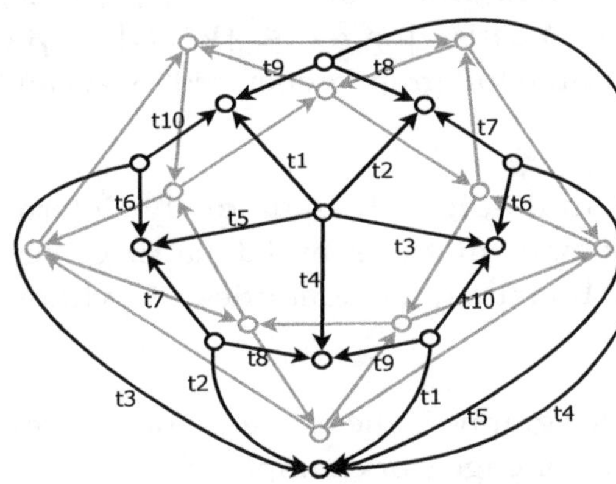

Figure 30. Constructing the dual $\boldsymbol{D'}$ of digraph $\boldsymbol{D}$.

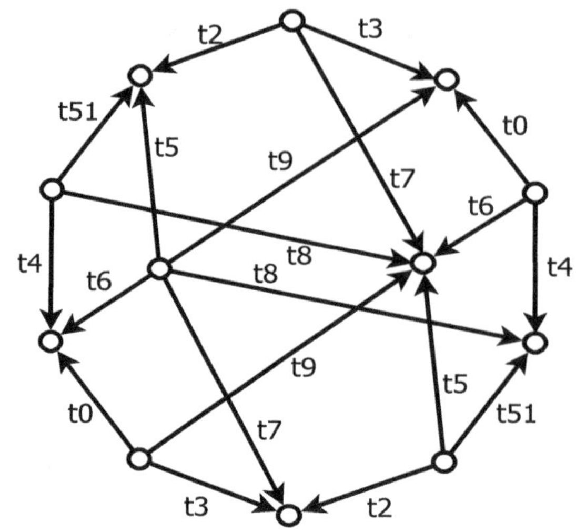

Figure 31. R^2-vector graph $\overline{K'}$, a Kirchhoff dual of $\overline{K}$.

7. Conclusions

In this paper, we have described a new type of graph, a vector graph, allowing us to define the notion of Kirchhoff graphs without any reference to matrices. In particular, any simple general vector graph is guaranteed to be Kirchhoff, whereas any simple vector graph having exactly one pair of identified edges is not Kirchhoff unless both such edges are bridges, or form a directed cycle, which is a cut set. Similar to R^2-Kirchhoff graphs, Maxwell's reciprocal figures also address cycles and vertex cuts in graphs having edges with a specified length and direction. Given any Maxwell diagram with no parallel edges, we showed that any pair of R^2-vector graphs corresponding to this diagram and its reciprocal are a pair of Kirchhoff duals. As a result, *most* Maxwell figures correspond to many pairs of dual R^2-Kirchhoff graphs. Given R^2-Kirchhoff graphs with identified edges, more careful treatment is required to construct Kirchhoff duals directly. Using R^2-vector graph representations of K_5 and $K_{3,3}$, which have no dual under standard notions, we illustrated R^2-vector graphs, which are dual to K_5 and $K_{3,3}$ in the Kirchhoff sense.

Moving forward, there are many unresolved questions in the study of Kirchhoff graphs. As mentioned above, the overarching open problem is the existence and construction of a Kirchhoff graph generated by any integer-valued matrix. The following is a (by no means exhaustive) set of open problems surrounding Kirchhoff graphs, which could lead to an existence proof.

Problem 1. Given any $A \in Z^{m\times n}$, determine constants $\alpha_A, \beta_A \in Z$, such that there always exists a Kirchhoff graph D generated by A satisfying $\alpha_A \leq |V(D)| \leq \beta_A$. Determining bounds on the number of vertices required for a Kirchhoff graph generated by A can lead to algorithmic approaches of Kirchhoff graph construction.

Problem 2. Though a standard multi-digraph is uniquely determined by its incidence matrix, a Kirchhoff graph is generated by many matrices, and a matrix can generate many Kirchhoff graphs. Define a matrix-like structure that both uniquely identifies a Kirchhoff graph and reflects which class of matrices generate this graph.

Problem 3. A unique facet of vector graphs is the existence of vertices that are incident to some edge, but are null with respect to that vector edge (for example, if vertex v is the initial vertex of some edge $\overline{e}_i$ and the terminal vertex of edge $\overline{e}_j$ where $\varphi(e_i) = \varphi(e_j)$). Determine the properties of matrix A that require that any Kirchhoff graph generated by A have such null vertices.

Problem 4. Classify the integer matrices that generate a Kirchhoff graph that is simple.

Problem 5. Determine a canonical or minimal form of Kirchhoff graph that is uniquely determined by its generating matrices.

Acknowledgments: Tyler Reese receives graduate funding from the National Science Foundation Graduate Research Fellowship Program. We wish to thank Brigitte Servatius for her ideas and guidance. We also thank the referees for their suggestions.

Author Contributions: Tyler Reese, Randy Paffenroth, Joseph Fehribach developed the theory, definitions and organization of the paper. Tyler Reese wrote the paper.

References

1. Fehribach, J.D. Vector-space methods and Kirchhoff Graphs for reaction networks. *SIAM J. Appl. Math.* **2009**, *70*, 543–562.
2. Fehribach, J.D. Matrices and their Kirchhoff Graphs. *Math. Contemp.* **2015**, *9*, 125–144.
3. Fishtik, I.; Callaghan, C.A.; Datta, R. Reaction route graphs I. Theory and algorithm. *J. Phys. Chem. B* **2004**, *108*, 5671–5682.
4. Fishtik, I.; Callaghan, C.A.; Datta, R. Reaction route graphs II. Examples of enzyme and surface catalyzed single overall reactions. *J. Phys. Chem. B* **2004**, *108*, 5683–5697.
5. Fishtik, I.; Callaghan, C.A.; Fehribach, J.D.; Datta, R. A reaction route graph analysis of the electrochemical hydrogen oxidation and evolution reactions. *J. Electroanal. Chem.* **2005**, *576*, 57–63.
6. Fishtik, I.; Datta, R. A reaction route network for hydrogen combustion. *Physica* **2007**, *373*, 777–784.
7. Fehribach, J.D. The use of graphs in the study of electrochemical reaction networks. In *Modern Aspects of Electrochemistry*; Vayenas, C., White, R.E., Gamboa-Alceco, M.E., Eds.; Springer-Verlag: New York, NY, USA, 2007; Chapter 3; Volume 41, pp. 197–219.
8. Craciun, G.; Tangy, Y.; Feinberg, M. Understanding bistability in complex enzyme-driven reaction routes. *Proc. Natl. Acad. Sci. USA* **2006**, *103*, 8697–8701.
9. Maxwell, J.C. On reciprocal figures, frames, and diagrams of forces. *Trans. R. Soc. Edinb.* **1870**, *26*, 1–40.
10. Baker, W.F.; Beghini, L.L.; Mazurek, A.; Carrion, J.; Beghini, A. Maxwell's reciprocal diagrams and discrete michell frames. *Struct. Multidiscip. Optim.* **2013**, *48*, 267–277.
11. Bollobás, B. *Modern Graph Theory. Springer Graduate Texts in Mathematics*; Springer-Verlag: New York, NY, USA, 1998.
12. Friedman, J. *Sheaves on Graphs, Their Homological Invariants, and a Proof of the Hanna Neumann Conjecture: With an Appendix by Warren Dicks*; Memoirs of the AMS: Providence, RI, USA, 2015; Volume 233.

3

Isoperimetric Numbers of Randomly Perturbed Intersection Graphs

Yilun Shang

Department of Computer and Information Sciences, Faculty of Engineering and Environment, Northumbria University, Newcastle NE1 8ST, UK; shylmath@hotmail.com.

Abstract: Social networks describe social interactions between people, which are often modeled by intersection graphs. In this paper, we propose an intersection graph model that is induced by adding a sparse random bipartite graph to a given bipartite graph. Under some mild conditions, we show that the vertex–isoperimetric number and the edge–isoperimetric number of the randomly perturbed intersection graph on n vertices are $\Omega(1/\ln n)$ asymptotically almost surely. Numerical simulations for small graphs extracted from two real-world social networks, namely, the board interlocking network and the scientific collaboration network, were performed. It was revealed that the effect of increasing isoperimetric numbers (i.e., expansion properties) on randomly perturbed intersection graphs is presumably independent of the order of the network.

Keywords: isoperimetric number; random graph; intersection graph; social network

---

1. Introduction

Complex large-scale network structures arise in a variety of natural and technological settings [1,2], and they pose numerous challenges to computer scientists and applied mathematicians. Many interesting ideas in this area come from the analysis of social networks [3], where each vertex (*actor*) is associated with a set of properties (*attributes*), and pairs of sets with nonempty intersections correspond to edges in the network. Complex and social networks represented by such *intersection graphs* are copious in the real world. Well-known examples include the film actor network [4], where actors are linked by an edge if they performed in the same movie, the academic co-authorship network [5], where two researchers are linked by an edge if they have a joint publication, the circle of friends in online social networks (e.g., Google+), where two users are declared adjacent if they share a common interest, and the Eschenauer–Gligor key predistribution scheme [6] in secure wireless sensor networks, where two sensors establish secure communication over a link if they have at least one common key. Remarkably, it was shown in Reference [7] that all graphs are indeed intersection graphs.

To understand statistical properties of intersection graphs, a probability model was introduced in References [8,9] as a generalization of the classical model $G(n,p)$ of Erdős and Rényi [10]. Formally, let n, m be positive integers and let $p \in [0,1]$. We start with a random bipartite graph $B(n,m,p)$ with independent vertex sets $V = \{v_1, \cdots, v_n\}$ and $W = \{w_1, \cdots, w_m\}$ and edges between V and W existing independently with probability p. In terms of social networks, V is interpreted as a set of actors and W a set of attributes. We then define the *random intersection graph* $G(n,m,p)$ with vertex set V and vertices $v_i, v_j \in V$ adjacent if and only if there exists some $w \in W$ such that both v_i and v_j are adjacent to w in $B(n,m,p)$. Several variant models of random intersection graphs have been proposed, and many graph-theoretic properties of $G(n,m,p)$, such as degree distribution, connected components, fixed subgraphs, independence number, clique number, diameter, Hamiltonicity and clustering, have been extensively studied [8,9,11–14]. We refer the reader to References [15,16] for an updated review of recent results in this prolific field.

In light of the above list of properties studied, it is, perhaps, surprising that there has been little work regarding *isoperimetric numbers* of random intersection graphs. The isoperimetric numbers, which measure the expansion properties of a graph (see Section 2 below for precise definitions), have a long history in random graph theory [17–19] and are strongly related to the graph spectrum and expanders [20]. They have found a wide range of applications in theoretical computer science, including algorithm design, data compression, rapid mixing, error correcting codes, and robust computer networks [21]. Social networks such as co-authorship networks are commonly believed to have poor expansion properties (i.e., small isoperimetric numbers), which indicate the existence of bottlenecks (e.g., cuts with small size) inside the networks, because of their modular and community organization [22,23]. In this paper, we hope to show that it is possible to increase the isoperimetric numbers by a gentle perturbation of the original bipartite graph structure underlying the intersection graphs.

In recent times, there has been an effort to study the effect of random perturbation on graphs. The most mathematically famous example is perhaps the Newman–Watts small-world network [1,24], which is a random instance obtained by adding random edges to a cycle, exhibiting short average distance and high clustering coefficient, namely, the so-called small-world phenomenon. A random graph model $G \cup R$ [25] with general connected base graph G on n vertices and R being a sparse Erdős-Rényi random graph $G(n, \varepsilon/n)$ where $\varepsilon > 0$ is some small constant has been introduced in [26], and its further properties, such as connectivity, fixed subgraphs, Hamiltonicity, diameter, mixing time, vertex and edge expansion, have been intensively examined; see, e.g., [27–34] and references therein. For instance, in Reference [29], a necessary condition for the base graph is given under which the perturbed graph $G \cup R$ is an expander a.a.s. (asymptomatically almost surely); for a connected base graph G, it is shown in Reference [30] that, a.a.s. the perturbed graph has an edge–isoperimetric number $\Omega(1/\ln n)$, diameter $O(\ln n)$, and vertex–isoperimetric number $\Omega(1/\ln n)$, where for the last property G is assumed to have bounded maximum degree. Here, we say that $G \cup R$ possesses a graph property $\mathcal{P}$ asymptotically almost surely, or a.a.s. for brevity, if the probability that $G \cup R$ possesses $\mathcal{P}$ tends to 1 as n goes to infinity. In this paper, to go a step further in this line of research, we investigate the bipartite graph type perturbation, where random edges are only added to the base (bipartite) graph between the two independent sets. We provide lower bounds for the isoperimetric numbers of random intersection graphs induced by such perturbations.

The rest of the paper is organized as follows. In Section 2, we state and discuss the main results, with proofs relegated to Section 4. In Section 3, we give numerical examples based upon real network data, complementing our theoretical results in small network sizes. Section 5 contains some concluding remarks.

2. Results

Let $G = (V, E)$ be a graph with vertex set V and edge set E. If $S \subseteq V$ is a set of vertices, then $\partial_G S$ denotes the set of edges of G having one end in S and the other end in $V \backslash S$. Given $S \subseteq V$, write $G[S]$ for the subgraph of G induced by S. We use $N_G(S)$ to denote the collection of vertices of $V \backslash S$ which are adjacent to some vertex of S. For a vertex $v \in V$, $N_G(v)$ is the neighborhood of v, and we denote by $N_G^2(v) = N_G(N_G(v))$ the second neighborhood of v. The above subscript G will be omitted when no ambiguity may arise. For a graph G, its edge–isoperimetric number, $c(G)$ (also called its Cheeger constant), is given by:

$$c(G) = \min_{\substack{S \subseteq V \\ 0 < |S| \leq |V|/2}} \frac{|\partial_G S|}{|S|}.$$

The vertex–isoperimetric number of G, $\iota(G)$, can be defined similarly as:

$$\iota(G) = \min_{\substack{S \subseteq V \\ 0 < |S| \leq |V|/2}} \frac{|N_G(S)|}{|S|}.$$

It is well-known that $c(G)/\Delta(G) \leq \iota(G) \leq c(G)$ [35], where $\Delta(G)$ is the maximum degree of G.

We will consider the following model of randomly perturbed intersection graphs. Given a fixed bipartite graph $B = B(V, W, E)$ with two independent vertex sets V ($|V| = n$) and W ($|W| = m$), the intersection graph derived from B is denoted by $G(B)$. That is, $G(B)$ is a graph on the vertex set V with two vertices adjacent if they have a common neighbor in B. For each pair of vertices $v \in V$ and $w \in W$, we add the edge $\{v, w\}$ to B independently with probability p. The resulting bipartite graph, denoted $B \cup R$, can be viewed as the union of B and a bipartite graph $R \sim B(n, m, p)$, meaning that R is a random graph distributed according to $B(n, m, p)$. We write $G(B \cup R)$, the intersection graph derived from $B \cup R$. Clearly, if the base graph $B(V, W, E)$ is taken to be the empty bipartite graph, our model $G(B \cup R)$ reduces to the random intersection graph $G(n, m, p)$.

Throughout the paper, the standard Landau asymptotic notations will be utilized (see, e.g., [10]). Let $\lfloor \cdot \rfloor$ be the round-down operator. As customary in the theory of random intersection graphs, we take $m = \lfloor n^\alpha \rfloor$ for a fixed real $\alpha \in (0, \infty)$, which allows for a natural progression from sparse to dense graphs. Recall that we say that $G(B \cup R)$ possesses a graph property $\mathcal{P}$ a.a.s. if the probability that $G(B \cup R)$ possesses $\mathcal{P}$ tends to 1 as n goes to infinity.

We are now ready to formulate the main results of this paper.

Theorem 1. *Let $B = B(V, W, E)$ be a bipartite graph with $|V| = n$ and $|W| = m = \lfloor n^\alpha \rfloor$ such that any two vertices in V are connected by a path and $\Delta := \max_{v \in V} N_B^2(v)$ is a constant (i.e., independent of n). For any $\varepsilon > 0$, let $R \sim B(n, m, p)$ with $p = \varepsilon/n$ if $\alpha \leq 1$ and $p = \varepsilon/\sqrt{nm}$ if $\alpha > 1$. Then there exists some constant $\delta > 0$ satisfying $\iota(G(B \cup R)) \geq \delta/\ln n$ a.a.s.*

A couple of remarks are in order.

Remark 1. *The local effects of the perturbation are quite mild, as a small ε is of interest. Nonetheless, the global influence on the vertex–isoperimetric number can be prominent. To see this, note that any connected (intersection) graph G has $\iota(G) = \Omega(1/n)$. In particular, if G is a tree, we have $\iota(G) = \Theta(1/n)$ (see e.g., [36]).*

Remark 2. *It is easy to check that the maximum degree of $G(B)$ is Δ. In fact, $v \in V$ and $v_1 \in V$ are adjacent in $G(B)$ if and only if they have a common neighbor $w \in W$, namely, $w \in N_B(v)$ and $v_1 \in N_B(w)$. Hence, the degree of v is $N_B(N_B(v))$. The assumption that Δ is a constant cannot be removed in general. Indeed, when $\alpha \geq 1$, consider the bipartite graph $B(V, W, E)$ with $V = \{v_1, \cdots, v_n\}$, $W = \{w_1, \cdots, w_m\}$, and the edge set $E = \{\{v_1, w_i\}, \{v_j, w_{j-1}\} | i = 1, \cdots, n-1, j = 2, \cdots, n\}$. It is clear that $G(B)$ is a star with center v_1 over the vertex set V. There are no more than $n^2 p$ edges over $V \backslash \{v_1\}$ in the graph $G(B \cup R)$, which covers at most $2n^2 p$ vertices. In $G(B \cup R)$, there will be an independent set S (meaning that $G(B \cup R)[S]$ is empty) of order at least:*

$$n - 2n^2 p = n\left(1 - 2\varepsilon\sqrt{\frac{n}{m}}\right)$$

and $N_{G(B \cup R)}(S) = 1$. Therefore, $\iota(G(B \cup R)) \leq 1/(n(1 - 2\varepsilon\sqrt{n/m})) = O(1/n)$. When $\alpha < 1$, consider the bipartite graph $B(V, W, E)$ with the edge set $E = \{\{v_1, w_i\}, \{v_j, w_{j-1}\}, \{v_l, w_m\} | \ i = 1, \cdots, m$, $j = 2, \cdots, m+1, l = m+2, \cdots, n\}$. Then $G(B)$ can be thought of as the joining of a star $K_{1,m}$ having center v_1 and a complete graph K_{n-m+1} by identifying v_1 with any vertex of K_{n-m+1}. After adding $nm\varepsilon/n = m\varepsilon$ edges to B, in $G(B \cup R)$, there will be an independent set S of order at least $m - 1 - 2m\varepsilon$ and $N_{G(B \cup R)}(S) = 1$. Therefore, $\iota(G(B \cup R)) \leq 1/(m - 1 - 2m\varepsilon) = O(1/m)$.

Recall that the inequality $c(G) \geq \iota(G)$ holds for any graph G. Therefore, a direct corollary of Theorem 1 reads $c(G(B \cup R)) \geq \delta/\ln n$ a.a.s. for some $\delta > 0$. The following theorem shows that this lower bound for edge–isoperimetric number actually holds without any assumption on Δ.

Theorem 2. *Let $B = B(V, W, E)$ be a bipartite graph with $|V| = n$ and $|W| = m = \lfloor n^\alpha \rfloor$ such that any two vertices in V are connected by a path. For any $\varepsilon > 0$, let $R \sim B(n, m, p)$ with $p = \varepsilon/\sqrt{nm}$. Then there exists some constant $\delta > 0$ satisfying $c(G(B \cup R)) \geq \delta/(1 + \ln n)$ a.a.s.*

Theorems 1 and 2 hold in the sense of large n limit. In the next section, we shall demonstrate that the isoperimetric numbers can be improved as well for small randomly perturbed intersection graphs based upon real network data.

3. Illustration on Small Networks

To find the exact isoperimetric numbers, one needs to calculate the minimum fraction of neighboring vertices or edges over the nodes inside the subset for all possible subsets of vertices with order at most $|V|/2$. Since this is an NP-hard problem, it is intractable to compute the exact values for general graphs [21,35]. It is well known that Cheeger's inequality, also known as the Alon–Milman inequality, provides bounds for the isoperimetric numbers using graph Laplacian eigenvalues. On the other hand, standard algorithms in linear algebra can be used to efficiently compute the spectrum of a given large graph. Here, instead of evaluating "approximate" values involving other parameters such as eigenvalues, we are interested in obtaining exact values of $\iota(G(B \cup R))$ and $c(G(B \cup R))$ for small networks.

Two intersection-based social networks are considered here: (i) The Norwegian interlocking directorate network Nor-Boards [37], where two directors are adjacent if they are sitting on the board of the same company based on the Norwegian Business Register on 5 August 2009. The underlying bipartite graph $\bar{B}(\bar{V}, \bar{W}, \bar{E})$ contains $|\bar{V}| = 1495$ directors, $|\bar{W}| = 367$ companies, and $|\bar{E}| = 1834$ edges indicating the affiliation relations; (ii) the co-authorship network ca-CondMat [5] based on preprints posted to the Condensed Matter Section of arXiv E-Print Archive between 1995 and 1999. The underlying bipartite graph $\bar{B}(\bar{V}, \bar{W}, \bar{E})$ contains $|\bar{V}|$ =16,726 authors, $|\bar{W}| =$ 22,016 papers, and $|\bar{E}| =$ 58,596 edges indicating authorship.

Figures 1 and 2 report the vertex–isoperimetric numbers and edge–isoperimetric numbers for subsets of Nor-Boards and ca-CondMat, respectively. For a given $n \in [20, 30]$, we first take a subgraph $B = B(V, W, E)$ from $\bar{B}(\bar{V}, \bar{W}, \bar{E})$ with $|V| = n$ so that $G(B)$ is connected and calculate its vertex–isoperimetric and edge–isoperimetric numbers. Each data point (blue square) in Figures 1 and 2 is obtained by means of an ensemble averaging of 30 independently taken graphs. For each chosen bipartite graph B, we then perturb it following the rules specified in Theorems 1 and 2 with $\varepsilon = 1$ to get the perturbed intersection graph $G(B \cup R)$. Each data point (red circle) in Figures 1 and 2 is obtained by means of a mixed ensemble averaging of 50 independently-implemented perturbations for 30 graphs. From a statistics viewpoint, it is clear that our random perturbation scheme increases both the vertex–isoperimetric and the edge–isoperimetric number for both cases. This, together with the theoretical results, suggests that the quantitative effect of random perturbations is independent of the order of the network.

Remark 3. *It is worth stressing that the theoretical results (Theorems 1 and 2) are in the large limit of the network size n. In other words, the form $\frac{1}{\ln n}$ only makes sense as $n \to \infty$. The simulation results presented in Figures 1 and 2 are for very small networks. Therefore, these results have no bearing on the $\frac{1}{\ln n}$ dependence (although a slight decline tendency for $\iota(G(B \cup R))$ can be seen in Figure 1a). The main phenomenon we observe from Figures 1 and 2 is that the random perturbation increases both vertex– and edge–isoperimetric numbers for all the cases considered. The numerical results (for small finite graphs) are a nice complement to the theoretical results (for infinite graphs). However, our numerical observations neither prove the $\frac{1}{\ln n}$ dependence would hold for small graphs nor show that such an increase of isoperimetric numbers would be universal in any sense. (A practical issue stems from graph sampling. To establish a proper model fit to the data, Akaike information criteria and Bayesian information criteria need to be applied.) The establishment of correlation between isoperimetric numbers and graph size n for finite intersection graphs is an interesting future work.*

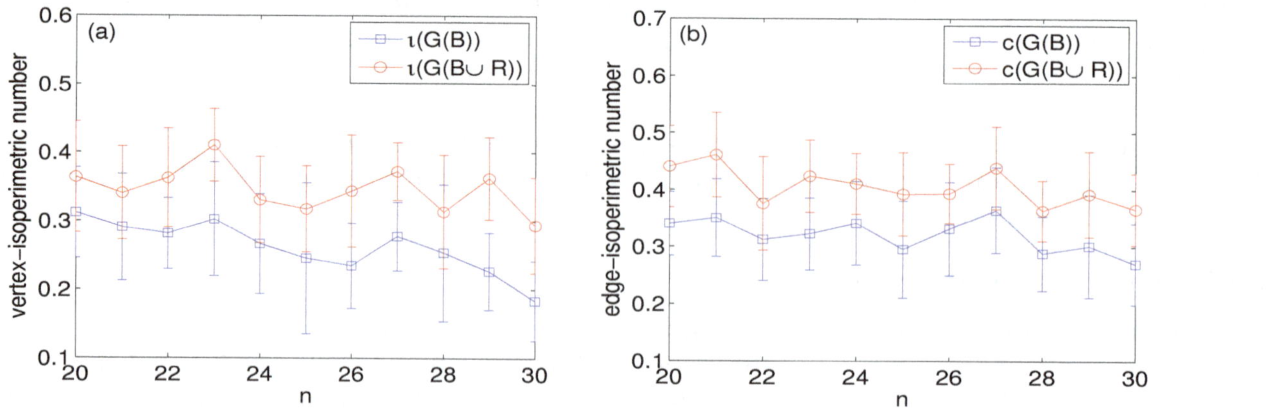

Figure 1. Vertex–isoperimetric number (panel (**a**)) and edge–isoperimetric number (panel (**b**)) versus $n = |G(B)|$ for subgraphs $G(B)$ (and its randomly perturbed version $G(B \cup R)$) taken from Nor-Boards.

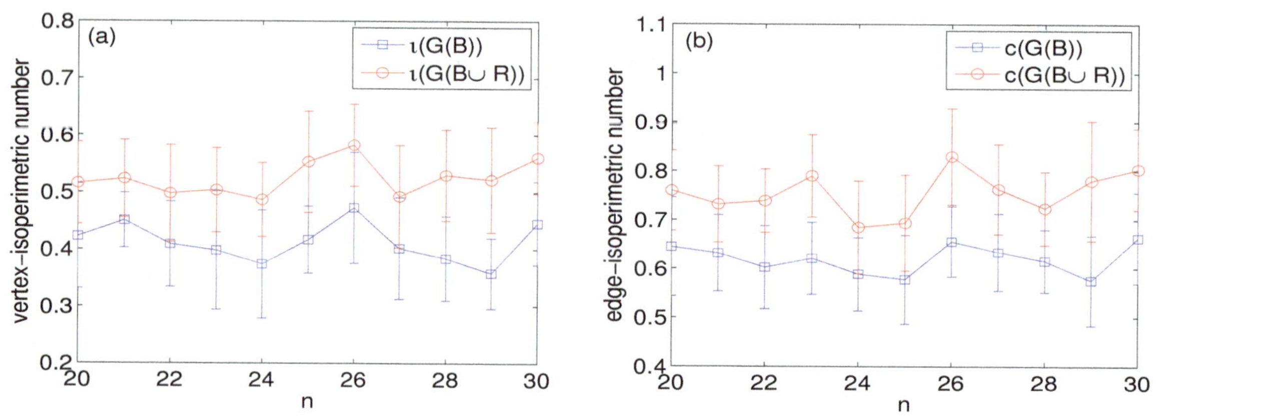

Figure 2. Vertex–isoperimetric number (panel (**a**)) and edge–isoperimetric number (panel (**b**)) versus $n = |G(B)|$ for subgraphs $G(B)$ (and its randomly perturbed version $G(B \cup R)$) taken from ca-CondMat.

4. Proofs

In this section, we prove Theorems 1 and 2. Our idea behind this is somewhat simple: If the network can be carefully decomposed into some subnetworks so that the resulting super-network (with these subnetworks being super-vertices) is sparse and highly connected, then its isoperimetric numbers are expected to be high. Similar approaches have been applied in, e.g., References [29–31].

Proof of Theorem 1. Set $s = C\Delta(\ln n)/\varepsilon$ for some constant $C = C(\varepsilon) > 0$ to be determined. By assumption, $G(B)$ is connected. Following Reference [38] (Proposition 4.5), we can divide the vertex set V into disjoint sets $V_1, V_2, \cdots, V_\theta$ satisfying $s \le |V_i| \le \Delta s$ and $G(B)[V_i]$ connected for each i. Clearly, $n/(\Delta s) \le \theta \le n/s$. Let $[\theta] = \{1, 2, \cdots, \theta\}$. For a graph $G = (V, E)$, we say two sets $S_1, S_2 \subseteq V$ have common neighbors in G if there exist $v_1 \in S_1$, $v_2 \in S_2$, and $v \in V$ such that $\{v_1, v\} \in E$ and $\{v_2, v\} \in E$ hold.

We will first show the following property for the random bipartite graph R holds a.a.s.: For every $\Theta \subseteq [\theta]$ with $0 < |\Theta| \le \theta/2$, there exist at least $|\Theta|/2$ many of V_i ($i \in [\theta] \backslash \Theta$) which have common neighbors with $\cup_{i\in\Theta} V_i$ in R.

Indeed, the probability that two sets V_i and V_j have no common neighbors in R can be computed as $\left\{1 - [1-(1-p)^{|V_i|}][1-(1-p)^{|V_j|}]\right\}^m$. Hence, the probability that there exists a set $\Theta \subseteq [\theta]$ with $0 < |\Theta| \le \theta/2$ such that no more than $|\Theta|/2$ many of V_i ($i \in [\theta] \backslash \Theta$) have common neighbors with $\cup_{i\in\Theta} V_i$ in R is upper bounded by:

$$\begin{aligned} &\sum_{1\le j\le \theta/2} \binom{\theta}{j}\binom{\theta-j}{\lfloor \frac{j}{2}\rfloor}\left\{1-[1-(1-p)^{js}][1-(1-p)^{(\theta-\lfloor 3j/2\rfloor)s}]\right\}^m, \\ \le &\sum_{1\le j\le \theta/2} \left(\frac{e\theta}{j}\right)^j \left(\frac{2e(\theta-j)}{j}\right)^{j/2}\left(1-\frac{s^2 j(\theta-3j/2)p^2}{4}\right)^m, \end{aligned}$$

where $\binom{\theta}{j}\binom{\theta-j}{\lfloor \frac{j}{2}\rfloor}$ counts the choice of Θ (with $|\Theta| = j$) and the corresponding sets $\{V_i\}$ described above, the estimate $|V_i| \ge s$ for all $i \in [\theta]$ is utilized in the multiplicative probabilities (i.e., there are at least $(\theta - \lfloor 3j/2\rfloor)$ sets in the union $\cup_{i\in\Theta}V_i$), and the upper bound comes from a direct application of inequalities ([10], p. 386). The above probability is further upper bounded by $(C(\ln n)/n)^m \sum_{1\le j\le\theta/2}(2\theta/j)^{3j/2} = o(1)$ when $\alpha \le 1$, and is upper bounded by $\sum_{1\le j\le \theta/2}\theta^{3j/2+2}\exp(-C\varepsilon\Delta j\ln n) = o(1)$ when $\alpha > 1$ for a sufficiently large C. Therefore, the above property for the random bipartite graph R holds a.a.s. In the following, we will condition on such an R.

Fix a set $S \subseteq V$ with $|S| \le n/2$. Define three sets of indices: $\Theta_0 = \{i \in [\theta] | V_i \subseteq S\}$, $\Theta_1 = \{i \in [\theta] | 0 < |V_i \cap S| < |V_i|\}$, and $\Theta_2 = \{i \in [\theta]\backslash\Theta_0 | N_{G(B\cup R)}(V_i) \cap S \neq \emptyset\}$. Note that Θ_0 and Θ_1 are deterministic, but Θ_2 is a random set. If $|\Theta_0| \le \theta/2$, $|\Theta_2| \ge |\Theta_0|/2$ a.a.s. by the above assumed property of R. Similarly, if $|\Theta_0| > \theta/2$, we have $|\Theta_2| \ge |\Theta|/2 = (\theta - |\Theta_0| - |\Theta_2|)/2$ a.a.s., where $\Theta = [\theta]\backslash(\Theta_0 \cup \Theta_2)$. Hence, $|\Theta_2| \ge \min\{|\Theta_0|/2, (\theta - |\Theta_0|)/3\}$ a.a.s. Recall that $|S| \le n/2$. We derive that $n/2 \le |V\backslash S| \le |\cup_{i\notin\Theta_0} V_i| \le (\theta - |\Theta_0|)\Delta s \le (\theta - |\Theta_0|)\Delta n/\theta$, and thus, $\theta - |\Theta_0| \ge \theta/(2\Delta)$. Therefore, we have a.a.s.:

$$|\Theta_2| \ge \min\left\{\frac{|\Theta_0|}{2}, \frac{\theta}{6\Delta}\right\} \ge \frac{|\Theta_0|}{6\Delta}.$$

By definition, we have $S \subseteq \cup_{i\in\Theta_0\cup\Theta_1} V_i$. Thus, $|S| \le (|\Theta_0| + |\Theta_1|)\Delta s$. Since $G(B)[V_i]$ for $i \in \Theta_1$ is connected, $|N_{G(B\cup R)}(S)| \ge |\Theta_1 \cup \Theta_2|$. Now we consider two cases. If $|\Theta_1| \ge |\Theta_0|$, then $|N_{G(B\cup R)}(S)| \ge |\Theta_1| \ge |S|/(2\Delta s)$. If $|\Theta_1| \le |\Theta_0|$, then $|N_{G(B\cup R)}(S)| \ge |\Theta_2| \ge |\Theta_0|/(6\Delta) \ge |S|/(12\Delta^2 s)$ a.a.s. Therefore:

$$\frac{|N_{G(B\cup R)}(S)|}{|S|} \ge \min\left\{\frac{1}{2\Delta s}, \frac{1}{12\Delta^2 s}\right\} \qquad a.a.s.$$

Recall the definition of s at the beginning of the proof, and we complete the proof by taking $\delta = \varepsilon/(12\Delta^3 C)$. □

We have made no attempt to optimize the constants in the proof. It is easy to check that the condition that $G(B)$ is connected in Theorem 1 can be weakened. For example, the above proof holds if each connected component of $G(B)$ is of order at least $C\Delta(\ln n)/\varepsilon$.

Let $G = (V, E)$ be a graph of order n. For integers a, b, and c, define $\mathcal{S}(a, b, c)$ as a collection of all sets $S \subseteq V$ such that $|S| = a$ and there exists a partition $S = S_1 \cup \cdots \cup S_b$, where each $G[S_i]$ is connected, there are no edges in E connecting different S_i, and $|N_G(S_1)| + \cdots + |N_G(S_b)| = c$. The next lemma gives an upper bound of the size of $\mathcal{S}(a, b, c)$.

Lemma 1. *([30])*

$$|\mathcal{S}(a,b,c)| \le \left(\frac{en}{b}\right)^b \left(\frac{ea}{b}\right)^b \left(\frac{ec}{b}\right)^b \left(\frac{e(a+c)}{c}\right)^c.$$

Proof of Theorem 2. Consider the family $\mathcal{S}(a, b, c)$ of sets defined in graph $G(B)$. Since $G(B)$ is connected, we have for each $S \in \mathcal{S}(a, b, c)$, $|\partial_{G(B)}S| \ge c \ge b$. Note that $|\partial_{G(B\cup R)}S| \ge |\partial_{G(R)}S|$

holds. It suffices to show that the following property for the random bipartite graph R holds a.a.s.: There are constants $K, \delta > 0$ such that for any $K \ln n \le a \le n/2$, we have:

$$|\partial_{G(R)} S| \ge \frac{\delta a}{1 + \ln n},$$

for each $S \in \mathcal{S}(a,b,c)$ with $b \le c \le \delta a/(1+\ln n)$. Indeed, when $|S| = a \le K \ln n$, we can choose a small δ such that $2K\delta \le 1$. Thus, $|\partial_{G(B \cup R)} S| \ge |\partial_{G(B)}(S)| \ge 1 \ge \delta a/(1 + \ln n)$.

It follows from Lemma 1 and $b \le c \le \delta a/(1+\ln n) \le a$ that:

$$|\mathcal{S}(a,b,c)| \le \left(\frac{2e^4 n a^2}{c^3}\right)^c \le \left(\frac{2e^4 n (1+\ln n)^3}{\delta^3 a}\right)^{\delta a/(1+\ln n)} \le e^{C\delta a \ln(1/\delta)},$$

for some constant $C > 0$, where the first inequality holds since $f(x) = (e\rho/x)^x$ is increasing on $(0, \rho]$ and the second inequality holds since $g(x) = (\rho/x^3)^x$ is increasing on $(0, \rho^{1/3}]$.

Note that $mp^2 \to 0$ and $1 - (1-p^2)^m \sim mp^2$. For a fixed S with $|S| = a \le n/2$, we obtain:

$$\begin{aligned} \mathbb{P}(|\partial_{G(R)} S| < \delta a) &\lesssim \mathbb{P}(\mathrm{Bin}(a(n-a), mp^2) < \delta a) \le \mathbb{P}\left(\mathrm{Bin}\left(\frac{na}{2}, mp^2\right) < \delta a\right) \\ &\lesssim \exp\left(-\frac{a\varepsilon^2}{16}\right), \end{aligned}$$

provided $\delta < \varepsilon^2/4$, where the first inequality relies on Reference [9] (Theorem 2.2) and the last line uses a standard Chernoff's bound (e.g., [10]). Hence:

$$\begin{aligned} &\mathbb{P}\left(|\partial_{G(R)} S| < \frac{\delta a}{1+\ln n}, \exists S \in \mathcal{S}(a,b,c), b \le c \le \frac{\delta a}{1+\ln n}, K\ln n \le a \le \frac{n}{2}\right) \\ \le\ & \mathbb{P}\left(|\partial_{G(R)} S| < \delta a, \exists S \in \mathcal{S}(a,b,c), b \le c \le n, K \ln n \le a \le n\right) \\ \le\ & n^3 \exp\left(C\delta a \ln\left(\frac{1}{\delta}\right) - \frac{a\varepsilon^2}{16}\right). \end{aligned}$$

By taking $C\delta \ln(1/\delta) \le \varepsilon^2/32$ and $K \ge 100/\varepsilon^2$, the last line above is upper bounded by $n^3 \exp(-\varepsilon^2 a/32) \le n^3 \exp(-\varepsilon^2 K (\ln n)/32) \le n^3 \exp(-25(\ln n)/8) = o(1)$ as $n \to \infty$. The proof is complete. □

5. Concluding Remarks

In this paper, we presented a model of randomly perturbed intersection graphs. The intersection graph is induced by a given bipartite graph (base graph) plus a binomial random bipartite graph. We proved that a.a.s., the vertex–isoperimetric number and the edge–isoperimetric number of the randomly perturbed intersection graphs are of order $\Omega(1/\ln n)$ under some mild conditions. It would be interesting to investigate path length, diameter, and clustering coefficient of this model, which are important characteristics of real-life complex and social networks.

Another intriguing direction is to examine more general intersection graph models, such as active and passive intersection graphs [39]. In particular, if two vertices in one independent set V are declared adjacent when they have at least $k \ge 1$ common neighbors in the other independent set W, what role will k play in estimating the isoperimetric numbers, clustering, and path length of the resulting perturbed intersection graphs? Other perturbation mechanisms are also of research interest.

Acknowledgments: I am very grateful to the editor and two anonymous referees for their valuable comments that have greatly improved the presentation of the paper. The author was supported by a Starting Grant of Northumbria University.

References

1. Newman, M.E.J. *Networks: An Introduction*; Oxford University Press: Oxford, UK, 2010.
2. Guillaume, J.L.; Latapy, M. Bipartite structure of all complex networks. *Inf. Process. Lett.* **2004**, *90*, 215–221. [CrossRef]
3. Kadushin, C. *Understanding Social Networks: Theories, Concepts, and Findings*; Oxford University Press: Oxford, UK, 2012
4. Watts, D.J.; Strogatz, S.H. Collective dynamics of 'small-world' networks. *Nature* **1998**, *393*, 440–442. [CrossRef]
5. Newman, M.E.J. The structure of scientific collaboration networks. *Proc. Natl. Acad. Sci. USA* **2001**, *98*, 404–409. [CrossRef] [PubMed]
6. Eschenauer, L.; Gligor, V. A key-management scheme for distributed sensor networks. In Proceedings of the 9th ACM Conference on Computer and Communications Security, Washington, DC, USA, 18–22 November 2002; pp. 41–47.
7. Marczewski, E. Sur deux propriétés des classes d'ensembles. *Fund. Math.* **1945**, *33*, 303–307. [CrossRef]
8. Karoński, M.; Scheinerman, E.R.; Singer-Cohen, K.B. On random intersection graphs: The subgraph problem. *Comb. Probab. Comput.* **1999**, *8*, 131–159. [CrossRef]
9. Singer-Cohen, K.B. Random Intersection Graphs. Ph.D. Thesis, Johns Hopkins University, Baltimore, MD, USA, 1995.
10. Frieze, A.; Karoński, M. *Introduction to Random Graphs*; Cambridge University Press: New York, NY, USA, 2016.
11. Shang, Y. Degree distributions in general random intersection graphs. *Electron. J. Combin.* **2010**, *17*, R23.
12. Kendall, M.; Martin, K.M. Graph-theoretic design and analysis of key predistribution schemes. *Des. Codes Cryptogr.* **2016**, *81*, 11–34. [CrossRef]
13. Zhao, J.; Yagan, O.; Gligor, V. On connectivity and robustness in random intersection graphs. *IEEE Trans. Autom. Contr.* **2017**, *62*, 2121–2136. [CrossRef]
14. Rybarczyk, K. The chromatic number of random intersection graphs. *Discuss. Math. Graph Theory* **2017**, *37*, 465–476. [CrossRef]
15. Bloznelis, M.; Godehardt, E.; Jaworski, J.; Kurauskas, V.; Rybarczyk, K. Recent progress in complex network analysis: Models of random intersection graphs. In *European Conference on Data Analysis*; Lausen, B., Krolak-Schwerdt, S., Boehmer, M., Eds.; Springer: Berlin, Germany, 2015; pp. 69–78.
16. Bloznelis, M.; Godehardt, E.; Jaworski, J.; Kurauskas, V.; Rybarczyk, K. Recent progress in complex network analysis: Properties of random intersection graphs. In *European Conference on Data Analysis*; Lausen, B., Krolak-Schwerdt, S., Boehmer, M., Eds.; Springer: Berlin, Germany, 2015; pp. 79–88.
17. Bollobás, B. The isoperimetric number of random regular graphs. *Eur. J. Comb.* **1988**, *9*, 241–244. [CrossRef]
18. Alon, N. On the edge-expansion of graphs. *Comb. Probab. Comput.* **1997**, *6*, 145–152. [CrossRef]
19. Benjamini, I.; Haber, S.; Krivelevich, M.; Lubetzky, E. The isoperimetric constant of the random graph process. *Random Struct. Algorithms* **2008**, *32*, 101–114. [CrossRef]
20. Puder, D. Expansion of random graphs: New proofs, new results. *Invent. Math.* **2015**, *201*, 845–908. [CrossRef]
21. Hoory, S.; Linial, N.; Wigderson, A. Expander graphs and their applications. *Bull. Am. Math. Soc.* **2006**, *43*, 439–561. [CrossRef]
22. Estrada, E. Spectral scaling and good expansion properties in complex networks. *Europhys. Lett.* **2006**, *73*, 649–655. [CrossRef]
23. Newman, M.E.J. Modularity and community structure in networks. *Proc. Natl. Acad. Sci. USA* **2006**, *103*, 8577–8582. [CrossRef] [PubMed]
24. Newman, M.E.J.; Watts, D.J. Renormalization group analysis of the small-world network model. *Phys. Lett. A* **1999**, *263*, 341–346. [CrossRef]
25. Shang, Y. A sharp threshold for rainbow connection in small-world networks. *Miskolc Math. Notes* **2012**, *13*, 493–497. [CrossRef]
26. Bohman, T.; Frieze, A.; Martin, R. How many random edges make a dense graph Hamiltonian? *Random Struct. Algorithms* **2003**, *22*, 33–42. [CrossRef]
27. Bohman, T.; Frieze, A.; Krivelevich, M.; Martin, R.; Adding random edges to dense graphs. *Random Struct. Algorithms* **2004**, *24*, 105–117. [CrossRef]

28. Krivelevich, M.; Sudakov, B.; Tetali, P. On smoothed analysis in dense graphs and formulas. *Random Struct. Algorithms* **2006**, *29*, 180–193. [CrossRef]
29. Flaxman, A.D. Expansion and lack thereof in randomly perturbed graphs. *Internet Math.* **2007**, *4*, 131–147. [CrossRef]
30. Krivelevich, M.; Reichman, D.; Samotij, W. Smoothed analysis on connected graphs. *SIAM J. Discrete Math.* **2015**, *29*, 1654–1669. [CrossRef]
31. Addario-Berry, L.; Lei, T. The mixing time of the Newman-Watts small-world model. *Adv. Appl. Probab.* **2015**, *47*, 37–56. [CrossRef]
32. Balogh, J.; Treglown, A.; Wagner, A.Z. Tilings in randomly perturbed dense graphs. *Combin. Probab. Comput.* **2019**, *28*, 159–176. [CrossRef]
33. Krivelevich, M.; Kwan, M.; Sudakov, B. Bounded-degree spanning trees in randomly perturbed graphs. *SIAM J. Discrete Math.* **2017**, *31*, 155–171. [CrossRef]
34. Böttcher, J.; Montgomery, R.; Parczyk, O.; Person, Y. Embedding spanning bounded degree subgraphs in randomly perturbed graphs. *Electron. Notes Discrete Math.* **2017**, *61*, 155–161. [CrossRef]
35. Mohar, B. Isoperimetric numbers of graphs. *J. Comb. Theory Ser. B* **1989**, *47*, 274–291. [CrossRef]
36. Grohe, M.; Marx, D. On tree width, bramble size, and expansion. *J. Comb. Theory Ser. B* **2009**, *99*, 218–228. [CrossRef]
37. Seierstad, C.; Opsahl, T. For the few not the many? The effects of affirmative action on presence, prominence, and social capital of women directors in Norway. *Scand. J. Manag.* **2011**, *27*, 44–54. [CrossRef]
38. Krivelevich, M.; Nachmias, A. Coloring complete bipartite graphs from random lists. *Random Struct. Algorithms* **2006**, *29*, 436–449. [CrossRef]
39. Shang, Y. Joint probability generating function for degrees of active/passive random intersection graphs. *Front. Math. China* **2012**, *7*, 117–124. [CrossRef]

4

Edge Even Graceful Labeling of Polar Grid Graphs

Salama Nagy Daoud [1,2]

1 Department of Mathematics, Faculty of Science, Taibah University, Al-Madinah 41411, Saudi Arabia; salamadaoud@gmail.com

2 Department of Mathematics and Computer Science, Faculty of Science, Menoufia University, Shebin El Kom 32511, Egypt

Abstract: Edge Even Graceful Labelingwas first defined byElsonbaty and Daoud in 2017. An edge even graceful labeling of a simple graph G with p vertices and q edges is a bijection f from the edges of the graph to the set $\{2, 4, \ldots, 2q\}$ such that, when each vertex is assigned the sum of all edges incident to it mod$2r$ where $r = \max\{p, q\}$, the resulting vertex labels are distinct. In this paper we proved necessary and sufficient conditions for the polar grid graph to be edge even graceful graph.

Keywords: graceful labeling; edge graceful labeling; edge even graceful labeling; polar grid graph

---

1. Introduction

The field of Graph Theory plays an important role in various areas of pure and applied sciences. One of the important areas in graph theory is Graph Labeling of a graph G which is an assignment of integers either to the vertices or edges or both subject to certain conditions. Graph labeling is a very powerful tool that eventually makes things in different fields very ease to be handled in mathematical way. Nowadays graph labeling has much attention from different brilliant researches ingraph theory which has rigorous applications in many disciplines, e.g., communication networks, coding theory, x-raycrystallography, radar, astronomy, circuit design, communication network addressing, data base management and graph decomposition problems. More interesting applications of graph labeling can be found in [1–10].

Let $G = (V(G), E(G))$ with $p = |V(G)|$ and $q = |E(G)|$ be a simple, connected, finite, undirected graph.

A function f is called a graceful labeling of a graph G if $f : V(G) \to \{0, 1, 2, \ldots, q\}$ is injective and the induced function $f^* : E(G) \to \{1, 2, \ldots, q\}$ defined as $f^*(e = uv) = |f(u) - f(v)|$ is bijective. This type of graph labeling first introducedby Rosa in 1967 [11] as a $\beta-$ valuation, later on Solomon W. Golomb [12] called as graceful labeling.

A function f is called an odd graceful labeling of a graph G if $f : V(G) \to \{0, 1, 2, \ldots, 2q-1\}$ is injective and the induced function $f^* : E(G) \to \{1, 3, \ldots, 2q-1\}$ defined as $f^*(e = uv) = |f(u) - f(v)|$ is bijective. This type of graph labeling first introducedby Gnanajothi in 1991 [13]. For more results on this type of labeling see [14,15].

A function f is called an edge graceful labeling of a graph G if $f : E(G) \to \{1, 2, \ldots, q\}$ is bijective and the induced function $f^* : V(G) \to \{0, 1, 2, \ldots, p-1\}$ defined as $f^*(u) = \sum_{e=uv \in E(G)} f(e) \bmod p$ is bijective. This type of graph labeling first introducedby Lo in 1985 [16]. For more results on this labeling see [17,18].

A function f is called an edge odd graceful labeling of a graph G if $f : E(G) \to \{1, 3, \ldots, 2q-1\}$ is bijective and the induced function $f^* : V(G) \to \{0, 1, 2, \ldots, 2q-1\}$ defined as $f^*(u) = \sum_{e=uv \in E(G)} f(e) \bmod 2q$ is injective. This type of graph labeling first introducedby Solairaju and Chithra in 2009 [19]. See also Daoud [20].

A function f is called an edge even graceful labeling of a graph G if $f : E(G) \rightarrow \{2, 4, \ldots, 2q\}$ is bijective and the induced function $f^* : V(G) \rightarrow \{0, 2, 4, \ldots, 2q-2\}$ defined as $f^*(u) = \sum_{e=uv \in E(G)} f(e) \mathrm{mod} 2r$, where $r = \max\{p, q\}$ is injective. This type of graph labeling first introduced by Elsonbaty and Daoud in 2017 [21].

For a summary of the results on these five types of graceful labels as well as all known labels so far, see [22].

2. Polar Grid Graph $P_{m,n}$

The polargrid graph $P_{m,n}$ is the graph consists of n copies of circles C_m which will be numbered from the inner most circle to the outer circle as $C_m^{(1)}, C_m^{(2)}, \ldots, C_m^{(n-1)}, C_m^{(n)}$ and m copies of paths P_{n+1} intersected at the center vertex v_0 which will be numbered as $P_{n+1}^{(1)}, P_{n+1}^{(2)}, \ldots, P_{n+1}^{(m-1)}, P_{n+1}^{(m)}$. See Figure 1.

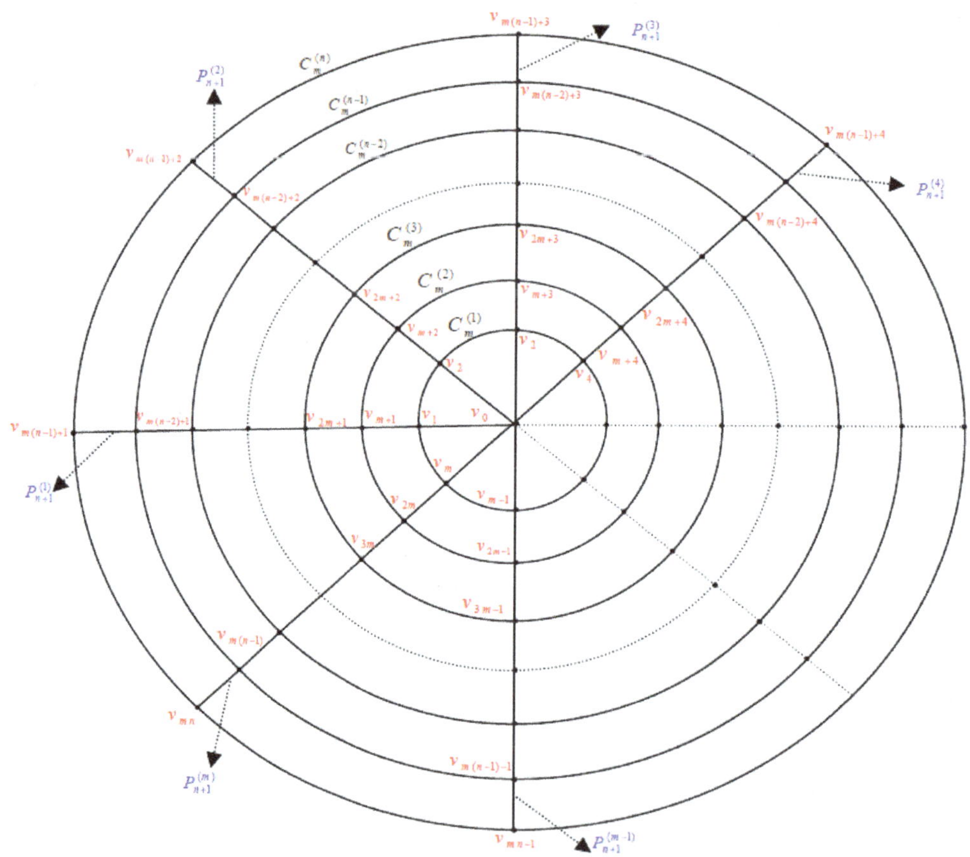

Figure 1. Polar grid graph $P_{m,n}$.

Theorem 1. *If m and n are even positive integes such that $m \geq 4$ and $n \geq 2$, then the polar grid graph $P_{m,n}$ is an edge even graceful graph.*

Proof. Using standard notation $p = |V(P_{m,n})| = mn + 1$, $q = |E(P_{m,n})| = 2mn$ and $r = \max\{p, q\} = 2mn$ Let the polar grid graph $P_{m,n}$ be labeled as in Figure 2. Let $f : E(G) \rightarrow \{2, 4, \ldots, 2q\}$. □

First we label the edges of paths $P_{n+1}^{(k)}, \leq k \leq m$ begin with the edges of the path $P_{n+1}^{(1)}$ to the edges of the path $P_{n+1}^{(m)}$ as follows: Move clockwise to label the edges $v_0v_1, v_0v_2, \ldots, v_0v_{m-1}, v_0v_m$ by $2, 4, \ldots, 2m-2, 2m$, then move anticlockwise to label the edges

$v_1v_{m+1}, v_mv_{2m}, v_{m-1}v_{2m-1}, \ldots, v_3v_{m+3}, v_2v_{m+2}$ by $2m+2, 2m+4, 2m+6, \ldots, 4m-2, 4m$, then move clockwise to label the edges $v_{m+1}v_{2m+1}, v_{m+2}v_{2m+2}, v_{m+3}v_{2m+3}, \ldots, v_{2m-1}v_{3m-1}, v_mv_{3m}$ by $4m+2, 4m+4, 4m+6, \ldots, 6m-2, 6m$ and so on. Finally move anticlockwise to label the edges $v_{m(n-2)+1}v_{m(n-1)+1}, v_{m(n-1)-1}v_{mn}, v_{m(n-1)-1}v_{mn-1}, \ldots, v_{m(n-2)+2}v_{m(n-1)+2}$ by $2m(n-1)+2, 2m(n-1)+4, 2m(n-1)+6, \ldots, 2mn-2, 2mn$. Second we label the edges of the circles $C_m^{(k)}, 1 \le k \le n$ begin with the edges of the inner most circle $C_m^{(1)}$ to the edges of the circle $C_m^{(\frac{n}{2})}$, then the edges of the outer circle $C_m^{(n)}$. Finally the edges of circles $C_m^{(\frac{n}{2}+1)}, C_m^{(\frac{n}{2}+2)}, \ldots, C_m^{(n-1)}$ respectively as follows: $f(v_{m(k-1)+i}v_{m(k-1)+i+1}) = 2mn + 2m(k-1) + 2i$, $f(v_{km}v_{m(k-1)+1}) = 2mn + 2mk$, $1 \le i \le m-1, 1 \le k \le \frac{n}{2}$; $f(v_{m(n-1)+i}v_{m(n-1)+i+1}) = 3mn + 2i$, $1 \le i \le m-1$; $f(v_{km}v_{m(k-1)+1}) = 2mn + (2k+1)m$, $\frac{n}{2} + 1 \le k \le n-1$.

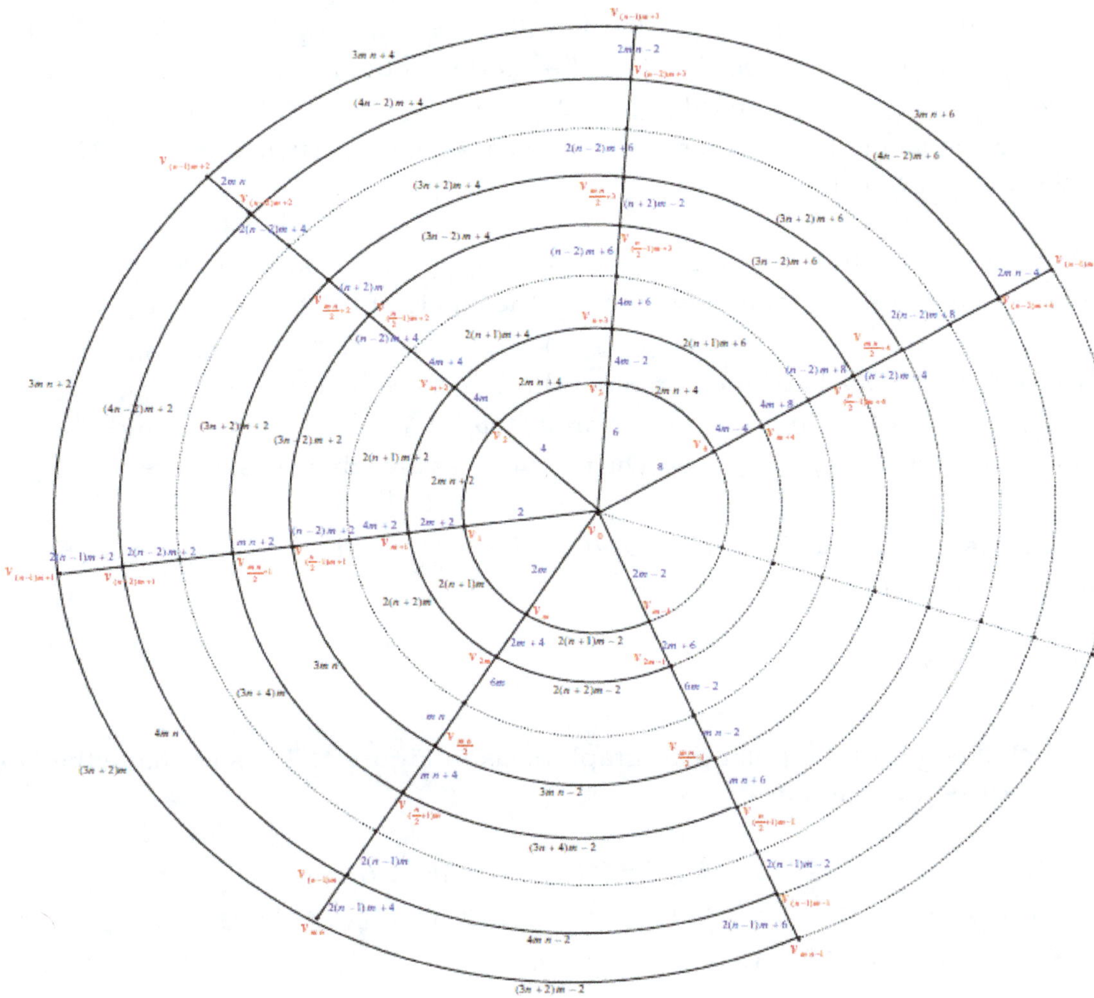

Figure 2. Labeling ofthe polar grid graph $P_{m,n}$ when n is even, $n \ge 2$.

Now the corresponding labels of vertices mod$4mn$ are assigned as follows:

Case (1) $m \equiv 4k \bmod 4n$, $1 \le k \le n-1$ and $m = 2 \bmod 4n$.

The labels of the vertices of the inner most circle $C_m^{(1)}$ to the circle $C_m^{(\frac{n}{2})}$ are given by $f^*(v_{(k-1)m+i}) \equiv 4m(2k-1) + 4i$, $1 \le k \le \frac{n}{2}, 1 \le i \le m$, the labels of the vertices of the outre circle $C_m^{(n)}$ are given by $f^*(v_{(n-1)m+i}) \equiv 2i + 2$, $1 \le i \le m$ and the labels of the vertices of the circles $C_m^{(\frac{n}{2}+1)}, \ldots, C_m^{(n-1)}$ are given by $f^*(v_{(k-1)m+i}) \equiv 8m(k - \frac{n}{2}) + 4i + 2$, $1 \le i \le m$, $\frac{n}{2} + 1 \le k \le n-1$.

The label of the center vertex v_0 is assigned as follows: when $m = 4k \bmod 4n$, $1 \le k \le n-1$, $f^*(v_0) \equiv \frac{m}{2}(2m+2) = m^2 + m$, since $m = 4k \bmod 4n$ then $m = 4nh + 4k$, thus $f^*(v_0) = m(4k+1)$ and when $m = 2 \bmod 4n$, we have $f^*(v_0) = 3m$.

Case (2) $m \equiv (8k-2) \bmod 4n$, $1 \leq k \leq \frac{n}{2}$. In this case the vertex $v_{km-(\frac{m+2}{4})}$ in the circle $C_m^{(k)}$ will repeat with the center vertex v_0. To avoid this problem we replace the labels of the two edges $v_{km-(\frac{m+2}{4})}v_{km-(\frac{m-2}{4})}$ and $v_{km-(\frac{m-2}{4})}v_{km-(\frac{m-6}{4})}$. That is $f(v_{km-(\frac{m+2}{4})}v_{km-(\frac{m-2}{4})}) = 2mn + m(2k-1) + \frac{m+2}{2}$ and $f(v_{km-(\frac{m-2}{4})}v_{km-(\frac{m-6}{4})}) = 2mn + m(2k-1) + \frac{m-2}{2}$ and we obtain the labels of the corresponding vertices as follows $f^*(v_{km-(\frac{m+2}{4})}) \equiv m(8k-1)+2$, $f^*(v_{km-(\frac{m-2}{4})}) \equiv m(8k-1)+4$, $f^*(v_{km-(\frac{m-6}{4})}) \equiv m(8k-1)+6$ and the label of the center vertex v_0 is assigened as $f^*(v_0) \equiv m(8k-1)$.The rest vertices are labeled as in case(1).

Case (3) $m \equiv (8k+2) \bmod 4n$, $1 \leq k \leq \frac{n}{2}-1$. In this case the vertex $v_{km-(\frac{m+2}{2})}$ in the circle $C_m^{(\frac{n}{2}+k)}$ will repeat with the center vertex v_0. To avoid this problem we replace the labels of the two edges $v_{km-(\frac{m+2}{2})}v_{km-(\frac{m}{2})}$ and $v_{km-(\frac{m}{2})}v_{km-(\frac{m-2}{2})}$. That is $f(v_{km-(\frac{m+2}{2})}v_{km-(\frac{m}{2})}) = 2mn + m(2k+1) + \frac{m+2}{2}$ and $f(v_{km-(\frac{m}{2})}v_{km-(\frac{m-2}{2})}) = 2mn + m(2k+1) + \frac{m-2}{2}$ and we obtain the labels of the corresponding vertices as follows $f^*(v_{km-(\frac{m+2}{2})}) \equiv m(8k+3)+2$, $f^*(v_{km-(\frac{m}{2})}) \equiv m(8k+3)+4$, $f^*(v_{km-(\frac{m-2}{2})}) \equiv m(8k+3)+6$ and the label of the center vertex v_0 is assigened $f^*(v_0) \equiv m(8k-29)$ as. The rest vertices are labeled as in case(1).

Case (4) $m \equiv 0 \bmod 4n$. In this case the vertex $v_{mn-(\frac{m+2}{2})}$ in the outer circle will repeat with the center vertex v_0. To avoid this problem we replace the labels of the two edges $v_{mn-(\frac{m+4}{2})}v_{mn-(\frac{m+2}{2})}$ and $v_{mn}v_{m(n-1)+1}$. That is $f(v_{mn-(\frac{m+4}{2})}v_{mn-(\frac{m+2}{2})}) = m(3n+2)$ and $f(v_{mn}v_{m(n-1)+1}) = 3mn + m - 4$ and we obtain the labels of the corresponding vertices as follows $f^*(v_{mn-(\frac{m+4}{2})}) \equiv 2m + 2$, $f^*(v_{mn-(\frac{m+2}{2})}) \equiv 2m+4$, $f^*(v_{mn}) \equiv m-2$ and $f^*(v_{m(n-1)+1}) \equiv 4mn - m$ and the label of the center vertex v_0 is assigened as $f^*(v_0) \equiv m$. The rest vertices are labeled as in case (1).

Illustration. *The edge even graceful labeling of the polar grid graphs $P_{14,6}$, $P_{16,6}$, $P_{18,6}$, $P_{24,6}$ and $P_{26,6}$ respectively are shown in Figure 3.*

Remark 1. *In case $m = 2$ and n is even, $n > 2$.*

Let the label of edges of the polar grid graph be as in Figure 4. Thus we have the label of the corresponding vertices are as follows:

$f^*(v_1) \equiv 4n+12$; $f^*(v_i) \equiv 16i-2,\ 2 \leq i \leq \frac{n}{2}$; $f^*(v_{\frac{n}{2}+i}) \equiv 16i+6,\ 1 \leq i \leq \frac{n}{2}-1$;
$f^*(v_n) \equiv 4$; $f^*(v'_i) \equiv 16i+2,\ 1 \leq i \leq \frac{n}{2}-1$; $f^*(v'_{\frac{n}{2}+i}) \equiv 2$; $f^*(v'_{\frac{n}{2}+i}) \equiv 16i+10,\ 1 \leq i \leq \frac{n}{2}-2$;
$f^*(v'_{n-1}) \equiv 4n-4$; $f^*(v'_n) \equiv 4n+8$ and $f^*(v_0) \equiv 4n+4$.

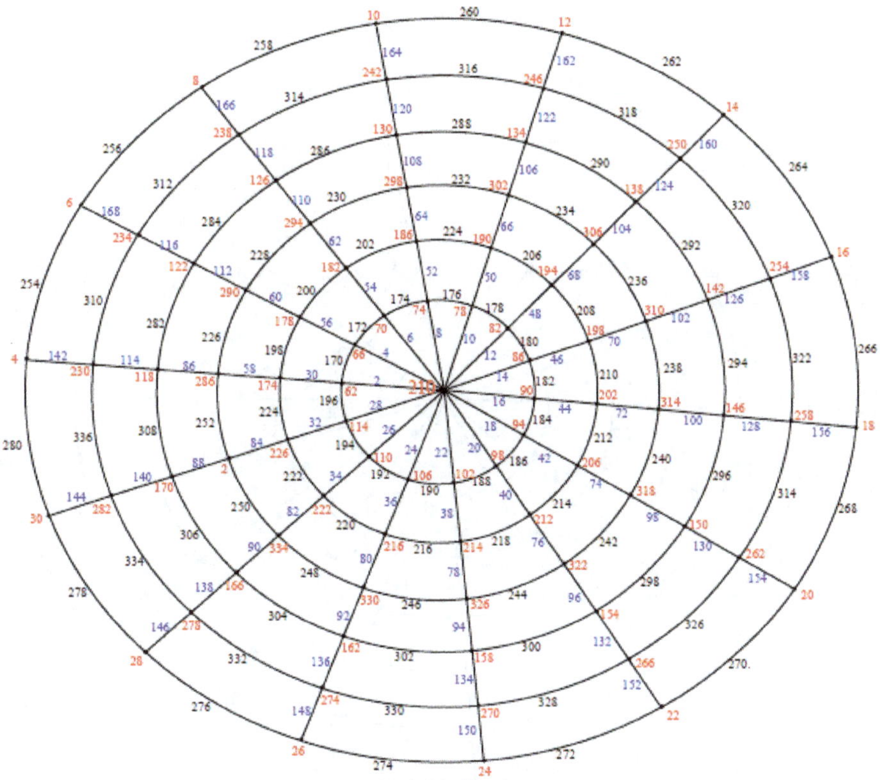

(a) $P_{14,6}$

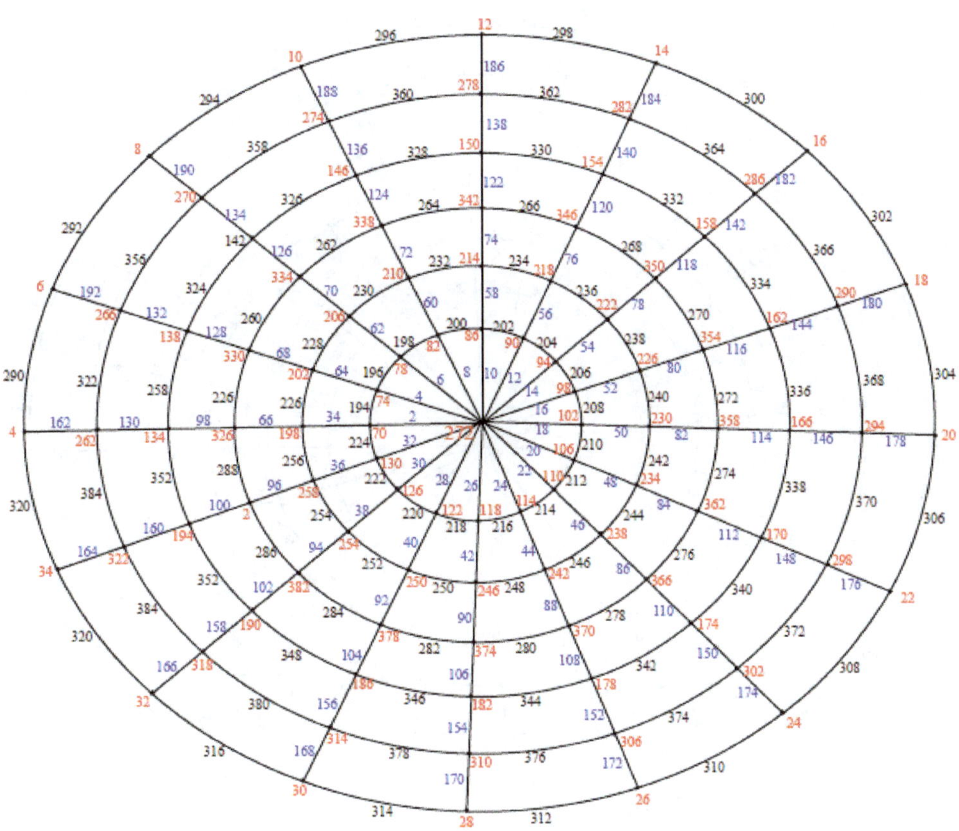

(b) $P_{16,6}$

Figure 3. *Cont.*

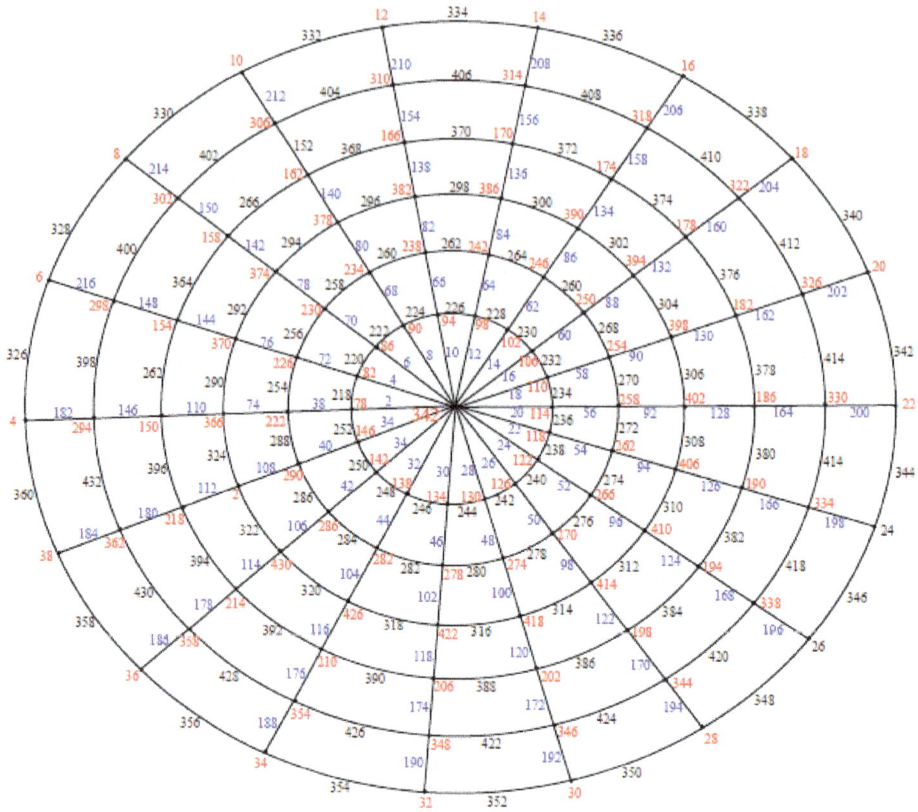

(c) $P_{18,6}$

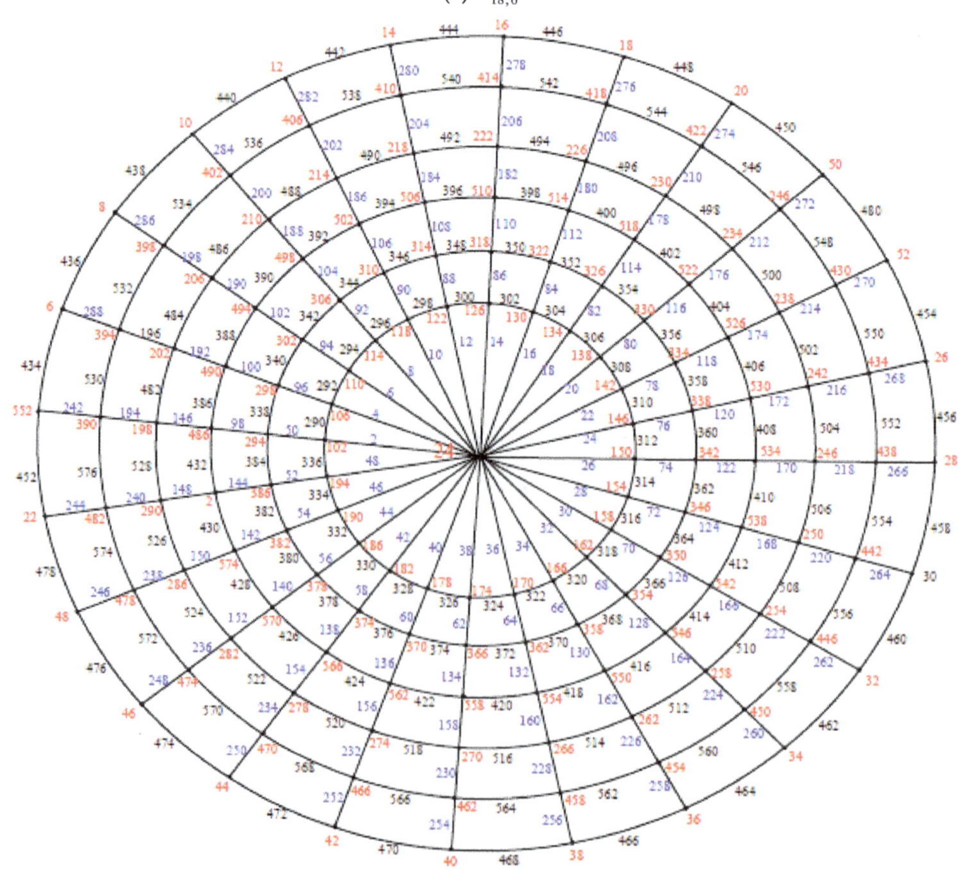

(d) $P_{24,6}$

Figure 3. *Cont.*

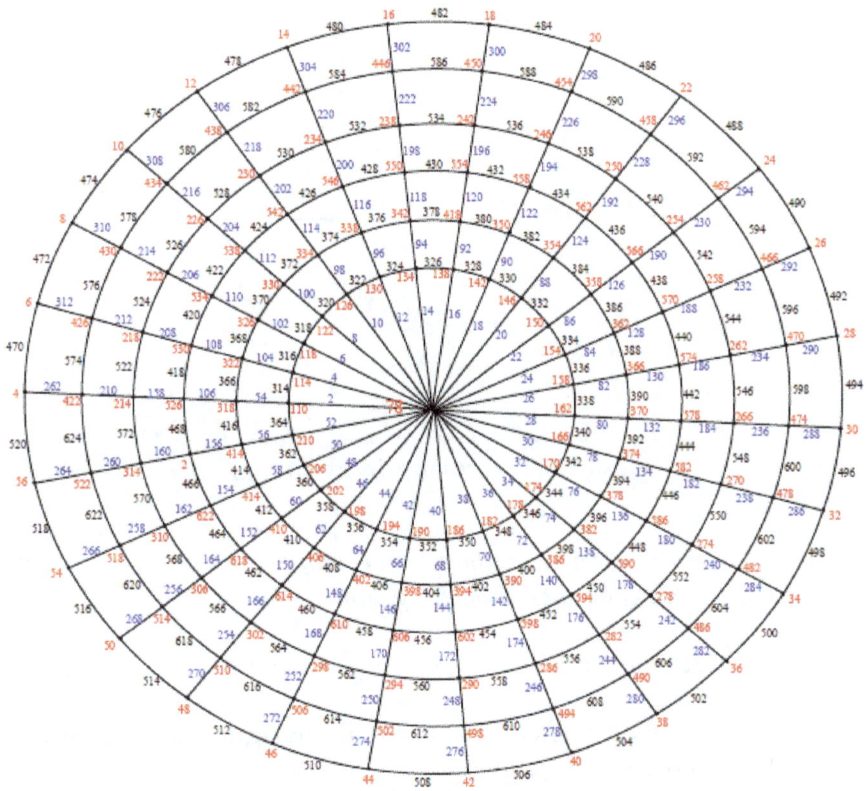

(e) $P_{26,6}$

Figure 3. The edge even graceful labeling of the polar grid graphs $P_{14,6}$, $P_{16,6}$, $P_{18,6}$, $P_{24,6}$ and $P_{26,6}$.

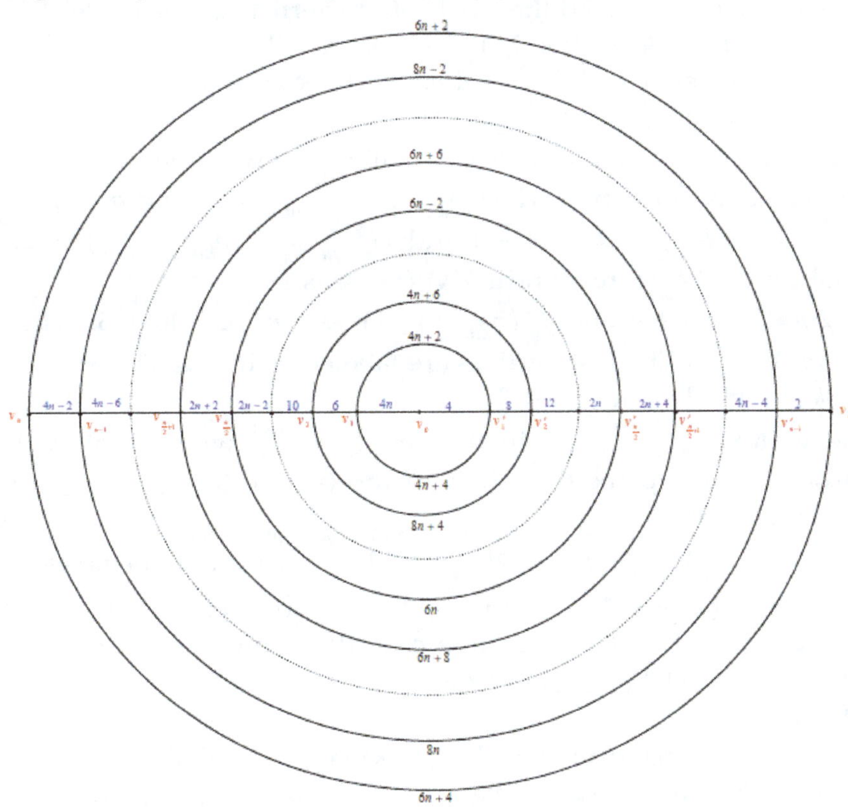

Figure 4. Labeling of the polar grid graph $P_{2,n}$, n is even integer greater than 2.

Note that $P_{2,2}$ is an edge even graceful graph but not follow this rule. See Figure 5.

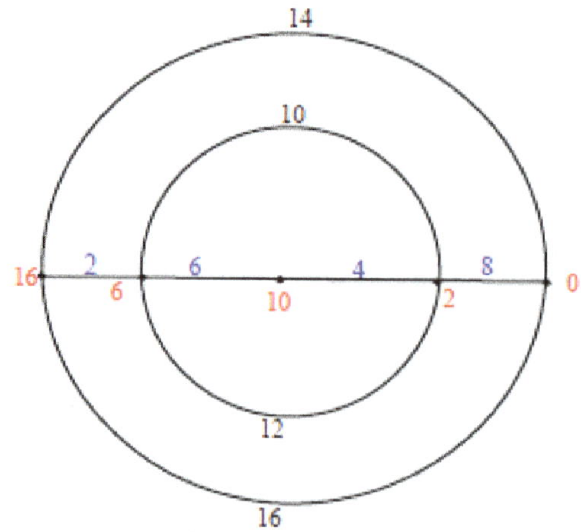

Figure 5. The polar grid graph $P_{2,2}$.

Theorem 2. *If m is an odd positive integer greater than 1 and n an is even positive integer greater than or equal 2, then the polar grid graph $P_{m,n}$ is an edge even graceful graph.*

Proof. Let the edges of the polar grid graph $P_{m,n}$ be labeled as in Figure 2. □

Now the corresponding labels of vertices mod$4mn$ are assigned as follows: There are four cases Case (1): $m \equiv (4k-1)\text{mod}4n,\ 1 \le k \le n$

The labels of the vertices of the inner most circle $C_m^{(1)}$ to the circle $C_m^{(\frac{n}{2})}$ are given by $f^*(v_{(k-1)m+i}) \equiv 4m(2k-1)+4i+2,\ 1 \le k \le \frac{n}{2},\ 1 \le i \le m$, the labels of the vertices of the outer circle $C_m^{(n)}$ are given by $f^*(v_{(n-1)m+i}) \equiv 2i+2,\ 1 \le i \le m$ and the labels of the vertices of the circles $C_m^{(\frac{n}{2}+1)}, \ldots, C_m^{(n-1)}$ are given by $f^*(v_{(k-1)m+i}) \equiv 8m(k-\frac{n}{2})+4i+2,\ 1 \le i \le m,\ \frac{n}{2}+1 \le k \le n-1$.

The center vertex v_0 is labeled as $f^*(v_0) \equiv 4mk$, and if $k=n$, we have $f^*(v_0) \equiv 0$.

Case (2): $m \equiv (8k-3)\text{mod}4n,\ 1 \le k \le \frac{n}{2}$.

In this case the vertex $v_{km(\frac{m+1}{2})}$ in the circle $C_m^{(k)}$ will repeat with the center vertex v_0. To avoid this problem we replace the labels of the two edges $v_{km-(\frac{m+1}{2})}v_{km-(\frac{m-1}{2})}$ and $v_{km-(\frac{m-1}{2})}v_{km-(\frac{m-3}{2})}$. That is $f(v_{km-(\frac{m+1}{2})}v_{km-(\frac{m-1}{2})}) = 2mn+m(2k-1)+1$ and $f(v_{km-(\frac{m-1}{2})}v_{km-(\frac{m-3}{2})}) = 2mn+m(2k-1)-1$ and we obtain the labels of the corresponding vertices as follows $f^*(v_{km-(\frac{m+1}{2})}) \equiv 2m(4k-1)+2$, $f^*(v_{km-(\frac{m-1}{2})}) \equiv 2m(4k-1)+4$ and $f^*(v_{km-(\frac{m-3}{2})}) \equiv 2m(4k-1)+6$. The center vertex v_0 is labeled as $f^*(v_0) = 2m(4k-1)$. The rest vertices are labeled as in case (1).

Case (3): $m \equiv (8k+1)\text{mod}4n,\ 1 \le k \le \frac{n}{2}-1$.

In this case the vertex $v_{m(\frac{n}{2}+k)(\frac{m+1}{2})}$ in the circle $C_m^{(\frac{n}{2}+k)}$ will repeat with the center vertex v_0. To avoid this problem we replace the labels of the two edges $v_{m(\frac{n}{2}+k)-(\frac{m+1}{2})}v_{m(\frac{n}{2}+k)(\frac{m-1}{2})}$ and $v_{m(\frac{n}{2}+k)-(\frac{m-1}{2})}v_{m(\frac{n}{2}+k)-(\frac{m-3}{2})}$. That is $f(v_{m(\frac{n}{2}+k)-(\frac{m+1}{2})}v_{m(\frac{n}{2}+k)-(\frac{m-1}{2})}) = 3mn+m(2k-1)+1$ and $f(v_{m(\frac{n}{2}+k)-(\frac{m-1}{2})}v_{m(\frac{n}{2}+k)-(\frac{m-3}{2})}) = 3mn+m(2k+1)-1$ and we obtain the labels of the corresponding vertices as follows $f^*(v_{m(\frac{n}{2}+k)-(\frac{m+1}{2})}) \equiv 2m(4k+1)+2$, $f^*(v_{m(\frac{n}{2}+k)-(\frac{m-1}{2})}) \equiv 2m(4k+1)+4$, and $f^*(v_{m(\frac{n}{2}+k)-(\frac{m-3}{2})}) \equiv 2m(4k+1)+6$ and in this case the center vertex v_0 is labeled as $f^*(v_0) = 2m(4k+1)$. The rest vertices are labeled as in case (1).

Case (4): $m \equiv 1\text{mod}4n$

In this case the vertex v_{mn-1} in the outer circle $C_m^{(n)}$ will repeat with the center vertex v_0. To avoid this problem we replace the labels of the two edges $v_{mn-2}v_{mn-1}$ and $v_{mn}v_{m(n-1)+1}$. That is $f(v_{mn-2}v_{mn-1}) = m(3n+2)$ and $f(v_{mn}v_{m(n-1)+1}) = m(3n+2)-4$ and we obtain the labels of the corresponding vertices as follows $f^*(v_{mn-2}) \equiv 2m+2$, $f^*(v_{mn-1}) \equiv 2m+4$, $f^*(v_{mn}) \equiv 2m-2$ and $f^*(v_{m(n-1)+1}) \equiv 0$, the center vertex v_0 is labeled as $f^*(v_0) \equiv 2m$. The rest vertices are labeled as in case (1).

Illustration. *The edge even graceful labeling of the polar grid graphs* $P_{13,6}$*,* $P_{15,6}$*,* $P_{17,6}$ *and* $P_{25,6}$ *respectively are shown in Figure 6.*

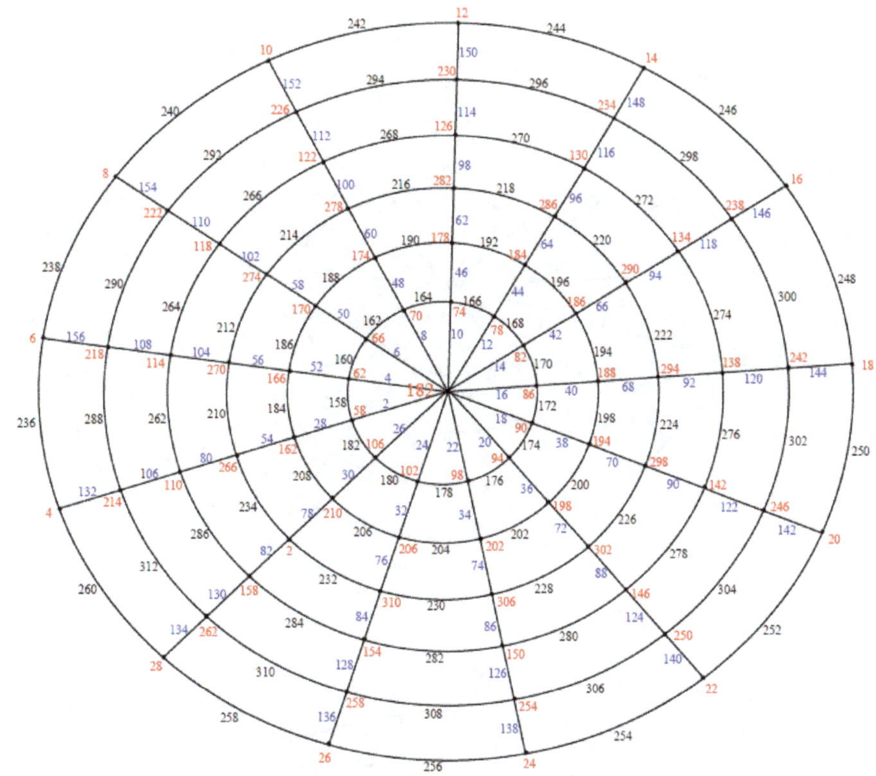

(a) $P_{13,6}$

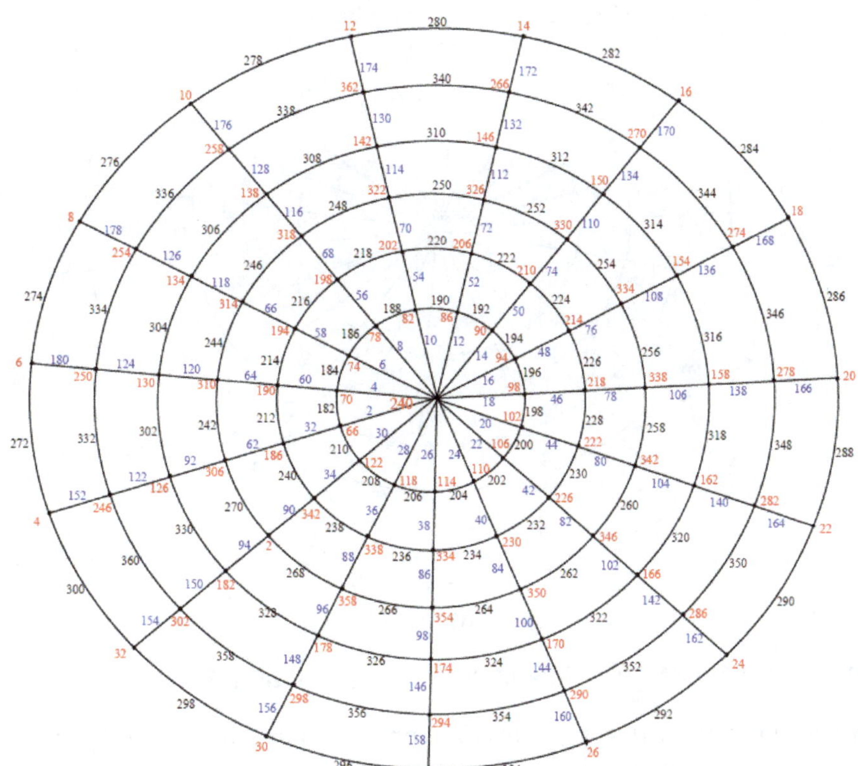

(b) $P_{15,6}$

Figure 6. *Cont.*

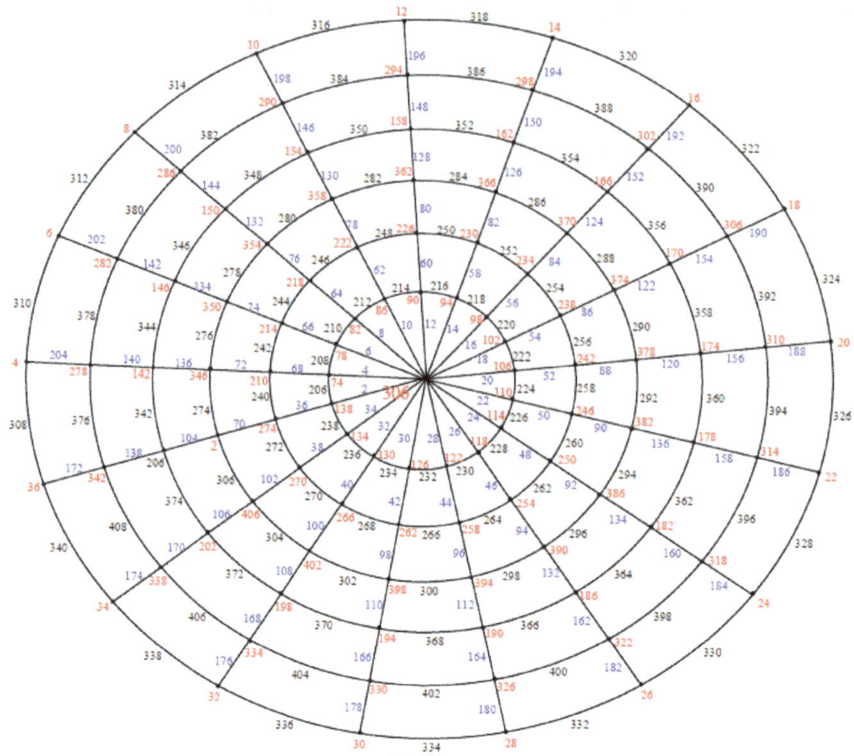

(c) $P_{17,6}$

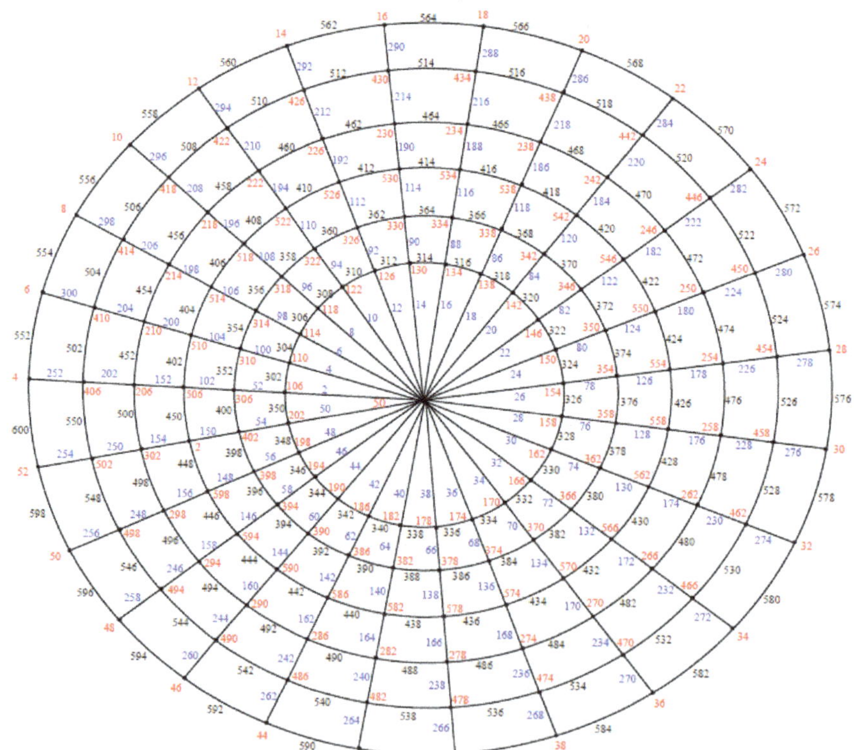

(d) $P_{25,6}$

Figure 6. The edge even graceful labeling of the polar grid graphs $P_{13,6}$, $P_{15,6}$, $P_{17,6}$ and $P_{25,6}$.

Theorem 3. *If m is an even positive integer greater than or equal 4 and n is an odd positive integer greater than or equal 3. Then the polar grid graph $P_{m,n}$ is an edge even graceful graph.*

Proof. Let the polar grid graph $P_{m,n}$ be labeled as in Figure 7. Let $f : E(G) \to \{2, 4, \ldots, 2q\}$. □

First we label the edges of the circles $C_m^{(k)}$, $1 \leq k \leq n$ begin with the edges of the inner most circle $C_m^{(1)}$ to edges of the outer circle $C_m^{(n)}$ as follows:

$$f(v_{m(k-1)+i}v_{m(k-1)+i+1}) = 2m(k-1) + 2i,\ f(v_{km}v_{(k-1)m+1}) = 2km,\ 1 \leq i \leq m-1,\ 1 \leq k \leq n.$$

Second we label the edges of paths $P_{n+1}^{(k)}$, $1 \leq k \leq m$ begin with the edges of the path $P_{n+1}^{(1)}$ as follows: Move anticlockwise to label the edges $v_0v_1, v_0v_m, v_0v_{m-1}, \ldots, v_0v_3, v_0v_2$ by $2mn+2$, $2mn+4, 2mn+6, \ldots, 2m(n+1)-2$, $2m(n+1)$, then move clockwise to label the edges $v_1v_{m+1}, v_2v_{m+2}, v_3v_{m+3}, \ldots, v_{m-1}v_{2m-1}, v_mv_{2m}$ by $2m(n+1)+2, 2m(n+1)+4, 2m(n+1)+6, \ldots, 2m(n+2)-2, 2m(n+2)$, then move anticlockwise to label the edges $v_{m+1}v_{2m+1}, v_{2m}v_{3m}, v_{2m-1}v_{3m-1}, \ldots, v_{m+3}v_{2m+3}, v_{m+2}v_{2m+2}$ by $2m(n+2)+2$, $2m(n+2)+4, 2m(n+2)+6, \ldots, 2m(n+3)-2$, $2m(n+3)$ and so on. Finally move anticlockwise to label the edges $v_{m(n-2)+1}v_{m(n-1)+1}, v_{m(n-1)+1}v_{mn}, v_{m(n-1)-1}v_{mn-1}, \ldots, v_{m(n-2)+3}v_{m(m-1)+3}, v_{m(n-2)+2}v_{m(m-1)+2}$ by $2m(2n-1)+2, 2m(2n-1)+4, 2m(2n-1)+6, \ldots, 4mn-2, 4mn$.

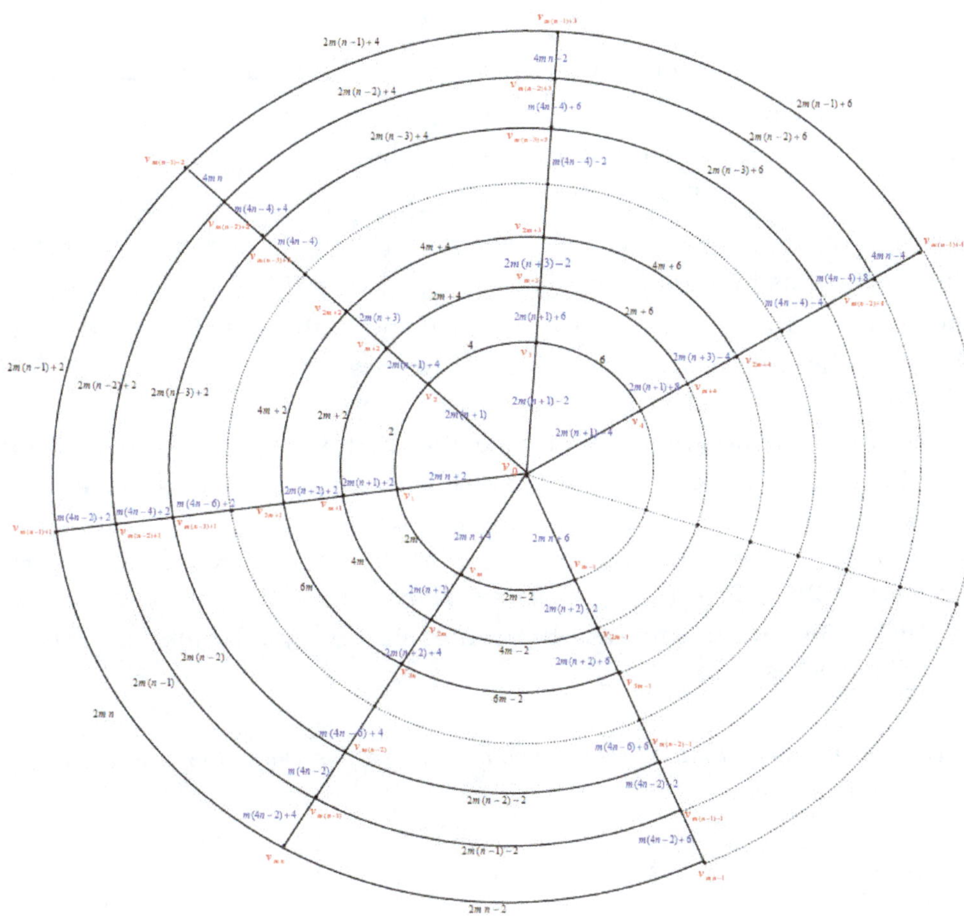

Figure 7. Labeling of the polar grid graph $P_{m,n}$ when n is odd and $n \geq 3$.

The corresponding labels of vertices mod$4mn$ are assigned as follows: There are four cases
Case (1) $m \equiv 4k \bmod 4n$, $1 \leq k \leq n-2$; $m \equiv (4n-2) \bmod 4n$ and $m \equiv 0 \bmod 4n$
$f^*(v_{(k-1)m+i}) \equiv 4m(2k-1) + 4i + 2$, $1 \leq i \leq m$, $1 \leq k \leq n-1$. That is the labels of the vertices in the most inner circle $C_m^{(1)}$ are assigned by $f^*(v_i) \equiv 4m + 4i + 2$, $1 \leq i \leq m$, the labels of the vertices in the circle $C_m^{(2)}$ are assigned by $f^*(v_{m+i}) \equiv 12m + 4i + 2$, the labels of vertices of the circle $C_m^{(\frac{n-1}{2})}$ are assigned by $f^*(\frac{v_{m(n-3)}}{2} + i) \equiv 4mn - 8m + 4i + 2$, the labels of vertices of the circle $C_m^{(\frac{n+1}{2})}$ are assigned by $f^*(\frac{v_{m(n-1)}}{2} + i) \equiv 4i + 2$, the labels of vertices of the circle $C_m^{(\frac{n+3}{2})}$ are assigned

by $f^*(\frac{v_{m(n+1)}}{2}+i) \equiv 8m+4i+2, \ldots,$ the labels of the vertices in the circle $C_m^{(n-1)}$ are assigned by $f^*(v_{m(n-2)+i}) \equiv 4mn-12m+4i+2,\ 1 \le i \le m$ and the labels of the vertices of the outer circle $C_m^{(n)}$ are assigned by $f^*(v_{m(n-1)+i}) \equiv 4mn-4m+2i+2,\ 1 \le i \le m$. The labels of the center vertex v_0 is assigned by $f^*(v_0) \equiv m^2(2n+1)$ when $m \equiv 4k \bmod 4n$, we have $f^*(v_0) \equiv m(4k+1)$, when $m \equiv (4n-2) \bmod 4n$, $f^*(v_0) \equiv 4mn-m$ and when $m \equiv 0 \bmod 4n$, $f^*(v_0) \equiv m$.

Case (2) $m \equiv (8k-2) \bmod 4n,\ 1 \le k \le \frac{n-1}{2}$.

In this case the vertex $v_{km-(\frac{m+2}{4})}$ in the circle $C_m^{(k)}$ will repeat with the center vertex v_0. To avoid this problem we replace the label of two edges $v_{km-(\frac{m+2}{4})}v_{km-(\frac{m-2}{4})}$ and $v_{km-(\frac{m-2}{4})}v_{km-(\frac{m-6}{4})}$. That is $f(v_{km-(\frac{m+2}{4})}v_{km-(\frac{m-2}{4})}) = m(2k-1)+\frac{m+2}{2}$ and $f(v_{km-(\frac{m-2}{4})}v_{km-(\frac{m-6}{2})}) = m(2k-1)+\frac{m-2}{2}$ and we obtain the labels of the corresponding vertices as follows $f^*(v_{km-(\frac{m+2}{4})}) \equiv m(8k-1)+2$, $f^*(v_{km-(\frac{m-2}{4})}) \equiv m(8k-1)+4$ and $f^*(v_{km-(\frac{m-6}{2})}) \equiv m(8k-1)+6$. In this case the center vertex v_0 is labeled as $f^*(v_0) \equiv m(2mn+m+1) \equiv m(8k-1)$. The rest vertices are labeled as in case (1).

Case (3) $m \equiv (8k-6) \bmod 4n,\ 1 \le k \le \frac{n-1}{2}$ and $m \ne 2$. In this case the vertex $v_{m(\frac{n+2k-1}{2})-(\frac{m+2}{4})}$ in the circle $C_m^{(\frac{n+2k-1}{2})}$ will repeat with the center vertex v_0. To avoid this problem we replace the labels of the two edges $v_{m(\frac{n+2k-1}{2})-(\frac{m+2}{4})}v_{m(\frac{n+2k-1}{2})-(\frac{m-2}{4})}$ and $v_{m(\frac{n+2k-1}{2})-(\frac{m-2}{4})}v_{m(\frac{n+2k-1}{2})-(\frac{m-6}{4})}$. That is $f(v_{m(\frac{n+2k-1}{2})-(\frac{m+2}{4})}v_{m(\frac{n+2k-1}{2})-(\frac{m-2}{4})}) = mn+2m(k-1)+\frac{m+2}{2}$ and $f(v_{m(\frac{n+2k-1}{2})-(\frac{m-2}{4})}v_{m(\frac{n+2k-1}{2})-(\frac{m-6}{4})}) = mn+2m(k-1)+\frac{m-2}{2}$ and we obtain the labels of the corresponding vertices as follows $f^*(v_{m(\frac{n+2k-1}{2})-(\frac{m+2}{4})}) \equiv m(8k-5)+2$, $f^*(v_{m(\frac{n+2k-1}{2})-(\frac{m-2}{4})}) \equiv m(8k-5)+4$ and $f^*(v_{m(\frac{n+2k-1}{2})-(\frac{m-6}{4})}) \equiv m(8k-5)+6$ and in this case the center vertex v_0 is labeled as $f^*(v_0) \equiv m(8k-5)$. The rest vertices are labeled as in case (1).

Case (4) $m \equiv (4n-4) \bmod 4n$. In this case the vertex $v_{mn-(\frac{m+2}{2})}$ in the outer circle $C_m^{(n)}$ will repeat with the center vertex v_0. To avoid this problem we replace the labels of the two edges $v_{mn-(\frac{m+6}{4})}v_{mn-(\frac{m+2}{4})}$ and $v_{mn}v_{m(n-1)+1}$. That is $f(v_{mn-(\frac{m+6}{4})}v_{mn-(\frac{m+2}{4})}) = 2mn$ and $f(v_{mn}v_{m(n-1)+1}) = m(2n-1)-4$ and we obtain the labels of the corresponding vertices as follows $f^*(v_{mn-(\frac{m+6}{4})}) \equiv 4mn-2m+2$, $f^*(v_{mn-(\frac{m+2}{4})}) \equiv 4mn-2m+8$, $f^*(v_{mn}) \equiv 4mn-3m-2$ and $f^*(v_{m(n-1)+1}) \equiv 4mn-5m$ and in this case the center vertex v_0 is labeled as $f^*(v_0) \equiv m(4n-3)$. The rest vertices are labeled as in case (1).

Illustration. *The edge even graceful labeling of the polar grid graphs $P_{10,5}$, $P_{12,5}$, $P_{14,5}$, $P_{16,5}$, $P_{18,5}$ and $P_{20,5}$ respectively are shown in Figure 8.*

Remark 2. *In case $m = 2$, n is odd, $n \ge 3$.*

Let the label of edges of the polar grid graph $P_{2,n}$ be as in Figure 9. Thus we have the labels of the corresponding vertices as follows: $f^*(v_1) \equiv 12$; $f^*(v_i) \equiv 16i-2,\ 2 \le i \le \frac{n-1}{2}$; $f^*(v_{\frac{n+2i-1}{2}}) \equiv 16i-10, 1 \le i \le \frac{n-1}{2}$; $f^*(v_n) \equiv 8n-2$; $f^*(v'_i) \equiv 16i+2,\ 1 \le i \le \frac{n-1}{2}$; $f^*(v'_{\frac{n+1}{2}}) \equiv 2$; $f^*(v'_{\frac{n+2i-1}{2}}) \equiv 16i-6,\ 2 \le i \le \frac{n-1}{2}$; $f^*(v'_n) \equiv 0$ and $f^*(v_0) \equiv 4$.

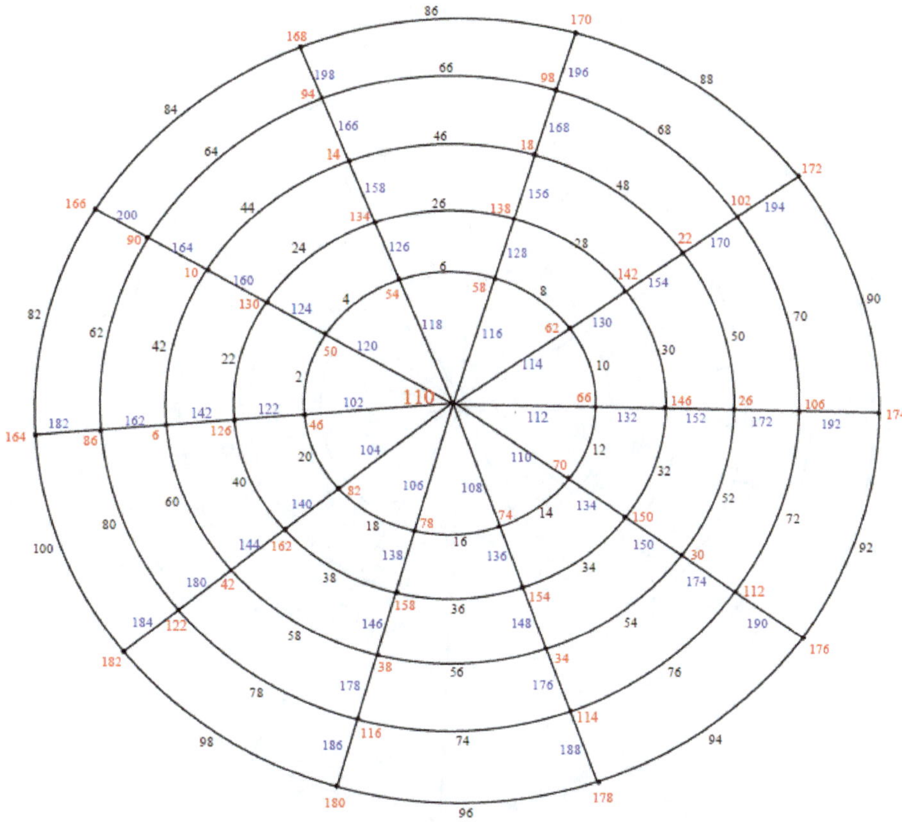

(a) $P_{10,5}$

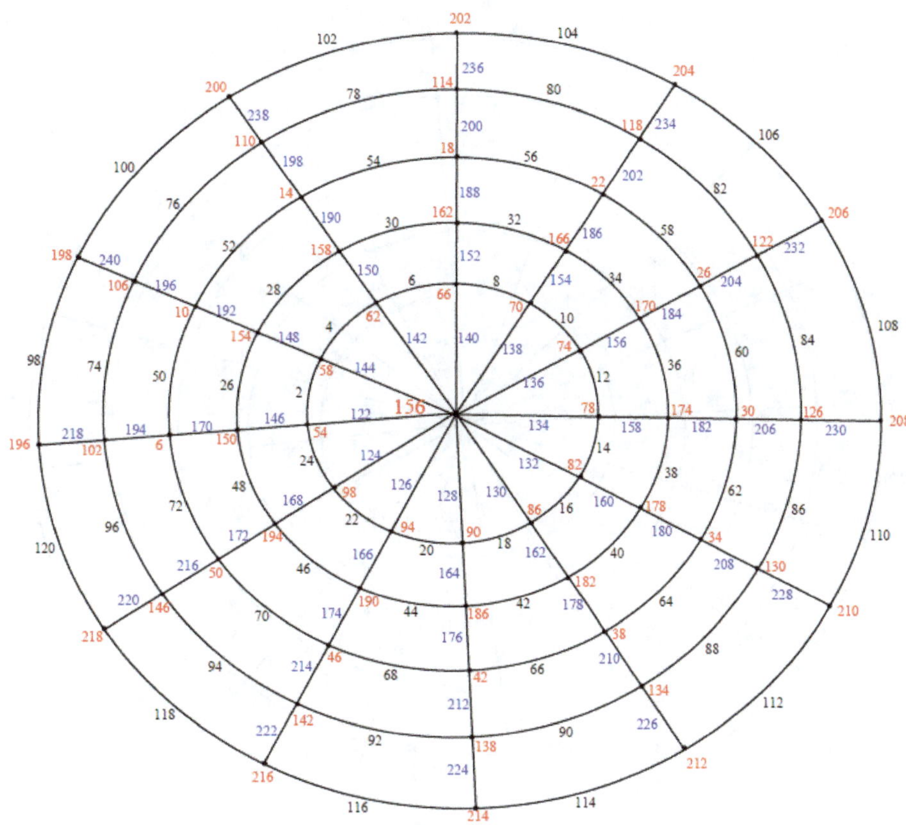

(b) $P_{12,5}$

Figure 8. *Cont.*

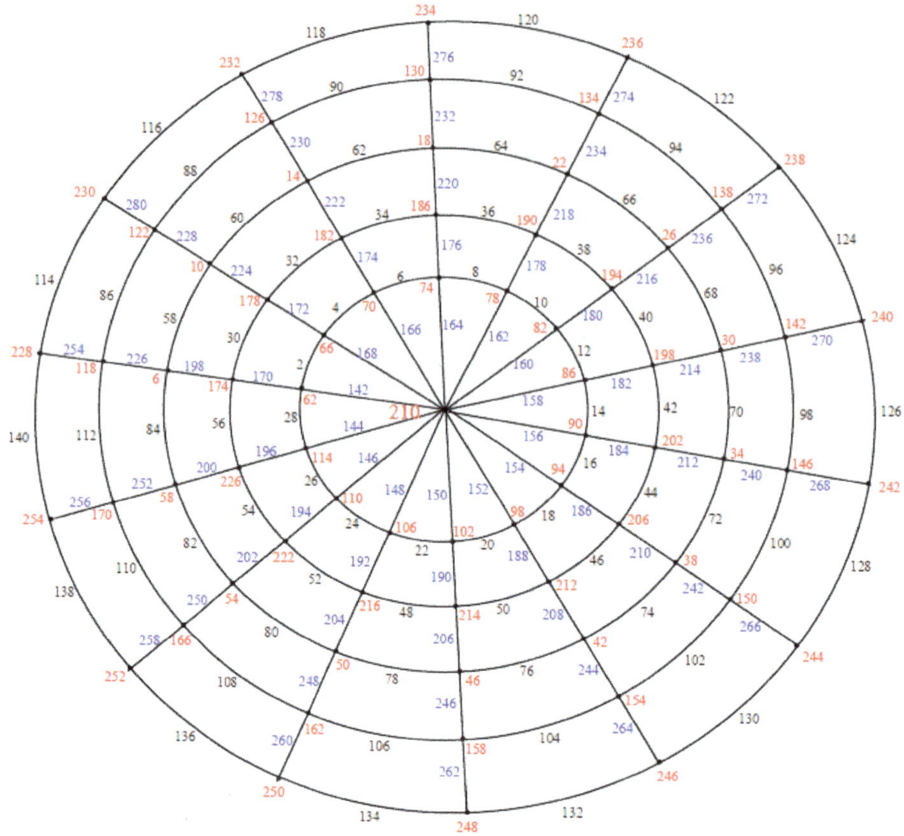

(c) $P_{14,5}$

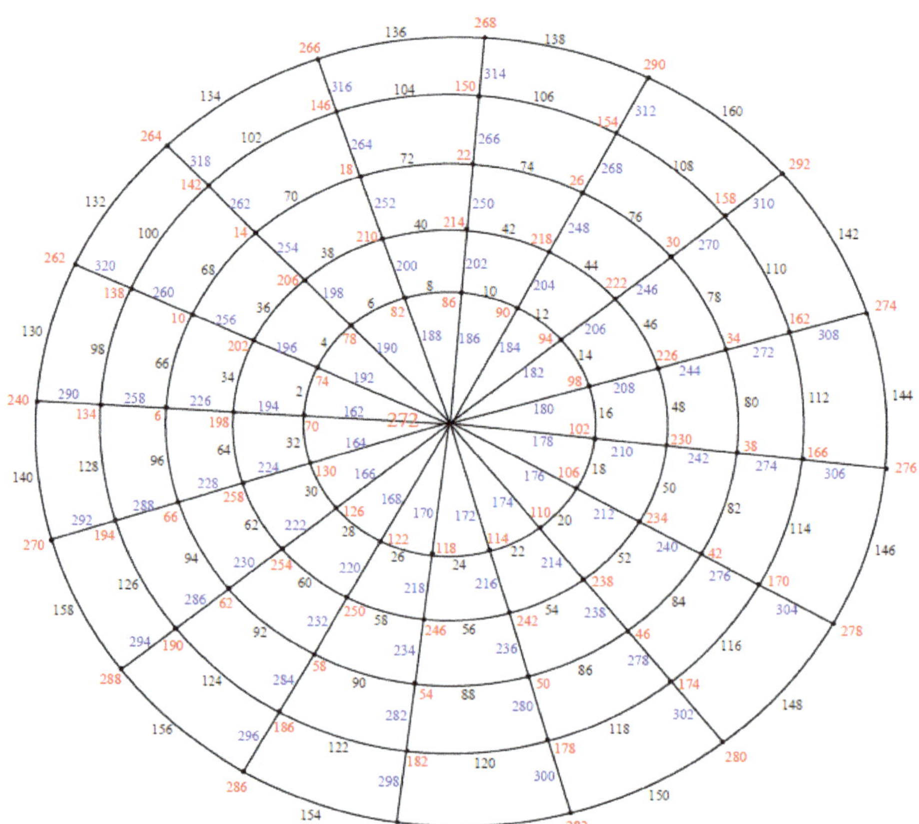

(d) $P_{16,5}$

Figure 8. *Cont.*

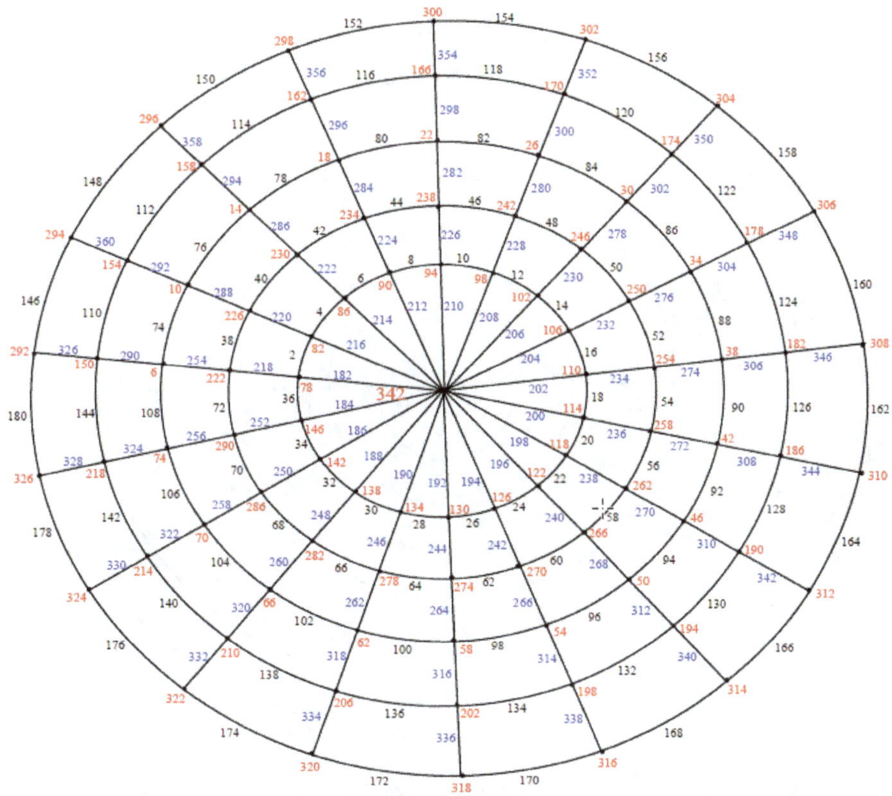

(e) $P_{18,5}$

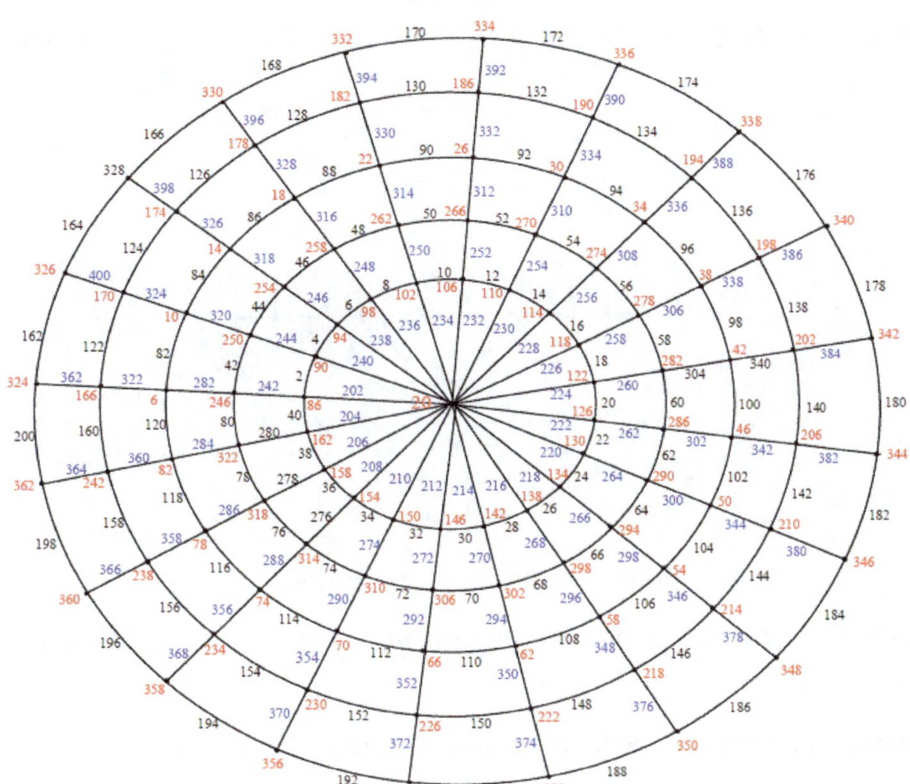

(f) $P_{20,5}$

Figure 8. The edge even graceful labeling of the polar grid graphs $P_{10,5}$, $P_{12,5}$, $P_{14,5}$, $P_{16,5}$, $P_{18,5}$ and $P_{20,5}$.

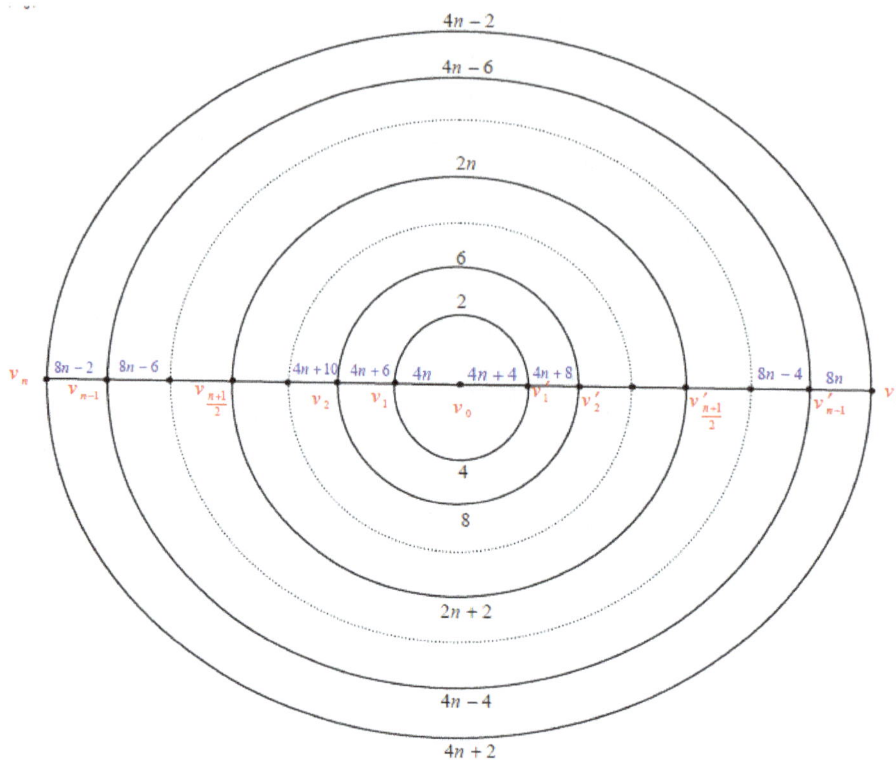

Figure 9. The labeling of the polar grid graph $P_{2,n}$, $n \geq 3$.

Illustration. *The edge even graceful labeling of the polar grid graphs $P_{2,5}$ is shown in Figure 10.*

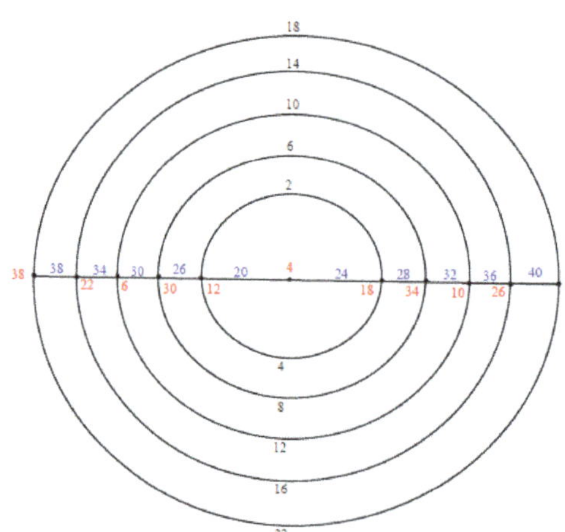

Figure 10. The labeling of the polar grid graph $P_{2,5}$.

Theorem 4. *If m and n are odd positive integers greater than 1. Then the polar grid graph $P_{m,n}$ is an edge even graceful graph.*

Proof. Let the polar grid graph $P_{m,n}$ be labeled as in Figure 7. Let $f : E(G) \rightarrow \{2, 4, \ldots, 2q\}$. □

The corresponding labels of vertices mod$4mn$ are assigned as follows: There are two cases:
Case (1) $n \equiv 1 \bmod 4$, this case contains five subcases as follows:
SubCase (i) $m \equiv (4k-3) \bmod 4n$, $2 \leq k \leq n$
$f^*(v_{(k-1)m+i}) \equiv 4m(2k-1) + 4i + 2, 1 \leq k \leq n-1, 1 \leq i \leq m$. That is the labels of vertices of the most inner circle $C_m^{(1)}$ are assigned by $f^*(v_i) = 4m + 4i + 2$, the labels of vertices of the

circle $C_m^{(2)}$ are assigned by $f^*(v_{m+i}) \equiv 12m + 4i + 2$, the labels of the vertices of the circle $C_m^{(\frac{n-1}{2})}$ are assigned by $f^*(v_{\frac{m(n-3)}{2}+i}) \equiv 4mn - 8m + 4i + 2$, the labels of the vertices of the circle $C_m^{(\frac{n+1}{2})}$ are assigned by $f^*(v_{\frac{m(n-1)}{2}+i}) \equiv 4i + 2$, the labels of the vertices of the circle $C_m^{(\frac{n+3}{2})}$ are assigned by $f^*(v_{\frac{m(n+1)}{2}+i}) \equiv 8m + 4i + 2, \ldots$, the labels of the vertices of the circle $C_m^{(n-1)}$ are assigned by $f^*(v_{m(n-2)+i}) \equiv 4mn - 12m + 4i + 2$, $1 \le i \le m$ and the labels of the vertices of the outer circle $C_m^{(n)}$ are assigned by $f^*(v_{m(n-1)+i}) \equiv 4mn - 12m + 2i + 2$, $1 \le i \le m$. The label of the center vertex v_0 is assigned by $f^*(v_0) \equiv 2mn + 2m(2k-1)$, when $k = \frac{n+1}{2}$, we have $f^*(v_0) \equiv 0$.

SubCase (ii) $m \equiv (8k-5) \bmod 4n$, $1 \le k \le \frac{n+1}{2}$, $m \ne 3$. In this subcase the vertex $v_{m(\frac{n+4k-1}{4})-(\frac{m+1}{2})}$ in the circle $C_m^{(\frac{n+4k-1}{4})}$ will repeat with the center vertex v_0. To avoid this problem we replace the labels of the two edges $v_{m(\frac{n+4k-1}{4})-(\frac{m+1}{2})}v_{m(\frac{n+4k-1}{4})-(\frac{m-1}{2})}$ and $v_{m(\frac{n+4k-1}{4})-(\frac{m-1}{2})}v_{m(\frac{n+4k-1}{4})-(\frac{m-3}{2})}$. That is $f(v_{m(\frac{n+4k-1}{4})-(\frac{m+1}{2})}v_{m(\frac{n+4k-1}{4})-(\frac{m-1}{2})}) = m[2k + \frac{n-3}{2}] + 1$ and $f(v_{m(\frac{n+4k-1}{4})-(\frac{m-1}{2})}v_{m(\frac{n+4k-1}{4})-(\frac{m-3}{2})}) = m[2k + \frac{n-3}{2}] - 1$ and we obtain the labels of the corresponding vertices as follows $f^*(v_{m(\frac{n+4k-1}{4})-(\frac{m+1}{2})}) \equiv 2mn + 4m(2k-1) + 2$, $f^*(v_{m(\frac{n+4k-1}{4})-(\frac{m-1}{2})}) \equiv 2mn + 4m(2k-1) + 4$, $f^*(v_{m(\frac{n+4k-1}{4})-(\frac{m-3}{2})}) \equiv 2mn + 4m(2k-1) + 6$, and in this case the center vertex v_0 is labeled as $f^*(v_0) \equiv 2mn + 4m(2k-1)$. The rest vertices will be labeled as in subCase (i).

Remark 3. *When $n \equiv 1 \bmod 4$ and $m = 3$, in this case the vertex $v_{3(\frac{n-1}{4})+1}$ in the circle $C_3^{(\frac{n-1}{4}+1)}$ will repeat with the center vertex v_0. To avoid this problem we replace the labels of the two edges $v_{3(\frac{n-1}{4})+2}v_{3(\frac{n-1}{4})+3}$ and $v_{3(\frac{n-1}{4})+1}v_{3(\frac{n-1}{4})+3}$. That is $f(v_{3(\frac{n-1}{4})+2}v_{3(\frac{n-1}{4})+3}) = 3(\frac{n-1}{2}) + 6$ and $f(v_{3(\frac{n-1}{4})+1}v_{3(\frac{n-1}{4})+3}) = 3(\frac{n-1}{2}) + 4$ and we obtain the labels of the corresponding vertices as follows $f^*(v_{3(\frac{n-1}{4})+1}) \equiv 6n + 10$, $f^*(v_{3(\frac{n-1}{4})+2}) \equiv 6n + 18$, $f^*(v_{3(\frac{n-1}{4})+3}) \equiv 6n + 16$ and the center vertex v_0 is labeld as $f^*(v_0) \equiv 6n + 12$.*

SubCase (iii) $m \equiv (8k-1) \bmod 4n$, $1 \le k \le \frac{n-5}{4}$. In this subcase the vertex $v_{m(\frac{3n+4k+1}{4})-(\frac{m+1}{2})}$ in the circle $C_m^{(\frac{3n+4k-1}{4})}$ will repeat with the center vertex v_0. To avoid this problem we replace the labels of the two edges $v_{m(\frac{3n+4k+1}{4})-(\frac{m+1}{2})}v_{m(\frac{3n+4k+1}{4})-(\frac{m-1}{2})}$ and $v_{m(\frac{3n+4k+1}{4})-(\frac{m-1}{2})}v_{m(\frac{3n+4k+1}{4})-(\frac{m-3}{2})}$. That is $f(v_{m(\frac{3n+4k+1}{4})-(\frac{m+1}{2})}v_{m(\frac{3n+4k+1}{4})-(\frac{m-1}{2})}) = 2mn + m[2k - \frac{n+1}{2}] + 1$ and $f(v_{m(\frac{3n+4k-1}{4})-(\frac{m-1}{2})}v_{m(\frac{3n+4k-1}{4})-(\frac{m-3}{2})}) = 2mn + m[2k - \frac{n+1}{2}] - 1$ and we obtain the labels of the corresponding vertices as follows $f^*(v_{m(\frac{3n+4k+1}{4})-(\frac{m+1}{2})}) \equiv 2mn + 8km + 2$, $f^*(v_{m(\frac{3n+4k-1}{4})-(\frac{m-1}{2})}) \equiv 2mn + 8km + 4$, $f^*(v_{m(\frac{3n+4k+1}{4})-(\frac{m-3}{2})}) \equiv 2mn + 8km + 6$, and in this case the center vertex v_0 is labeled as $f^*(v_0) \equiv 2mn + 8km$. The rest vertices will be labeled as in subCase (i).

SubCase (iv) $m \equiv (8k-1) \bmod 4n$, $\frac{n+3}{4} \le k \le \frac{n-1}{2}$. In this case the vertex $v_{m(\frac{4k-n+1}{4})-(\frac{m+1}{2})}$ in the circle $C_m^{(\frac{4n-k+1}{4})}$ will repeat with the center vertex v_0. To avoid this problem we replace the labels of the two edges $v_{m(\frac{4k-n+1}{4})-(\frac{m+1}{2})}v_{m(\frac{4k-n+1}{4})-(\frac{m-1}{2})}$ and $v_{m(\frac{4k-n+1}{4})-(\frac{m-1}{2})}v_{m(\frac{4k-n+1}{4})-(\frac{m-3}{2})}$. That is $f(v_{m(\frac{4k-n+1}{4})-(\frac{m+1}{2})}v_{m(\frac{4k-n+1}{4})-(\frac{m-1}{2})}) = m[2k - \frac{n+1}{2}] + 1$ and $f(v_{m(\frac{4k-n+1}{4})-(\frac{m-1}{2})}v_{m(\frac{4k-n+1}{4})-(\frac{m-3}{2})}) = m[2k - \frac{n+1}{2}] - 1$ and we obtain the labels of the corresponding vertices as follows $f^*(v_{m(\frac{4k-n+1}{4})-(\frac{m+1}{2})}) \equiv 2mn + 8km + 2$, $f^*(v_{m(\frac{4k-n+1}{4})-(\frac{m-1}{2})}) \equiv 8km - 2mn + 4$, $f^*(v_{m(\frac{4k-n+1}{4})-(\frac{m-3}{2})}) \equiv 8km - 2mn + 6$, and in this case the center vertex v_0 is labeled as $f^*(v_0) \equiv 8km - 2mn$. The rest vertices will be labeled as in subCase (i).

SubCase (v) $m \equiv (2n-3) \bmod 4n$. In this case the vertex v_{mn-1} in the outer circle $C_m^{(n)}$ will repeat with the center vertex v_0. To avoid this problem we replace the labels of the two edges $v_{mn-2}v_{mn-1}$ and $v_{mn}v_{m(n-1)+1}$. That is $f(v_{mn-2}v_{mn-1}) = 2mn - 4$, $f(v_{mn}v_{m(n-1)+1}) = 2mn$ and we obtain the labels of the corresponding vertices are as follows $f^*(v_{mn-2}) \equiv 4mn - 2m + 2$, $f^*(v_{mn-1}) \equiv 4mn - 2m + 4$,

$f^*(v_{mn}) \equiv 4mn - 2m - 2$ and $f^*(v_{m(n-1)+1}) \equiv 4mn - 4m$, and in this case the center vertex v_0 is labeled as $f^*(v_0) \equiv 4mn - 2m$. The rest vertices will be labeled as in subCase (i).

Illustration. *The edge odd graceful labeling of the polar grid graphs* $P_{3,5}, P_{13,5}, P_{11,5}, P_{7,9}, P_{15,5}$ *and* $P_{7,5}$ *respectively are shown in Figure 11.*

Case (2) $n \equiv 3 \bmod 4$. This case contains also five subcases as follows:

SubCase (i) $m \equiv (4k-3) \bmod 4n$, $2 \le k \le n$

$f^*(v_{(k-1)m+i}) \equiv 4m(2k-1) + 4i + 2, 1 \le k \le n-1, 1 \le i \le m$. That is the labels of vertices of the most inner circle $C_m^{(1)}$ are assigned by $f^*(v_i) \equiv 4m + 4i + 2$, the label of vertices of the circle $C_m^{(2)}$ are assigned by $f^*(v_{m+i}) \equiv 12m + 4i + 2$, the labels of vertices of the circle $C_m^{(\frac{n-1}{2})}$ are assigned by $f^*(v_{\frac{m(n-1)}{2}+i}) \equiv 4mn - 8m + 4i + 2$, the labels of vertices of the circle $C_m^{(\frac{n+1}{2})}$ are assigned by $f^*(v_{\frac{m(n-3)}{2}+i}) \equiv 4i + 2$, the labels of vertices of the circle $C_m^{(\frac{n+3}{2})}$ are assigned by $f^*(v_{\frac{m(n+1)}{2}+i}) \equiv 8m + 4i + 2, \ldots$, the labels of vertices of the circle $C_m^{(n-1)}$ are assigned by $f^*(v_{m(n-2)+i}) \equiv 4mn - 12m + 4i + 2$, $1 \le i \le m$ and the labels of the vertices of the outer circle $C_m^{(n)}$ are assigned by $f^*(v_{m(n-1)+i}) \equiv 4mn - 12m + 2i + 2$, $1 \le i \le m$. The label of the center vertex v_0 is assigned by $f^*(v_0) \equiv 2mn + 2m(2k-1)$, when $k = \frac{n+1}{2}$, we have $f^*(v_0) \equiv 0$.

SubCase (ii) $m \equiv (8k-5) \bmod 4n$, $1 \le k \le \frac{n-3}{4}$, $m \ne 3$

In this subcase the vertex $v_{m(\frac{3n+4k-1}{4})-(\frac{m+1}{2})}$ in the circle $C_m^{(\frac{3n+4k-1}{4})}$ will repeat with the center vertex v_0. To avoid this problem we replace the labels of the two edges $v_{m(\frac{3n+4k-1}{4})-(\frac{m+1}{2})}v_{m(\frac{3n+4k-1}{4})-(\frac{m-1}{2})}$ and $v_{m(\frac{3n+4k-1}{4})-(\frac{m-1}{2})}v_{m(\frac{3n+4k-1}{4})-(\frac{m-3}{2})}$. That is $f(v_{m(\frac{3n+4k-1}{4})-(\frac{m+1}{2})}v_{m(\frac{3n+4k-1}{4})-(\frac{m-1}{2})}) = 2mn + m[2k - \frac{n+3}{2}] + 1$ and $f(v_{m(\frac{3n+4k-1}{4})-(\frac{m-1}{2})}v_{m(\frac{3n+4k-1}{4})-(\frac{m-3}{2})}) = 2mn + m[2k - \frac{n+3}{2}] - 1$ and we obtain the labels of the corresponding vertices as follows $f^*(v_{m(\frac{3n+4k-1}{4})-(\frac{m+1}{2})}) \equiv 2mn + 4m(2k-1) + 2$, $f^*(v_{m(\frac{3n+4k-1}{4})-(\frac{m-1}{2})}) \equiv 2mn + 4m(2k-1) + 4$, $f^*(v_{m(\frac{3n+4k-1}{4})-(\frac{m-3}{2})}) \equiv 2mn + 4m(2k-1) + 6$, and the label of the center vertex v_0 is assigned by $f^*(v_0) \equiv 2mn + 4m(2k-1)$. That rest vertices will be labeled as in subcase (i).

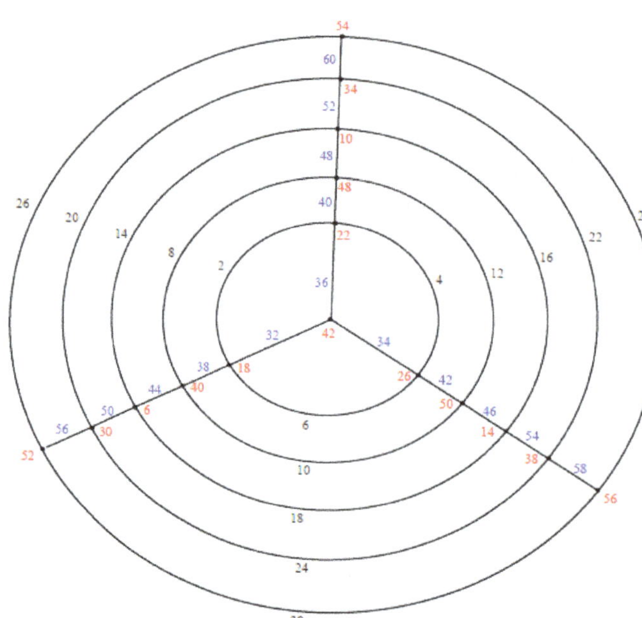

(a) $P_{3,5}$

Figure 11. *Cont.*

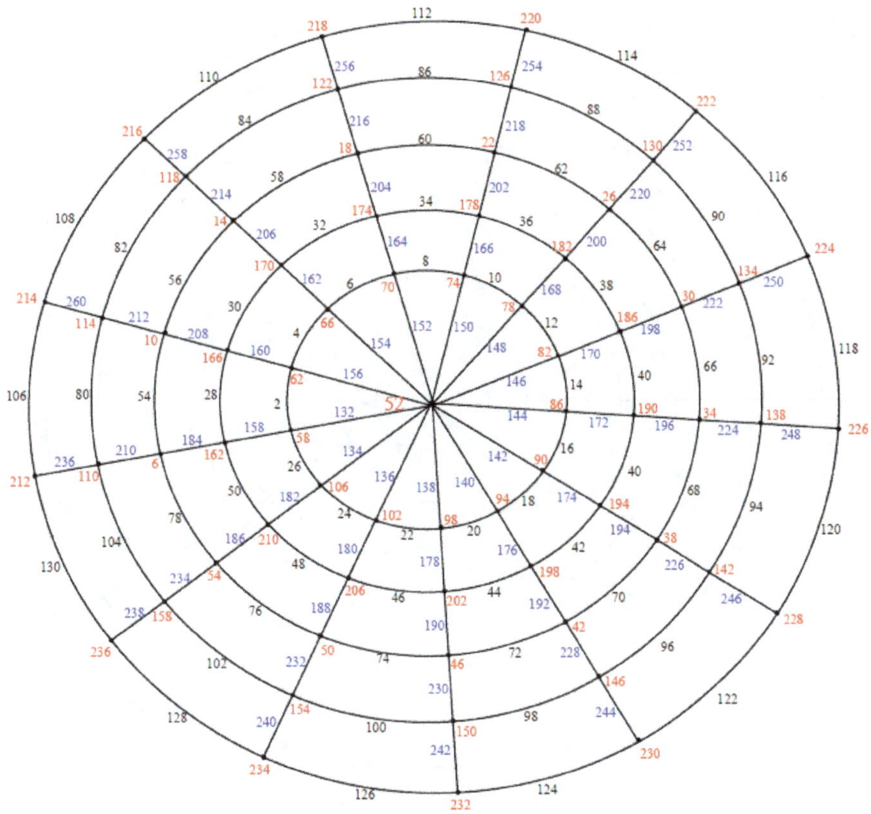

(b) $P_{13,5}$

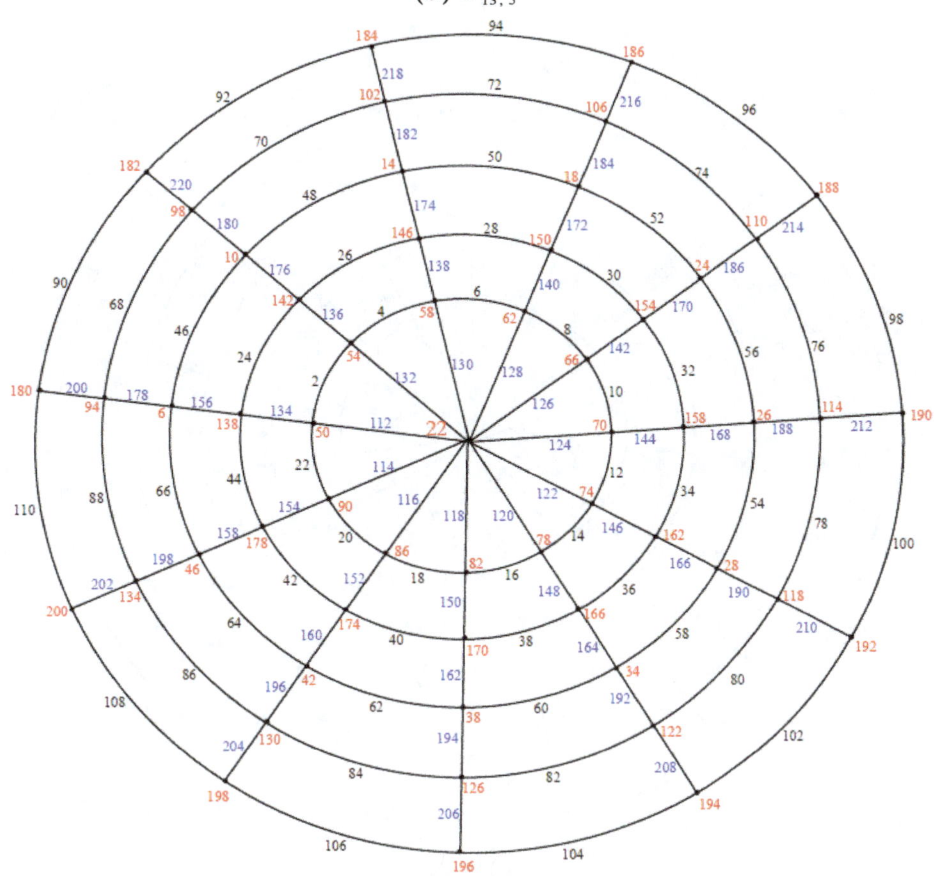

(c) $P_{11,5}$

Figure 11. *Cont.*

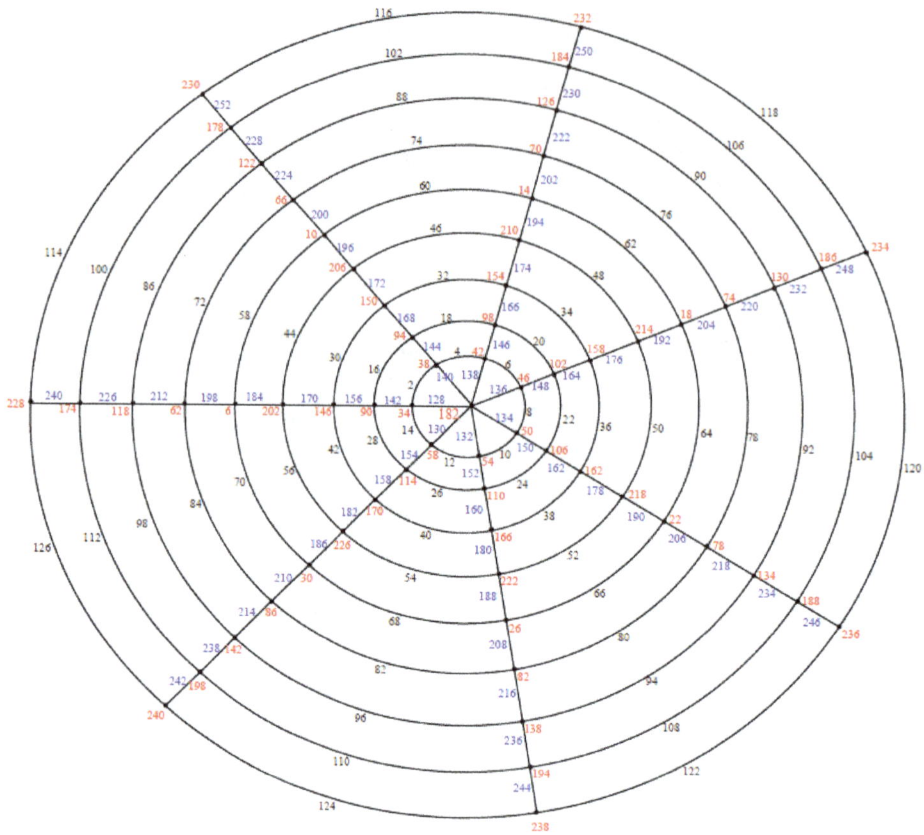

(d) $P_{7,9}$

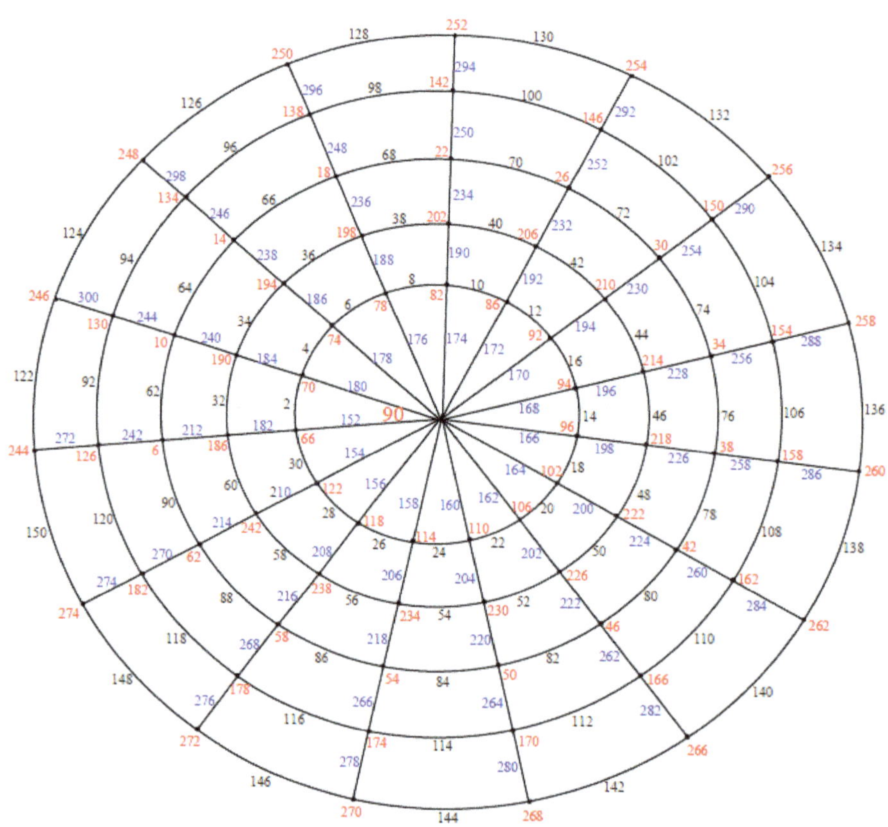

(e) $P_{15,5}$

Figure 11. *Cont.*

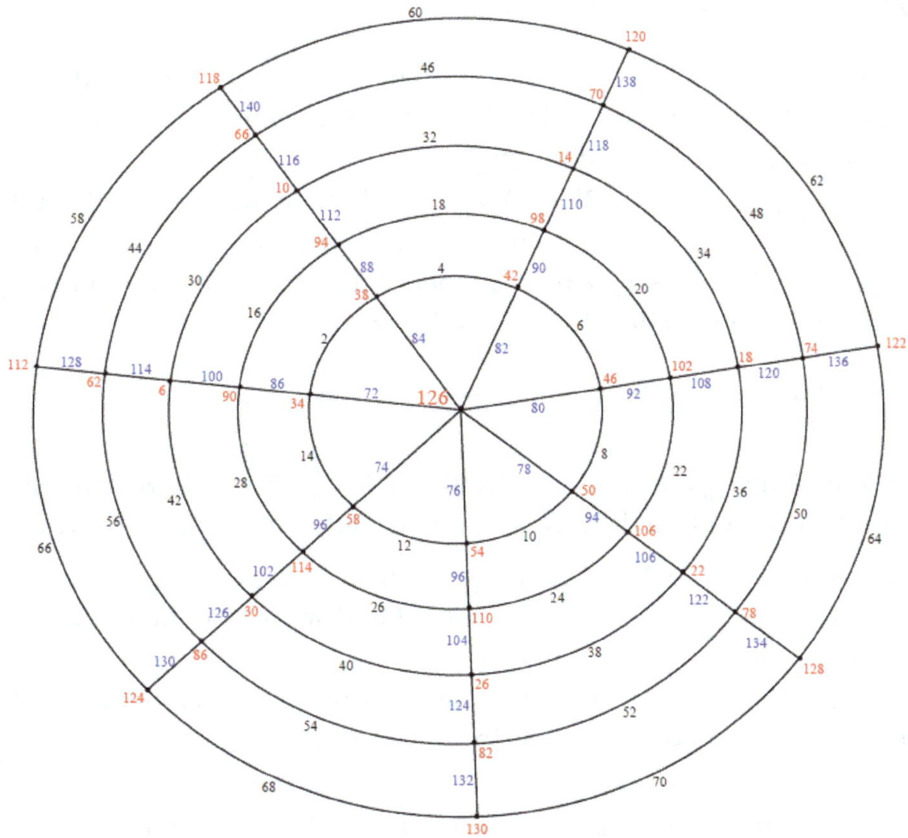

(f) $P_{7,5}$

Figure 11. The polar grid graphs $P_{3,5}$, $P_{13,5}$, $P_{11,5}$, $P_{7,9}$, $P_{15,5}$ and $P_{7,5}$.

Remark 4. *When* $n \equiv 3 \bmod 4$ *and* $m = 3$, *we have the vertex* $v_{3(\frac{3n-1}{4})+1}$ *in the circle* $C_m^{(\frac{3n-1}{4}+1)}$ *will repeat with the center vertex* v_0. *To avoid this problem we replace the labels of the two edges* $v_{3(\frac{3n-1}{4})+2}v_{3(\frac{3n-1}{4})+3}$ *and* $v_{3(\frac{3n-1}{4})+1}v_{3(\frac{3n-1}{4})+3}$. *That is* $f(v_{3(\frac{3n-1}{4})+2}v_{3(\frac{3n-1}{4})+3}) = 3(\frac{3n-1}{2}) + 6$ *and* $f(v_{3(\frac{3n-1}{4})+1}v_{3(\frac{3n-1}{4})+3}) = 3(\frac{3n-1}{2}) + 4$ *and we obtain the labes of the corresponding vertices mod4mn are as follows:* $f^*(v_{3(\frac{3n-1}{4})+1}) \equiv 6n+10$, $f^*(v_{3(\frac{3n-1}{4})+2}) \equiv 6n+18$, $f^*(v_{3(\frac{3n-1}{4})+3}) \equiv 6n+20$ *and the label of the center vertex* v_0 *is assigned by* $f^*(v_0) \equiv 6n+12$.

Note that $P_{3,3}$ is an edge even graceful grapg but not follow this rule. See Figure 12.

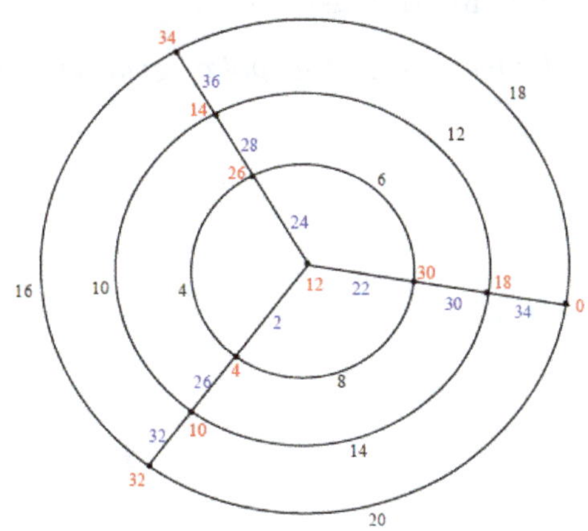

Figure 12. The polar grid graphs $P_{3,3}$.

SubCase (iii) $m \equiv (8k-1)\mathrm{mod}4n$, $\frac{n+5}{4} \leq k \leq \frac{n+1}{2}$,

In this subcase the vertex $v_{m(\frac{4k-n-1}{4})-(\frac{m+1}{2})}$ in the circle $C_m^{(\frac{4k-n-1}{4})}$ will repeat with the center vertex v_0. To avoid this problem we replace the labels of the two edges $v_{m(\frac{4k-n-1}{4})-(\frac{m+1}{2})}v_{m(\frac{4k-n-1}{4})-(\frac{m-1}{2})}$ and $v_{m(\frac{4k-n-1}{4})-(\frac{m-1}{2})}v_{m(\frac{4k-n-1}{4})-(\frac{m-3}{2})}$ That is $f(v_{m(\frac{4k-n-1}{4})-(\frac{m+1}{2})}v_{m(\frac{4k-n-1}{4})-(\frac{m-1}{2})}) = (2k-1)m+1$ and $f(v_{m(\frac{4k-n-1}{4})-(\frac{m-1}{2})}v_{m(\frac{4k-n-1}{4})-(\frac{m-3}{2})}) = (2k-1)m-1$ and we obtain the labels of the corresponding vertices as follows: $f^*(v_{m(\frac{4k-n-1}{4})-(\frac{m+1}{2})}) \equiv 2m(4k-1)+2$, $f^*(v_{m(\frac{4k-n-1}{4})-(\frac{m-1}{2})}) \equiv 2m(4k-1)+4$ and $f^*(v_{m(\frac{4k-n-1}{4})-(\frac{m-3}{2})}) \equiv 2m(4k-1)+6$. The label of the center vertex v_0 is assigned by $f^*(v_0) \equiv 2m(4k-1)$. The rest vertices will be labeled as in subCase (i).

SubCase (iv) $m \equiv (8k-1)\mathrm{mod}4n$, $1 \leq k \leq \frac{n-1}{2}$

In this subcase the vertex $v_{m(\frac{n+4k+1}{4})-(\frac{m+1}{2})}$ in the circle $C_m^{(\frac{n+4k+1}{4})}$ will repeat with the center vertex v_0. To avoid this problem we replace the labels of the two edges $v_{m(\frac{n+4k+1}{4})-(\frac{m+1}{2})}v_{m(\frac{n+4k+1}{4})-(\frac{m-1}{2})}$ and $v_{m(\frac{n+4k+1}{4})-(\frac{m-1}{2})}v_{m(\frac{n+4k+1}{4})-(\frac{m-3}{2})}$ That is $f(v_{m(\frac{n+4k+1}{4})-(\frac{m+1}{2})}v_{m(\frac{n+4k+1}{4})-(\frac{m-1}{2})}) = (2k+\frac{n-1}{2})m+1$ and $f(v_{m(\frac{n+4k+1}{4})-(\frac{m-1}{2})}v_{m(\frac{4k-n-1}{4})-(\frac{m-3}{2})}) = (2k+\frac{n-1}{2})m-1$ and we obtain the labels of the corresponding vertices as follows $f^*(v_{m(\frac{n+4k+1}{4})-(\frac{m+1}{2})}) = 2mn+8km+2$, $f^*(v_{m(\frac{n+4k+1}{4})-(\frac{m-1}{2})}) = 2mn+8km+4$, $f^*(v_{m(\frac{n+4k+1}{4})-(\frac{m-3}{2})}) = 2mn+8km+6$ and the label of the center vertex v_0 is labeled as $f^*(v_0) \equiv 2mn+8km$. The rest vertices will be labeled as in subCase (i).

Remark 5. *If* $k = \frac{n+1}{4}$ *we have* $f^*(v_{m(\frac{n+4k+1}{4})-(\frac{m+1}{2})}) = 8km-2mn+2$, $f^*(v_{m(\frac{n+4k+1}{4})-(\frac{m-1}{2})}) = 8km-2mn+4$, $f^*(v_{m(\frac{n+4k+1}{4})-(\frac{m-3}{2})}) = 8km-2mn+6$ *and the center vertex* v_0 *is labeled as* $f^*(v_0) \equiv 8km-2mn$.

SubCase (v) $m \equiv (2n-3)\mathrm{mod}4n$

In this subcase the vertex v_{mn-1} in the outer circle $C_m^{(n)}$ will repeat with the center vertex v_0. To avoid this problem we replace the labels of the two edges $v_{mn-2}v_{mn-1}$ and $v_{mn}v_{m(n-1)+1}$. That is $f(v_{mn-2}v_{mn-1}) = 2mn-4$, $f(v_{mn}v_{m(n-1)+1}) = 2mn$ and we obtain the labes of the corresponding vertices as follows:

$f^*(v_{mn-2}) = 4mn-2m+2$, $f^*(v_{mn-1}) = 4mn-2m+4$, $f^*(v_{mn}) = 4mn-2m-2$ and $f^*(v_{m(n-1)+1}) = 4m(n-1)$ and the label of the center vertex v_0 is assigned by $f^*(v_0) \equiv 2m(2n-1)$.

The rest vertices will be labeled as in subCase (i).

Illustration. *The edge odd graceful labeling of the polar grid graphs* $P_{3,7}$, $P_{13,7}$, $P_{19,7}$, $P_{11,7}$ *and* $P_{15,7}$ *respectively are shown in Figure 13.*

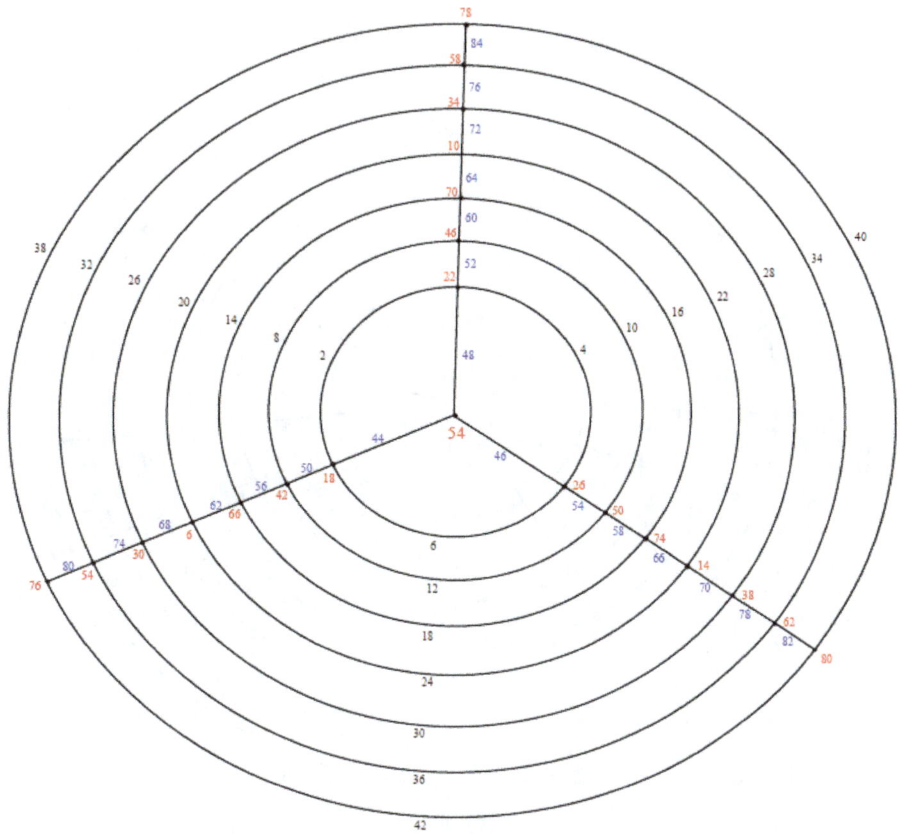

(a) $P_{3,7}$

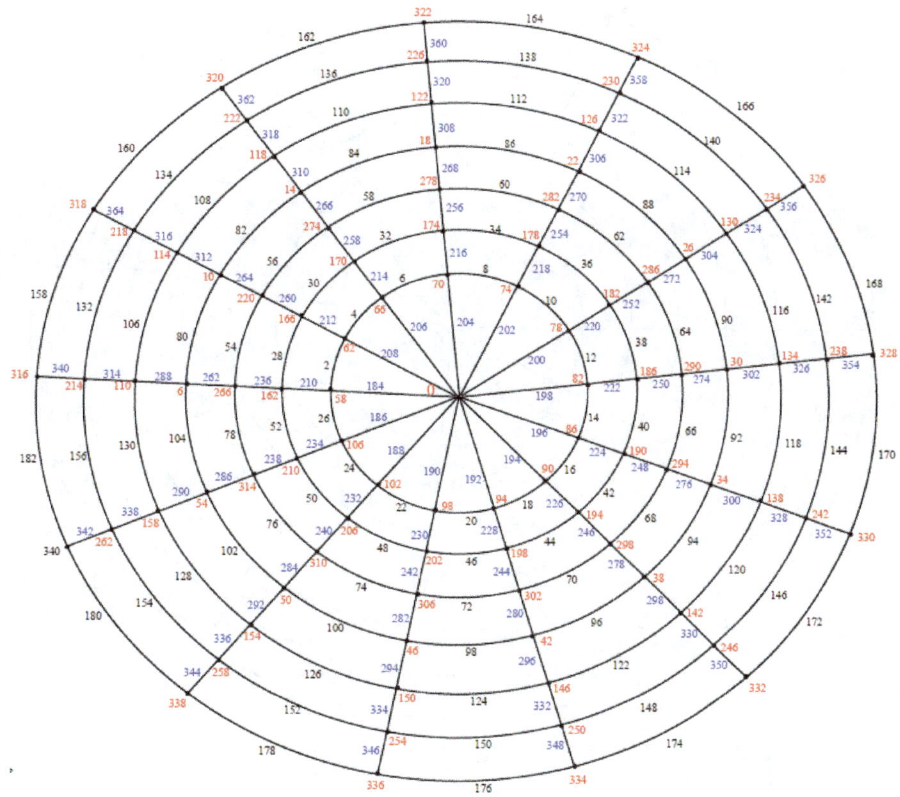

(b) $P_{13,7}$

Figure 13. *Cont.*

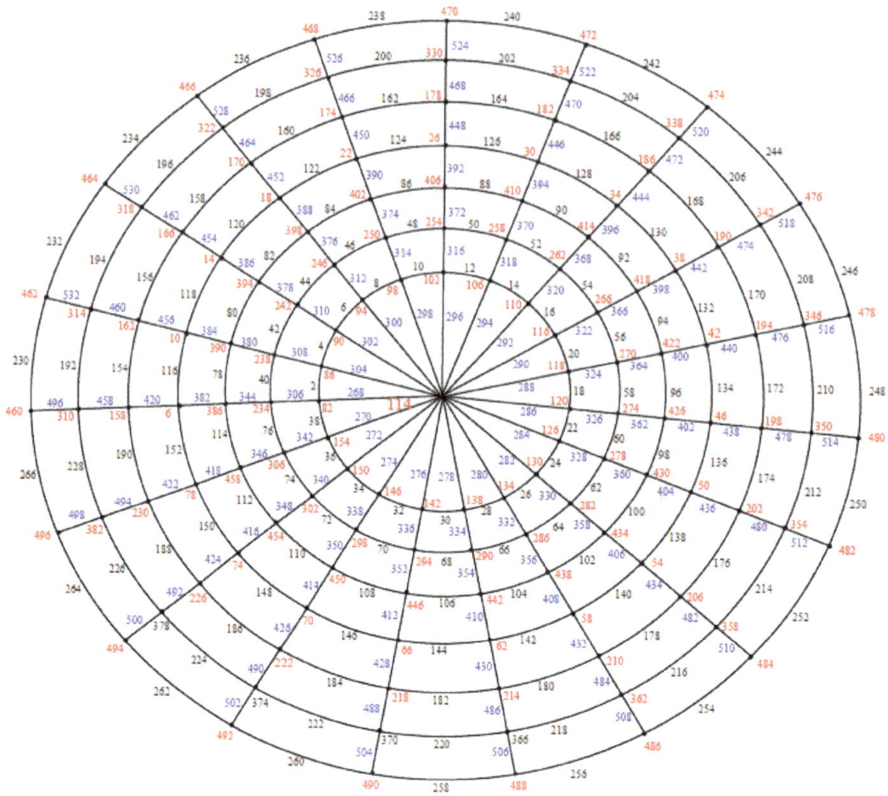

(c) $P_{19,7}$

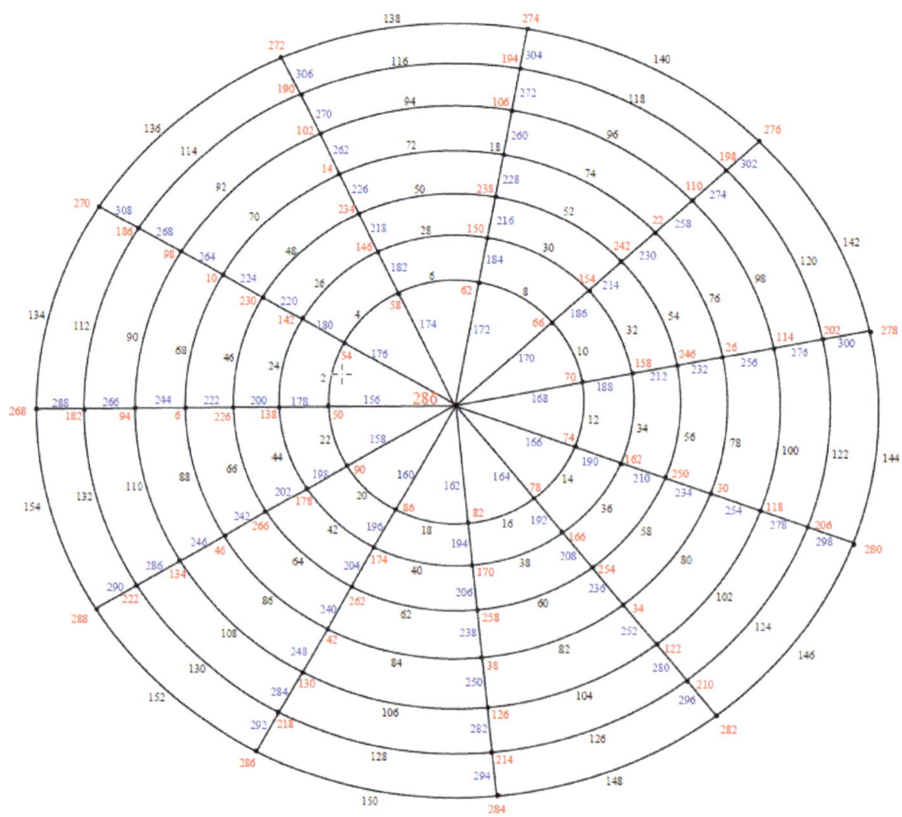

(d) $P_{11,7}$

Figure 13. *Cont.*

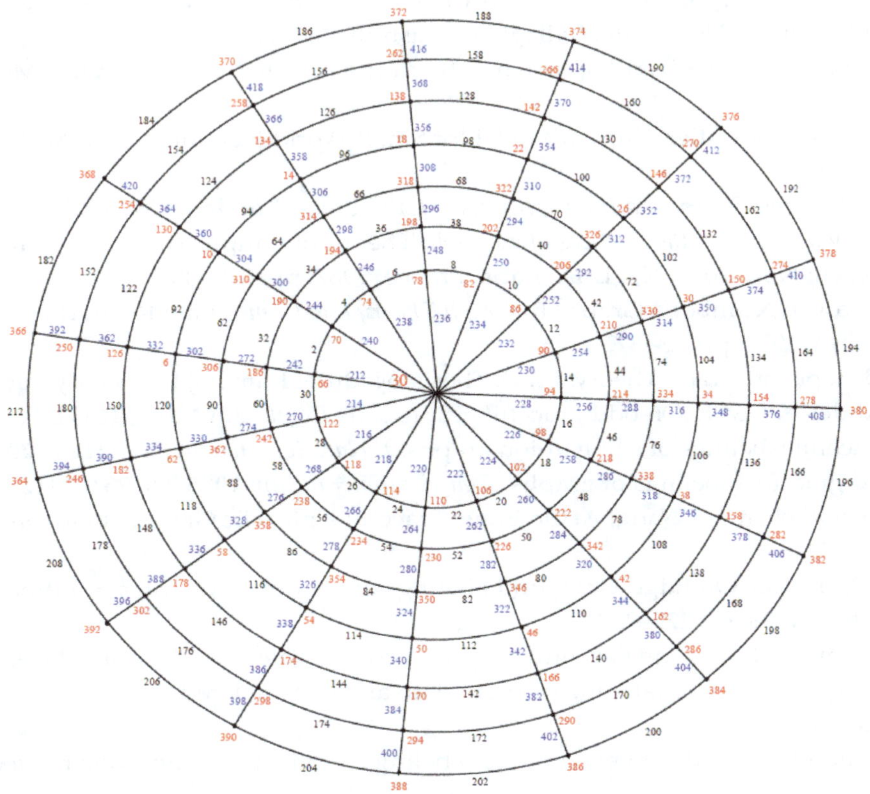

(e) $P_{15,7}$

Figure 13. The polar grid graphs $P_{3,7}$, $P_{13,7}$, $P_{19,7}$, $P_{11,7}$ $P_{15,7}$.

3. Conclusions

This paper gives some basic knowledge about the application of Graph labeling and Graph Theory in real life which is the one branch of mathematics. It is designed for the researcher who research in graph labeling and graph Theory. In this paper, we give necessary and sufficient conditions for a polar grid graph to admit edge even labeling. In future work we will study the necessary and sufficient conditions for the cylinder $P_m \times C_n$, torus $C_m \times C_n$ and rectangular $P_m \times P_n$ grid graphs to be edge even graceful.

References

1. Acharya, B.D.; Arumugam, S.; Rosa, A. *Labeling of Discrete Structures and Applications;* Narosa Publishing House: New Delhi, India, 2008; pp. 1–14.
2. Bloom, G.S. Numbered Undirected Graphs and Their Uses, a Survey of a Unifying Scientific and Engineering Concept and Its Use in Developing a Theory of Non-Redundant Homometric Sets Relating to Some Ambiguities in X-ray Diffraction Analysis. Ph.D. Thesis, University Southern California, Los Angeles, CA, USA, 1975.
3. Bloom, G.S.; Golomb, S.W. Numbered complete graphs, unusual rulers, and assorted applications. In *Theory and Applications of Graphs;* Lecture Notes in Math; Springer: New York, NY, USA, 1978; Volume 642, pp. 53–65.
4. Bloom, G.S.; Golomb, S.W. Applications of numbered undirected graphs. *Proc. IEEE* **1977**, *65*, 562–570. [CrossRef]
5. Bloom, G.S.; Hsu, D.F. On graceful digraphs and a problem in network addressing. *Congr. Numer.* **1982**, *35*, 91–103.
6. Graham, R.L.; Pollak, H.O. On the addressing problem for loop switching. *Bell Syst. Tech. J.* **1971**, *50*, 2495–2519. [CrossRef]

7. Sutton, M. Summable Graphs Labellings and their Applications. Ph.D. Thesis, Department Computer Science, The University of Newcastle, Callaghan, Australia, 2001.
8. Shang, Y. More on the normalized Laplacian Estrada index. *Appl. Anal. Discret. Math.* **2014**, *8*, 346–357. [CrossRef]
9. Shang, Y. Geometric Assortative Growth Model for Small-World Networks. *Sci. World J.* **2014**, *2014*, 759391. [CrossRef] [PubMed]
10. Gross, J.; Yellen, J. *Graph Theory and Its Applications*; CRC Press: London, UK, 1999.
11. Rosa, A. *Certain Valuations of the Vertices of a Graph*; Theory of Graphs (Internat. Symp, Rome, July 1966); Gordan and Breach: New York, NY, USA; Paris, France, 1967; pp. 349–355.
12. Golomb, S.W. How to Number a Graph. In *Graph Theory and Computing*; Read, R.C., Ed.; Academic Press: NewYork, NY, USA, 1972; pp. 23–37.
13. Gnanajothi, R.B. Topics in Graph Theory. Ph.D. Thesis, Madurai Kamaraj University, Tamil Nadu, India, 1991.
14. Seoud, M.A.; Abdel-Aal, M.E. On odd graceful graphs. *Ars Comb.* **2013**, *108*, 161–185.
15. Gao, Z. Odd graceful labelings of some union graphs. *J. Nat. Sci. Heilongjiang Univ.* **2007**, *24*, 35–39.
16. Lo, S.P. On edge-graceful1abelings of graphs. *Congr. Numerantium* **1985**, *50*, 231–241.
17. Kuan, Q.; Lee, S.; Mitchem, J.; Wang, A. On Edge-Graceful Unicyclic Graphs. *Congr. Numerantium* **1988**, *61*, 65–74.
18. Lee, L.; Lee, S.; Murty, G. On Edge-Graceful Labelings of Complete Graphs: Solutions of Lo's Conjecture. *Congr. Numerantium* **1988**, *62*, 225–233.
19. Solairaju, A.; Chithra, K. Edge-Odd Graceful Graphs. *Electron. Notes Discret. Math.* **2009**, *33*, 15–20. [CrossRef]
20. Daoud, S.N. Edge odd graceful labeling of some path and cycle related graphs. *AKCE Int. J. Graphs Comb.* **2017**, *14*, 178–203. [CrossRef]
21. Elsonbaty, A.; Daoud, S.N. Edge even graceful labeling of some path and cycle related graphs. *Ars Comb.* **2017**, *130*, 79–96.
22. Gallian, J.A. A Dynamic Survey of Graph Labeling. *Electron. J. Comb.* **2017**, *22*, #DS6.

5

The Complexity of Some Classes of Pyramid Graphs Created from a Gear Graph

Jia-Bao Liu [1] and Salama Nagy Daoud [2,3,*]

1 School of Mathematics and Physics, Anhui Jianzhu University, Hefei 230601, China; liujiabaoad@163.com
2 Department of Mathematics, Faculty of Science, Taibah University, Al-Madinah 41411, Saudi Arabia
3 Department of Mathematics and Computer Science, Faculty of Science, Menoufia University, Shebin El Kom 32511, Egypt
* Correspondence: sdaoud@taibahu.edu.sa.

Abstract: The methods of measuring the complexity (spanning trees) in a finite graph, a problem related to various areas of mathematics and physics, have been inspected by many mathematicians and physicists. In this work, we defined some classes of pyramid graphs created by a gear graph then we developed the Kirchhoff's matrix tree theorem method to produce explicit formulas for the complexity of these graphs, using linear algebra, matrix analysis techniques, and employing knowledge of Chebyshev polynomials. Finally, we gave some numerical results for the number of spanning trees of the studied graphs.

Keywords: complexity; Chebyshev polynomials; gear graph; pyramid graphs

MSC: 05C05, 05C50

---

1. Introduction

The graph theory is a theory that combines computer science and mathematics, which can solve considerable problems in several fields (telecom, social network, molecules, computer network, genetics, etc.) by designing graphs and facilitating them through idealistic cases such as the spanning trees, see [1–10].

A spanning tree of a finite connected graph G is a maximal subset of the edges that contains no cycle, or equivalently a minimal subset of the edges that connects all the vertices. The history of enumerating the number of spanning trees $\tau(G)$ of a graph G dates back to 1842 when the physicist Kirchhoff [11] offered the matrix tree theorem established on the determinants of a certain matrix gained from the Laplacian matrix L defined by the difference between the degree matrix D and adjacency matrix A, where D is a diagonal matrix, $D = dig\,(d_1, d_2, \ldots, d_n)$ corresponding to a graph G with n vertices that has the vertex degree of d_i in the ith position of a graph G and A is a matrix with rows and columns labeled by graph vertices, with a 1 or 0 in position (u_i, u_j) according to whether u_i and u_j are adjacent or not. That is

$$L_{ij} = \begin{cases} a_i & \text{if } i = j \\ -1 & \text{if } i \neq j \text{ and } i \text{ is adjacent to } j \\ 0 & \text{otherwise} \end{cases},$$

where a_i denotes the degree of the vertex i.

This method allows beneficial results for a graph comprising a small number of vertices, but is not feasible for large graphs. There is one more method for calculating $\tau(G)$. Let $\lambda_1 \geq \lambda_2 \geq \ldots \geq \lambda_k = 0$

denote the eigenvalues of the matrix L of a graph G with n vertices. "Kelmans" and "Chelnokov" [12] have derived that

$$\tau(G) = \frac{1}{k}\prod_{i=1}^{k-1} \lambda_k.$$

One of the favorite methods of calculating the complexity is the contraction–deletion theorem. For any graph G, the complexity $\tau(G)$ of G is equal to $\tau(G) = \tau(G-e) + \tau(G/e)$, where e is any edge of G, and where $G - e$ is the deletion of e from G, and G/e is the contraction of e in G. This gives a recursive method to calculate the complexity of a graph [13,14].

Another important method is using electrically equivalent transformations of networks. Yilun Shang [15] derived a closed-form formula for the enumeration of spanning trees the subdivided-line graph of a simple connected graph using the theory of electrical networks.

Many works have conceived techniques to derive the number of spanning trees of a graph, some of which can be found at [16–18].

Now, we give the following Lemma:

Lemma 1 [19]. $\tau(G) = \frac{1}{k^2}\det(kI - D^c + A^c)$ *where* A^c *and* D^c *are the adjacency and degree matrices of* G^c, *the complement of* G, *respectively, and* I *is the* $k \times k$ *identity matrix.*

The characteristic of this formula is to express $\tau(G)$ straightway as a determinant rather than in terms of cofactors as in Kirchhoff theorem or eigenvalues as in Kelmans and Chelnokov formula.

2. Chebyshev Polynomial

In this part we insert some relations regarding Chebyshev polynomials of the first and second types which we use in our calculations.

We start from their definitions, see Yuanping, et al. [20].

Let $A_n(x)$ be $n \times n$ matrix such that

$$A_n(x) = \begin{pmatrix} 2x & -1 & 0 & \cdots & 0 \\ -1 & 2x & -1 & \ddots & \vdots \\ 0 & \ddots & \ddots & \ddots & 0 \\ \vdots & \ddots & \ddots & \ddots & -1 \\ 0 & \cdots & 0 & -1 & 2x \end{pmatrix}.$$

Furthermore, we render that the Chebyshev polynomials of the first type are defined by

$$T_n(x) = \cos(n\ \cos^{-1} x) \tag{1}$$

The Chebyshev polynomials of the second type are defined by

$$U_{n-1}(x) = \frac{1}{n}\frac{d}{dx}T_n(x) = \frac{\sin(n\ \cos^{-1} x)}{\sin(\cos^{-1} x)} \tag{2}$$

It is easily confirmed that

$$U_n(x) - 2xU_{n-1}(x) + U_{n-2}(x) = 0 \tag{3}$$

It can then be shown from this recursion that by expanding $\det A_n(x)$ one obtains

$$U_n(x) = \det(A_n(x)),\ n \geq 1 \tag{4}$$

Moreover, by solving the recursion (3), one gets the straightforward formula

$$U_n(x) = \frac{(x+\sqrt{x^2-1})^{n+1} - (x-\sqrt{x^2-1})^{n+1}}{2\sqrt{x^2-1}}, \quad n \geq 1, \tag{5}$$

where the conformity is valid for all complex x (except at $x = \pm 1$, where the function can be taken as the limit).

The definition of $U_n(x)$ easily yields its zeros and it can therefore be confirmed that

$$U_{n-1}(x) = 2^{n-1} \prod_{j=1}^{n-1} (x - \cos \frac{j\pi}{n}) \tag{6}$$

One further notes that

$$U_{n-1}(-x) = (-1)^{n-1} U_{n-1}(x) \tag{7}$$

From Equations (6) and (7), we have:

$$U_{n-1}{}^2(x) = 4^{n-1} \prod_{j=1}^{n-1} (x^2 - \cos^2 \frac{j\pi}{n}) \tag{8}$$

Finally, straightforward manipulation of the above formula produces the following formula (9), which is highly beneficial to us later:

$$U_{n-1}{}^2(\sqrt{\frac{x+2}{4}}) = \prod_{j=1}^{n-1} (x - 2\cos \frac{2j\pi}{n}) \tag{9}$$

Moreover, one can see that

$$U_{n-1}{}^2(x) = \frac{1 - T_{2n}(x)}{2(1-x^2)} = \frac{1 - T_n(2x^2-1)}{2(1-x^2)} \tag{10}$$

$$T_n(x) = \frac{1}{2}[(x+\sqrt{x^2-1})^n + ((x-\sqrt{x^2-1})^n] \tag{11}$$

Now we introduce the following important two Lemmas.

Lemma 2 [21]. *Let $B_n(x)$ be $n \times n$ Circulant matrix such that*

$$B_n(x) = \begin{pmatrix} x & 0 & 1 & \cdots & 1 & 0 \\ 0 & \ddots & \ddots & \ddots & \ddots & 1 \\ 1 & \ddots & \ddots & \ddots & \ddots & \vdots \\ \vdots & \ddots & \ddots & \ddots & \ddots & 1 \\ 1 & \ddots & \ddots & \ddots & \ddots & 0 \\ 0 & 1 & \cdots & 1 & 0 & x \end{pmatrix}.$$

Then for $n \geq 3,\ x \geq 4$, one has

$$\det(B_n(x)) = \frac{2(x+n-3)}{x-3}[T_n(\frac{x-1}{2}) - 1].$$

Lemma 3 [22]. *If $A \in F^{n\times n}$, $B \in F^{n\times m}$, $C \in F^{m\times n}$ and $D \in F^{m\times m}$. Suppose that A and D are nonsingular matrices, then*:

$$\det\begin{pmatrix} A & B \\ C & D \end{pmatrix} = \det(A - BD^{-1}C)\det D = \det A \det(D - CA^{-1}B).$$

This Lemma gives a type of symmetry for some matrices which simplify our calculations of the complexity of graphs studied in this paper.

3. Main Results

Definition 1. *The pyramid graph* $A_n^{(m)}$ is the graph created from the gear graph G_{m+1} with vertices $\{u_0;\ u_1, u_2, \ldots, u_m;\ w_1, w_2, \ldots, w_m\}$ and m sets of vertices, say, $\{v_1^1, v_2^1, \ldots, v_n^1\}$, $\{v_1^2, v_2^2, \ldots, v_n^2\}, \ldots, \{v_1^m, v_2^m, \ldots, v_n^m\}$, such that for all $i = 1, 2, \ldots, n$ the vertex v_i^j is adjacent to u_j and u_{j+1},where $j = 1, 2, \ldots, m-1$, and v_i^m is adjacent to u_1 and u_m. See Figure 1.

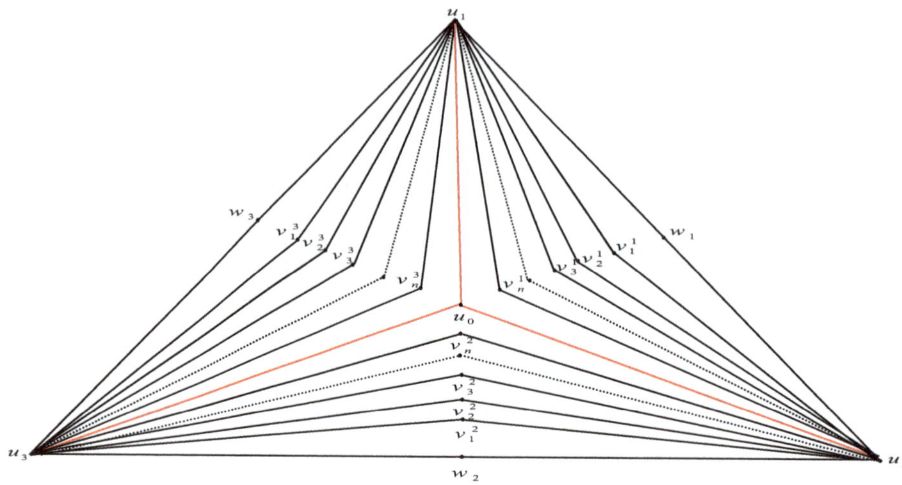

Figure 1. The pyramid graph $A_n^{(3)}$.

Theorem 1. *For* $n \geq 0$, $m \geq 3$, $\tau(A_n^{(m)}) = 2^{mn}[(n+2+\sqrt{2n+3}\,)^m + (n+2-\sqrt{2n+3}\,)^m - 2(n+1)^m]$.

Proof. Using Lemma 1, we have

$$\tau(A_n^{(m)}) = \frac{1}{(mn+2m+1)^2} \times \det((mn+2m+1)\, I - D^c + A^c) = \frac{1}{(mn+2m+1)^2} \times$$

$$\det\left(\begin{array}{cccccccccccccccccccccccccccccc}
(m+1) & 0 & 0 & \cdots & \cdots & \cdots & 0 & 1 & \cdots & \cdots & \cdots & \cdots & 1 & 1 & \cdots & \cdots & \cdots & \cdots & \cdots & \cdots & \cdots & \cdots & \cdots & \cdots & \cdots & \cdots & \cdots & \cdots & \cdots & 1 \\
0 & 2(n+2) & 1 & \cdots & \cdots & \cdots & 1 & 0 & 1 & \cdots & \cdots & 1 & 0 & 0 & \cdots & 0 & 1 & \cdots & 1 & \cdots & \cdots & 1 & \cdots & 1 & 1 & \cdots & 1 & 0 & \cdots & 0 \\
\vdots & 1 & \ddots & \ddots & \ddots & \ddots & \vdots & 0 & \ddots & \ddots & \ddots & \ddots & 1 & 0 & \cdots & 0 & 0 & \cdots & 0 & 1 & \cdots & 1 & \cdots & \cdots & 1 & \cdots & 1 & 1 & \cdots & 1 \\
\vdots & \vdots & \ddots & \ddots & \ddots & \ddots & \vdots & 1 & \ddots & \ddots & \ddots & \vdots & \vdots & 1 & \cdots & 1 & 0 & \cdots & 0 & 0 & \cdots & 0 & 1 & \cdots & 1 & \cdots & \cdots & 1 & \cdots & 1 \\
\vdots & \vdots & \ddots & \ddots & \ddots & \ddots & \vdots & \vdots & \ddots & \ddots & \ddots & \vdots & \vdots & 1 & \cdots & 1 & 1 & \cdots & 1 & 0 & \cdots & 0 & 0 & \cdots & 0 & 1 & \cdots & 1 & \cdots & \cdots \\
\vdots & \vdots & \ddots & \ddots & \ddots & \ddots & 1 & \vdots & \ddots & \ddots & \ddots & \ddots & 1 & \vdots & \cdots & \cdots & \cdots & \cdots & \cdots & \cdots & \cdots & \cdots & \cdots & \cdots & \cdots & \cdots & \cdots & \cdots & \cdots & \vdots \\
0 & 1 & \cdots & \cdots & \cdots & 1 & 2(n+2) & 1 & \cdots & \cdots & 1 & 0 & 0 & 1 & \cdots & 1 & 1 & \cdots & 1 & \cdots & \cdots & 1 & \cdots & 1 & 0 & \cdots & 0 & 0 & \cdots & 0 \\
1 & 0 & 0 & 1 & \cdots & \cdots & 1 & 3 & 1 & \cdots & \cdots & \cdots & 1 & 1 & \cdots & 1 & 1 & \cdots & 1 & \cdots & \cdots & \cdots & \cdots & \cdots & 1 & \cdots & 1 & 1 & \cdots & 1 \\
\vdots & 1 & \ddots & \ddots & \ddots & \ddots & \vdots & 1 & \ddots & \ddots & \ddots & \ddots & \vdots & 1 & \cdots & 1 & 1 & \cdots & 1 & \cdots & \cdots & \cdots & \cdots & \cdots & 1 & \cdots & 1 & 1 & \cdots & 1 \\
\vdots & \vdots & \ddots & \ddots & \ddots & \ddots & \vdots & \vdots & \ddots & \ddots & \ddots & \ddots & \vdots & \vdots & \cdots & \vdots & \vdots & \cdots & \vdots & \cdots & \cdots & \cdots & \cdots & \cdots & \vdots & \cdots & \vdots & \vdots & \cdots & 1 \\
\vdots & \vdots & \ddots & \ddots & \ddots & \ddots & 1 & \vdots & \ddots & \ddots & \ddots & \ddots & \vdots & \vdots & \cdots & \vdots & \vdots & \cdots & \vdots & \cdots & \cdots & \cdots & \cdots & \cdots & \vdots & \cdots & \vdots & \vdots & \cdots & \vdots \\
\vdots & 1 & \ddots & \ddots & \ddots & \ddots & 0 & \vdots & \ddots & \ddots & \ddots & \ddots & \vdots & 1 & \cdots & 1 & 1 & \cdots & 1 & \cdots & \cdots & \cdots & \cdots & \cdots & 1 & \cdots & 1 & 1 & \cdots & 1 \\
1 & 0 & 1 & \cdots & \cdots & 1 & 0 & 1 & \cdots & \cdots & \cdots & 1 & 3 & 1 & \cdots & 1 & 1 & \cdots & 1 & \cdots & \cdots & \cdots & \cdots & \cdots & 1 & \cdots & 1 & 1 & \cdots & 1 \\
1 & 0 & 0 & 1 & \cdots & 1 & 0 & 1 & 1 & \cdots & \cdots & 1 & 1 & 3 & 1 & \cdots & \cdots & \cdots & \cdots & \cdots & \cdots & \cdots & \cdots & \cdots & \cdots & \cdots & \cdots & \cdots & \cdots & 1 \\
\vdots & \vdots & \vdots & \vdots & \vdots & \vdots & \vdots & \vdots & \vdots & \vdots & \vdots & \vdots & \vdots & 1 & \ddots & \ddots & \ddots & \ddots & \ddots & \ddots & \ddots & \ddots & \ddots & \ddots & \ddots & \ddots & \ddots & \ddots & \ddots & \vdots \\
\vdots & 0 & 0 & 1 & \vdots & 1 & \vdots & 1 & 1 & \cdots & \cdots & 1 & 1 & \vdots & \ddots & \ddots & \ddots & \ddots & \ddots & \ddots & \ddots & \ddots & \ddots & \ddots & \ddots & \ddots & \ddots & \ddots & \ddots & \vdots \\
\vdots & 1 & 0 & 0 & \vdots & 1 & \vdots & 1 & 1 & \cdots & \cdots & 1 & 1 & \vdots & \ddots & \ddots & \ddots & \ddots & \ddots & \ddots & \ddots & \ddots & \ddots & \ddots & \ddots & \ddots & \ddots & \ddots & \ddots & \vdots \\
\vdots & \vdots & \vdots & \vdots & \vdots & \vdots & \vdots & \vdots & \vdots & \vdots & \vdots & \vdots & \vdots & \vdots & \ddots & \ddots & \ddots & \ddots & \ddots & \ddots & \ddots & \ddots & \ddots & \ddots & \ddots & \ddots & \ddots & \ddots & \ddots & \vdots \\
\vdots & 1 & 0 & 0 & \vdots & 1 & \vdots & 1 & 1 & \cdots & \cdots & 1 & 1 & \vdots & \ddots & \ddots & \ddots & \ddots & \ddots & \ddots & \ddots & \ddots & \ddots & \ddots & \ddots & \ddots & \ddots & \ddots & \ddots & \vdots \\
\vdots & 1 & 1 & 0 & \vdots & \vdots & \vdots & \vdots & \vdots & \vdots & \vdots & \vdots & \vdots & \vdots & \ddots & \ddots & \ddots & \ddots & \ddots & \ddots & \ddots & \ddots & \ddots & \ddots & \ddots & \ddots & \ddots & \ddots & \ddots & \vdots \\
\vdots & \vdots & \vdots & \vdots & \vdots & \vdots & \vdots & \vdots & \vdots & \vdots & \vdots & \vdots & \vdots & \vdots & \ddots & \ddots & \ddots & \ddots & \ddots & \ddots & \ddots & \ddots & \ddots & \ddots & \ddots & \ddots & \ddots & \ddots & \ddots & \vdots \\
\vdots & 1 & 1 & 0 & \vdots & 1 & \vdots & \vdots & \vdots & \vdots & \vdots & \vdots & \vdots & \vdots & \ddots & \ddots & \ddots & \ddots & \ddots & \ddots & \ddots & \ddots & \ddots & \ddots & \ddots & \ddots & \ddots & \ddots & \ddots & \vdots \\
\vdots & \vdots & \vdots & 1 & \vdots & \vdots & \vdots & \vdots & \vdots & \vdots & \vdots & \vdots & \vdots & \vdots & \ddots & \ddots & \ddots & \ddots & \ddots & \ddots & \ddots & \ddots & \ddots & \ddots & \ddots & \ddots & \ddots & \ddots & \ddots & \vdots \\
\vdots & \vdots & \vdots & \vdots & \vdots & 1 & \vdots & \vdots & \vdots & \vdots & \vdots & \vdots & \vdots & \vdots & \ddots & \ddots & \ddots & \ddots & \ddots & \ddots & \ddots & \ddots & \ddots & \ddots & \ddots & \ddots & \ddots & \ddots & \ddots & \vdots \\
\vdots & 1 & 1 & 1 & \vdots & 0 & \vdots & \vdots & \vdots & \vdots & \vdots & \vdots & \vdots & \vdots & \ddots & \ddots & \ddots & \ddots & \ddots & \ddots & \ddots & \ddots & \ddots & \ddots & \ddots & \ddots & \ddots & \ddots & \ddots & \vdots \\
\vdots & \vdots & \vdots & \vdots & \vdots & \vdots & \vdots & \vdots & \vdots & \vdots & \vdots & \vdots & \vdots & \vdots & \ddots & \ddots & \ddots & \ddots & \ddots & \ddots & \ddots & \ddots & \ddots & \ddots & \ddots & \ddots & \ddots & \ddots & \ddots & \vdots \\
\vdots & 1 & 1 & \vdots & \vdots & 0 & \vdots & 1 & 1 & \cdots & \cdots & 1 & 1 & \vdots & \ddots & \ddots & \ddots & \ddots & \ddots & \ddots & \ddots & \ddots & \ddots & \ddots & \ddots & \ddots & \ddots & \ddots & \ddots & \vdots \\
\vdots & 0 & 1 & 1 & \vdots & 0 & \vdots & 1 & 1 & \cdots & \cdots & 1 & 1 & \vdots & \ddots & \ddots & \ddots & \ddots & \ddots & \ddots & \ddots & \ddots & \ddots & \ddots & \ddots & \ddots & \ddots & \ddots & \ddots & \vdots \\
\vdots & \vdots & \vdots & \vdots & \vdots & \vdots & \vdots & \vdots & \vdots & \vdots & \vdots & \vdots & \vdots & \vdots & \ddots & \ddots & \ddots & \ddots & \ddots & \ddots & \ddots & \ddots & \ddots & \ddots & \ddots & \ddots & \ddots & \ddots & \ddots & 1 \\
1 & 0 & 1 & 1 & \cdots & 0 & \cdots & 1 & 1 & \cdots & \cdots & 1 & 1 & 1 & \cdots & \cdots & \cdots & \cdots & \cdots & \cdots & \cdots & \cdots & \cdots & \cdots & \cdots & \cdots & \cdots & \cdots & 1 & 3
\end{array}\right)$$

Let $j = (1 \cdots 1)$ be the $1 \times n$ matrix with all one, and J_n be the $n \times n$ matrix with all one. Set $a = 2n + 4$ and $b = mn + 2m + 1$. Then we obtain:

$$
\tau\left(A_n^{(m)}\right) = \frac{1}{b^2} \times \det \begin{pmatrix}
m+1 & 0 & \cdots & \cdots & \cdots & \cdots & 0 & 1 & \cdots & \cdots & \cdots & \cdots & 1 & j & \cdots & \cdots & \cdots & \cdots & j \\
0 & a & 1 & \cdots & \cdots & \cdots & 1 & 0 & 1 & \cdots & \cdots & 1 & 0 & 0 & j & \cdots & \cdots & j & 0 \\
\vdots & 1 & \ddots & \ddots & \ddots & \ddots & \vdots & 0 & 0 & \ddots & \ddots & \ddots & 1 & 0 & \ddots & \ddots & \ddots & \ddots & j \\
\vdots & \vdots & \ddots & \ddots & \ddots & \ddots & \vdots & 1 & 0 & \ddots & \ddots & \ddots & \vdots & j & \ddots & \ddots & \ddots & \ddots & \vdots \\
\vdots & \vdots & \ddots & \ddots & \ddots & \ddots & \vdots & \vdots & \ddots & \ddots & \ddots & \ddots & \vdots & \vdots & \ddots & \ddots & \ddots & \ddots & \vdots \\
\vdots & \vdots & \ddots & \ddots & \ddots & \ddots & 1 & \vdots & \ddots & \ddots & 0 & 0 & 1 & \vdots & \ddots & \ddots & \ddots & \ddots & j \\
0 & 1 & \cdots & \cdots & \cdots & 1 & a & 1 & \cdots & \cdots & 1 & 0 & 0 & j & \cdots & \cdots & j & 0 & 0 \\
1 & 0 & 0 & 1 & \cdots & \cdots & 1 & 3 & 1 & \cdots & \cdots & \cdots & 1 & j & \cdots & \cdots & \cdots & \cdots & j \\
\vdots & 1 & 0 & \ddots & \ddots & \ddots & 1 & 1 & \ddots & \ddots & \ddots & \ddots & \vdots & \vdots & \ddots & \ddots & \ddots & \ddots & \vdots \\
\vdots & \vdots & \ddots & \ddots & \ddots & \ddots & \vdots & \vdots & \ddots & \ddots & \ddots & \ddots & \vdots & \vdots & \ddots & \ddots & \ddots & \ddots & \vdots \\
\vdots & \vdots & \ddots & \ddots & \ddots & \ddots & 1 & \vdots & \ddots & \ddots & \ddots & \ddots & \vdots & \vdots & \ddots & \ddots & \ddots & \ddots & \vdots \\
\vdots & 1 & \ddots & \ddots & \ddots & \ddots & 0 & \vdots & \ddots & \ddots & \ddots & \ddots & 1 & \vdots & \ddots & \ddots & \ddots & \ddots & \vdots \\
1 & 0 & 1 & \cdots & \cdots & 1 & 0 & 1 & \cdots & \cdots & \cdots & 1 & 3 & j & \cdots & \cdots & \cdots & \cdots & j \\
j^t & 0 & 0 & j^t & \cdots & \cdots & j^t & j^t & \cdots & \cdots & \cdots & \cdots & j^t & & & & & & \\
\vdots & j^t & 0 & \ddots & \ddots & \ddots & \vdots & \vdots & \ddots & \ddots & \ddots & \ddots & \vdots & & & & & & \\
\vdots & \vdots & \ddots & \ddots & \ddots & \ddots & \vdots & \vdots & \ddots & \ddots & \ddots & \ddots & \vdots & & & 2I_{mn} + J_{mn} & & & \\
\vdots & \vdots & \ddots & \ddots & \ddots & \ddots & j^t & \vdots & \ddots & \ddots & \ddots & \ddots & \vdots & & & & & & \\
\vdots & j^t & \ddots & \ddots & \ddots & \ddots & 0 & \vdots & \ddots & \ddots & \ddots & \ddots & \vdots & & & & & & \\
j^t & 0 & j^t & \cdots & \cdots & j^t & 0 & j^t & \cdots & \cdots & \cdots & \cdots & j^t & & & & & &
\end{pmatrix}
$$

$$
= \frac{1}{b^2} \times \det \begin{pmatrix}
b & 0 & \cdots & \cdots & \cdots & \cdots & 0 & 1 & \cdots & \cdots & \cdots & \cdots & 1 & j & \cdots & \cdots & \cdots & \cdots & j \\
b & a & 1 & \cdots & \cdots & \cdots & 1 & 0 & 1 & \cdots & \cdots & 1 & 0 & 0 & j & \cdots & \cdots & j & 0 \\
\vdots & 1 & \ddots & \ddots & \ddots & \ddots & \vdots & 0 & 0 & \ddots & \ddots & \ddots & 1 & 0 & \ddots & \ddots & \ddots & \ddots & j \\
\vdots & \vdots & \ddots & \ddots & \ddots & \ddots & \vdots & 1 & 0 & \ddots & \ddots & \ddots & \vdots & j & \ddots & \ddots & \ddots & \ddots & \vdots \\
\vdots & \vdots & \ddots & \ddots & \ddots & \ddots & \vdots & \vdots & \ddots & \ddots & \ddots & \ddots & \vdots & \vdots & \ddots & \ddots & \ddots & \ddots & \vdots \\
\vdots & \vdots & \ddots & \ddots & \ddots & \ddots & 1 & \vdots & \ddots & \ddots & 0 & 0 & 1 & \vdots & \ddots & \ddots & \ddots & \ddots & j \\
b & 1 & \cdots & \cdots & \cdots & 1 & a & 1 & \cdots & \cdots & 1 & 0 & 0 & j & \cdots & \cdots & j & 0 & 0 \\
b & 0 & 0 & 1 & \cdots & \cdots & 1 & 3 & 1 & \cdots & \cdots & \cdots & 1 & j & \cdots & \cdots & \cdots & \cdots & j \\
\vdots & 1 & 0 & \ddots & \ddots & \ddots & 1 & 1 & \ddots & \ddots & \ddots & \ddots & \vdots & \vdots & \ddots & \ddots & \ddots & \ddots & \vdots \\
\vdots & \vdots & \ddots & \ddots & \ddots & \ddots & \vdots & \vdots & \ddots & \ddots & \ddots & \ddots & \vdots & \vdots & \ddots & \ddots & \ddots & \ddots & \vdots \\
\vdots & \vdots & \ddots & \ddots & \ddots & \ddots & 1 & \vdots & \ddots & \ddots & \ddots & \ddots & \vdots & \vdots & \ddots & \ddots & \ddots & \ddots & \vdots \\
\vdots & 1 & \ddots & \ddots & \ddots & \ddots & 0 & \vdots & \ddots & \ddots & \ddots & \ddots & 1 & \vdots & \ddots & \ddots & \ddots & \ddots & \vdots \\
b & 0 & 1 & \cdots & \cdots & 1 & 0 & 1 & \cdots & \cdots & \cdots & 1 & 3 & j & \cdots & \cdots & \cdots & \cdots & j \\
bj^t & 0 & 0 & j^t & \cdots & \cdots & j^t & j^t & \cdots & \cdots & \cdots & \cdots & j^t & & & & & & \\
\vdots & j^t & 0 & \ddots & \ddots & \ddots & \vdots & \vdots & \ddots & \ddots & \ddots & \ddots & \vdots & & & & & & \\
\vdots & \vdots & \ddots & \ddots & \ddots & \ddots & \vdots & \vdots & \ddots & \ddots & \ddots & \ddots & \vdots & & & 2I_{mn} + J_{mn} & & & \\
\vdots & \vdots & \ddots & \ddots & \ddots & \ddots & j^t & \vdots & \ddots & \ddots & \ddots & \ddots & \vdots & & & & & & \\
\vdots & j^t & \ddots & \ddots & \ddots & \ddots & 0 & \vdots & \ddots & \ddots & \ddots & \ddots & \vdots & & & & & & \\
bj^t & 0 & j^t & \cdots & \cdots & j^t & 0 & j^t & \cdots & \cdots & \cdots & \cdots & j^t & & & & & &
\end{pmatrix}
$$

$$
= \frac{1}{b} \times \det \begin{pmatrix}
1 & 0 & \cdots & \cdots & \cdots & \cdots & 0 & 1 & \cdots & \cdots & \cdots & \cdots & 1 & j & \cdots & \cdots & \cdots & \cdots & j \\
1 & a & 1 & \cdots & \cdots & \cdots & 1 & 0 & 1 & \cdots & \cdots & 1 & 0 & 0 & j & \cdots & \cdots & j & 0 \\
\vdots & 1 & \ddots & \ddots & \ddots & \ddots & \vdots & 0 & 0 & \ddots & \ddots & \ddots & 1 & 0 & \ddots & \ddots & \ddots & \ddots & j \\
\vdots & \vdots & \ddots & \ddots & \ddots & \ddots & \vdots & 1 & 0 & \ddots & \ddots & \ddots & \vdots & j & \ddots & \ddots & \ddots & \ddots & \vdots \\
\vdots & \vdots & \ddots & \ddots & \ddots & \ddots & \vdots & \vdots & \ddots & \ddots & \ddots & \ddots & \vdots & \vdots & \ddots & \ddots & \ddots & \ddots & \vdots \\
\vdots & \vdots & \ddots & \ddots & \ddots & \ddots & 1 & \vdots & \ddots & \ddots & 0 & 0 & 1 & \vdots & \ddots & \ddots & \ddots & \ddots & j \\
1 & 1 & \cdots & \cdots & \cdots & 1 & a & 1 & \cdots & \cdots & 1 & 0 & 0 & j & \cdots & \cdots & j & 0 & 0 \\
1 & 0 & 0 & 1 & \cdots & \cdots & 1 & 3 & 1 & \cdots & \cdots & \cdots & 1 & j & \cdots & \cdots & \cdots & \cdots & j \\
\vdots & 1 & 0 & \ddots & \ddots & \ddots & 1 & 1 & \ddots & \ddots & \ddots & \ddots & \vdots & \vdots & \ddots & \ddots & \ddots & \ddots & \vdots \\
\vdots & \vdots & \ddots & \ddots & \ddots & \ddots & \vdots & \vdots & \ddots & \ddots & \ddots & \ddots & \vdots & \vdots & \ddots & \ddots & \ddots & \ddots & \vdots \\
\vdots & \vdots & \ddots & \ddots & \ddots & \ddots & 1 & \vdots & \ddots & \ddots & \ddots & \ddots & \vdots & \vdots & \ddots & \ddots & \ddots & \ddots & \vdots \\
\vdots & 1 & \ddots & \ddots & \ddots & \ddots & 0 & \vdots & \ddots & \ddots & \ddots & \ddots & 1 & \vdots & \ddots & \ddots & \ddots & \ddots & \vdots \\
1 & 0 & 1 & \cdots & \cdots & 1 & 0 & 1 & \cdots & \cdots & \cdots & 1 & 3 & j & \cdots & \cdots & \cdots & \cdots & j \\
1j^t & 0 & 0 & j^t & \cdots & \cdots & j^t & j^t & \cdots & \cdots & \cdots & \cdots & j^t & & & & & & \\
\vdots & j^t & 0 & \ddots & \ddots & \ddots & \vdots & \vdots & \ddots & \ddots & \ddots & \ddots & \vdots & & & & & & \\
\vdots & \vdots & \ddots & \ddots & \ddots & \ddots & \vdots & \vdots & \ddots & \ddots & \ddots & \ddots & \vdots & & & 2I_{mn} + J_{mn} & & & \\
\vdots & \vdots & \ddots & \ddots & \ddots & \ddots & j^t & \vdots & \ddots & \ddots & \ddots & \ddots & \vdots & & & & & & \\
\vdots & j^t & \ddots & \ddots & \ddots & \ddots & 0 & \vdots & \ddots & \ddots & \ddots & \ddots & \vdots & & & & & & \\
1j^t & 0 & j^t & \cdots & \cdots & j^t & 0 & j^t & \cdots & \cdots & \cdots & \cdots & j^t & & & & & &
\end{pmatrix}
$$

$$
= \tfrac{1}{b} \times \det\begin{pmatrix}
1 & 0 & \cdots & \cdots & \cdots & \cdots & 0 & 1 & \cdots & \cdots & \cdots & \cdots & 1 & j & \cdots & \cdots & \cdots & \cdots & j \\
0 & a & 1 & \cdots & \cdots & \cdots & 1 & -1 & 0 & \cdots & \cdots & 0 & -1 & -j & 0 & \cdots & \cdots & 0 & -j \\
\vdots & 1 & \ddots & \ddots & \ddots & \ddots & \vdots & -1 & -1 & \ddots & \ddots & \ddots & 0 & -j & \ddots & \ddots & \ddots & \ddots & 0 \\
\vdots & \vdots & \ddots & \ddots & \ddots & \ddots & \vdots & 0 & \ddots & \ddots & \ddots & \ddots & \vdots & 0 & \ddots & \ddots & \ddots & \ddots & \vdots \\
\vdots & \vdots & \ddots & \ddots & \ddots & \ddots & \vdots & \vdots & \ddots & \ddots & \ddots & \ddots & \vdots & \vdots & \ddots & \ddots & \ddots & \ddots & \vdots \\
\vdots & \vdots & \ddots & \ddots & \ddots & \ddots & 1 & \vdots & \ddots & \ddots & \ddots & -1 & 0 & \vdots & \ddots & \ddots & \ddots & \ddots & 0 \\
0 & 1 & \cdots & \cdots & \cdots & 1 & a & 0 & \cdots & \cdots & 0 & -1 & -1 & 0 & \cdots & \cdots & 0 & -j & -j \\
0 & 0 & 0 & 1 & \cdots & \cdots & 1 & 2 & 0 & \cdots & \cdots & \cdots & 0 & 0 & \cdots & \cdots & \cdots & \cdots & 0 \\
\vdots & 1 & 0 & \ddots & \ddots & \ddots & 1 & 0 & \ddots & \ddots & \ddots & \ddots & \vdots & \vdots & \ddots & \ddots & \ddots & \ddots & \vdots \\
\vdots & \vdots & \ddots & \ddots & \ddots & \ddots & \vdots & \vdots & \ddots & \ddots & \ddots & \ddots & \vdots & \vdots & \ddots & \ddots & \ddots & \ddots & \vdots \\
\vdots & \vdots & \ddots & \ddots & \ddots & \ddots & 1 & \vdots & \ddots & \ddots & \ddots & \ddots & \vdots & \vdots & \ddots & \ddots & \ddots & \ddots & \vdots \\
\vdots & 1 & \ddots & \ddots & \ddots & \ddots & 0 & \vdots & \ddots & \ddots & \ddots & \ddots & 0 & \vdots & \ddots & \ddots & \ddots & \ddots & \vdots \\
0 & 0 & 1 & \cdots & \cdots & 1 & 0 & 0 & \cdots & \cdots & \cdots & 0 & 2 & 0 & \cdots & \cdots & \cdots & \cdots & 0 \\
0 & 0 & 0 & j^t & \cdots & \cdots & j^t & 0 & \cdots & \cdots & \cdots & \cdots & 0 & & & & & & \\
\vdots & j^t & 0 & \ddots & \ddots & \ddots & \vdots & \vdots & \ddots & \ddots & \ddots & \ddots & \vdots & & & & & & \\
\vdots & \vdots & \ddots & \ddots & \ddots & \ddots & \vdots & \vdots & \ddots & \ddots & \ddots & \ddots & \vdots & & & & 2I_{mn} & & \\
\vdots & \vdots & \ddots & \ddots & \ddots & \ddots & j^t & \vdots & \ddots & \ddots & \ddots & \ddots & \vdots & & & & & & \\
\vdots & j^t & \ddots & \ddots & \ddots & \ddots & 0 & \vdots & \ddots & \ddots & \ddots & \ddots & \vdots & & & & & & \\
0 & 0 & j^t & \cdots & \cdots & j^t & 0 & 0 & \cdots & \cdots & \cdots & \cdots & 0 & & & & & &
\end{pmatrix}
$$

$$
= \tfrac{1}{b} \times \det\begin{pmatrix}
a & 1 & \cdots & \cdots & \cdots & 1 & -1 & 0 & \cdots & \cdots & 0 & -1 & -j & 0 & \cdots & \cdots & 0 & -j \\
1 & \ddots & \ddots & \ddots & \ddots & \vdots & -1 & -1 & \ddots & \ddots & \ddots & 0 & -j & \ddots & \ddots & \ddots & \ddots & 0 \\
\vdots & \ddots & \ddots & \ddots & \ddots & \vdots & 0 & \ddots & \ddots & \ddots & \ddots & \vdots & 0 & \ddots & \ddots & \ddots & \ddots & \vdots \\
\vdots & \ddots & \ddots & \ddots & \ddots & \vdots & \vdots & \ddots & \ddots & \ddots & \ddots & \vdots & \vdots & \ddots & \ddots & \ddots & \ddots & \vdots \\
\vdots & \ddots & \ddots & \ddots & \ddots & 1 & \vdots & \ddots & \ddots & \ddots & -1 & 0 & \vdots & \ddots & \ddots & \ddots & \ddots & 0 \\
1 & \cdots & \cdots & \cdots & 1 & a & 0 & \cdots & \cdots & 0 & -1 & -1 & 0 & \cdots & \cdots & 0 & -j & -j \\
0 & 0 & 1 & \cdots & \cdots & 1 & 2 & 0 & \cdots & \cdots & \cdots & 0 & 0 & \cdots & \cdots & \cdots & \cdots & 0 \\
1 & 0 & \ddots & \ddots & \ddots & 1 & 0 & \ddots & \ddots & \ddots & \ddots & \vdots & \vdots & \ddots & \ddots & \ddots & \ddots & \vdots \\
\vdots & \ddots & \ddots & \ddots & \ddots & \vdots & \vdots & \ddots & \ddots & \ddots & \ddots & \vdots & \vdots & \ddots & \ddots & \ddots & \ddots & \vdots \\
\vdots & \ddots & \ddots & \ddots & \ddots & 1 & \vdots & \ddots & \ddots & \ddots & \ddots & \vdots & \vdots & \ddots & \ddots & \ddots & \ddots & \vdots \\
1 & \ddots & \ddots & \ddots & \ddots & 0 & \vdots & \ddots & \ddots & \ddots & \ddots & 0 & \vdots & \ddots & \ddots & \ddots & \ddots & \vdots \\
0 & 1 & \cdots & \cdots & 1 & 0 & 0 & \cdots & \cdots & \cdots & 0 & 2 & 0 & \cdots & \cdots & \cdots & \cdots & 0 \\
0 & 0 & j^t & \cdots & \cdots & j^t & 0 & \cdots & \cdots & \cdots & \cdots & 0 & & & & & & \\
j^t & 0 & \ddots & \ddots & \ddots & \vdots & \vdots & \ddots & \ddots & \ddots & \ddots & \vdots & & & & & & \\
\vdots & \ddots & \ddots & \ddots & \ddots & \vdots & \vdots & \ddots & \ddots & \ddots & \ddots & \vdots & & & & 2I_{mn} & & \\
\vdots & \ddots & \ddots & \ddots & \ddots & j^t & \vdots & \ddots & \ddots & \ddots & \ddots & \vdots & & & & & & \\
j^t & \ddots & \ddots & \ddots & \ddots & 0 & \vdots & \ddots & \ddots & \ddots & \ddots & \vdots & & & & & & \\
0 & j^t & \cdots & \cdots & j^t & 0 & 0 & \cdots & \cdots & \cdots & \cdots & 0 & & & & & &
\end{pmatrix}
$$

Using Lemma 3, yields

$$
\tau\,(A_n^{(m)}) = \tfrac{1}{b} \times \det\begin{pmatrix} A & B \\ C & 2I_{mn} \end{pmatrix} = \tfrac{1}{b} \times \det(A - B\tfrac{1}{2I_{mn}}\ C) \times 2^{mn}
$$

$$
= \tfrac{1}{b}2^{mn} \times 2^{-2m} \times \det\begin{pmatrix}
2a & n+2 & 2(n+1) & \cdots & 2(n+1) & n+2 & -2 & 0 & \cdots & \cdots & 0 & -2 \\
n+2 & 2a & n+2 & 2(n+1) & \cdots & 2(n+1) & -2 & \ddots & \ddots & \ddots & \ddots & 0 \\
2(n+1) & n+2 & \ddots & \ddots & \ddots & \vdots & 0 & \ddots & \ddots & \ddots & \ddots & \vdots \\
\vdots & \ddots & \ddots & \ddots & \ddots & 2(n+1) & \vdots & \ddots & \ddots & \ddots & \ddots & \vdots \\
2(n+1) & \ddots & \ddots & \ddots & \ddots & n+2 & \vdots & \ddots & \ddots & \ddots & \ddots & 0 \\
n+2 & 2(n+1) & \cdots & 2(n+1) & n+2 & 2a & 0 & \cdots & \cdots & 0 & -2 & -2 \\
0 & 0 & 2 & \cdots & \cdots & 2 & 4 & 0 & \cdots & \cdots & \cdots & 0 \\
2 & \ddots & \ddots & \ddots & \ddots & \vdots & 0 & \ddots & \ddots & \ddots & \ddots & \vdots \\
\vdots & \ddots & \ddots & \ddots & \ddots & \vdots & \vdots & \ddots & \ddots & \ddots & \ddots & \vdots \\
\vdots & \ddots & \ddots & \ddots & \ddots & 2 & \vdots & \ddots & \ddots & \ddots & \ddots & \vdots \\
2 & \ddots & \ddots & \ddots & \ddots & 0 & \vdots & \ddots & \ddots & \ddots & \ddots & 0 \\
0 & 2 & \cdots & \cdots & 2 & 0 & 0 & \cdots & \cdots & \cdots & 0 & 4
\end{pmatrix}
$$

Using Lemma 3 again, yields

$$
\tau\,(A_n^{(m)}) = \frac{2^{mn-2m}}{b} \times \det\begin{pmatrix} D & E \\ F & 4I_m \end{pmatrix} = \frac{2^{mn}}{b} \times \det(D - E\frac{1}{4I_m}\ F)
$$

$$\tau(A_n^{(m)}) = \frac{2^{mn}}{b} \times \det\begin{pmatrix} 2a & (n+3) & 2(n+2) & \cdots & 2(n+2) & (n+3) \\ (n+3) & 2a & (n+3) & \ddots & \cdots & 2(n+2) \\ 2(n+2) & \ddots & \ddots & \ddots & \ddots & \vdots \\ \vdots & \ddots & \ddots & \ddots & \ddots & 2(n+2) \\ 2(n+2) & \ddots & \ddots & \ddots & \ddots & (n+3) \\ (n+3) & 2(n+2) & \cdots & 2(n+2) & (n+3) & 2a \end{pmatrix}$$

Straightforward inducement using the properties of determinants, one can obtain

$$\tau(A_n^{(m)}) = \frac{2^{mn}}{b} \times \frac{2b}{mn+m+2} \times \det\begin{pmatrix} (2a-n-3) & 0 & (n+1) & \cdots & (n+1) & 0 \\ 0 & (2a-n-3) & 0 & \ddots & \cdots & (n+1) \\ (n+1) & \ddots & \ddots & \ddots & \ddots & \vdots \\ \vdots & \ddots & \ddots & \ddots & \ddots & (n+1) \\ (n+1) & \ddots & \ddots & \ddots & \ddots & 0 \\ 0 & (n+1) & \cdots & (n+1) & 0 & (2a-n-3) \end{pmatrix}$$

$$= \frac{2^{mn+1}(n+1)^m}{mn+m+2} \times \det\begin{pmatrix} \frac{(2a-n-3)}{(n+1)} & 0 & 1 & \cdots & 1 & 0 \\ 0 & \frac{(2a-n-3)}{(n+1)} & 0 & \ddots & \ddots & 1 \\ 1 & \ddots & \ddots & \ddots & \ddots & \vdots \\ \vdots & \ddots & \ddots & \ddots & \ddots & 1 \\ 1 & \ddots & \ddots & \ddots & \ddots & 0 \\ 0 & 1 & \cdots & 1 & 0 & \frac{(2a-n-3)}{(n+1)} \end{pmatrix}$$

Using Lemma 2, yields

$$\tau(A_n^{(m)}) = 2^{mn+1} \times \frac{(n+1)^m}{mn+m+2} \times \frac{2(\frac{2a-n-3}{n+1}+m-3)}{\frac{2a-n-3}{n+1}-3} \times [T_m(\frac{\frac{2a-n-3}{n+1}-1}{2}) - 1]$$
$$= 2^{mn+1} \times (n+1)^m \times [T_m(\tfrac{n+2}{n+1}) - 1].$$

Using Equation (11), yields the result. □

Definition 2. *The pyramid graph $B_n^{(m)}$ is the graph created from the gear graph G_{m+1} with vertices $\{u_0;\ u_1, u_2, \ldots, u_m;\ w_1, w_2, \ldots, w_m\}$ with double internal edges and m sets of vertices, say, $\{v_1^1, v_2^1, \ldots, v_n^1\}$, $\{v_1^2, v_2^2, \ldots, v_n^2\}, \ldots, \{v_1^m, v_2^m, \ldots, v_n^m\}$, such that for all $i = 1, 2, \ldots, n$ the vertex v_i^j is adjacent to u_j and u_{j+1}, where $j = 1, 2, \ldots, m-1$, and v_i^m is adjacent to u_1 and u_m. See Figure 2.*

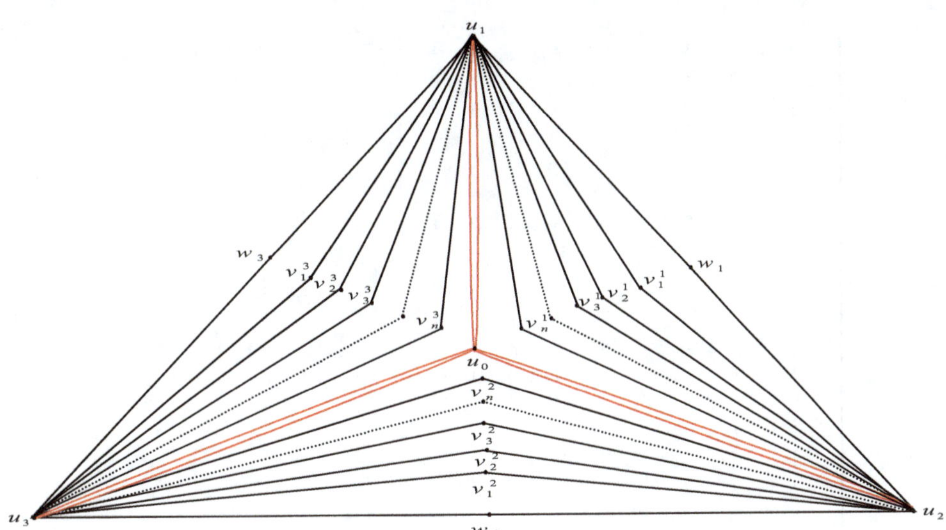

Figure 2. The pyramid graph $B_n^{(3)}$.

Theorem 2. *For* $n \geq 0$, $m \geq 3$, $\tau(B_n^{(m)}) = 2^{mn}[(n+3+2\sqrt{n+2}\,)^m + (n+3-2\sqrt{n+2}\,)^m - 2(n+1)^m]$.

Proof. Using Lemma 1, we get:

$$\tau(B_n^{(m)}) = \frac{1}{(mn+2m+1)^2} \times \det((mn+2m+1)\,I - D^c + A^c) = \frac{1}{(mn+2m+1)^2} \times$$

$$\det\left(\begin{array}{cccccccccccccccccccccccccccccc}
(2m+1) & -1 & -1 & \cdots & \cdots & \cdots & -1 & 1 & \cdots & \cdots & \cdots & \cdots & 1 & 1 & \cdots & \cdots & \cdots & \cdots & \cdots & \cdots & \cdots & \cdots & \cdots & \cdots & \cdots & \cdots & \cdots & \cdots & \cdots & 1 \\
-1 & (2n+5) & 1 & \cdots & \cdots & \cdots & 1 & 0 & 1 & \cdots & \cdots & 1 & 0 & 0 & \cdots & 0 & 1 & \cdots & 1 & \cdots & \cdots & 1 & \cdots & 1 & 1 & \cdots & 1 & 0 & \cdots & 0 \\
\vdots & 1 & \ddots & \ddots & \ddots & \ddots & \vdots & 0 & \ddots & \ddots & \ddots & \ddots & 1 & 0 & \cdots & 0 & 0 & \cdots & 0 & 1 & \cdots & 1 & \cdots & \cdots & 1 & \cdots & 1 & 1 & \cdots & 1 \\
\vdots & \vdots & \ddots & \ddots & \ddots & \ddots & \vdots & 1 & \ddots & \ddots & \ddots & \vdots & \vdots & 1 & \cdots & 1 & 0 & \cdots & 0 & 0 & \cdots & 0 & 1 & \cdots & 1 & \cdots & \cdots & 1 & \cdots & 1 \\
\vdots & \vdots & \ddots & \ddots & \ddots & \ddots & \vdots & \vdots & \ddots & \ddots & \ddots & \vdots & \vdots & 1 & \cdots & 1 & 1 & \cdots & 1 & 0 & \cdots & 0 & 0 & \cdots & 0 & 1 & \cdots & 1 & \cdots & \cdots \\
\vdots & \vdots & \ddots & \ddots & \ddots & \ddots & 1 & \vdots & \ddots & \ddots & \ddots & \ddots & 1 & \vdots & \cdots & \cdots & \cdots & \cdots & \cdots & \cdots & \cdots & \cdots & \cdots & \cdots & \cdots & \cdots & \cdots & \cdots & \cdots & \vdots \\
-1 & 1 & \cdots & \cdots & \cdots & 1 & (2n+5) & 1 & \cdots & \cdots & 1 & 0 & 0 & 1 & \cdots & 1 & 1 & \cdots & 1 & \cdots & \cdots & 1 & \cdots & 1 & 0 & \cdots & 0 & 0 & \cdots & 0 \\
1 & 0 & 0 & 1 & \cdots & \cdots & 1 & 3 & 1 & \cdots & \cdots & \cdots & 1 & 1 & \cdots & 1 & 1 & \cdots & 1 & \cdots & \cdots & \cdots & \cdots & \cdots & 1 & \cdots & 1 & 1 & \cdots & 1 \\
\vdots & 1 & \ddots & \ddots & \ddots & \ddots & \vdots & 1 & \ddots & \ddots & \ddots & \ddots & \vdots & 1 & \cdots & 1 & 1 & \cdots & 1 & \cdots & \cdots & \cdots & \cdots & \cdots & 1 & \cdots & 1 & 1 & \cdots & 1 \\
\vdots & \vdots & \ddots & \ddots & \ddots & \ddots & \vdots & \vdots & \ddots & \ddots & \ddots & \ddots & \vdots & \vdots & \cdots & \vdots & \vdots & \cdots & \vdots & \cdots & \cdots & \cdots & \cdots & \cdots & \vdots & \cdots & \vdots & \vdots & \cdots & 1 \\
\vdots & \vdots & \ddots & \ddots & \ddots & \ddots & 1 & \vdots & \ddots & \ddots & \ddots & \ddots & \vdots & \vdots & \cdots & \vdots & \vdots & \cdots & \vdots & \cdots & \cdots & \cdots & \cdots & \cdots & \vdots & \cdots & \vdots & \vdots & \cdots & \vdots \\
\vdots & 1 & \ddots & \ddots & \ddots & \ddots & 0 & \vdots & \ddots & \ddots & \ddots & \ddots & \vdots & 1 & \cdots & 1 & 1 & \cdots & 1 & \cdots & \cdots & \cdots & \cdots & \cdots & 1 & \cdots & 1 & 1 & \cdots & 1 \\
1 & 0 & 1 & \cdots & \cdots & 1 & 0 & 1 & \cdots & \cdots & \cdots & 1 & 3 & 1 & \cdots & 1 & 1 & \cdots & 1 & \cdots & \cdots & \cdots & \cdots & \cdots & 1 & \cdots & 1 & 1 & \cdots & 1 \\
1 & 0 & 0 & 1 & \cdots & 1 & 0 & 1 & 1 & \cdots & \cdots & 1 & 1 & 3 & 1 & \cdots & \cdots & \cdots & \cdots & \cdots & \cdots & \cdots & \cdots & \cdots & \cdots & \cdots & \cdots & \cdots & \cdots & 1 \\
\vdots & \vdots & \vdots & \vdots & \vdots & \vdots & \vdots & \vdots & \vdots & \vdots & \vdots & \vdots & \vdots & 1 & \ddots & \ddots & \ddots & \ddots & \ddots & \ddots & \ddots & \ddots & \ddots & \ddots & \ddots & \ddots & \ddots & \ddots & \ddots & \vdots \\
\vdots & 0 & 0 & 1 & \vdots & 1 & \vdots & 1 & 1 & \cdots & \cdots & 1 & 1 & \vdots & \ddots & \ddots & \ddots & \ddots & \ddots & \ddots & \ddots & \ddots & \ddots & \ddots & \ddots & \ddots & \ddots & \ddots & \ddots & \vdots \\
\vdots & 1 & 0 & 0 & \vdots & 1 & \vdots & 1 & 1 & \cdots & \cdots & 1 & 1 & \vdots & \ddots & \ddots & \ddots & \ddots & \ddots & \ddots & \ddots & \ddots & \ddots & \ddots & \ddots & \ddots & \ddots & \ddots & \ddots & \vdots \\
\vdots & \vdots & \vdots & \vdots & \vdots & \vdots & \vdots & \vdots & \vdots & \vdots & \vdots & \vdots & \vdots & \vdots & \ddots & \ddots & \ddots & \ddots & \ddots & \ddots & \ddots & \ddots & \ddots & \ddots & \ddots & \ddots & \ddots & \ddots & \ddots & \vdots \\
\vdots & 1 & 0 & 0 & \vdots & 1 & \vdots & 1 & 1 & \cdots & \cdots & 1 & 1 & \vdots & \ddots & \ddots & \ddots & \ddots & \ddots & \ddots & \ddots & \ddots & \ddots & \ddots & \ddots & \ddots & \ddots & \ddots & \ddots & \vdots \\
\vdots & 1 & 1 & 0 & \vdots & \vdots & \vdots & \vdots & \vdots & \vdots & \vdots & \vdots & \vdots & \vdots & \ddots & \ddots & \ddots & \ddots & \ddots & \ddots & \ddots & \ddots & \ddots & \ddots & \ddots & \ddots & \ddots & \ddots & \ddots & \vdots \\
\vdots & \vdots & \vdots & \vdots & \vdots & \vdots & \vdots & \vdots & \vdots & \vdots & \vdots & \vdots & \vdots & \vdots & \ddots & \ddots & \ddots & \ddots & \ddots & \ddots & \ddots & \ddots & \ddots & \ddots & \ddots & \ddots & \ddots & \ddots & \ddots & \vdots \\
\vdots & 1 & 1 & 0 & \vdots & 1 & \vdots & \vdots & \vdots & \vdots & \vdots & \vdots & \vdots & \vdots & \ddots & \ddots & \ddots & \ddots & \ddots & \ddots & \ddots & \ddots & \ddots & \ddots & \ddots & \ddots & \ddots & \ddots & \ddots & \vdots \\
\vdots & \vdots & \vdots & 1 & \vdots & \vdots & \vdots & \vdots & \vdots & \vdots & \vdots & \vdots & \vdots & \vdots & \ddots & \ddots & \ddots & \ddots & \ddots & \ddots & \ddots & \ddots & \ddots & \ddots & \ddots & \ddots & \ddots & \ddots & \ddots & \vdots \\
\vdots & \vdots & \vdots & \vdots & \vdots & 1 & \vdots & \vdots & \vdots & \vdots & \vdots & \vdots & \vdots & \vdots & \ddots & \ddots & \ddots & \ddots & \ddots & \ddots & \ddots & \ddots & \ddots & \ddots & \ddots & \ddots & \ddots & \ddots & \ddots & \vdots \\
\vdots & 1 & 1 & 1 & \vdots & 0 & \vdots & \vdots & \vdots & \vdots & \vdots & \vdots & \vdots & \vdots & \ddots & \ddots & \ddots & \ddots & \ddots & \ddots & \ddots & \ddots & \ddots & \ddots & \ddots & \ddots & \ddots & \ddots & \ddots & \vdots \\
\vdots & \vdots & \vdots & \vdots & \vdots & \vdots & \vdots & \vdots & \vdots & \vdots & \vdots & \vdots & \vdots & \vdots & \ddots & \ddots & \ddots & \ddots & \ddots & \ddots & \ddots & \ddots & \ddots & \ddots & \ddots & \ddots & \ddots & \ddots & \ddots & \vdots \\
\vdots & 1 & 1 & \vdots & \vdots & 0 & \vdots & 1 & 1 & \cdots & \cdots & 1 & 1 & \vdots & \ddots & \ddots & \ddots & \ddots & \ddots & \ddots & \ddots & \ddots & \ddots & \ddots & \ddots & \ddots & \ddots & \ddots & \ddots & \vdots \\
\vdots & 0 & 1 & 1 & \vdots & 0 & \vdots & 1 & 1 & \cdots & \cdots & 1 & 1 & \vdots & \ddots & \ddots & \ddots & \ddots & \ddots & \ddots & \ddots & \ddots & \ddots & \ddots & \ddots & \ddots & \ddots & \ddots & \ddots & \vdots \\
\vdots & \vdots & \vdots & \vdots & \vdots & \vdots & \vdots & \vdots & \vdots & \vdots & \vdots & \vdots & \vdots & \vdots & \ddots & \ddots & \ddots & \ddots & \ddots & \ddots & \ddots & \ddots & \ddots & \ddots & \ddots & \ddots & \ddots & \ddots & \ddots & 1 \\
1 & 0 & 1 & 1 & \cdots & 0 & \cdots & 1 & 1 & \cdots & \cdots & 1 & 1 & 1 & \cdots & \cdots & \cdots & \cdots & \cdots & \cdots & \cdots & \cdots & \cdots & \cdots & \cdots & \cdots & \cdots & \cdots & 1 & 3
\end{array}\right)$$

Let $j = (1 \cdots 1)$ be the $1 \times n$ matrix with all one, and J_n be the $n \times n$ matrix with all one. Set $a = 2n+5$ and $b = mn+2m+1$. Then we get:

$$\tau\left(B_n^{(m)}\right) = \frac{1}{b^2} \times \det\left(\begin{array}{ccccccccccccccccccc}
2m+1 & -1 & \cdots & \cdots & \cdots & \cdots & -1 & 1 & \cdots & \cdots & \cdots & \cdots & 1 & j & \cdots & \cdots & \cdots & \cdots & j \\
-1 & a & 1 & \cdots & \cdots & \cdots & 1 & 0 & 1 & \cdots & \cdots & 1 & 0 & 0 & j & \cdots & \cdots & j & 0 \\
\vdots & 1 & \ddots & \ddots & \ddots & \ddots & \vdots & 0 & 0 & \ddots & \ddots & \ddots & 1 & 0 & \ddots & \ddots & \ddots & \ddots & j \\
\vdots & \vdots & \ddots & \ddots & \ddots & \ddots & \vdots & 1 & 0 & \ddots & \ddots & \ddots & \vdots & j & \ddots & \ddots & \ddots & \ddots & \vdots \\
\vdots & \vdots & \ddots & \ddots & \ddots & \ddots & \vdots & \vdots & \ddots & \ddots & \ddots & \ddots & \vdots & \vdots & \ddots & \ddots & \ddots & \ddots & \vdots \\
\vdots & \vdots & \ddots & \ddots & \ddots & \ddots & 1 & \vdots & \ddots & \ddots & 0 & 0 & 1 & \vdots & \ddots & \ddots & \ddots & \ddots & j \\
-1 & 1 & \cdots & \cdots & \cdots & 1 & a & 1 & \cdots & \cdots & 1 & 0 & 0 & j & \cdots & \cdots & j & 0 & 0 \\
1 & 0 & 0 & 1 & \cdots & \cdots & 1 & 3 & 1 & \cdots & \cdots & \cdots & 1 & j & \cdots & \cdots & \cdots & \cdots & j \\
\vdots & 1 & 0 & \ddots & \ddots & \ddots & 1 & 1 & \ddots & \ddots & \ddots & \ddots & \vdots & \vdots & \ddots & \ddots & \ddots & \ddots & \vdots \\
\vdots & \vdots & \ddots & \ddots & \ddots & \ddots & \vdots & \vdots & \ddots & \ddots & \ddots & \ddots & \vdots & \vdots & \ddots & \ddots & \ddots & \ddots & \vdots \\
\vdots & \vdots & \ddots & \ddots & \ddots & \ddots & 1 & \vdots & \ddots & \ddots & \ddots & \ddots & \vdots & \vdots & \ddots & \ddots & \ddots & \ddots & \vdots \\
\vdots & 1 & \ddots & \ddots & \ddots & \ddots & 0 & \vdots & \ddots & \ddots & \ddots & \ddots & 1 & \vdots & \ddots & \ddots & \ddots & \ddots & \vdots \\
1 & 0 & 1 & \cdots & \cdots & 1 & 0 & 1 & \cdots & \cdots & \cdots & 1 & 3 & j & \cdots & \cdots & \cdots & \cdots & j \\
j^t & 0 & 0 & j^t & \cdots & \cdots & j^t & j^t & \cdots & \cdots & \cdots & \cdots & j^t & & & & & & \\
\vdots & j^t & 0 & \ddots & \ddots & \ddots & \vdots & \vdots & \ddots & \ddots & \ddots & \ddots & \vdots & & & & & & \\
\vdots & \vdots & \ddots & \ddots & \ddots & \ddots & \vdots & \vdots & \ddots & \ddots & \ddots & \ddots & \vdots & & & 2I_{mn}+J_{mn} & & & \\
\vdots & \vdots & \ddots & \ddots & \ddots & \ddots & j^t & \vdots & \ddots & \ddots & \ddots & \ddots & \vdots & & & & & & \\
\vdots & j^t & \ddots & \ddots & \ddots & \ddots & 0 & \vdots & \ddots & \ddots & \ddots & \ddots & \vdots & & & & & & \\
j^t & 0 & j^t & \cdots & \cdots & j^t & 0 & j^t & \cdots & \cdots & \cdots & \cdots & j^t & & & & & &
\end{array}\right)$$

$$
= \frac{1}{b^2} \times \det \begin{pmatrix}
b & -1 & \cdots & \cdots & \cdots & \cdots & -1 & 1 & \cdots & \cdots & \cdots & \cdots & 1 & j & \cdots & \cdots & \cdots & \cdots & j \\
b & a & 1 & \cdots & \cdots & \cdots & 1 & 0 & 1 & \cdots & \cdots & 1 & 0 & 0 & j & \cdots & \cdots & j & 0 \\
\vdots & 1 & \ddots & \ddots & \ddots & \ddots & \vdots & 0 & 0 & \ddots & \ddots & \ddots & 1 & 0 & \ddots & \ddots & \ddots & \ddots & j \\
\vdots & \vdots & \ddots & \ddots & \ddots & \ddots & \vdots & 1 & 0 & \ddots & \ddots & \ddots & \vdots & j & \ddots & \ddots & \ddots & \ddots & \vdots \\
\vdots & \vdots & \ddots & \ddots & \ddots & \ddots & \vdots & \vdots & \ddots & \ddots & \ddots & \ddots & \vdots & \vdots & \ddots & \ddots & \ddots & \ddots & \vdots \\
\vdots & \vdots & \ddots & \ddots & \ddots & \ddots & 1 & \vdots & \ddots & \ddots & 0 & 0 & 1 & \vdots & \ddots & \ddots & \ddots & \ddots & j \\
b & 1 & \cdots & \cdots & \cdots & 1 & a & 1 & \cdots & \cdots & 1 & 0 & 0 & j & \cdots & \cdots & j & 0 & 0 \\
b & 0 & 0 & 1 & \cdots & \cdots & 1 & 3 & 1 & \cdots & \cdots & \cdots & 1 & j & \cdots & \cdots & \cdots & \cdots & j \\
\vdots & 1 & 0 & \ddots & \ddots & \ddots & 1 & 1 & \ddots & \ddots & \ddots & \ddots & \vdots & \vdots & \ddots & \ddots & \ddots & \ddots & \vdots \\
\vdots & \vdots & \ddots & \ddots & \ddots & \ddots & \vdots & \vdots & \ddots & \ddots & \ddots & \ddots & \vdots & \vdots & \ddots & \ddots & \ddots & \ddots & \vdots \\
\vdots & \vdots & \ddots & \ddots & \ddots & \ddots & 1 & \vdots & \ddots & \ddots & \ddots & \ddots & \vdots & \vdots & \ddots & \ddots & \ddots & \ddots & \vdots \\
\vdots & 1 & \ddots & \ddots & \ddots & \ddots & 0 & \vdots & \ddots & \ddots & \ddots & \ddots & 1 & \vdots & \ddots & \ddots & \ddots & \ddots & \vdots \\
b & 0 & 1 & \cdots & \cdots & 1 & 0 & 1 & \cdots & \cdots & \cdots & 1 & 3 & j & \cdots & \cdots & \cdots & \cdots & j \\
bj^t & 0 & 0 & j^t & \cdots & \cdots & j^t & j^t & \cdots & \cdots & \cdots & \cdots & j^t & & & & & & \\
\vdots & j^t & 0 & \ddots & \ddots & \ddots & \vdots & \vdots & \ddots & \ddots & \ddots & \ddots & \vdots & & & & & & \\
\vdots & \vdots & \ddots & \ddots & \ddots & \ddots & \vdots & \vdots & \ddots & \ddots & \ddots & \ddots & \vdots & & & 2I_{mn}+J_{mn} & & & \\
\vdots & \vdots & \ddots & \ddots & \ddots & \ddots & j^t & \vdots & \ddots & \ddots & \ddots & \ddots & \vdots & & & & & & \\
\vdots & j^t & \ddots & \ddots & \ddots & \ddots & 0 & \vdots & \ddots & \ddots & \ddots & \ddots & \vdots & & & & & & \\
bj^t & 0 & j^t & \cdots & \cdots & j^t & 0 & j^t & \cdots & \cdots & \cdots & \cdots & j^t & & & & & &
\end{pmatrix}
$$

$$
= \frac{1}{b} \times \det \begin{pmatrix}
1 & -1 & \cdots & \cdots & \cdots & \cdots & -1 & 1 & \cdots & \cdots & \cdots & \cdots & 1 & j & \cdots & \cdots & \cdots & \cdots & j \\
1 & a & 1 & \cdots & \cdots & \cdots & 1 & 0 & 1 & \cdots & \cdots & 1 & 0 & 0 & j & \cdots & \cdots & j & 0 \\
\vdots & 1 & \ddots & \ddots & \ddots & \ddots & \vdots & 0 & 0 & \ddots & \ddots & \ddots & 1 & 0 & \ddots & \ddots & \ddots & \ddots & j \\
\vdots & \vdots & \ddots & \ddots & \ddots & \ddots & \vdots & 1 & 0 & \ddots & \ddots & \ddots & \vdots & j & \ddots & \ddots & \ddots & \ddots & \vdots \\
\vdots & \vdots & \ddots & \ddots & \ddots & \ddots & \vdots & \vdots & \ddots & \ddots & \ddots & \ddots & \vdots & \vdots & \ddots & \ddots & \ddots & \ddots & \vdots \\
\vdots & \vdots & \ddots & \ddots & \ddots & \ddots & 1 & \vdots & \ddots & \ddots & 0 & 0 & 1 & \vdots & \ddots & \ddots & \ddots & \ddots & j \\
1 & 1 & \cdots & \cdots & \cdots & 1 & a & 1 & \cdots & \cdots & 1 & 0 & 0 & j & \cdots & \cdots & j & 0 & 0 \\
1 & 0 & 0 & 1 & \cdots & \cdots & 1 & 3 & 1 & \cdots & \cdots & \cdots & 1 & j & \cdots & \cdots & \cdots & \cdots & j \\
\vdots & 1 & 0 & \ddots & \ddots & \ddots & 1 & 1 & \ddots & \ddots & \ddots & \ddots & \vdots & \vdots & \ddots & \ddots & \ddots & \ddots & \vdots \\
\vdots & \vdots & \ddots & \ddots & \ddots & \ddots & \vdots & \vdots & \ddots & \ddots & \ddots & \ddots & \vdots & \vdots & \ddots & \ddots & \ddots & \ddots & \vdots \\
\vdots & \vdots & \ddots & \ddots & \ddots & \ddots & 1 & \vdots & \ddots & \ddots & \ddots & \ddots & \vdots & \vdots & \ddots & \ddots & \ddots & \ddots & \vdots \\
\vdots & 1 & \ddots & \ddots & \ddots & \ddots & 0 & \vdots & \ddots & \ddots & \ddots & \ddots & 1 & \vdots & \ddots & \ddots & \ddots & \ddots & \vdots \\
1 & 0 & 1 & \cdots & \cdots & 1 & 0 & 1 & \cdots & \cdots & \cdots & 1 & 3 & j & \cdots & \cdots & \cdots & \cdots & j \\
1j^t & 0 & 0 & j^t & \cdots & \cdots & j^t & j^t & \cdots & \cdots & \cdots & \cdots & j^t & & & & & & \\
\vdots & j^t & 0 & \ddots & \ddots & \ddots & \vdots & \vdots & \ddots & \ddots & \ddots & \ddots & \vdots & & & & & & \\
\vdots & \vdots & \ddots & \ddots & \ddots & \ddots & \vdots & \vdots & \ddots & \ddots & \ddots & \ddots & \vdots & & & 2I_{mn}+J_{mn} & & & \\
\vdots & \vdots & \ddots & \ddots & \ddots & \ddots & j^t & \vdots & \ddots & \ddots & \ddots & \ddots & \vdots & & & & & & \\
\vdots & j^t & \ddots & \ddots & \ddots & \ddots & 0 & \vdots & \ddots & \ddots & \ddots & \ddots & \vdots & & & & & & \\
1j^t & 0 & j^t & \cdots & \cdots & j^t & 0 & j^t & \cdots & \cdots & \cdots & \cdots & j^t & & & & & &
\end{pmatrix}
$$

$$
= \frac{1}{b} \times \det \begin{pmatrix}
1 & -1 & \cdots & \cdots & \cdots & \cdots & -1 & 1 & \cdots & \cdots & \cdots & \cdots & 1 & j & \cdots & \cdots & \cdots & \cdots & j \\
0 & (a+1) & 2 & \cdots & \cdots & \cdots & 2 & -1 & 0 & \cdots & \cdots & 0 & -1 & -j & 0 & \cdots & \cdots & 0 & -j \\
\vdots & 2 & \ddots & \ddots & \ddots & \ddots & \vdots & -1 & -1 & \ddots & \ddots & \ddots & 0 & -j & \ddots & \ddots & \ddots & \ddots & 0 \\
\vdots & \vdots & \ddots & \ddots & \ddots & \ddots & \vdots & 0 & \ddots & \ddots & \ddots & \ddots & \vdots & 0 & \ddots & \ddots & \ddots & \ddots & \vdots \\
\vdots & \vdots & \ddots & \ddots & \ddots & \ddots & \vdots & \vdots & \ddots & \ddots & \ddots & \ddots & \vdots & \vdots & \ddots & \ddots & \ddots & \ddots & \vdots \\
\vdots & \vdots & \ddots & \ddots & \ddots & \ddots & 2 & \vdots & \ddots & \ddots & \ddots & -1 & 0 & \vdots & \ddots & \ddots & \ddots & \ddots & 0 \\
0 & 2 & \cdots & \cdots & \cdots & 2 & (a+1) & 0 & \cdots & \cdots & 0 & -1 & -1 & 0 & \cdots & \cdots & 0 & -j & -j \\
0 & 1 & 1 & 2 & \cdots & \cdots & 2 & 2 & 0 & \cdots & \cdots & \cdots & 0 & 0 & \cdots & \cdots & \cdots & \cdots & 0 \\
\vdots & 2 & 1 & \ddots & \ddots & \ddots & 2 & 0 & \ddots & \ddots & \ddots & \ddots & \vdots & \vdots & \ddots & \ddots & \ddots & \ddots & \vdots \\
\vdots & \vdots & \ddots & \ddots & \ddots & \ddots & \vdots & \vdots & \ddots & \ddots & \ddots & \ddots & \vdots & \vdots & \ddots & \ddots & \ddots & \ddots & \vdots \\
\vdots & \vdots & \ddots & \ddots & \ddots & \ddots & 2 & \vdots & \ddots & \ddots & \ddots & \ddots & \vdots & \vdots & \ddots & \ddots & \ddots & \ddots & \vdots \\
\vdots & 2 & \ddots & \ddots & \ddots & \ddots & 1 & \vdots & \ddots & \ddots & \ddots & \ddots & 0 & \vdots & \ddots & \ddots & \ddots & \ddots & \vdots \\
0 & 1 & 2 & \cdots & \cdots & 2 & 1 & 0 & \cdots & \cdots & \cdots & 0 & 2 & 0 & \cdots & \cdots & \cdots & \cdots & 0 \\
0 & j^t & j^t & 2j^t & \cdots & \cdots & 2j^t & 0 & \cdots & \cdots & \cdots & \cdots & 0 & & & & & & \\
\vdots & 2j^t & j^t & \ddots & \ddots & \ddots & \vdots & \vdots & \ddots & \ddots & \ddots & \ddots & \vdots & & & & & & \\
\vdots & \vdots & \ddots & \ddots & \ddots & \ddots & \vdots & \vdots & \ddots & \ddots & \ddots & \ddots & \vdots & & & 2I_{mn} & & & \\
\vdots & \vdots & \ddots & \ddots & \ddots & \ddots & 2j^t & \vdots & \ddots & \ddots & \ddots & \ddots & \vdots & & & & & & \\
\vdots & 2j^t & \ddots & \ddots & \ddots & \ddots & j^t & \vdots & \ddots & \ddots & \ddots & \ddots & \vdots & & & & & & \\
0 & j^t & 2j^t & \cdots & \cdots & 2j^t & j^t & 0 & \cdots & \cdots & \cdots & \cdots & 0 & & & & & &
\end{pmatrix}
$$

$$
= \tfrac{1}{b} \times \det \begin{pmatrix}
(a+1) & 2 & \cdots & \cdots & \cdots & 2 & -1 & 0 & \cdots & \cdots & 0 & -1 & -j & 0 & \cdots & \cdots & 0 & -j \\
2 & \ddots & \ddots & \ddots & \ddots & \vdots & -1 & -1 & \ddots & \ddots & \ddots & 0 & -j & \ddots & \ddots & \ddots & \ddots & 0 \\
\vdots & \ddots & \ddots & \ddots & \ddots & \vdots & 0 & \ddots & \ddots & \ddots & \ddots & \vdots & 0 & \ddots & \ddots & \ddots & \ddots & \vdots \\
\vdots & \ddots & \ddots & \ddots & \ddots & \vdots & \vdots & \ddots & \ddots & \ddots & \ddots & \vdots & \vdots & \ddots & \ddots & \ddots & \ddots & \vdots \\
\vdots & \ddots & \ddots & \ddots & \ddots & 2 & \vdots & \ddots & \ddots & \ddots & -1 & 0 & \vdots & \ddots & \ddots & \ddots & \ddots & 0 \\
2 & \cdots & \cdots & \cdots & 2 & (a+1) & 0 & \cdots & \cdots & 0 & -1 & -1 & 0 & \cdots & \cdots & 0 & -j & -j \\
1 & 1 & 2 & \cdots & \cdots & 2 & 2 & 0 & \cdots & \cdots & \cdots & 0 & 0 & \cdots & \cdots & \cdots & \cdots & 0 \\
2 & 1 & \ddots & \ddots & \ddots & 2 & 0 & \ddots & \ddots & \ddots & \ddots & \vdots & \vdots & \ddots & \ddots & \ddots & \ddots & \vdots \\
\vdots & \ddots & \ddots & \ddots & \ddots & \vdots & \vdots & \ddots & \ddots & \ddots & \ddots & \vdots & \vdots & \ddots & \ddots & \ddots & \ddots & \vdots \\
\vdots & \ddots & \ddots & \ddots & \ddots & 2 & \vdots & \ddots & \ddots & \ddots & \ddots & \vdots & \vdots & \ddots & \ddots & \ddots & \ddots & \vdots \\
2 & \ddots & \ddots & \ddots & \ddots & 1 & \vdots & \ddots & \ddots & \ddots & \ddots & 0 & \vdots & \ddots & \ddots & \ddots & \ddots & \vdots \\
1 & 2 & \cdots & \cdots & 2 & 1 & 0 & \cdots & \cdots & \cdots & 0 & 2 & 0 & \cdots & \cdots & \cdots & \cdots & 0 \\
j^t & j^t & 2j^t & \cdots & \cdots & 2j^t & 0 & \cdots & \cdots & \cdots & \cdots & 0 & & & & & & \\
2j^t & j^t & \ddots & \ddots & \ddots & \vdots & \vdots & \ddots & \ddots & \ddots & \ddots & \vdots & & & & & & \\
\vdots & \ddots & \ddots & \ddots & \ddots & \vdots & \vdots & \ddots & \ddots & \ddots & \ddots & \vdots & & & & 2I_{mn} & & \\
\vdots & \ddots & \ddots & \ddots & \ddots & 2j^t & \vdots & \ddots & \ddots & \ddots & \ddots & \vdots & & & & & & \\
2j^t & \ddots & \ddots & \ddots & \ddots & j^t & \vdots & \ddots & \ddots & \ddots & \ddots & \vdots & & & & & & \\
j^t & 2j^t & \cdots & \cdots & 2j^t & j^t & 0 & \cdots & \cdots & \cdots & \cdots & 0 & & & & & &
\end{pmatrix}
$$

Using Lemma 3, yields

$$
\tau\,(B_n^{(m)}) = \tfrac{1}{b} \times \det \begin{pmatrix} A & B \\ C & 2I_{mn} \end{pmatrix} = \tfrac{1}{b} \times \det(A - B\tfrac{1}{2I_{mn}}\, C) \times 2^{mn}
$$

$$
= \tfrac{1}{b} 2^{mn} \times 2^{-2m} \times \det \begin{pmatrix}
(2a+2n+2) & 3n+4 & 4(n+1) & \cdots & 4(n+1) & 3n+4 & -2 & 0 & \cdots & \cdots & 0 & -2 \\
3n+4 & (2a+2n+2) & 3n+4 & 4(n+1) & \cdots & 4(n+1) & -2 & \ddots & \ddots & \ddots & \ddots & 0 \\
4(n+1) & 3n+4 & \ddots & \ddots & \ddots & \vdots & 0 & \ddots & \ddots & \ddots & \ddots & \vdots \\
\vdots & \ddots & \ddots & \ddots & \ddots & 4(n+1) & \vdots & \ddots & \ddots & \ddots & \ddots & \vdots \\
4(n+1) & \ddots & \ddots & \ddots & \ddots & 3n+4 & \vdots & \ddots & \ddots & \ddots & \ddots & 0 \\
3n+4 & 4(n+1) & \cdots & 4(n+1) & 3n+4 & (2a+2n+2) & 0 & \cdots & \cdots & 0 & -2 & -2 \\
2 & 2 & 4 & \cdots & \cdots & 4 & 4 & 0 & \cdots & \cdots & \cdots & 0 \\
4 & \ddots & \ddots & \ddots & \ddots & \vdots & 0 & \ddots & \ddots & \ddots & \ddots & \vdots \\
\vdots & \ddots & \ddots & \ddots & \ddots & \vdots & \vdots & \ddots & \ddots & \ddots & \ddots & \vdots \\
\vdots & \ddots & \ddots & \ddots & \ddots & 4 & \vdots & \ddots & \ddots & \ddots & \ddots & \vdots \\
4 & \ddots & \ddots & \ddots & \ddots & 2 & \vdots & \ddots & \ddots & \ddots & \ddots & 0 \\
2 & 4 & \cdots & \cdots & 4 & 2 & 0 & \cdots & \cdots & \cdots & 0 & 4
\end{pmatrix}
$$

Using Lemma 3 again, yields

$$
\tau\,(B_n^{(m)}) = \tfrac{2^{mn-2m}}{b} \times \det \begin{pmatrix} D & E \\ F & 4I_m \end{pmatrix} = \tfrac{2^{mn}}{b} \times \det(D - E\tfrac{1}{4I_m}\, F)
$$

$$
\tau\,(B_n^{(m)}) = \tfrac{2^{mn}}{b} \times \det \begin{pmatrix}
(2a+2n+4) & (3n+7) & 4(n+2) & \cdots & 4(n+2) & (3n+7) \\
(3n+7) & (2a+2n+4) & (3n+7) & \ddots & \cdots & 4(n+2) \\
4(n+2) & \ddots & \ddots & \ddots & \ddots & \vdots \\
\vdots & \ddots & \ddots & \ddots & \ddots & 4(n+2) \\
4(n+2) & \ddots & \ddots & \ddots & \ddots & (3n+7) \\
(3n+7) & 4(n+2) & \cdots & 4(n+2) & (3n+7) & (2a+2n+4)
\end{pmatrix}
$$

With a straightforward inducement using properties of determinants, we obtain

$$
\tau\,(B_n^{(m)}) = \tfrac{2^{mn}}{b} \times \tfrac{4b}{mn+m+4} \times \det \begin{pmatrix}
(2a-n-3) & 0 & (n+1) & \cdots & (n+1) & 0 \\
0 & (2a-n-3) & 0 & \ddots & \cdots & (n+1) \\
(n+1) & \ddots & \ddots & \ddots & \ddots & \vdots \\
\vdots & \ddots & \ddots & \ddots & \ddots & (n+1) \\
(n+1) & \ddots & \ddots & \ddots & \ddots & 0 \\
0 & (n+1) & \cdots & (n+1) & 0 & (2a-n-3)
\end{pmatrix}
$$

$$= \frac{2^{mn+2} \times (n+1)^m}{mn+m+4} \times \det \begin{pmatrix} \frac{(2a-n-3)}{(n+1)} & 0 & 1 & \cdots & 1 & 0 \\ 0 & \frac{(2a-n-3)}{(n+1)} & 0 & \ddots & \ddots & 1 \\ 1 & \ddots & \ddots & \ddots & \ddots & \vdots \\ \vdots & \ddots & \ddots & \ddots & \ddots & 1 \\ 1 & \ddots & \ddots & \ddots & \ddots & 0 \\ 0 & 1 & \cdots & 1 & 0 & \frac{(2a-n-3)}{(n+1)} \end{pmatrix}$$

Using Lemma 2, yields

$$\begin{aligned} \tau(B_n^{(m)}) &= 2^{mn+2} \times \frac{(n+1)^m}{mn+m+4} \times \frac{2(\frac{2a-n-3}{n+1}+m-3)}{\frac{2a-n-3}{n+1}-3} \times [T_m(\frac{\frac{2a-n-3}{n+1}-1}{2}) - 1] \\ &= 2^{mn+1} \times (n+1)^m \times [T_m(\tfrac{n+3}{n+1}) - 1]. \end{aligned}$$

Using Equation (11), yields the result. □

Definition 3. *The pyramid graph $C_n^{(m)}$ is the graph created from the gear graph G_{m+1} with vertices $\{u_0; u_1, u_2, \ldots, u_m; w_1, w_2, \ldots, w_m\}$ with double external edges and m sets of vertices, say, $\{v_1^1, v_2^1, \ldots, v_n^1\}$, $\{v_1^2, u_2^2, \ldots, v_n^2\}, \ldots, \{v_1^m, v_2^m, \ldots, v_n^m\}$, such that for all $i = 1, 2, \ldots, n$ the vertex v_i^j is adjacent to u_j and u_{j+1}, where $j = 1, 2, \ldots, m-1$, and v_i^m is adjacent to u_1 and u_m. See Figure 3.*

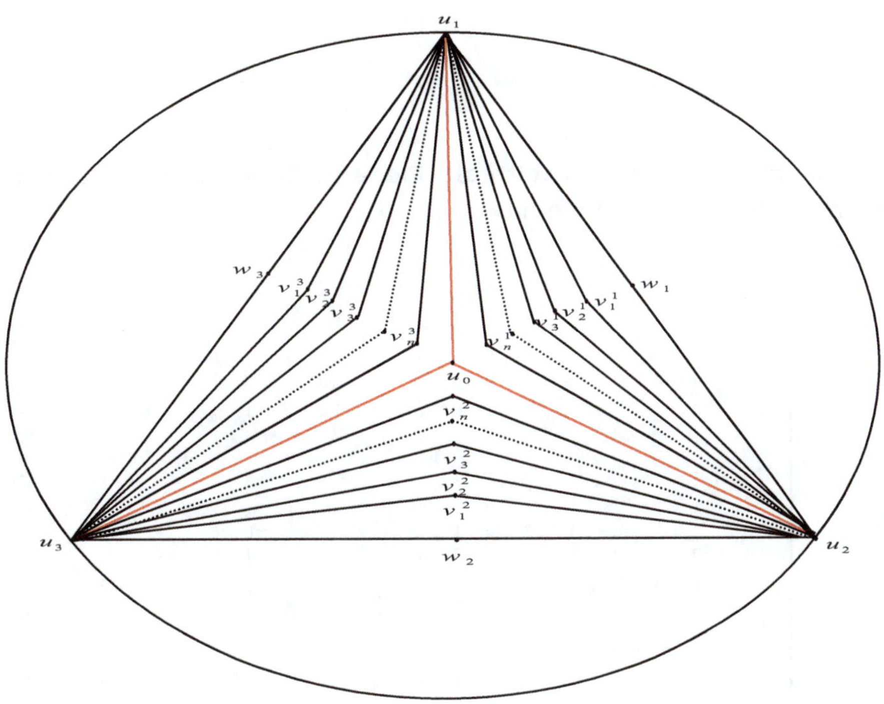

Figure 3. The pyramid graph $C_n^{(3)}$.

Theorem 3. *For $n \geq 0$, $m \geq 3$, $\tau(C_n^{(m)}) = 2^{mn}[(n+4+\sqrt{2n+7}\,)^m + (n+4-\sqrt{2n+7}\,)^m - 2(n+3)^m]$.*

Proof. Using Lemma 1, we have:

$$\tau(C_n^{(m)}) = \frac{1}{(mn+2m+1)^2} \times \det((mn+2m+1)\,I - D^c + A^c) = \frac{1}{(mn+2m+1)^2} \times$$

$$\det\begin{pmatrix}
(m+1) & 0 & 0 & \cdots & \cdots & \cdots & 0 & 1 & \cdots & \cdots & \cdots & \cdots & 1 & 1 & \cdots & \cdots & \cdots & \cdots & \cdots & \cdots & \cdots & \cdots & \cdots & \cdots & \cdots & \cdots & \cdots & \cdots & \cdots & 1 \\
0 & 2(n+3) & 0 & 1 & \cdots & 1 & 0 & 0 & 1 & \cdots & \cdots & 1 & 0 & 0 & \cdots & 0 & 1 & \cdots & 1 & \cdots & \cdots & 1 & \cdots & 1 & 1 & \cdots & 1 & 0 & \cdots & 0 \\
\vdots & 0 & \ddots & \ddots & \ddots & \ddots & 1 & 0 & \ddots & \ddots & \ddots & \ddots & 1 & 0 & \cdots & 0 & 0 & \cdots & 0 & 1 & \cdots & 1 & \cdots & \cdots & 1 & \cdots & 1 & 1 & \cdots & 1 \\
\vdots & 1 & \ddots & \ddots & \ddots & \ddots & \vdots & 1 & \ddots & \ddots & \ddots & \vdots & \vdots & 1 & \cdots & 1 & 0 & \cdots & 0 & 0 & \cdots & 0 & 1 & \cdots & 1 & \cdots & \cdots & 1 & \cdots & 1 \\
\vdots & \vdots & \ddots & \ddots & \ddots & \ddots & 1 & \vdots & \ddots & \ddots & \ddots & \vdots & \vdots & 1 & \cdots & 1 & 1 & \cdots & 1 & 0 & \cdots & 0 & 0 & \cdots & 0 & 1 & \cdots & 1 & \cdots & \cdots \\
\vdots & 1 & \ddots & \ddots & \ddots & \ddots & 0 & \vdots & \ddots & \ddots & \ddots & \ddots & 1 & \vdots & \cdots & \cdots & \cdots & \cdots & \cdots & \cdots & \cdots & \cdots & \cdots & \cdots & \cdots & \cdots & \cdots & \cdots & \cdots & \vdots \\
0 & 0 & 1 & \cdots & 1 & 0 & 2(n+3) & 1 & \cdots & \cdots & 1 & 0 & 0 & 1 & \cdots & 1 & 1 & \cdots & 1 & \cdots & \cdots & 1 & \cdots & 1 & 0 & \cdots & 0 & 0 & \cdots & 0 \\
1 & 0 & 0 & 1 & \cdots & \cdots & 1 & 3 & 1 & \cdots & \cdots & \cdots & 1 & 1 & \cdots & 1 & 1 & \cdots & 1 & \cdots & \cdots & \cdots & \cdots & \cdots & 1 & \cdots & 1 & 1 & \cdots & 1 \\
\vdots & 1 & \ddots & \ddots & \ddots & \ddots & \vdots & 1 & \ddots & \ddots & \ddots & \ddots & \vdots & 1 & \cdots & 1 & 1 & \cdots & 1 & \cdots & \cdots & \cdots & \cdots & \cdots & 1 & \cdots & 1 & 1 & \cdots & 1 \\
\vdots & \vdots & \ddots & \ddots & \ddots & \ddots & \vdots & \vdots & \ddots & \ddots & \ddots & \ddots & \vdots & \vdots & \cdots & \vdots & \vdots & \cdots & \vdots & \cdots & \cdots & \cdots & \cdots & \cdots & \vdots & \cdots & \vdots & \vdots & \cdots & 1 \\
\vdots & \vdots & \ddots & \ddots & \ddots & \ddots & 1 & \vdots & \ddots & \ddots & \ddots & \ddots & \vdots & \vdots & \cdots & \vdots & \vdots & \cdots & \vdots & \cdots & \cdots & \cdots & \cdots & \cdots & \vdots & \cdots & \vdots & \vdots & \cdots & \vdots \\
\vdots & 1 & \ddots & \ddots & \ddots & \ddots & 0 & \vdots & \ddots & \ddots & \ddots & \ddots & \vdots & 1 & \cdots & 1 & 1 & \cdots & 1 & \cdots & \cdots & \cdots & \cdots & \cdots & 1 & \cdots & 1 & 1 & \cdots & 1 \\
1 & 0 & 1 & \cdots & \cdots & 1 & 0 & 1 & \cdots & \cdots & \cdots & 1 & 3 & 1 & \cdots & 1 & 1 & \cdots & 1 & \cdots & \cdots & \cdots & \cdots & \cdots & 1 & \cdots & 1 & 1 & \cdots & 1 \\
1 & 0 & 0 & 1 & \cdots & 1 & 0 & 1 & 1 & \cdots & \cdots & 1 & 1 & 3 & 1 & \cdots & \cdots & \cdots & \cdots & \cdots & \cdots & \cdots & \cdots & \cdots & \cdots & \cdots & \cdots & \cdots & \cdots & 1 \\
\vdots & \vdots & \vdots & \vdots & \vdots & \vdots & \vdots & \vdots & \vdots & \vdots & \vdots & \vdots & \vdots & 1 & \ddots & \ddots & \ddots & \ddots & \ddots & \ddots & \ddots & \ddots & \ddots & \ddots & \ddots & \ddots & \ddots & \ddots & \ddots & \vdots \\
\vdots & 0 & 0 & 1 & \vdots & 1 & \vdots & 1 & 1 & \cdots & \cdots & 1 & 1 & \vdots & \ddots & \ddots & \ddots & \ddots & \ddots & \ddots & \ddots & \ddots & \ddots & \ddots & \ddots & \ddots & \ddots & \ddots & \ddots & \vdots \\
\vdots & 1 & 0 & 0 & \vdots & 1 & \vdots & 1 & 1 & \cdots & \cdots & 1 & 1 & \vdots & \ddots & \ddots & \ddots & \ddots & \ddots & \ddots & \ddots & \ddots & \ddots & \ddots & \ddots & \ddots & \ddots & \ddots & \ddots & \vdots \\
\vdots & \vdots & \vdots & \vdots & \vdots & \vdots & \vdots & \vdots & \vdots & \vdots & \vdots & \vdots & \vdots & \vdots & \ddots & \ddots & \ddots & \ddots & \ddots & \ddots & \ddots & \ddots & \ddots & \ddots & \ddots & \ddots & \ddots & \ddots & \ddots & \vdots \\
\vdots & 1 & 0 & 0 & \vdots & 1 & \vdots & 1 & 1 & \cdots & \cdots & 1 & 1 & \vdots & \ddots & \ddots & \ddots & \ddots & \ddots & \ddots & \ddots & \ddots & \ddots & \ddots & \ddots & \ddots & \ddots & \ddots & \ddots & \vdots \\
\vdots & 1 & 1 & 0 & \vdots & \vdots & \vdots & \vdots & \vdots & \vdots & \vdots & \vdots & \vdots & \vdots & \ddots & \ddots & \ddots & \ddots & \ddots & \ddots & \ddots & \ddots & \ddots & \ddots & \ddots & \ddots & \ddots & \ddots & \ddots & \vdots \\
\vdots & \vdots & \vdots & \vdots & \vdots & \vdots & \vdots & \vdots & \vdots & \vdots & \vdots & \vdots & \vdots & \vdots & \ddots & \ddots & \ddots & \ddots & \ddots & \ddots & \ddots & \ddots & \ddots & \ddots & \ddots & \ddots & \ddots & \ddots & \ddots & \vdots \\
\vdots & 1 & 1 & 0 & \vdots & 1 & \vdots & \vdots & \vdots & \vdots & \vdots & \vdots & \vdots & \vdots & \ddots & \ddots & \ddots & \ddots & \ddots & \ddots & \ddots & \ddots & \ddots & \ddots & \ddots & \ddots & \ddots & \ddots & \ddots & \vdots \\
\vdots & \vdots & \vdots & 1 & \vdots & \vdots & \vdots & \vdots & \vdots & \vdots & \vdots & \vdots & \vdots & \vdots & \ddots & \ddots & \ddots & \ddots & \ddots & \ddots & \ddots & \ddots & \ddots & \ddots & \ddots & \ddots & \ddots & \ddots & \ddots & \vdots \\
\vdots & \vdots & \vdots & \vdots & \vdots & 1 & \vdots & \vdots & \vdots & \vdots & \vdots & \vdots & \vdots & \vdots & \ddots & \ddots & \ddots & \ddots & \ddots & \ddots & \ddots & \ddots & \ddots & \ddots & \ddots & \ddots & \ddots & \ddots & \ddots & \vdots \\
\vdots & 1 & 1 & 1 & \vdots & 0 & \vdots & \vdots & \vdots & \vdots & \vdots & \vdots & \vdots & \vdots & \ddots & \ddots & \ddots & \ddots & \ddots & \ddots & \ddots & \ddots & \ddots & \ddots & \ddots & \ddots & \ddots & \ddots & \ddots & \vdots \\
\vdots & \vdots & \vdots & \vdots & \vdots & \vdots & \vdots & \vdots & \vdots & \vdots & \vdots & \vdots & \vdots & \vdots & \ddots & \ddots & \ddots & \ddots & \ddots & \ddots & \ddots & \ddots & \ddots & \ddots & \ddots & \ddots & \ddots & \ddots & \ddots & \vdots \\
\vdots & 1 & 1 & \vdots & \vdots & 0 & \vdots & 1 & 1 & \cdots & \cdots & 1 & 1 & \vdots & \ddots & \ddots & \ddots & \ddots & \ddots & \ddots & \ddots & \ddots & \ddots & \ddots & \ddots & \ddots & \ddots & \ddots & \ddots & \vdots \\
\vdots & 0 & 1 & 1 & \vdots & 0 & \vdots & 1 & 1 & \cdots & \cdots & 1 & 1 & \vdots & \ddots & \ddots & \ddots & \ddots & \ddots & \ddots & \ddots & \ddots & \ddots & \ddots & \ddots & \ddots & \ddots & \ddots & \ddots & \vdots \\
\vdots & \vdots & \vdots & \vdots & \vdots & \vdots & \vdots & \vdots & \vdots & \vdots & \vdots & \vdots & \vdots & \vdots & \ddots & \ddots & \ddots & \ddots & \ddots & \ddots & \ddots & \ddots & \ddots & \ddots & \ddots & \ddots & \ddots & \ddots & \ddots & 1 \\
1 & 0 & 1 & 1 & \cdots & 0 & \cdots & 1 & 1 & \cdots & \cdots & 1 & 1 & 1 & \cdots & \cdots & \cdots & \cdots & \cdots & \cdots & \cdots & \cdots & \cdots & \cdots & \cdots & \cdots & \cdots & \cdots & 1 & 3
\end{pmatrix}$$

Let $j = (1 \cdots 1)$ be the $1 \times n$ matrix with all one, and J_n be the $n \times n$ matrix with all one. Set $a = 2n + 6$ and $b = mn + 2m + 1$. Then we have:

$$\tau\left(C_n^{(m)}\right) = \frac{1}{b^2} \times \det\begin{pmatrix}
m+1 & 0 & \cdots & \cdots & \cdots & \cdots & 0 & 1 & \cdots & \cdots & \cdots & \cdots & 1 & j & \cdots & \cdots & \cdots & \cdots & j \\
0 & a & 0 & 1 & \cdots & 1 & 0 & 0 & 1 & \cdots & \cdots & 1 & 0 & 0 & j & \cdots & \cdots & j & 0 \\
\vdots & 0 & \ddots & \ddots & \ddots & \ddots & 1 & 0 & 0 & \ddots & \ddots & \ddots & 1 & 0 & \ddots & \ddots & \ddots & \ddots & j \\
\vdots & 1 & \ddots & \ddots & \ddots & \ddots & \vdots & 1 & 0 & \ddots & \ddots & \ddots & \vdots & j & \ddots & \ddots & \ddots & \ddots & \vdots \\
\vdots & \vdots & \ddots & \ddots & \ddots & \ddots & 1 & \vdots & \ddots & \ddots & \ddots & \ddots & \vdots & \vdots & \ddots & \ddots & \ddots & \ddots & \vdots \\
\vdots & 1 & \ddots & \ddots & \ddots & \ddots & 0 & \vdots & \ddots & \ddots & 0 & 0 & 1 & \vdots & \ddots & \ddots & \ddots & \ddots & j \\
0 & 0 & 1 & \cdots & 1 & 0 & a & 1 & \cdots & \cdots & 1 & 0 & 0 & j & \cdots & \cdots & j & 0 & 0 \\
1 & 0 & 0 & 1 & \cdots & \cdots & 1 & 3 & 1 & \cdots & \cdots & \cdots & 1 & j & \cdots & \cdots & \cdots & \cdots & j \\
\vdots & 1 & 0 & \ddots & \ddots & \ddots & 1 & 1 & \ddots & \ddots & \ddots & \ddots & \vdots & \vdots & \ddots & \ddots & \ddots & \ddots & \vdots \\
\vdots & \vdots & \ddots & \ddots & \ddots & \ddots & \vdots & \vdots & \ddots & \ddots & \ddots & \ddots & \vdots & \vdots & \ddots & \ddots & \ddots & \ddots & \vdots \\
\vdots & \vdots & \ddots & \ddots & \ddots & \ddots & 1 & \vdots & \ddots & \ddots & \ddots & \ddots & \vdots & \vdots & \ddots & \ddots & \ddots & \ddots & \vdots \\
\vdots & 1 & \ddots & \ddots & \ddots & \ddots & 0 & \vdots & \ddots & \ddots & \ddots & \ddots & 1 & \vdots & \ddots & \ddots & \ddots & \ddots & \vdots \\
1 & 0 & 1 & \cdots & \cdots & 1 & 0 & 1 & \cdots & \cdots & \cdots & 1 & 3 & j & \cdots & \cdots & \cdots & \cdots & j \\
j^t & 0 & 0 & j^t & \cdots & \cdots & j^t & j^t & \cdots & \cdots & \cdots & \cdots & j^t & & & & & & \\
\vdots & j^t & 0 & \ddots & \ddots & \ddots & \vdots & \vdots & \ddots & \ddots & \ddots & \ddots & \vdots & & & & & & \\
\vdots & \vdots & \ddots & \ddots & \ddots & \ddots & \vdots & \vdots & \ddots & \ddots & \ddots & \ddots & \vdots & & & 2I_{m\,n} + J_{m\,n} & & & \\
\vdots & \vdots & \ddots & \ddots & \ddots & \ddots & j^t & \vdots & \ddots & \ddots & \ddots & \ddots & \vdots & & & & & & \\
\vdots & j^t & \ddots & \ddots & \ddots & \ddots & 0 & \vdots & \ddots & \ddots & \ddots & \ddots & \vdots & & & & & & \\
j^t & 0 & j^t & \cdots & \cdots & j^t & 0 & j^t & \cdots & \cdots & \cdots & \cdots & j^t & & & & & &
\end{pmatrix}$$

$$
= \frac{1}{b^2} \times \det \begin{pmatrix}
b & 0 & \cdots & \cdots & \cdots & \cdots & 0 & 1 & \cdots & \cdots & \cdots & \cdots & 1 & j & \cdots & \cdots & \cdots & \cdots & j \\
b & a & 0 & 1 & \cdots & 1 & 0 & 0 & 1 & \cdots & \cdots & 1 & 0 & 0 & j & \cdots & \cdots & j & 0 \\
\vdots & 1 & \ddots & \ddots & \ddots & \ddots & 1 & 0 & 0 & \ddots & \ddots & \ddots & 1 & 0 & \ddots & \ddots & \ddots & \ddots & j \\
\vdots & \vdots & \ddots & \ddots & \ddots & \ddots & \vdots & 1 & 0 & \ddots & \ddots & \ddots & \vdots & j & \ddots & \ddots & \ddots & \ddots & \vdots \\
\vdots & \vdots & \ddots & \ddots & \ddots & \ddots & 1 & \vdots & \ddots & \ddots & \ddots & \ddots & \vdots & \vdots & \ddots & \ddots & \ddots & \ddots & \vdots \\
\vdots & 1 & \ddots & \ddots & \ddots & \ddots & 0 & \vdots & \ddots & \ddots & 0 & 0 & 1 & \vdots & \ddots & \ddots & \ddots & \ddots & j \\
b & 0 & 1 & \cdots & 1 & 0 & a & 1 & \cdots & \cdots & 1 & 0 & 0 & j & \cdots & \cdots & j & 0 & 0 \\
b & 0 & 0 & 1 & \cdots & \cdots & 1 & 3 & 1 & \cdots & \cdots & \cdots & 1 & j & \cdots & \cdots & \cdots & \cdots & j \\
\vdots & 1 & 0 & \ddots & \ddots & \ddots & 1 & 1 & \ddots & \ddots & \ddots & \ddots & \vdots & \vdots & \ddots & \ddots & \ddots & \ddots & \vdots \\
\vdots & \vdots & \ddots & \ddots & \ddots & \ddots & \vdots & \vdots & \ddots & \ddots & \ddots & \ddots & \vdots & \vdots & \ddots & \ddots & \ddots & \ddots & \vdots \\
\vdots & \vdots & \ddots & \ddots & \ddots & \ddots & 1 & \vdots & \ddots & \ddots & \ddots & \ddots & \vdots & \vdots & \ddots & \ddots & \ddots & \ddots & \vdots \\
\vdots & 1 & \ddots & \ddots & \ddots & \ddots & 0 & \vdots & \ddots & \ddots & \ddots & \ddots & 1 & \vdots & \ddots & \ddots & \ddots & \ddots & \vdots \\
b & 0 & 1 & \cdots & \cdots & 1 & 0 & 1 & \cdots & \cdots & \cdots & 1 & 3 & j & \cdots & \cdots & \cdots & \cdots & j \\
bj^t & 0 & 0 & j^t & \cdots & \cdots & j^t & j^t & \cdots & \cdots & \cdots & \cdots & j^t & & & & & & \\
\vdots & j^t & 0 & \ddots & \ddots & \ddots & \vdots & \vdots & \ddots & \ddots & \ddots & \ddots & \vdots & & & & & & \\
\vdots & \vdots & \ddots & \ddots & \ddots & \ddots & \vdots & \vdots & \ddots & \ddots & \ddots & \ddots & \vdots & & & 2I_{mn}+J_{mn} & & & \\
\vdots & \vdots & \ddots & \ddots & \ddots & \ddots & j^t & \vdots & \ddots & \ddots & \ddots & \ddots & \vdots & & & & & & \\
\vdots & j^t & \ddots & \ddots & \ddots & \ddots & 0 & \vdots & \ddots & \ddots & \ddots & \ddots & \vdots & & & & & & \\
bj^t & 0 & j^t & \cdots & \cdots & j^t & 0 & j^t & \cdots & \cdots & \cdots & \cdots & j^t & & & & & &
\end{pmatrix}
$$

$$
= \frac{1}{b} \times \det \begin{pmatrix}
1 & 0 & \cdots & \cdots & \cdots & \cdots & 0 & 1 & \cdots & \cdots & \cdots & \cdots & 1 & j & \cdots & \cdots & \cdots & \cdots & j \\
1 & a & 0 & 1 & \cdots & 1 & 0 & 0 & 1 & \cdots & \cdots & 1 & 0 & 0 & j & \cdots & \cdots & j & 0 \\
\vdots & 0 & \ddots & \ddots & \ddots & \ddots & 1 & 0 & 0 & \ddots & \ddots & \ddots & 1 & 0 & \ddots & \ddots & \ddots & \ddots & j \\
\vdots & 1 & \ddots & \ddots & \ddots & \ddots & \vdots & 1 & 0 & \ddots & \ddots & \ddots & \vdots & j & \ddots & \ddots & \ddots & \ddots & \vdots \\
\vdots & \vdots & \ddots & \ddots & \ddots & \ddots & 1 & \vdots & \ddots & \ddots & \ddots & \ddots & \vdots & \vdots & \ddots & \ddots & \ddots & \ddots & \vdots \\
\vdots & 1 & \ddots & \vdots & \ddots & \ddots & 0 & \vdots & \ddots & \ddots & 0 & 0 & 1 & \vdots & \ddots & \ddots & \ddots & \ddots & j \\
1 & 0 & 1 & \cdots & 1 & 0 & a & 1 & \cdots & \cdots & 1 & 0 & 0 & j & \cdots & \cdots & j & 0 & 0 \\
1 & 0 & 0 & 1 & \cdots & \cdots & 1 & 3 & 1 & \cdots & \cdots & \cdots & 1 & j & \cdots & \cdots & \cdots & \cdots & j \\
\vdots & 1 & 0 & \ddots & \ddots & \ddots & 1 & 1 & \ddots & \ddots & \ddots & \ddots & \vdots & \vdots & \ddots & \ddots & \ddots & \ddots & \vdots \\
\vdots & \vdots & \ddots & \ddots & \ddots & \ddots & \vdots & \vdots & \ddots & \ddots & \ddots & \ddots & \vdots & \vdots & \ddots & \ddots & \ddots & \ddots & \vdots \\
\vdots & \vdots & \ddots & \ddots & \ddots & \ddots & 1 & \vdots & \ddots & \ddots & \ddots & \ddots & \vdots & \vdots & \ddots & \ddots & \ddots & \ddots & \vdots \\
\vdots & 1 & \ddots & \ddots & \ddots & \ddots & 0 & \vdots & \ddots & \ddots & \ddots & \ddots & 1 & \vdots & \ddots & \ddots & \ddots & \ddots & \vdots \\
1 & 0 & 1 & \cdots & \cdots & 1 & 0 & 1 & \cdots & \cdots & \cdots & 1 & 3 & j & \cdots & \cdots & \cdots & \cdots & j \\
1j^t & 0 & 0 & j^t & \cdots & \cdots & j^t & j^t & \cdots & \cdots & \cdots & \cdots & j^t & & & & & & \\
\vdots & j^t & 0 & \ddots & \ddots & \ddots & \vdots & \vdots & \ddots & \ddots & \ddots & \ddots & \vdots & & & & & & \\
\vdots & \vdots & \ddots & \ddots & \ddots & \ddots & \vdots & \vdots & \ddots & \ddots & \ddots & \ddots & \vdots & & & 2I_{mn}+J_{mn} & & & \\
\vdots & \vdots & \ddots & \ddots & \ddots & \ddots & j^t & \vdots & \ddots & \ddots & \ddots & \ddots & \vdots & & & & & & \\
\vdots & j^t & \ddots & \ddots & \ddots & \ddots & 0 & \vdots & \ddots & \ddots & \ddots & \ddots & \vdots & & & & & & \\
1j^t & 0 & j^t & \cdots & \cdots & j^t & 0 & j^t & \cdots & \cdots & \cdots & \cdots & j^t & & & & & &
\end{pmatrix}
$$

$$
= \frac{1}{b} \times \det \begin{pmatrix}
1 & 0 & \cdots & \cdots & \cdots & \cdots & 0 & 1 & \cdots & \cdots & \cdots & \cdots & 1 & j & \cdots & \cdots & \cdots & \cdots & j \\
0 & a & 0 & 1 & \cdots & 1 & 0 & -1 & 0 & \cdots & \cdots & 0 & -1 & -j & 0 & \cdots & \cdots & 0 & -j \\
\vdots & 1 & \ddots & \ddots & \ddots & \ddots & 1 & -1 & -1 & \ddots & \ddots & \ddots & 0 & -j & \ddots & \ddots & \ddots & \ddots & 0 \\
\vdots & \vdots & \ddots & \ddots & \ddots & \ddots & \vdots & 0 & \ddots & \ddots & \ddots & \ddots & \vdots & 0 & \ddots & \ddots & \ddots & \ddots & \vdots \\
\vdots & \vdots & \ddots & \ddots & \ddots & \ddots & 1 & \vdots & \ddots & \ddots & \ddots & \ddots & \vdots & \vdots & \ddots & \ddots & \ddots & \ddots & \vdots \\
\vdots & 1 & \ddots & \ddots & \ddots & \ddots & 0 & \vdots & \ddots & \ddots & \ddots & -1 & 0 & \vdots & \ddots & \ddots & \ddots & \ddots & 0 \\
0 & 0 & 1 & \cdots & 1 & 0 & a & 0 & \cdots & \cdots & 0 & -1 & -1 & 0 & \cdots & \cdots & 0 & -j & -j \\
0 & 0 & 0 & 1 & \cdots & \cdots & 1 & 2 & 0 & \cdots & \cdots & \cdots & 0 & 0 & \cdots & \cdots & \cdots & \cdots & 0 \\
\vdots & 1 & 0 & \ddots & \ddots & \ddots & 1 & 0 & \ddots & \ddots & \ddots & \ddots & \vdots & \vdots & \ddots & \ddots & \ddots & \ddots & \vdots \\
\vdots & \vdots & \ddots & \ddots & \ddots & \ddots & \vdots & \vdots & \ddots & \ddots & \ddots & \ddots & \vdots & \vdots & \ddots & \ddots & \ddots & \ddots & \vdots \\
\vdots & \vdots & \ddots & \ddots & \ddots & \ddots & 1 & \vdots & \ddots & \ddots & \ddots & \ddots & \vdots & \vdots & \ddots & \ddots & \ddots & \ddots & \vdots \\
\vdots & 1 & \ddots & \ddots & \ddots & \ddots & 0 & \vdots & \ddots & \ddots & \ddots & \ddots & 0 & \vdots & \ddots & \ddots & \ddots & \ddots & \vdots \\
0 & 0 & 1 & \cdots & \cdots & 1 & 0 & 0 & \cdots & \cdots & \cdots & 0 & 2 & 0 & \cdots & \cdots & \cdots & \cdots & 0 \\
0 & 0 & 0 & j^t & \cdots & \cdots & j^t & 0 & \cdots & \cdots & \cdots & \cdots & 0 & & & & & & \\
\vdots & j^t & 0 & \ddots & \ddots & \ddots & \vdots & \vdots & \ddots & \ddots & \ddots & \ddots & \vdots & & & & & & \\
\vdots & \vdots & \ddots & \ddots & \ddots & \ddots & \vdots & \vdots & \ddots & \ddots & \ddots & \ddots & \vdots & & & 2I_{mn} & & & \\
\vdots & \vdots & \ddots & \ddots & \ddots & \ddots & j^t & \vdots & \ddots & \ddots & \ddots & \ddots & \vdots & & & & & & \\
\vdots & j^t & \ddots & \ddots & \ddots & \ddots & 0 & \vdots & \ddots & \ddots & \ddots & \ddots & \vdots & & & & & & \\
0 & 0 & j^t & \cdots & \cdots & j^t & 0 & 0 & \cdots & \cdots & \cdots & \cdots & 0 & & & & & &
\end{pmatrix}
$$

$$
= \tfrac{1}{b} \times \det \begin{pmatrix}
a & 0 & 1 & \cdots & 1 & 0 & -1 & 0 & \cdots & \cdots & 0 & -1 & -j & 0 & \cdots & \cdots & 0 & -j \\
0 & \ddots & \ddots & \ddots & \ddots & 1 & -1 & -1 & \ddots & \ddots & \ddots & 0 & -j & \ddots & \ddots & \ddots & \ddots & 0 \\
1 & \ddots & \ddots & \ddots & \ddots & \vdots & 0 & \ddots & \ddots & \ddots & \ddots & \vdots & 0 & \ddots & \ddots & \ddots & \ddots & \vdots \\
\vdots & \ddots & \ddots & \ddots & \ddots & 1 & \vdots & \ddots & \ddots & \ddots & \ddots & \vdots & \vdots & \ddots & \ddots & \ddots & \ddots & \vdots \\
1 & \ddots & \ddots & \ddots & \ddots & 0 & \vdots & \ddots & \ddots & \ddots & -1 & 0 & \vdots & \ddots & \ddots & \ddots & \ddots & 0 \\
0 & 1 & \cdots & 1 & 0 & a & 0 & \cdots & \cdots & 0 & -1 & -1 & 0 & \cdots & \cdots & 0 & -j & -j \\
0 & 0 & 1 & \cdots & \cdots & 1 & 2 & 0 & \cdots & \cdots & \cdots & 0 & 0 & \cdots & \cdots & \cdots & \cdots & 0 \\
1 & 0 & \ddots & \ddots & \ddots & 1 & 0 & \ddots & \ddots & \ddots & \ddots & \vdots & \vdots & \ddots & \ddots & \ddots & \ddots & \vdots \\
\vdots & \ddots & \ddots & \ddots & \ddots & \vdots & \vdots & \ddots & \ddots & \ddots & \ddots & \vdots & \vdots & \ddots & \ddots & \ddots & \ddots & \vdots \\
\vdots & \ddots & \ddots & \ddots & \ddots & 1 & \vdots & \ddots & \ddots & \ddots & \ddots & \vdots & \vdots & \ddots & \ddots & \ddots & \ddots & \vdots \\
1 & \ddots & \ddots & \ddots & \ddots & 0 & \vdots & \ddots & \ddots & \ddots & \ddots & 0 & \vdots & \ddots & \ddots & \ddots & \ddots & \vdots \\
0 & 1 & \cdots & \cdots & 1 & 0 & 0 & \cdots & \cdots & \cdots & 0 & 2 & 0 & \cdots & \cdots & \cdots & \cdots & 0 \\
0 & 0 & j^t & \cdots & \cdots & j^t & 0 & \cdots & \cdots & \cdots & \cdots & 0 & & & & & & \\
j^t & 0 & \ddots & \ddots & \ddots & \vdots & \vdots & \ddots & \ddots & \ddots & \ddots & \vdots & & & & & & \\
\vdots & \ddots & \ddots & \ddots & \ddots & \vdots & \vdots & \ddots & \ddots & \ddots & \ddots & \vdots & & & & 2I_{m\,n} & & \\
\vdots & \ddots & \ddots & \ddots & \ddots & j^t & \vdots & \ddots & \ddots & \ddots & \ddots & \vdots & & & & & & \\
j^t & \ddots & \ddots & \ddots & \ddots & 0 & \vdots & \ddots & \ddots & \ddots & \ddots & \vdots & & & & & & \\
0 & j^t & \cdots & \cdots & j^t & 0 & 0 & \cdots & \cdots & \cdots & \cdots & 0 & & & & & &
\end{pmatrix}
$$

Using Lemma 3, yields

$$
\tau\,(C_n^{(m)}) = \tfrac{1}{b} \times \det\begin{pmatrix} A & B \\ C & 2I_{mn} \end{pmatrix} = \tfrac{1}{b} \times \det(A - B\tfrac{1}{2I_{mn}}\ C) \times 2^{mn}
$$

$$
= \tfrac{1}{b} \times 2^{mn} \times 2^{-2m} \times \det \begin{pmatrix}
2a & n & 2(n+1) & \cdots & 2(n+1) & n & -2 & 0 & \cdots & \cdots & 0 & -2 \\
n & 2a & n+2 & 2(n+1) & \cdots & 2(n+1) & -2 & \ddots & \ddots & \ddots & \ddots & 0 \\
2(n+1) & n & \ddots & \ddots & \ddots & \vdots & 0 & \ddots & \ddots & \ddots & \ddots & \vdots \\
\vdots & \ddots & \ddots & \ddots & \ddots & 2(n+1) & \vdots & \ddots & \ddots & \ddots & \ddots & \vdots \\
2(n+1) & \ddots & \ddots & \ddots & \ddots & n & \vdots & \ddots & \ddots & \ddots & \ddots & 0 \\
n & 2(n+1) & \cdots & 2(n+1) & n & 2a & 0 & \cdots & \cdots & 0 & -2 & -2 \\
0 & 0 & 2 & \cdots & \cdots & 2 & 4 & 0 & \cdots & \cdots & \cdots & 0 \\
2 & \ddots & \ddots & \ddots & \ddots & \vdots & 0 & \ddots & \ddots & \ddots & \ddots & \vdots \\
\vdots & \ddots & \ddots & \ddots & \ddots & \vdots & \vdots & \ddots & \ddots & \ddots & \ddots & \vdots \\
\vdots & \ddots & \ddots & \ddots & \ddots & 2 & \vdots & \ddots & \ddots & \ddots & \ddots & \vdots \\
2 & \ddots & \ddots & \ddots & \ddots & 0 & \vdots & \ddots & \ddots & \ddots & \ddots & 0 \\
0 & 2 & \cdots & \cdots & 2 & 0 & 0 & \cdots & \cdots & \cdots & 0 & 4
\end{pmatrix}
$$

Using Lemma 3 again, yields

$$
\tau\,(C_n^{(m)}) = \tfrac{2^{m\,n-2m}}{b} \times \det\begin{pmatrix} D & E \\ F & 4I_m \end{pmatrix} = \tfrac{2^{m\,n}}{b} \times \det(D - E\tfrac{1}{4I_m}\ F)
$$

$$
\tau\,(C_n^{(m)}) = \tfrac{2^{mn}}{b} \times \det \begin{pmatrix}
2a & (n+1) & 2(n+2) & \cdots & 2(n+2) & (n+1) \\
(n+1) & 2a & (n+3) & \ddots & \cdots & 2(n+2) \\
2(n+2) & \ddots & \ddots & \ddots & \ddots & \vdots \\
\vdots & \ddots & \ddots & \ddots & \ddots & 2(n+2) \\
2(n+2) & \ddots & \ddots & \ddots & \ddots & (n+1) \\
(n+1) & 2(n+2) & \cdots & 2(n+2) & (n+1) & 2a
\end{pmatrix}
$$

Using properties of determinants, we have:

$$\tau(C_n^{(m)}) = \frac{2^{mn}}{b} \times \frac{2b}{mn+3m+2} \times \det \begin{pmatrix} (2a-n-1) & 0 & (n+3) & \cdots & (n+3) & 0 \\ 0 & (2a-n-1) & 0 & \ddots & \cdots & (n+3) \\ (n+3) & \ddots & \ddots & \ddots & \ddots & \vdots \\ \vdots & \ddots & \ddots & \ddots & \ddots & (n+3) \\ (n+3) & \ddots & \ddots & \ddots & \ddots & 0 \\ 0 & (n+3) & \cdots & (n+3) & 0 & (2a-n-1) \end{pmatrix}$$

$$= \frac{2^{mn+1}(n+3)^m}{mn+3m+2} \times \det \begin{pmatrix} \frac{(2a-n-1)}{(n+3)} & 0 & 1 & \cdots & 1 & 0 \\ 0 & \frac{(2a-n-1)}{(n+3)} & 0 & \ddots & \ddots & 1 \\ 1 & \ddots & \ddots & \ddots & \ddots & \vdots \\ \vdots & \ddots & \ddots & \ddots & \ddots & 1 \\ 1 & \ddots & \ddots & \ddots & \ddots & 0 \\ 0 & 1 & \cdots & 1 & 0 & \frac{(2a-n-1)}{(n+3)} \end{pmatrix}$$

Using Lemma 2, yields:

$$\begin{aligned} \tau(C_n^{(m)}) &= 2^{mn+1} \times \frac{(n+3)^m}{mn+3m+2} \times \frac{2(\frac{2a-n-1}{n+3}+m-3)}{\frac{2a-n-1}{n+3}-3} \times [T_m(\frac{\frac{2a-n-1}{n+3}-1}{2}) - 1] \\ &= 2^{mn+1} \times (n+3)^m \times [T_m(\tfrac{n+4}{n+3}) - 1]. \end{aligned}$$

Using Equation (11), yields the result. □

Definition 4. *The pyramid graph $D_n^{(m)}$ is the graph created from the gear graph G_{m+1} with vertices $\{u_0;\ u_1, u_2, \ldots, u_m;\ w_1, w_2, \ldots, w_m\}$ with double internal and external edges and m sets of vertices, say, $\{v_1^1, v_2^1, \ldots, v_n^1\}$, $\{v_1^2, v_2^2, \ldots, v_n^2\}, \ldots, \{v_1^m, v_2^m, \ldots, v_n^m\}$, such that for all $i = 1, 2, \ldots, n$ the vertex v_i^j is adjacent to u_j and u_{j+1}, where $j = 1, 2, \ldots, m-1$, and v_i^m is adjacent to u_1 and u_m. See Figure 4.*

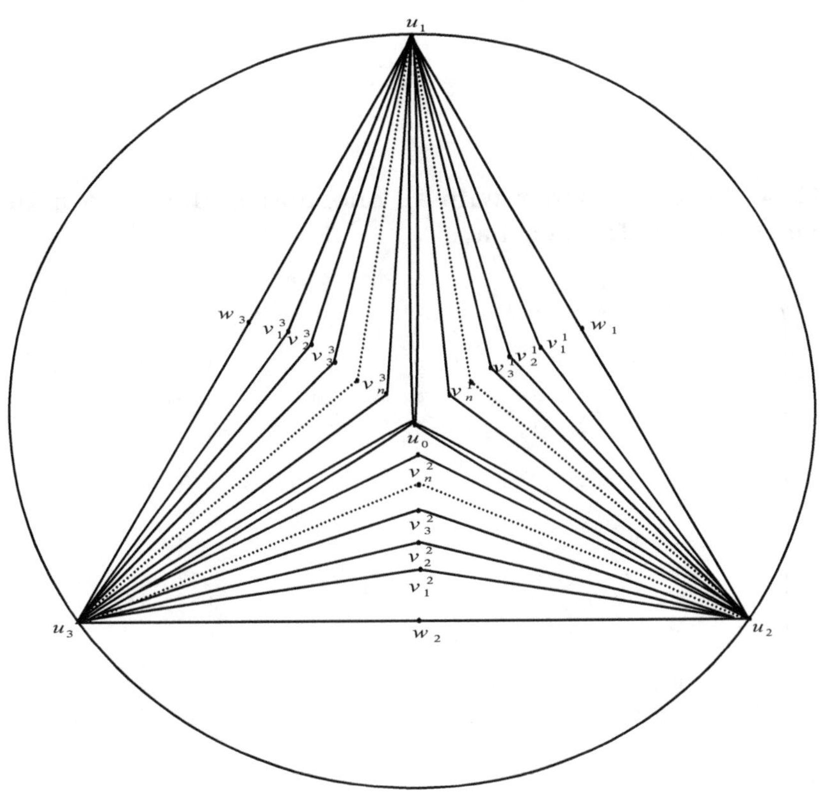

Figure 4. The pyramid graph $D_n^{(3)}$.

Theorem 4. *For* $n \geq 0$, $m \geq 3$, $\tau(D_n^{(m)}) = 2^{mn}[(n+5+2\sqrt{n+4}\,)^m + (n+5-2\sqrt{n+4}\,)^m - 2(n+3)^m]$.

Proof. Applying Lemma 1, we have:

$$\tau(D_n^{(m)}) = \frac{1}{(mn+2m+1)^2} \times \det((mn+2m+1)\,I - D^c + A^c) = \frac{1}{(mn+2m+1)^2} \times$$

$$\det\left(\begin{array}{cccccccccccccccccccccccccccccc}
(2m+1) & -1 & -1 & \cdots & \cdots & \cdots & -1 & 1 & \cdots & \cdots & \cdots & \cdots & 1 & 1 & \cdots & \cdots & \cdots & \cdots & \cdots & \cdots & \cdots & \cdots & \cdots & \cdots & \cdots & \cdots & \cdots & \cdots & \cdots & 1 \\
-1 & (2n+7) & 0 & 1 & \cdots & 1 & 0 & 0 & 1 & \cdots & \cdots & 1 & 0 & 0 & \cdots & 0 & 1 & \cdots & 1 & \cdots & \cdots & 1 & \cdots & 1 & 1 & \cdots & 1 & 0 & \cdots & 0 \\
\vdots & 0 & \ddots & \ddots & \ddots & \ddots & 1 & 0 & \ddots & \ddots & \ddots & \ddots & 1 & 0 & \cdots & 0 & 0 & \cdots & 0 & 1 & \cdots & 1 & \cdots & \cdots & 1 & \cdots & 1 & 1 & \cdots & 1 \\
\vdots & 1 & \ddots & \ddots & \ddots & \ddots & \vdots & 1 & \ddots & \ddots & \ddots & \vdots & \vdots & 1 & \cdots & 1 & 0 & \cdots & 0 & 0 & \cdots & 0 & 1 & \cdots & 1 & \cdots & \cdots & 1 & \cdots & 1 \\
\vdots & \vdots & \ddots & \ddots & \ddots & \ddots & 1 & \vdots & \ddots & \ddots & \ddots & \vdots & \vdots & 1 & \cdots & 1 & 1 & \cdots & 1 & 0 & \cdots & 0 & 0 & \cdots & 0 & 1 & \cdots & 1 & \cdots & \cdots \\
\vdots & 1 & \ddots & \ddots & \ddots & \ddots & 0 & \vdots & \ddots & \ddots & \ddots & \ddots & 1 & \vdots & \cdots & \cdots & \cdots & \cdots & \cdots & \cdots & \cdots & \cdots & \cdots & \cdots & \cdots & \cdots & \cdots & \cdots & \cdots & \vdots \\
-1 & 0 & 1 & \cdots & 1 & 0 & (2n+7) & 1 & \cdots & \cdots & 1 & 0 & 0 & 1 & \cdots & 1 & 1 & \cdots & 1 & \cdots & \cdots & 1 & \cdots & 1 & 0 & \cdots & 0 & 0 & \cdots & 0 \\
1 & 0 & 0 & 1 & \cdots & \cdots & 1 & 3 & 1 & \cdots & \cdots & \cdots & 1 & 1 & \cdots & 1 & 1 & \cdots & 1 & \cdots & \cdots & \cdots & \cdots & \cdots & 1 & \cdots & 1 & 1 & \cdots & 1 \\
\vdots & 1 & \ddots & \ddots & \ddots & \ddots & \vdots & 1 & \ddots & \ddots & \ddots & \ddots & \vdots & 1 & \cdots & 1 & 1 & \cdots & 1 & \cdots & \cdots & \cdots & \cdots & \cdots & 1 & \cdots & 1 & 1 & \cdots & 1 \\
\vdots & \vdots & \ddots & \ddots & \ddots & \ddots & \vdots & \vdots & \ddots & \ddots & \ddots & \ddots & \vdots & \vdots & \cdots & \vdots & \vdots & \cdots & \vdots & \cdots & \cdots & \cdots & \cdots & \cdots & \vdots & \cdots & \vdots & \vdots & \cdots & 1 \\
\vdots & \vdots & \ddots & \ddots & \ddots & \ddots & 1 & \vdots & \ddots & \ddots & \ddots & \ddots & \vdots & \vdots & \cdots & \vdots & \vdots & \cdots & \vdots & \cdots & \cdots & \cdots & \cdots & \cdots & \vdots & \cdots & \vdots & \vdots & \cdots & \vdots \\
1 & 1 & \ddots & \ddots & \ddots & \ddots & 0 & \vdots & \ddots & \ddots & \ddots & \ddots & \vdots & 1 & \cdots & 1 & 1 & \cdots & 1 & \cdots & \cdots & \cdots & \cdots & \cdots & 1 & \cdots & 1 & 1 & \cdots & 1 \\
1 & 0 & 1 & \cdots & \cdots & 1 & 0 & 1 & \cdots & \cdots & \cdots & 1 & 3 & 1 & \cdots & 1 & 1 & \cdots & 1 & \cdots & \cdots & \cdots & \cdots & \cdots & 1 & \cdots & 1 & 1 & \cdots & 1 \\
\vdots & 0 & 0 & 1 & \cdots & 1 & 0 & 1 & 1 & \cdots & \cdots & 1 & 1 & 3 & 1 & \cdots & \cdots & \cdots & \cdots & \cdots & \cdots & \cdots & \cdots & \cdots & \cdots & \cdots & \cdots & \cdots & \cdots & 1 \\
\vdots & \vdots & \vdots & \vdots & \vdots & \vdots & \vdots & \vdots & \vdots & \vdots & \vdots & \vdots & \vdots & 1 & \ddots & \ddots & \ddots & \ddots & \ddots & \ddots & \ddots & \ddots & \ddots & \ddots & \ddots & \ddots & \ddots & \ddots & \ddots & \vdots \\
\vdots & 0 & 0 & 1 & \vdots & 1 & \vdots & 1 & 1 & \cdots & \cdots & 1 & 1 & \vdots & \ddots & \ddots & \ddots & \ddots & \ddots & \ddots & \ddots & \ddots & \ddots & \ddots & \ddots & \ddots & \ddots & \ddots & \ddots & \vdots \\
\vdots & 1 & 0 & 0 & \vdots & 1 & \vdots & 1 & 1 & \cdots & \cdots & 1 & 1 & \vdots & \ddots & \ddots & \ddots & \ddots & \ddots & \ddots & \ddots & \ddots & \ddots & \ddots & \ddots & \ddots & \ddots & \ddots & \ddots & \vdots \\
\vdots & \vdots & \vdots & \vdots & \vdots & \vdots & \vdots & \vdots & \vdots & \vdots & \vdots & \vdots & \vdots & \vdots & \ddots & \ddots & \ddots & \ddots & \ddots & \ddots & \ddots & \ddots & \ddots & \ddots & \ddots & \ddots & \ddots & \ddots & \ddots & \vdots \\
\vdots & 1 & 0 & 0 & \vdots & 1 & \vdots & 1 & 1 & \cdots & \cdots & 1 & 1 & \vdots & \ddots & \ddots & \ddots & \ddots & \ddots & \ddots & \ddots & \ddots & \ddots & \ddots & \ddots & \ddots & \ddots & \ddots & \ddots & \vdots \\
\vdots & 1 & 1 & 0 & \vdots & \vdots & \vdots & \vdots & \vdots & \vdots & \vdots & \vdots & \vdots & \vdots & \ddots & \ddots & \ddots & \ddots & \ddots & \ddots & \ddots & \ddots & \ddots & \ddots & \ddots & \ddots & \ddots & \ddots & \ddots & \vdots \\
\vdots & \vdots & \vdots & \vdots & \vdots & \vdots & \vdots & \vdots & \vdots & \vdots & \vdots & \vdots & \vdots & \vdots & \ddots & \ddots & \ddots & \ddots & \ddots & \ddots & \ddots & \ddots & \ddots & \ddots & \ddots & \ddots & \ddots & \ddots & \ddots & \vdots \\
\vdots & 1 & 1 & 0 & \vdots & 1 & \vdots & \vdots & \vdots & \vdots & \vdots & \vdots & \vdots & \vdots & \ddots & \ddots & \ddots & \ddots & \ddots & \ddots & \ddots & \ddots & \ddots & \ddots & \ddots & \ddots & \ddots & \ddots & \ddots & \vdots \\
\vdots & \vdots & \vdots & 1 & \vdots & \vdots & \vdots & \vdots & \vdots & \vdots & \vdots & \vdots & \vdots & \vdots & \ddots & \ddots & \ddots & \ddots & \ddots & \ddots & \ddots & \ddots & \ddots & \ddots & \ddots & \ddots & \ddots & \ddots & \ddots & \vdots \\
\vdots & \vdots & \vdots & \vdots & \vdots & 1 & \vdots & \vdots & \vdots & \vdots & \vdots & \vdots & \vdots & \vdots & \ddots & \ddots & \ddots & \ddots & \ddots & \ddots & \ddots & \ddots & \ddots & \ddots & \ddots & \ddots & \ddots & \ddots & \ddots & \vdots \\
\vdots & 1 & 1 & 1 & \vdots & 0 & \vdots & \vdots & \vdots & \vdots & \vdots & \vdots & \vdots & \vdots & \ddots & \ddots & \ddots & \ddots & \ddots & \ddots & \ddots & \ddots & \ddots & \ddots & \ddots & \ddots & \ddots & \ddots & \ddots & \vdots \\
\vdots & \vdots & \vdots & \vdots & \vdots & \vdots & \vdots & \vdots & \vdots & \vdots & \vdots & \vdots & \vdots & \vdots & \ddots & \ddots & \ddots & \ddots & \ddots & \ddots & \ddots & \ddots & \ddots & \ddots & \ddots & \ddots & \ddots & \ddots & \ddots & \vdots \\
\vdots & 1 & 1 & \vdots & \vdots & 0 & \vdots & 1 & 1 & \cdots & \cdots & 1 & 1 & \vdots & \ddots & \ddots & \ddots & \ddots & \ddots & \ddots & \ddots & \ddots & \ddots & \ddots & \ddots & \ddots & \ddots & \ddots & \ddots & \vdots \\
\vdots & 0 & 1 & 1 & \vdots & 0 & \vdots & 1 & 1 & \cdots & \cdots & 1 & 1 & \vdots & \ddots & \ddots & \ddots & \ddots & \ddots & \ddots & \ddots & \ddots & \ddots & \ddots & \ddots & \ddots & \ddots & \ddots & \ddots & \vdots \\
\vdots & \vdots & \vdots & \vdots & \vdots & \vdots & \vdots & \vdots & \vdots & \vdots & \vdots & \vdots & \vdots & \vdots & \ddots & \ddots & \ddots & \ddots & \ddots & \ddots & \ddots & \ddots & \ddots & \ddots & \ddots & \ddots & \ddots & \ddots & \ddots & 1 \\
1 & 0 & 1 & 1 & \cdots & 0 & \cdots & 1 & 1 & \cdots & \cdots & 1 & 1 & 1 & \cdots & \cdots & \cdots & \cdots & \cdots & \cdots & \cdots & \cdots & \cdots & \cdots & \cdots & \cdots & \cdots & \cdots & 1 & 3
\end{array}\right)$$

Let $j = (1 \cdots 1)$ be the $1 \times n$ matrix with all one, and J_n the $n \times n$ matrix with all one. Set $a = 2n+7$ and $b = mn+2m+1$. Then we have:

$$\tau\left(D_n^{(m)}\right) = \frac{1}{b^2}\det\left(\begin{array}{ccccccccccccccccccc}
2m+1 & -1 & \cdots & \cdots & \cdots & \cdots & -1 & 1 & \cdots & \cdots & \cdots & \cdots & 1 & j & \cdots & \cdots & \cdots & \cdots & j \\
-1 & a & 0 & 1 & \cdots & 1 & 0 & 0 & 1 & \cdots & \cdots & 1 & 0 & 0 & j & \cdots & \cdots & j & 0 \\
\vdots & 0 & \ddots & \ddots & \ddots & \ddots & 1 & 0 & 0 & \ddots & \ddots & \ddots & 1 & 0 & \ddots & \ddots & \ddots & \ddots & j \\
\vdots & 1 & \ddots & \ddots & \ddots & \ddots & \vdots & 1 & 0 & \ddots & \ddots & \ddots & \vdots & j & \ddots & \ddots & \ddots & \ddots & \vdots \\
\vdots & \vdots & \ddots & \ddots & \ddots & \ddots & 1 & \vdots & \ddots & \ddots & \ddots & \ddots & \vdots & \vdots & \ddots & \ddots & \ddots & \ddots & \vdots \\
\vdots & 1 & \ddots & \ddots & \ddots & \ddots & 0 & \vdots & \ddots & \ddots & 0 & 0 & 1 & \vdots & \ddots & \ddots & \ddots & \ddots & j \\
-1 & 0 & 1 & \cdots & 1 & 0 & a & 1 & \cdots & \cdots & 1 & 0 & 0 & j & \cdots & \cdots & j & 0 & 0 \\
1 & 0 & 0 & 1 & \cdots & \cdots & 1 & 3 & 1 & \cdots & \cdots & \cdots & 1 & j & \cdots & \cdots & \cdots & \cdots & j \\
\vdots & 1 & 0 & \ddots & \ddots & \ddots & 1 & 1 & \ddots & \ddots & \ddots & \ddots & \vdots & \vdots & \ddots & \ddots & \ddots & \ddots & \vdots \\
\vdots & \vdots & \ddots & \ddots & \ddots & \ddots & \vdots & \vdots & \ddots & \ddots & \ddots & \ddots & \vdots & \vdots & \ddots & \ddots & \ddots & \ddots & \vdots \\
\vdots & \vdots & \ddots & \ddots & \ddots & \ddots & 1 & \vdots & \ddots & \ddots & \ddots & \ddots & \vdots & \vdots & \ddots & \ddots & \ddots & \ddots & \vdots \\
\vdots & 1 & \ddots & \ddots & \ddots & \ddots & 0 & \vdots & \ddots & \ddots & \ddots & \ddots & 1 & \vdots & \ddots & \ddots & \ddots & \ddots & \vdots \\
1 & 0 & 1 & \cdots & \cdots & 1 & 0 & 1 & \cdots & \cdots & \cdots & 1 & 3 & j & \cdots & \cdots & \cdots & \cdots & j \\
j^t & 0 & 0 & j^t & \cdots & \cdots & j^t & j^t & \cdots & \cdots & \cdots & \cdots & j^t & & & & & & \\
\vdots & j^t & 0 & \ddots & \ddots & \ddots & \vdots & \vdots & \ddots & \ddots & \ddots & \ddots & \vdots & & & & & & \\
\vdots & \vdots & \ddots & \ddots & \ddots & \ddots & \vdots & \vdots & \ddots & \ddots & \ddots & \ddots & \vdots & & & 2I_{mn}+J_{mn} & & & \\
\vdots & \vdots & \ddots & \ddots & \ddots & \ddots & j^t & \vdots & \ddots & \ddots & \ddots & \ddots & \vdots & & & & & & \\
\vdots & j^t & \ddots & \ddots & \ddots & \ddots & 0 & \vdots & \ddots & \ddots & \ddots & \ddots & \vdots & & & & & & \\
j^t & 0 & j^t & \cdots & \cdots & j^t & 0 & j^t & \cdots & \cdots & \cdots & \cdots & j^t & & & & & &
\end{array}\right)$$

$$
= \frac{1}{b^2}\det\left(\begin{array}{ccccccccccccccccccc}
b & -1 & \cdots & \cdots & \cdots & \cdots & -1 & 1 & \cdots & \cdots & \cdots & \cdots & 1 & j & \cdots & \cdots & \cdots & \cdots & j \\
b & a & 0 & 1 & \cdots & 1 & 0 & 0 & 1 & \cdots & \cdots & 1 & 0 & 0 & j & \cdots & \cdots & j & 0 \\
\vdots & 0 & \ddots & \ddots & \ddots & \ddots & 1 & 0 & 0 & \ddots & \ddots & \ddots & 1 & 0 & \ddots & \ddots & \ddots & \ddots & j \\
\vdots & 1 & \ddots & \ddots & \ddots & \ddots & \vdots & 1 & 0 & \ddots & \ddots & \ddots & \vdots & j & \ddots & \ddots & \ddots & \ddots & \vdots \\
\vdots & \vdots & \ddots & \ddots & \ddots & \ddots & 1 & \vdots & \ddots & \ddots & \ddots & \ddots & \vdots & \vdots & \ddots & \ddots & \ddots & \ddots & \vdots \\
\vdots & 1 & \ddots & \ddots & \ddots & \ddots & 0 & \vdots & \ddots & \ddots & 0 & 0 & 1 & \vdots & \ddots & \ddots & \ddots & \ddots & j \\
b & 0 & 1 & \cdots & 1 & 0 & a & 1 & \cdots & \cdots & 1 & 0 & 0 & j & \cdots & \cdots & j & 0 & 0 \\
b & 0 & 0 & 1 & \cdots & \cdots & 1 & 3 & 1 & \cdots & \cdots & \cdots & 1 & j & \cdots & \cdots & \cdots & \cdots & j \\
\vdots & 1 & 0 & \ddots & \ddots & \ddots & 1 & 1 & \ddots & \ddots & \ddots & \ddots & \vdots & \vdots & \ddots & \ddots & \ddots & \ddots & \vdots \\
\vdots & \vdots & \ddots & \ddots & \ddots & \ddots & \vdots & \vdots & \ddots & \ddots & \ddots & \ddots & \vdots & \vdots & \ddots & \ddots & \ddots & \ddots & \vdots \\
\vdots & \vdots & \ddots & \ddots & \ddots & \ddots & 1 & \vdots & \ddots & \ddots & \ddots & \ddots & \vdots & \vdots & \ddots & \ddots & \ddots & \ddots & \vdots \\
\vdots & 1 & \ddots & \ddots & \ddots & \ddots & 0 & \vdots & \ddots & \ddots & \ddots & \ddots & 1 & \vdots & \ddots & \ddots & \ddots & \ddots & \vdots \\
b & 0 & 1 & \cdots & \cdots & 1 & 0 & 1 & \cdots & \cdots & \cdots & 1 & 3 & j & \cdots & \cdots & \cdots & \cdots & j \\
bj^t & 0 & 0 & j^t & \cdots & \cdots & j^t & j^t & \cdots & \cdots & \cdots & \cdots & j^t & & & & & & \\
\vdots & j^t & 0 & \ddots & \ddots & \ddots & \vdots & \vdots & \ddots & \ddots & \ddots & \ddots & \vdots & & & & & & \\
\vdots & \vdots & \ddots & \ddots & \ddots & \ddots & \vdots & \vdots & \ddots & \ddots & \ddots & \ddots & \vdots & & & 2I_{mn}+J_{mn} & & & \\
\vdots & \vdots & \ddots & \ddots & \ddots & \ddots & j^t & \vdots & \ddots & \ddots & \ddots & \ddots & \vdots & & & & & & \\
\vdots & j^t & \ddots & \ddots & \ddots & \ddots & 0 & \vdots & \ddots & \ddots & \ddots & \ddots & \vdots & & & & & & \\
bj^t & 0 & j^t & \cdots & \cdots & j^t & 0 & j^t & \cdots & \cdots & \cdots & \cdots & j^t & & & & & &
\end{array}\right)
$$

$$
= \frac{1}{b}\det\left(\begin{array}{cccccccccccccccccccc}
 & 1 & -1 & \cdots & \cdots & \cdots & \cdots & -1 & 1 & \cdots & \cdots & \cdots & \cdots & 1 & j & \cdots & \cdots & \cdots & \cdots & j \\
 & 1 & a & 0 & 1 & \cdots & 1 & 0 & 0 & 1 & \cdots & \cdots & 1 & 0 & 0 & j & \cdots & \cdots & j & 0 \\
 & \vdots & 0 & \ddots & \ddots & \ddots & \ddots & 1 & 0 & 0 & \ddots & \ddots & \ddots & 1 & 0 & \ddots & \ddots & \ddots & \ddots & j \\
 & \vdots & 1 & \ddots & \ddots & \ddots & \ddots & \vdots & 1 & 0 & \ddots & \ddots & \ddots & \vdots & j & \ddots & \ddots & \ddots & \ddots & \vdots \\
 & \vdots & \vdots & \ddots & \ddots & \ddots & \ddots & 1 & \vdots & \ddots & \ddots & \ddots & \ddots & \vdots & \vdots & \ddots & \ddots & \ddots & \ddots & \vdots \\
 & \vdots & 1 & \ddots & \ddots & \ddots & \ddots & 0 & \vdots & \ddots & \ddots & 0 & 0 & 1 & \vdots & \ddots & \ddots & \ddots & \ddots & j \\
 & 1 & 0 & 1 & \cdots & 1 & 0 & a & 1 & \cdots & \cdots & 1 & 0 & 0 & j & \cdots & \cdots & j & 0 & 0 \\
 & 1 & 0 & 0 & 1 & \cdots & \cdots & 1 & 3 & 1 & \cdots & \cdots & \cdots & 1 & j & \cdots & \cdots & \cdots & \cdots & j \\
 & \vdots & 1 & 0 & \ddots & \ddots & \ddots & 1 & 1 & \ddots & \ddots & \ddots & \ddots & \vdots & \vdots & \ddots & \ddots & \ddots & \ddots & \vdots \\
\ddots & \vdots & \vdots & \ddots & \ddots & \ddots & \ddots & \vdots & \vdots & \ddots & \ddots & \ddots & \ddots & \vdots & \vdots & \ddots & \ddots & \ddots & \ddots & \vdots \\
 & \vdots & \vdots & \ddots & \ddots & \ddots & \ddots & 1 & \vdots & \ddots & \ddots & \ddots & \ddots & \vdots & \vdots & \ddots & \ddots & \ddots & \ddots & \vdots \\
 & \vdots & 1 & \ddots & \ddots & \ddots & \ddots & 0 & \vdots & \ddots & \ddots & \ddots & \ddots & 1 & \vdots & \ddots & \ddots & \ddots & \ddots & \vdots \\
 & 1 & 0 & 1 & \cdots & \cdots & 1 & 0 & 1 & \cdots & \cdots & \cdots & 1 & 3 & j & \cdots & \cdots & \cdots & \cdots & j \\
 & 1j^t & 0 & 0 & j^t & \cdots & \cdots & j^t & j^t & \cdots & \cdots & \cdots & \cdots & j^t & & & & & & \\
 & \vdots & j^t & 0 & \ddots & \ddots & \ddots & \vdots & \vdots & \ddots & \ddots & \ddots & \ddots & \vdots & & & & & & \\
 & \vdots & \vdots & \ddots & \ddots & \ddots & \ddots & \vdots & \vdots & \ddots & \ddots & \ddots & \ddots & \vdots & & & 2I_{mn}+J_{mn} & & & \\
 & \vdots & \vdots & \ddots & \ddots & \ddots & \ddots & j^t & \vdots & \ddots & \ddots & \ddots & \ddots & \vdots & & & & & & \\
 & \vdots & j^t & \ddots & \ddots & \ddots & \ddots & 0 & \vdots & \ddots & \ddots & \ddots & \ddots & \vdots & & & & & & \\
 & 1j^t & 0 & j^t & \cdots & \cdots & j^t & 0 & j^t & \cdots & \cdots & \cdots & \cdots & j^t & & & & & &
\end{array}\right)
$$

$$
= \frac{1}{b}\det\left(\begin{array}{ccccccccccccccccccc}
1 & -1 & \cdots & \cdots & \cdots & \cdots & -1 & 1 & \cdots & \cdots & \cdots & \cdots & 1 & j & \cdots & \cdots & \cdots & \cdots & j \\
0 & (a+1) & 1 & 2 & \cdots & 2 & 1 & -1 & 0 & \cdots & \cdots & 0 & -1 & -j & 0 & \cdots & \cdots & 0 & -j \\
\vdots & 1 & \ddots & \ddots & \ddots & \ddots & 2 & -1 & -1 & \ddots & \ddots & \ddots & 0 & -j & \ddots & \ddots & \ddots & \ddots & 0 \\
\vdots & 2 & \ddots & \ddots & \ddots & \ddots & \vdots & 0 & \ddots & \ddots & \ddots & \ddots & \vdots & 0 & \ddots & \ddots & \ddots & \ddots & \vdots \\
\vdots & \vdots & \ddots & \ddots & \ddots & \ddots & 2 & \vdots & \ddots & \ddots & \ddots & \ddots & \vdots & \vdots & \ddots & \ddots & \ddots & \ddots & \vdots \\
\vdots & 2 & \ddots & \ddots & \ddots & \ddots & 1 & \vdots & \ddots & \ddots & \ddots & -1 & 0 & \vdots & \ddots & \ddots & \ddots & \ddots & 0 \\
0 & 1 & 2 & \cdots & 2 & 1 & (a+1) & 0 & \cdots & \cdots & 0 & -1 & -1 & 0 & \cdots & \cdots & 0 & -j & -j \\
0 & 1 & 1 & 2 & \cdots & \cdots & 2 & 2 & 0 & \cdots & \cdots & \cdots & 0 & 0 & \cdots & \cdots & \cdots & \cdots & 0 \\
\vdots & 2 & 1 & \ddots & \ddots & \ddots & 2 & 0 & \ddots & \ddots & \ddots & \ddots & \vdots & \vdots & \ddots & \ddots & \ddots & \ddots & \vdots \\
\vdots & \vdots & \ddots & \ddots & \ddots & \ddots & \vdots & \vdots & \ddots & \ddots & \ddots & \ddots & \vdots & \vdots & \ddots & \ddots & \ddots & \ddots & \vdots \\
\vdots & \vdots & \ddots & \ddots & \ddots & \ddots & 2 & \vdots & \ddots & \ddots & \ddots & \ddots & \vdots & \vdots & \ddots & \ddots & \ddots & \ddots & \vdots \\
\vdots & 2 & \ddots & \ddots & \ddots & \ddots & 1 & \vdots & \ddots & \ddots & \ddots & \ddots & 0 & \vdots & \ddots & \ddots & \ddots & \ddots & \vdots \\
0 & 1 & 2 & \cdots & \cdots & 2 & 1 & 0 & \cdots & \cdots & \cdots & 0 & 2 & 0 & \cdots & \cdots & \cdots & \cdots & 0 \\
0 & j^t & j^t & 2j^t & \cdots & \cdots & 2j^t & 0 & \cdots & \cdots & \cdots & \cdots & 0 & & & & & & \\
\vdots & 2j^t & j^t & \ddots & \ddots & \ddots & \vdots & \vdots & \ddots & \ddots & \ddots & \ddots & \vdots & & & & & & \\
\vdots & \vdots & \ddots & \ddots & \ddots & \ddots & \vdots & \vdots & \ddots & \ddots & \ddots & \ddots & \vdots & & & 2I_{mn} & & & \\
\vdots & \vdots & \ddots & \ddots & \ddots & \ddots & 2j^t & \vdots & \ddots & \ddots & \ddots & \ddots & \vdots & & & & & & \\
\vdots & 2j^t & \ddots & \ddots & \ddots & \ddots & j^t & \vdots & \ddots & \ddots & \ddots & \ddots & \vdots & & & & & & \\
0 & j^t & 2j^t & \cdots & \cdots & 2j^t & j^t & 0 & \cdots & \cdots & \cdots & \cdots & 0 & & & & & &
\end{array}\right)
$$

$$
= \tfrac{1}{b}\det\begin{pmatrix}
(a+1) & 1 & 2 & \cdots & 2 & 1 & -1 & 0 & \cdots & \cdots & 0 & -1 & -j & 0 & \cdots & \cdots & 0 & -j \\
1 & \ddots & \ddots & \ddots & \ddots & 2 & -1 & -1 & \ddots & \ddots & \ddots & 0 & -j & \ddots & \ddots & \ddots & \ddots & 0 \\
2 & \ddots & \ddots & \ddots & \ddots & \vdots & 0 & \ddots & \ddots & \ddots & \ddots & \vdots & 0 & \ddots & \ddots & \ddots & \ddots & \vdots \\
\vdots & \ddots & \ddots & \ddots & \ddots & 2 & \vdots & \ddots & \ddots & \ddots & \ddots & \vdots & \vdots & \ddots & \ddots & \ddots & \ddots & \vdots \\
2 & \ddots & \ddots & \ddots & \ddots & 1 & \vdots & \ddots & \ddots & \ddots & -1 & 0 & \vdots & \ddots & \ddots & \ddots & \ddots & 0 \\
1 & 2 & \cdots & 2 & 1 & (a+1) & 0 & \cdots & \cdots & 0 & -1 & -1 & 0 & \cdots & \cdots & 0 & -j & -j \\
1 & 1 & 2 & \cdots & \cdots & 2 & 2 & 0 & \cdots & \cdots & \cdots & 0 & 0 & \cdots & \cdots & \cdots & \cdots & 0 \\
2 & 1 & \ddots & \ddots & \ddots & 2 & 0 & \ddots & \ddots & \ddots & \ddots & \vdots & \vdots & \ddots & \ddots & \ddots & \ddots & \vdots \\
\vdots & \ddots & \ddots & \ddots & \ddots & \vdots & \vdots & \ddots & \ddots & \ddots & \ddots & \vdots & \vdots & \ddots & \ddots & \ddots & \ddots & \vdots \\
\vdots & \ddots & \ddots & \ddots & \ddots & 2 & \vdots & \ddots & \ddots & \ddots & \ddots & \vdots & \vdots & \ddots & \ddots & \ddots & \ddots & \vdots \\
2 & \ddots & \ddots & \ddots & \ddots & 1 & \vdots & \ddots & \ddots & \ddots & \ddots & 0 & \vdots & \ddots & \ddots & \ddots & \ddots & \vdots \\
1 & 2 & \cdots & \cdots & 2 & 1 & 0 & \cdots & \cdots & \cdots & 0 & 2 & 0 & \cdots & \cdots & \cdots & \cdots & 0 \\
j^t & j^t & 2j^t & \cdots & \cdots & 2j^t & 0 & \cdots & \cdots & \cdots & \cdots & 0 & & & & & & \\
2j^t & j^t & \ddots & \ddots & \ddots & \vdots & \vdots & \ddots & \ddots & \ddots & \ddots & \vdots & & & & & & \\
\vdots & \ddots & \ddots & \ddots & \ddots & \vdots & \vdots & \ddots & \ddots & \ddots & \ddots & \vdots & & & & 2I_{mn} & & \\
\vdots & \ddots & \ddots & \ddots & \ddots & 2j^t & \vdots & \ddots & \ddots & \ddots & \ddots & \vdots & & & & & & \\
2j^t & \ddots & \ddots & \ddots & \ddots & j^t & \vdots & \ddots & \ddots & \ddots & \ddots & \vdots & & & & & & \\
j^t & 2j^t & \cdots & \cdots & 2j^t & j^t & 0 & \cdots & \cdots & \cdots & \cdots & 0 & & & & & &
\end{pmatrix}
$$

Using Lemma 3, yields

$$
\tau\,(D_n^{(m)}) = \tfrac{1}{b} \times \det\begin{pmatrix} A & B \\ C & 2I_{mn} \end{pmatrix} = \tfrac{1}{b} \times \det(A - B\tfrac{1}{2I_{mn}}\ C) \times 2^{mn}
$$

$$
= \tfrac{1}{b}2^{mn} \times 2^{-2m} \times \det\begin{pmatrix}
(2a+2n+2) & 3n+2 & 4(n+1) & \cdots & 4(n+1) & 3n+2 & -2 & 0 & \cdots & \cdots & 0 & -2 \\
3n+2 & (2a+2n+2) & 3n+2 & 4(n+1) & \cdots & 4(n+1) & -2 & \ddots & \ddots & \ddots & \ddots & 0 \\
4(n+1) & 3n+4 & \ddots & \ddots & \ddots & \vdots & 0 & \ddots & \ddots & \ddots & \ddots & \vdots \\
\vdots & \ddots & \ddots & \ddots & \ddots & 4(n+1) & \vdots & \ddots & \ddots & \ddots & \ddots & \vdots \\
4(n+1) & \ddots & \ddots & \ddots & \ddots & 3n+2 & \vdots & \ddots & \ddots & \ddots & \ddots & 0 \\
3n+2 & 4(n+1) & \cdots & 4(n+1) & 3n+2 & (2a+2n+2) & 0 & \cdots & \cdots & 0 & -2 & -2 \\
2 & 2 & 4 & \cdots & \cdots & 4 & 4 & 0 & \cdots & \cdots & \cdots & 0 \\
4 & \ddots & \ddots & \ddots & \ddots & \vdots & 0 & \ddots & \ddots & \ddots & \ddots & \vdots \\
\vdots & \ddots & \ddots & \ddots & \ddots & \vdots & \vdots & \ddots & \ddots & \ddots & \ddots & \vdots \\
\vdots & \ddots & \ddots & \ddots & \ddots & 4 & \vdots & \ddots & \ddots & \ddots & \ddots & \vdots \\
4 & \ddots & \ddots & \ddots & \ddots & 2 & \vdots & \ddots & \ddots & \ddots & \ddots & 0 \\
2 & 4 & \cdots & \cdots & 4 & 2 & 0 & \cdots & \cdots & \cdots & 0 & 4
\end{pmatrix}
$$

Using Lemma 3, yields

$$
\tau\,(D_n^{(m)}) = \tfrac{2^{mn-2m}}{b} \times \det\begin{pmatrix} A & B \\ C & 4I_m \end{pmatrix} = \tfrac{2^{mn}}{b} \times \det(A - B\tfrac{1}{4I_m}\ C)
$$

$$
\tau\,(D_n^{(m)}) = \tfrac{2^{mn}}{b} \times \det\begin{pmatrix}
(2a+2n+4) & (3n+5) & 4(n+2) & \cdots & 4(n+2) & (3n+5) \\
(3n+5) & (2a+2n+4) & (3n+5) & \ddots & \cdots & 4(n+2) \\
4(n+2) & \ddots & \ddots & \ddots & \ddots & \vdots \\
\vdots & \ddots & \ddots & \ddots & \ddots & 4(n+2) \\
4(n+2) & \ddots & \ddots & \ddots & \ddots & (3n+5) \\
(3n+5) & 4(n+2) & \cdots & 4(n+2) & (3n+5) & (2a+2n+4)
\end{pmatrix}
$$

Straightforward inducement using properties of determinants, we get:

$$
\tau\,(D_n^{(m)}) = \tfrac{2^{mn}}{b} \times \tfrac{4b}{mn+3m+4} \times \det\begin{pmatrix}
(2a-n-1) & 0 & (n+3) & \cdots & (n+3) & 0 \\
0 & (2a-n-1) & 0 & \ddots & \cdots & (n+3) \\
(n+3) & \ddots & \ddots & \ddots & \ddots & \vdots \\
\vdots & \ddots & \ddots & \ddots & \ddots & (n+3) \\
(n+3) & \ddots & \ddots & \ddots & \ddots & 0 \\
0 & (n+3) & \cdots & (n+3) & 0 & (2a-n-1)
\end{pmatrix}
$$

$$= \frac{2^{mn+2}(n+3)^m}{mn+3m+4} \times \det \begin{pmatrix} \frac{(2a-n-1)}{(n+3)} & 0 & 1 & \cdots & 1 & 0 \\ 0 & \frac{(2a-n-1)}{(n+3)} & 0 & \ddots & \ddots & 1 \\ 1 & \ddots & \ddots & \ddots & \ddots & \vdots \\ \vdots & \ddots & \ddots & \ddots & \ddots & 1 \\ 1 & \ddots & \ddots & \ddots & \ddots & 0 \\ 0 & 1 & \cdots & 1 & 0 & \frac{(2a-n-1)}{(n+3)} \end{pmatrix}$$

Using Lemma 2, yields:

$$\tau(D_n^{(m)}) = 2^{mn+2} \times \frac{(n+3)^m}{mn+3m+4} \times \frac{2(\frac{2a-n-1}{n+3}+m-3)}{\frac{2a-n-1}{n+3}-3} \times [T_m(\frac{\frac{2a-n-1}{n+3}-1}{2}) - 1] = 2^{mn+1} \times (n+3)^m \times [T_m(\frac{n+5}{n+3}) - 1].$$

Using Equation (11), yields the result. □

4. Numerical Results

The following Table 1 illustrates some values of the number of spanning trees of studied pyramid graphs.

Table 1. Some values of the number of spanning trees of studied pyramid graphs.

m	n	$\tau(P_n^{(m)})$	$\tau(A_n^{(m)})$	$\tau(B_n^{(m)})$	$\tau(C_n^{(m)})$
3	0	50	196	242	676
3	1	1024	3200	3136	8192
3	2	15,488	43,264	36,992	92,416
3	3	200,704	524,288	409,600	991,232
3	4	2,367,488	5,914,624	4,333,568	10,240,000
3	5	26,214,400	63,438,848	44,302,336	102,760,448
4	0	192	1152	1792	6400
4	1	11,520	49,152	57,600	184,320
4	2	458,752	1,638,400	1,622,016	4,816,896
4	3	14,745,600	47,185,920	41,746,432	117,440,512
4	4	415,236,096	1,233,125,376	1,006,632,960	2,717,908,992
4	5	10,687,086,592	30,064,771,072	23,102,226,432	60,397,977,600
5	0	722	6724	12,482	58,564
5	1	123,904	739,328	984,064	3,964,928
5	2	12,781,568	59,969,536	65,619,968	237,899,776
5	3	1,007,681,536	4,060,086,272	3,901,751,296	13,088,325,632
5	4	67,194,847,232	243,609,370,624	213,408,284,672	674,448,277,504
5	5	3,995,393,327,104	243,609,370,624	10,953,240,346,624	33,019,708,571,648

5. Conclusions

The number of spanning trees $\tau(G)$ in graphs (networks) is an important invariant. The computation of this number is not only beneficial from a mathematical (computational) standpoint, but it is also an important measure of the reliability of a network and electrical circuit layout. Some computationally laborious problems such as the traveling salesman problem can be resolved approximately by using spanning trees. In this paper, we define some classes of pyramid graphs created by a gear graph and we have studied the problem of computing the number of spanning trees of these graphs.

Author Contributions: All authors contributed equally to this work. Funding Acquisition, J.-B.L.; Methodology, J.-B.L. and S.N.D. Daoud; Writing—Original Draft, J.-B.L. and S.N.D. Daoud; All authors read and approved the final manuscript.

Acknowledgments: The authors are grateful to the anonymous reviewers for their helpful comments and suggestions for improving the original version of the paper.

References

1. Applegate, D.L.; Bixby, R.E.; Chvátal, V.; Cook, W.J. *The Traveling Salesman Problem: A Computational Study*; Princeton University Press: Princeton, NJ, USA, 2006.
2. Cvetkoviĕ, D.; Doob, M.; Sachs, H. *Spectra of Graphs: Theory and Applications*, 3rd ed.; Johann Ambrosius Barth: Heidelberg, Germany, 1995.
3. Kirby, E.C.; Klein, D.J.; Mallion, R.B.; Pollak, P.; Sachs, H. A theorem for counting spanning trees in general chemical graphs and its particular application to toroidal fullerenes. *Croat. Chem. Acta* **2004**, *77*, 263–278.
4. Boesch, F.T.; Satyanarayana, A.; Suffel, C.L. A survey of some network reliability analysis and synthesis results. *Networks* **2009**, *54*, 99–107. [CrossRef]
5. Boesch, F.T. On unreliability polynomials and graph connectivity in reliable network synthesis. *J. Graph Theory* **1986**, *10*, 339–352. [CrossRef]
6. Wu, F.Y. Number of spanning trees on a Lattice. *J. Phys. A* **1977**, *10*, 113–115. [CrossRef]
7. Zhang, F.; Yong, X. Asymptotic enumeration theorems for the number of spanning trees and Eulerian trail in circulant digraphs & graphs. *Sci. China Ser. A* **1999**, *43*, 264–271.
8. Chen, G.; Wu, B.; Zhang, Z. Properties and applications of Laplacian spectra for Koch networks. *J. Phys. A Math. Theor.* **2012**, *45*, 025102.
9. Atajan, T.; Inaba, H. Network reliability analysis by counting the number of spanning trees. In Proceedings of the IEEE International Symposium on Communications and Information Technology, ISCIT 2004, Sapporo, Japan, 26–29 October 2004; pp. 601–604.
10. Brown, T.J.N.; Mallion, R.B.; Pollak, P.; Roth, A. Some methods for counting the spanning trees in labelled molecular graphs, examined in relation to certain fullerenes. *Discret. Appl. Math.* **1996**, *67*, 51–66. [CrossRef]
11. Kirchhoff, G.G. Uber die Auflosung der Gleichungen, auf welche man be ider Untersuchung der Linearen Verteilung galvanischer Storme gefuhrt wird. *Ann. Phys. Chem.* **1847**, *72*, 497–508. [CrossRef]
12. Kelmans, A.K.; Chelnokov, V.M. A certain polynomials of a graph and graphs with an extermal number of trees. *J. Comb. Theory B* **1974**, *16*, 197–214. [CrossRef]
13. Biggs, N.L. *Algebraic Graph Theory*, 2nd ed.; Cambridge University Press: Cambridge, UK, 1993; p. 205.
14. Daoud, S.N. The deletion-contraction method for counting the number of spanning trees of graphs. *Eur. Phys. J. Plus* **2015**, *130*, 217. [CrossRef]
15. Shang, Y. On the number of spanning trees, the Laplacian eigenvalues, and the Laplacian Estrada index of subdivided-line graphs. *Open Math.* **2016**, *14*, 641–648. [CrossRef]
16. Bozkurt, Ş.B.; Bozkurt, D. On the Number of Spanning Trees of Graphs. *Sci. World J.* **2014**, *2014*, 294038. [CrossRef] [PubMed]
17. Daoud, S.N. Number of Spanning Trees in Different Product of Complete and Complete Tripartite Graphs. *ARS Comb.* **2018**, *139*, 85–103.
18. Daoud, S.N. Complexity of Graphs Generated by Wheel Graph and Their Asymptotic Limits. *J. Egypt. Math. Soc.* **2017**, *25*, 424–433. [CrossRef]
19. Daoud, S.N. Chebyshev polynomials and spanning tree formulas. *Int. J. Math. Comb.* **2012**, *4*, 68–79.
20. Zhang, Y.; Yong, X.; Golin, M.J. Chebyshev polynomials and spanning trees formulas for circulant and related graphs. *Discret. Math.* **2005**, *298*, 334–364. [CrossRef]
21. Daoud, S.N. On a class of some pyramid graphs and Chebyshev polynomials. *J. Math. Probl. Eng. Hindawi Publ. Corp.* **2013**, *2013*, 820549.
22. Marcus, M. *A Survey of Matrix Theory and Matrix Inequalities*; University Allyn and Bacon. Inc.: Boston, MA, USA, 1964.

6

Four Operators of Rough Sets Generalized to Matroids and a Matroidal Method for Attribute Reduction

Jingqian Wang [1] and Xiaohong Zhang [2,*]

[1] College of Electrical & Information Engineering, Shaanxi University of Science & Technology, Xi'an 710021, China; wangjingqianw@163.com

[2] School of Arts and Sciences, Shaanxi University of Science & Technology, Xi'an 710021, China

* Correspondence: zhangxiaohong@sust.edu.cn or zxhonghz@263.net

Abstract: Rough sets provide a useful tool for data preprocessing during data mining. However, many algorithms related to some problems in rough sets, such as attribute reduction, are greedy ones. Matroids propose a good platform for greedy algorithms. Therefore, it is important to study the combination between rough sets and matroids. In this paper, we investigate rough sets and matroids through their operators, and provide a matroidal method for attribute reduction in information systems. Firstly, we generalize four operators of rough sets to four operators of matroids through the interior, closure, exterior and boundary axioms, respectively. Thus, there are four matroids induced by these four operators of rough sets. Then, we find that these four matroids are the same one, which implies the relationship about operators between rough sets and matroids. Secondly, a relationship about operations between matroids and rough sets is presented according to the induced matroid. Finally, the girth function of matroids is used to compute attribute reduction in information systems.

Keywords: rough set; matroid; operator; attribute reduction

---

1. Introduction

Rough set theory was proposed by Pawlak [1,2] in 1982 as a mathematical tool to deal with various types of data in data mining. There are many practical problems have been solved by it, such as rule extraction [3,4], attribute reduction [5–7], feature selection [8–10] and knowledge discovery [11]. In Pawlak's rough sets, the relationships of objects are equivalence relations. However, it is well known that this requirement is excessive in practice [12,13]. Hence, Pawlak's rough sets have been extended by relations [14,15], coverings [16–18] and neighborhoods [6,19]. They have been combined with other theories including topology [20], lattice theory [21,22], graph theory [23,24] and fuzzy set theory [25,26].

However, many optimization issues related to rough sets, including attribute reduction, are NP-hard. Therefore, the algorithms to deal with them are often greedy ones [27]. Matroid theory [28–30] is a generalization of graph and linear algebra theories. It has been used in information coding [31] and cryptology [32]. Recently, the combination between rough sets and matroids has attracted many interesting research. For example, Zhu and Wang [33] established a matroidal structure through the upper approximation number and studied generalized rough sets with matroidal approaches. Liu and Zhu [34] established a parametric matroid through the lower approximation operator of rough sets. Li et al. [35,36] used matroidal approaches to investigate rough sets through closure operators. Su and Zhu [37] presented three types of matroidal structures of covering-based rough sets. Wang et al. [38] induced a matroid named 2-circuit matroid by equivalence relations, and equivalently formulated attribute reduction with matroidal approaches. Wang and Zhu used matrix approaches to study the 2-circuit matroid [39], and used contraction operation in matroids to study some relationships between a subset and the upper approximation of this subset in rough sets [40]. Unfortunately, all of these papers never study matroids and rough sets through the positive, negative and boundary operators of rough sets. Thus, it is necessary to further study rough sets and matroids by these operators in this paper. In addition, only Wang et al. [38] presented two equivalent descriptions of attribute reduction by closure operators and rank functions of matroids, respectively. We consider presenting a novel approach to attribute reduction through the girth function of matroids in this paper.

In this paper, we mainly use the positive operator, the negative operator and the boundary operator to study matroids and rough sets, and propose a method to compute attribute reduction in information systems through the girth function of matroids. Firstly, we generalize the positive (the lower approximation operator), upper approximation, negative and boundary operators of rough sets to the interior, closure, exterior and boundary operators of matroids respectively. Among them, the upper and lower approximation operators have been studied in [35]. Thus, there are four matroids induced by these four operators of rough sets. Then, the relationship between these four matroids is studied, which implies the relationship about operators between rough sets and matroids. In fact, these four matroids are the same one. Secondly, a relationship about the restriction operation both in matroids and rough sets is proposed. Finally, a matroidal approach is proposed to compute attribute reduction in information systems through the girth function of matroids, and an example about attribute reduction is solved. Using this matroidal approach, we can compute attribute reduction through their results "2" and "∞".

The rest of this paper is organized as follows. Section 2 recalls some basic notions about rough sets, information systems and matroids. In Section 3, we generalize four operators of rough sets to four operators of matroids, respectively. In addition, we study the relationship between four matroids induced by these four operators of rough sets. Moreover, a relationship about operations between matroids and rough sets is presented. In Section 4, an equivalent formulation of attribute reduction through the girth function is presented. Based on the equivalent formulation, a novel method is proposed to compute attribute reduction in information systems. Finally, Section 5 concludes this paper and indicates further works.

2. Basic Definitions

In this section, we review some notions in Pawlak's rough sets, information systems and matroids.

2.1. Pawlak's Rough Sets and Information Systems

The definition of approximation operators is presented in [1,41].

Let R an equivalence relation on U. For any $X \subseteq U$, a pair of approximation $\overline{R}(X)$ and $\underline{R}(X)$ of X are defined by

$$\overline{R}(X) = \{x \in U : RN(x) \bigcap X \neq \emptyset\},$$
$$\underline{R}(X) = \{x \in U : RN(x) \subseteq X\},$$

where $RN(x) = \{y \in U : xRy\}$. $\overline{R}$ and $\underline{R}$ are called the upper and lower approximation operators with respect to R, respectively.

In this paper, U is a nonempty and finite set called universe. Let $-X$ be the complement of X in U and $\emptyset$ be the empty set. We have the following conclusions about $\overline{R}$ and $\underline{R}$.

Proposition 1. *Refs. [1,41] Let R be an equivalence relation on U. For any $X, Y \subseteq U$,*

(1L) $\underline{R}(U) = U$,
(2L) $\underline{R}(\phi) = \phi$,
(3L) $\underline{R}(X) \subseteq X$,
(4L) $\underline{R}(X \bigcap Y) = \underline{R}(X) \bigcap \underline{R}(Y)$,
(5L) $\underline{R}(\underline{R}(X)) = \underline{R}(X)$,
(6L) $X \subseteq Y \Rightarrow \underline{R}(X) \subseteq \underline{R}(Y)$,
(7L) $\underline{R}(-\underline{R}(X)) = -\underline{R}(X)$,
(8LH) $\underline{R}(-X) = -\overline{R}(X)$,

(1H) $\overline{R}(U) = U$,
(2H) $\overline{R}(\phi) = \phi$,
(3H) $X \subseteq \overline{R}(X)$,
(4H) $\overline{R}(X \bigcup Y) = \overline{R}(X) \bigcup \overline{R}(Y)$,
(5H) $\overline{R}(\overline{R}(X)) = \overline{R}(X)$,
(6H) $X \subseteq Y \Rightarrow \overline{R}(X) \subseteq \overline{R}(Y)$,
(7H) $\overline{R}(-\overline{R}(X)) = -\overline{R}(X)$,
(9LH) $\underline{R}(X) \subseteq \overline{R}(X)$.

On the basis of the upper and lower approximation operators with respect to R, one can define three operators to divide the universe, namely, the negative operator NEG_R, the positive operator POS_R and the boundary operator BND_R:

$$NEG_R(X) = U - \overline{R}(X),$$
$$POS_R(X) = \underline{R}(X),$$
$$BND_R(X) = \overline{R}(X) - \underline{R}(X).$$

An information system [38] is an ordered pair $IS = (U, A)$, where U is a nonempty finite set of objects and A is a nonempty finite set of attributes such that $a : U \to V_a$ for any $a \in A$, where V_a is called the value set of a. For all $B \subseteq A$, the indiscernibility relation induced by B is defined as follows:

$$IND(B) = \{(x, y) \in U \times U : \forall b \in B, b(x) = b(y)\}.$$

Definition 1. *(Reduct [38]) Let $IS = (U, A)$ be an information system. For all $B \subseteq A$, B is called a reduct of IS, if the following two conditions hold:*

(1) $IND(B) \neq IND(B - b)$ for any $b \in B$,
(2) $IND(B) = IND(A)$.

2.2. Matroids

Definition 2. *(Matroid [29,30]) Let U is a finite set, and $\mathbf{I}$ is a nonempty subset of 2^U (the set of all subsets of U). $(U, \mathbf{I})$ is called a matroid, if the following conditions hold:*
(I1) If $I \in \mathbf{I}$ and $I' \subseteq I$, then $I' \in \mathbf{I}$.
(I2) If $I_1, I_2 \in \mathbf{I}$ and $|I_1| < |I_2|$, then there exists $e \in I_2 - I_1$ such that $I_1 \bigcup \{e\} \in \mathbf{I}$, where $|I|$ denotes the cardinality of I.

Let $M = (U, \mathbf{I})$ be a matroid. We shall often write $U(M)$ for U and $\mathbf{I}(M)$ for $\mathbf{I}$, particularly when several matroids are being considered. The members of $\mathbf{I}$ are the independent sets of M.

Example 1. *Let $U = \{a_1, a_2, a_3, a_4, a_5\}$ and $\mathbf{I} = \{\emptyset, \{a_1\}, \{a_2\}, \{a_3\}, \{a_4\}, \{a_5\}, \{a_1, a_3\}, \{a_1, a_4\}, \{a_1, a_5\}, \{a_2, a_3\}, \{a_2, a_4\}, \{a_2, a_5\}, \{a_3, a_4\}, \{a_3, a_5\}, \{a_4, a_5\}, \{a_1, a_3, a_4\}, \{a_1, a_3, a_5\}, \{a_1, a_4, a_5\}, \{a_2, a_3, a_4\}, \{a_2, a_3, a_5\}, \{a_2, a_4, a_5\}\}$. Then, $M = (U, \mathbf{I})$ is a matroid.*

In order to make some expressions brief, some denotations are presented. Let $\mathbf{A} \subseteq 2^U$. Then,

$$Min(\mathbf{A}) = \{X \in \mathbf{A} : \forall Y \in \mathbf{A}, Y \subseteq X \Rightarrow X = Y\},$$
$$Max(\mathbf{A}) = \{X \in \mathbf{A} : \forall Y \in \mathbf{A}, X \subseteq Y \Rightarrow X = Y\},$$
$$Opp(\mathbf{A}) = \{X \subseteq U : X \notin \mathbf{A}\}.$$

The set of all circuits of M is defined as $\mathbf{C}(M) = Min(Opp(\mathbf{I}))$. The rank function r_M of M is denoted by $r_M(X) = max\{|I| : I \subseteq X, I \in \mathbf{I}\}$ for any $X \subseteq U$. $r_M(X)$ is called the rank of X in M. The closure operator cl_M of M is defined as

$$cl_M(X) = \{u \in U : r_M(X) = r_M(X \bigcup \{u\})\} \text{ for all } X \subseteq U.$$

We call $cl_M(X)$ the closure of X in M. X is called a closed set if $cl_M(X) = X$, and we denote the family of all closed sets of M by $\mathbf{F}(M)$. The closure axiom of a matroid is introduced in the following proposition.

Proposition 2. *(Closure axiom [29,30]) Let cl be an operator of U. Then, there exists one and only one matroid M such that $cl = cl_M$ iff cl satisfies the following four conditions:*
$(CL1)$ $X \subseteq cl(X)$ for any $X \subseteq U$;
$(CL2)$ If $X \subseteq Y \subseteq U$, then $cl(X) \subseteq cl(Y)$;
$(CL3)$ $cl(cl(X)) = cl(X)$ for any $X \subseteq U$;
$(CL4)$ For any $x, y \in U$, if $y \in cl(X \bigcup \{x\}) - cl(X)$, then $x \in cl(X \bigcup \{y\})$.

Example 2. *(Continued from Example 1) Let $X = \{a_3, a_4\}$. Then,*
$\mathbf{C}(M) = Min(Opp(\mathbf{I})) = \{\{a_1, a_2\}, \{a_3, a_4, a_5\}\}$,
$r_M(X) = max\{|I| : I \subseteq X, I \in \mathbf{I}\} = 2$,
$cl_M(X) = \{u \in U : r_M(X) = r_M(X \bigcup \{u\})\} = \{a_3, a_4, a_5\}$,
$\mathbf{F}(M) = \{\emptyset, \{a_3\}, \{a_4\}, \{a_5\}, \{a_1, a_2\}, \{a_1, a_2, a_3\}, \{a_1, a_2, a_4\}, \{a_1, a_2, a_5\}, \{a_3, a_4, a_5\}, \{a_1, a_2, a_3, a_4, a_5\}\}$.

Based on $\mathbf{F}(M)$, the interior operator int_M of M is defined as

$$int_M(X) = \bigcup \{Y \subseteq X : U - Y \in \mathbf{F}(M)\} \text{ for any } X \subseteq U.$$

$int_M(X)$ is called the interior of X in M. X is called a open set if $int_M(X) = X$. The following proposition shows the interior axiom of a matroid.

Proposition 3. *(Interior axiom [29,30]) Let int be an operator of U. Then, there exists one and only one matroid M such that $int = int_M$ iff int satisfies the following four conditions:*
$(INT1)$ $int(X) \subseteq X$ for any $X \subseteq U$,
$(INT2)$ If $X \subseteq Y \subseteq U$, then $int(X) \subseteq int(Y)$,
$(INT3)$ $int(int(X)) = int(X)$ for any $X \subseteq U$,
$(INT4)$ For any $x, y \in U$, if $y \in int(X) - int(X - \{x\})$, then $x \notin int(X - \{y\})$.

Example 3. *(Continued from Example 2) $int_M(X) = \bigcup \{Y \subseteq X : U - Y \in \mathbf{F}(M)\} = \{a_3, a_4\}$.*

Based on the closure operator cl_M, the exterior operator ex_M and the boundary operator bo_M of M are defined as

$$ex_M(X) = -cl_M(X) \text{ and } bo_M(X) = cl_M(X) \bigcap cl_M(-X) \text{ for all } X \subseteq U.$$

$ex_M(X)$ is called the exterior of X in M, and $bo_M(X)$ is called the boundary of X in M. The following two propositions present the exterior and boundary axioms, respectively.

Proposition 4. *(Exterior axiom [42]) Let ex be an operator of U. Then, there exists one and only one matroid M such that $ex = ex_M$ iff ex_M satisfies the following four conditions:*
$(EX1)$ $X \bigcap ex(X) = \emptyset$ for any $X \subseteq U$;
$(EX2)$ If $X \subseteq Y \subseteq U$, then $ex(Y) \subseteq ex(X)$;
$(EX3)$ $ex(-ex(X)) = ex(X)$ for any $X \subseteq U$;
$(EX4)$ For any $x, y \in U$, if $y \in ex(X) - ex(X \bigcup \{x\})$, then $x \notin ex(X \bigcup \{y\})$.

Proposition 5. *(Boundary axiom [42]) Let bo be an operator of U. Then, there exists one and only one matroid M such that $bo = bo_M$ iff bo satisfies the following five conditions:*
$(BO1)$ $bo(X) = bo(-X)$ *for any* $X \subseteq U$;
$(BO2)$ $bo(bo(X)) \subseteq bo(X)$ *for any* $X \subseteq U$;
$(BO3)$ $X \bigcap Y \bigcap (bo(X) \bigcup bo(Y)) \subseteq X \bigcap Y \bigcap bo(X \bigcap Y)$ *for any* $X, Y \subseteq U$;
$(BO4)$ *For any* $x, y \in U$, *if* $y \in bo(X \bigcup \{x\}) - bo(X)$, *then* $x \in bo(X \bigcup \{y\})$;
$(BO5)$ $bo(X \bigcup bo(X)) \subseteq bo(X)$ *for any* $X \subseteq U$.

Example 4. *(Continued from Example 2)* $ex_M(X) = U - \{a_3, a_4, a_5\} = \{a_1, a_2\}$, $bo_M(X) = cl_M(X) \bigcap cl_M(-X) = \{a_3, a_4, a_5\} \bigcap \{a_1, a_2, a_5\} = \{a_5\}$.

The following proposition shows some relationships between these above four operators, namely cl_M, int_M, ex_M and bo_M.

Proposition 6. *Ref. [42] Let* $M = (U, \mathbf{I})$ *be a matroid. For all* $X \subseteq U$, *the following statements hold:*
(1) $int_M(X) = -cl_M(-X)$ *and* $cl_M(X) = -int_M(-X)$;
(2) $cl_M(bo_M(X)) = bo_M(X)$;
(3) $bo_M(ex_M(X)) = bo_M(-X)$.

3. The Relationship about Operators between Rough Sets and Matroids

In this section, four matroids are induced by four operators of rough sets. These four matroids are induced by the lower approximation operator $\underline{R}$ (because $\underline{R} = POS_R$, we only consider $\underline{R}$), the upper approximation operator $\overline{R}$, the negative operator NEG_R and the boundary operator BND_R through the interior axiom, the closure axiom, the exterior axiom and the boundary axiom, respectively. Among them, the upper approximation operator $\overline{R}$ has been studied in [35]. Then, the relationship between these four matroids are studied, and we find that these four are the same one. According to this work, we present the relationship about operators between rough sets and matroids.

3.1. Four Matroids Induced by Four Operators of Rough Sets

In this subsection, we generalize the positive operator (the lower approximation operator), the upper approximation operator, the negative operator and the boundary operator of rough sets to the interior operator, the closure operator, the exterior operator and the boundary operator of matroids, respectively. Firstly, the following lemma is proposed.

Lemma 1. *Refs. [1,41] Let R be an equivalence relation on U. For any* $x, y \in U$, *if* $x \in RN(y)$, *then* $y \in RN(x)$.

The following proposition shows that the lower approximation operator $\underline{R}$ satisfies the interior axiom of matroids.

Proposition 7. *Let R be an equivalence relation on U. Then,* $\underline{R}$ *satisfies* $(INT1)$, $(INT2)$, $(INT3)$ *and* $(INT4)$ *of Proposition 3.*

Proof. By $(1L)$, $(6L)$ and $(5L)$ of Proposition 1, $\underline{R}$ satisfies $(INT1)$, $(INT2)$ and $(INT3)$, respectively. $(INT4)$: For any $x, y \in U$, if $y \in \underline{R}(X) - \underline{R}(X - \{x\})$, then $y \in \underline{R}(X)$ but $y \notin \underline{R}(X - \{x\})$. Hence, $RN(y) \subseteq X$ but $RN(y) \nsubseteq X - \{x\}$. Therefore, $x \in RN(y)$. According to Lemma 1, $y \in RN(x)$. Hence, $RN(x) \nsubseteq X - \{y\}$, i.e., $x \notin \underline{R}(X - \{y\})$. □

Inspired by Proposition 7, there is a matroid such that $\underline{R}$ is its interior operator.

Definition 3. *Let R be an equivalence relation on U. The matroid whose interior operator is $\underline{R}$ is denoted by $M(\underline{R})$. We say $M(\underline{R})$ is the matroid induced by $\underline{R}$.*

Corollary 1. *Let R be an equivalence relation on U. Then, $int_{M(\underline{R})} = POS_R$.*

Proof. According to Definition 3, $int_{M(\underline{R})} = \underline{R}$. Since $POS_R = \underline{R}$, so $int_{M(\underline{R})} = POS_R$. □

The upper approximation operator $\overline{R}$ satisfies the closure axiom in [35,38].

Proposition 8. *Refs. [35,38] Let R be an equivalence relation on U. Then, $\overline{R}$ satisfies $(CL1)$, $(CL2)$, $(CL3)$ and $(CL4)$ of Proposition 2.*

Proposition 8 determines the second matroid induced by $\overline{R}$.

Definition 4. *Let R be an equivalence relation on U. The matroid whose closure operator is $\overline{R}$ is denoted by $M(\overline{R})$. We say that $M(\overline{R})$ is the matroid induced by $\overline{R}$.*

The negative operator NEG_R satisfies the exterior axiom.

Proposition 9. *Let R be an equivalence relation on U. Then, NEG_R satisfies $(EX1)$, $(EX2)$, $(EX3)$ and $(EX4)$ of Proposition 4.*

Proof. $(EX1)$: For any $X \subseteq U$, $NEG_R(X) = U - \overline{R}(X)$. According to $(3H)$ of Proposition 1, $X \subseteq \overline{R}(X)$. Therefore, $X \bigcap NEG_R(X) = \emptyset$;
$(EX2)$: According to $(6H)$ of Proposition 1, if $X \subseteq Y \subseteq U$, then $\overline{R}(X) \subseteq \overline{R}(Y)$. Therefore, $U - \overline{R}(Y) \subseteq U - \overline{R}(Y)$, i.e., $NEG_R(Y) \subseteq NEG_R(X)$;
$(EX3)$: For any $X \subseteq U$, $NEG_R(X) = U - \overline{R}(X)$. Hence, $-NEG_R(X) = U - NEG_R(X) = U - (U - \overline{R}(X)) = \overline{R}(X)$. Therefore, $NEG_R(-NEG_R(X)) = NEG_R(\overline{R}(X)) = U - \overline{R}(\overline{R}(X))$. According to $(5H)$ of Proposition 1, $\overline{R}(\overline{R}(X)) = \overline{R}(X)$. Hence, $NEG_R(-NEG_R(X)) = U - \overline{R}(X) = NEG_R(X)$;
$(EX4)$: For any $x, y \in U$, if $y \in NEG_R(X) - NEG_R(X \bigcup \{x\})$, then $y \in NEG_R(X)$ but $y \notin NEG_R(X \bigcup \{x\})$, i.e., $y \in U - \overline{R}(X)$ but $y \notin U - \overline{R}(X \bigcup \{x\})$. Since $\overline{R}(X) \subseteq U$ and $\overline{R}(X \bigcup \{x\}) \subseteq U$, so $y \in \overline{R}(X \bigcup \{x\})$ but $y \notin \overline{R}(X)$. Hence, $RN(y) \bigcap (X \bigcup \{x\}) \neq \emptyset$ but $RN(y) \bigcap X = \emptyset$. Therefore, $RN(y) \bigcap \{x\} \neq \emptyset$, i.e., $x \in RN(y)$. According to Lemma 1, $y \in RN(x)$. Hence, $RN(x) \bigcap (X \bigcup \{y\}) \neq \emptyset$, i.e., $x \in \overline{R}(X \bigcup \{y\})$. Therefore, $x \notin U - \overline{R}(X \bigcup \{y\})$, i.e., $x \notin NEG_R(X \bigcup \{y\})$. □

Proposition 9 determines the third matroid such that NEG_R is its exterior operator.

Definition 5. *Let R be an equivalence relation on U. The matroid whose exterior operator is NEG_R is denoted by $M(NEG_R)$. We say $M(NEG_R)$ is the matroid induced by NEG_R.*

In order to certify the boundary operator BND_R satisfies the boundary axiom, the following two lemmas are proposed.

Lemma 2. *Refs. [1,41] Let R be an equivalence relation on U. For all $X, Y \subseteq U$, $\overline{R}(X \bigcap Y) \subseteq \overline{R}(X) \bigcap \overline{R}(Y)$.*

Lemma 3. *Let R be an equivalence relation on U. If $X \subseteq Y \subseteq U$, then $X \bigcap BND_R(Y) \subseteq BND_R(X)$.*

Proof. For any $x \in X \bigcap BND_R(Y)$, $X \bigcap BND_R(Y) = X \bigcap (\overline{R}(X) - \underline{R}(X)) = X \bigcap \overline{R}(X) \bigcap \overline{R}(-X))$. Since $X \subseteq Y \subseteq U$, so $-Y \subseteq -X \subseteq U$. According to $(6H)$ of Proposition 1, $X \bigcap \overline{R}(X) \bigcap \overline{R}(-X)) = X \bigcap \overline{R}(-X) \subseteq \overline{R}(X) \bigcap \overline{R}(-X) = BND_R(X)$. Hence, $x \in BND_R(X)$, i.e., $X \bigcap BND_R(Y) \subseteq BND_R(X)$. □

The boundary operator BND_R satisfies the boundary axiom.

Proposition 10. *Let R be an equivalence relation on U. Then, BND_R satisfies $(BO1)$, $(BO2)$, $(BO3)$, $(BO4)$ and $(BO5)$ of Proposition 5.*

Proof. $(BO1)$: According to $(8LH)$ of Proposition 1, $\underline{R}(-X) = -\overline{R}(X)$. For any $X \subseteq U$,

$$\begin{aligned} BND_R(-X) &= \overline{R}(-X) - \underline{R}(-X) \\ &= \overline{R}(-X) \bigcap (U - \underline{R}(-X)) \\ &= (-\underline{R}(X)) \bigcap \overline{R}(X) \\ &= \overline{R}(X) \bigcap (-\underline{R}(X)) \\ &= \overline{R}(X) - \underline{R}(X) \\ &= BND_R(X). \end{aligned}$$

$(BO2)$: For any $X \subseteq U$,

$$\begin{aligned} BND_R(BND_R(X)) &= \overline{R}(BND_R(X)) - \underline{R}(BND_R(X)) \\ &= \overline{R}(BND_R(X)) \bigcap (U - \underline{R}(BND_R(X))) \\ &= \overline{R}(BND_R(X)) \bigcap (-\underline{R}(BND_R(X))) \\ &= \overline{R}(BND_R(X)) \bigcap (\overline{R}(-BND_R(X))) \\ &\subseteq \overline{R}(BND_R(X)) \\ &= \overline{R}(\overline{R}(X) - \underline{R}(X)) \\ &= \overline{R}(\overline{R}(X) \bigcap (-\underline{R}(X))). \end{aligned}$$

According to Lemma 1, we know

$$\begin{aligned} \overline{R}(\overline{R}(X) \bigcap (-\underline{R}(X))) &\subseteq \overline{R}(\overline{R}(X)) \bigcap \overline{R}(-\underline{R}(X)) \\ &= \overline{R}(X) \bigcap \overline{R}(-\underline{R}(X)) \\ &= \overline{R}(X) - \underline{R}(X) \\ &= BND_R(X). \end{aligned}$$

Hence, $BND_R(BND_R(X)) \subseteq BND_R(X)$;

$(BO3)$: For any $X, Y \subseteq U$, $X \bigcap Y \bigcap (BND_R(X) \bigcup BND_R(Y)) = X \bigcap Y \bigcap ((\overline{R}(X) - \underline{R}(X)) \bigcup (\overline{R}(Y) - \underline{R}(Y))) = X \bigcap Y \bigcap ((\overline{R}(X) \bigcap \overline{R}(-X)) \bigcup (\overline{R}(Y) \bigcap \overline{R}(-Y))) \subseteq X \bigcap Y \bigcap (\overline{R}(-X) \bigcup \overline{R}(-Y))$. According to $(4H)$ of Proposition 1, we know $X \bigcap Y \bigcap (\overline{R}(-X) \bigcup \overline{R}(-Y)) = X \bigcap Y \bigcap \overline{R}((-X) \bigcup (-Y)) = X \bigcap Y \bigcap \overline{R}(-(X \bigcap Y))$. According to $(6H)$ of Proposition 1, we know $X \bigcap Y \subseteq \overline{R}(X \bigcap Y)$. Therefore, $X \bigcap Y \bigcap \overline{R}(-(X \bigcap Y)) = X \bigcap Y \bigcap \overline{R}(-(X \bigcap Y)) \bigcap \overline{R}(X \bigcap Y) = X \bigcap Y \bigcap BND_R(X \bigcap Y)$.

$(BO4)$: When $x = y$ or $x \in X$, it is straightforward. When $y \in X$, it does not hold. (In fact, we suppose $y \in X$. If $y \in BND_R(X \bigcup \{x\})$, according to Lemma 3, we know $y \in X \bigcap BND_R(X \bigcup \{x\}) \subseteq BND_R(X)$, which is contradictory with $y \in BND_R(X\bigcup\{x\}) - BND_R(X)$. Hence, $y \notin X$.) We only need to prove it for $x \neq y$ and $x, y \notin X$. If $y \in BND_R(X\bigcup\{x\}) - BND_R(X)$, since

$$\begin{aligned}
& BND_R(X\bigcup\{x\}) - BND_R(X) \\
= & (\overline{R}(X\bigcup\{x\}) - \underline{R}(X\bigcup\{x\})) - (\overline{R}(X) - \underline{R}(X)) \\
= & (\overline{R}(X\bigcup\{x\}) \bigcap \overline{R}(-(X\bigcup\{x\}))) - (\overline{R}(X) \bigcap \overline{R}(-X)) \\
= & (\overline{R}(X\bigcup\{x\}) \bigcap \overline{R}(-(X\bigcup\{x\}))) \bigcap ((-\overline{R}(X)) \bigcup (-\overline{R}(-X))) \\
= & (\overline{R}(X\bigcup\{x\}) \bigcap \overline{R}(-(X\bigcup\{x\})) \bigcap (-\overline{R}(X))) \bigcup (\overline{R}(X\bigcup\{x\}) \\
& \bigcap (\overline{R}(-(X\bigcup\{x\}))) \bigcap (-\overline{R}(-X))) \\
= & \overline{R}(X\bigcup\{x\}) \bigcap \overline{R}(-(X\bigcup\{x\})) \bigcap (-\overline{R}(X)) \\
= & \overline{R}(X\bigcup\{x\}) \bigcap (-\overline{R}(X)) \bigcap \overline{R}(-(X\bigcup\{x\})) \\
= & (\overline{R}(X\bigcup\{x\}) - \overline{R}(X)) \bigcap \overline{R}(-(X\bigcup\{x\})),
\end{aligned}$$

then $y \in \overline{R}(X\bigcup\{x\}) - \overline{R}(X)$ and $y \in \overline{R}(-(X\bigcup\{x\}))$. According to Proposition 8, we have $x \in \overline{R}(X\bigcup\{y\})$. Since $y \in \overline{R}(-(X\bigcup\{x\}))$, so $x \in \overline{R}(-(X\bigcup\{y\}))$. Hence, $y \in \overline{R}(X\bigcup\{y\}) \bigcap \overline{R}(-(X\bigcup\{y\}))$, i.e., $y \in BND_R(X\bigcup\{y\})$.
$(BO5)$: For any $X, Y \subseteq U$,

$$\begin{aligned}
BND_R(X\bigcup BND_R(X)) & = \overline{R}(X\bigcup BND_R(X)) - \underline{R}(X\bigcup BND_R(X)) \\
& = \overline{R}(X\bigcup BND_R(X)) \bigcap \overline{R}(-(X\bigcup BND_R(X))) \\
& = \overline{R}(X\bigcup BND_R(X)) \bigcap \overline{R}((-X) \bigcap (BND_R(-X))) \\
& \subseteq \overline{R}(X\bigcup BND_R(X)) \bigcap \overline{R}(-X) \\
& = \overline{R}(X\bigcup(\overline{R}(X) \bigcap \overline{R}(-X))) \bigcap \overline{R}(-X) \\
& = \overline{R}(\overline{R}(X) \bigcap U) \bigcap \overline{R}(-X) \\
& = \overline{R}(\overline{R}(X)) \bigcap \overline{R}(-X).
\end{aligned}$$

According to $(5H)$ and $(8LH)$ of Proposition 1, $\overline{R}(\overline{R}(X)) \bigcap \overline{R}(-X) = \overline{R}(X) \bigcap \overline{R}(-X) = \overline{R}(X) - \underline{R}(X) = BND_R(X)$. Therefore, $BND_R(X\bigcup BND_R(X)) \subseteq BND_R(X)$. □

Proposition 8 determines the fourth matroid such that BND_R is its boundary operator.

Definition 6. *Let R be an equivalence relation on U. The matroid whose boundary operator is BND_R is denoted by $M(BND_R)$. We say that $M(BND_R)$ is the matroid induced by BND_R.*

3.2. The Relationship between These Four Matroids

This subsection studies the relationship between these four matroids in the above subsection. In fact, these four matroids are the same one.

Theorem 1. *Let R be an equivalence relation on U. Then,*

$$M(\underline{R}) = M(\overline{R}) = M(NEG_R) = M(BND_R).$$

Proof. (1) On one hand, $M(\underline{R})$ and $M(\overline{R})$ have the same grand U. On the other hand, according to Definition 3, we know $int_{M(\underline{R})}(X) = \underline{R}(X)$ for any $X \subseteq U$. By Proposition 6, $cl_{M(\underline{R})}(X) = -int_{M(\underline{R})}(-X) = -\underline{R}(-X)$. According to $(8LH)$ of Proposition 1, $-\underline{R}(-X) = \overline{R}(X)$. Hence, $cl_{M(\underline{R})}(X) = \overline{R}(X)$. According to Definition 4, $cl_{M(\overline{R})}(X) = \overline{R}(X)$. Therefore, $cl_{M(\underline{R})}(X) = cl_{M(\overline{R})}(X)$, i.e., $M(\underline{R}) = M(\overline{R})$.
(2) On one hand, $M(\overline{R})$ and $M(NEG_R)$ have the same grand U. On the other hand, according to Definition 4, we know $cl_{M(\overline{R})} = \overline{R}$. For any $X \subseteq U$, $ex_{M(\overline{R})}(X) = -cl_{M(\overline{R})}(X) = -\overline{R}(X) = U -$

$\overline{R}(X) = NEG_R(X)$. By Definition 5, $ex_{M(NEG_R)}(X) = NEG_R(X)$. Hence, $ex_{M(\overline{R})}(X) = ex_{M(NEG_R)}(X)$, i.e., $M(\overline{R}) = M(NEG_R)$.
(3) On one hand, $M(\overline{R})$ and $M(NEG_R)$ have the same grand U. On the other hand, according to Definition 4, we have $cl_{M(\overline{R})} = \overline{R}$. For all $X \subseteq U$, $bo_{M(\overline{R})}(X) = cl_{M(\overline{R})}(X) \bigcap cl_{M(\overline{R})}(-X) = \overline{R}(X) \bigcap \overline{R}(-X) = \overline{R}(X) \bigcap \overline{R}(-X) = \overline{R}(X) - \underline{R}(X) = BND_R(X)$. According to Definition 6, $bo_{M(NEG_R)}(X) = BND_R(X)$. Therefore, $bo_{M(\overline{R})}(X) = bo_{M(NEG_R)}(X)$, i.e., $M(\overline{R}) = M(NEG_R)$. □

Definition 7. *Let R be an equivalence relation on U. The matroid whose interior operator, closure operator, exterior operator and boundary operator are $\underline{R}$, $\overline{R}$, NEG_R and BND_R is defined as $M(R)$. We say that $M(R)$ is the matroid induced by R.*

According to the above definition, we have the relationship about operators between rough sets and matroids as Table 1:

Table 1. The relationship about operators between rough sets and matroids.

$M(R)$ Is the Matroid Induced by R
$int_{M(R)} = \underline{R} = POS_R$
$cl_{M(R)} = \overline{R}$
$ex_{M(R)} = NEG_R$
$bo_{M(R)} = BND_R$

3.3. The Relationship about Operations between Matroids and Rough Sets

In this subection, a relationship about the restriction operation both in matroids and rough sets is proposed. First of all, two definitions of these two operations are presented in the following two definitions.

Definition 8. *(Restriction [29,30]) Let $M = (U, \mathbf{I})$ be a matroid. For $X \subseteq U$, the restriction of M to X is defined as $M|X = (X, \mathbf{I}_X)$, where $\mathbf{I}_X = \{I \subseteq X : I \in \mathbf{I}\}$.*

Not that $\mathbf{C}(M|X) = \{C \subseteq X : C \in \mathbf{C}(M)\}$. For an equivalence relation R on U, there is also a definition of restriction of R. For any $X \subseteq U$, $R|X$ is an equivalence relation called the restriction of R to X, where $R|X = \{(x,y) \in X \times X : (x,y) \in R\}$, $X \times X$ is the product set of X and X. According to Definition 7, $M(R|X)$ is a matroid on X.

In [38], the set of independent sets of $M(R)$ is proposed in the following lemma.

Lemma 4. *Ref. [38] Let R be an equivalence relation on U. Then,*

$$\mathbf{I}(M(R)) = \{X \subseteq U : \forall x, y \in X, x \neq y \Rightarrow (x,y) \notin R\}.$$

Example 5. *Let R be an equivalence relation on U with $U = \{a,b,c,d,e\}$, and $U/R = \{\{a,b\},\{c,d,e\}\}$. According to Lemma 4, $\mathbf{I}(M(R)) = \{\emptyset, \{a\}, \{b\}, \{c\}, \{d\}, \{e\}, \{a,c\}, \{b,c\}, \{a,d\}, \{b,d\}, \{a,e\}, \{b,e\}\}$.*

Proposition 11. *Let R be an equivalence relation on U and $X \subseteq U$. Then, $M(R|X) = M(R)|X$.*

Proof. For any $X \subseteq U$, $R|X$ is an equivalence relation on X. Thus, $M(R|X)$ is a matroid on X. By Definition 8, $M(R)|X$ is a matroid on X. Therefore, we need to prove only $\mathbf{I}(M(R|X)) = \mathbf{I}(M(R)|X)$. According to Lemma 4, $\mathbf{I}(M(R|X)) = \{Y \subseteq X : \forall x, y \in Y, x \neq y \Rightarrow (x,y) \notin R|X\}$, $\mathbf{I}(M(R)|X) = \{Y \subseteq X : \forall x, y \in Y, x \neq y \Rightarrow (x,y) \notin R\}$. On one hand, since $R|X \subseteq R$, $\mathbf{I}(M(R)|X) \subseteq \mathbf{I}(M(R|X))$. On the other hand, suppose $Y \in \mathbf{I}(M(R|X)) - \mathbf{I}(M(R)|X)$. For any $x, y \in Y$, if $x \neq y$, then $(x,y) \notin R|X$ but $(x,y) \in R$. Therefore, $x, y \notin X$ but $x, y \in U$, i.e., $x, y \in U - X$. Hence, $Y \subseteq U - X$, which is contradictory with $Y \subseteq X$, i.e., $Y \in \mathbf{I}(M(R|X)) - \mathbf{I}(M(R)|X)$. Thus, $\mathbf{I}(M(R|X)) \subseteq \mathbf{I}(M(R)|X)$. □

Example 6. *(Continued from Example 5) Let* $X = \{a,b,c\}$. *According to Definition 8,* $\boldsymbol{I}(M(R)|X) = \{\emptyset, \{a\}, \{b\}, \{c\}, \{a,c\}, \{b,c\}\}$, *and* $M(R)|X = (X, \boldsymbol{I}(M(R)|\ X))$. *Since* $R|X = \{(a,a),(b,b),(c,c),(a,b),(b,a)\}$, *so* $X/(R|X) = \{\{a,b\},\{c\}\}$. *According to Lemma 4,* $\boldsymbol{I}(M(R|X)) = \{\emptyset, \{a\}, \{b\}, \{c\}, \{a,c\}, \{b,c\}\}$, *and* $M(R|X) = (X, \boldsymbol{I}(M(R|X))$. *Therefore,* $M(R|X) = M(R)|X$.

4. A Matroidal Approach to Attribute Reduction through the Girth Function

In this section, a matroidal approach is proposed to compute attribute reduction in information systems through the girth function of matroids.

4.1. An Equivalent Formulation of Attribute Reduction through the Girth Function

Lemma 5. *Ref. [15] Let* R_1 *and* R_2 *be two equivalence relations on* U, *respectively. Then,* $\underline{R_1} = \underline{R_2}$ *if and only if* $R_1 = R_2$.

Based on Lemma 5, we propose a necessary and sufficient condition for two equivalence relations induce the same matroids.

Proposition 12. *Let* R_1 *and* R_2 *be two equivalence relations on* U, *respectively. Then,* $M(R_1) = M(R_2)$ *if and only if* $R_1 = R_2$.

Proof. According to Definition 7, $M(R_1)$ and $M(R_2)$ have the same grand U. According to Proposition 3, Proposition 7 and Lemma 5,

$$\begin{aligned} M(R_1) = M(R_2) &\Leftrightarrow int_{M(R_1)} = int_{M(R_2)} \\ &\Leftrightarrow \underline{R_1} = \underline{R_2} \\ &\Leftrightarrow R_1 = R_2. \end{aligned}$$

□

An equivalent formulation of attribute reduction in information systems is presented from the viewpoint of matroids.

Proposition 13. *Let* $IS = (U, A)$ *be an information system. For all* $B \subseteq A$, B *is a reduct of* IS *if and only if it satisfies the following two conditions:*

(1) For all $b \in B$, $M(IND(B)) \neq M(IND(B-b))$;
(2) $M(IND(B)) = M(IND(A))$.

Proof. Since $IND(A)$, $IND(B)$ and $IND(B-b)$ are equivalence relations on U, $M(IND(A))$, $M(IND(B))$ and $M(IND(B-b))$ are matroids on U. According to Proposition 12,
(1) For all $b \in B$, $M(IND(B)) \neq M(IND(B-b)) \Leftrightarrow IND(B) \neq IND(B-b)$;
(2) $M(IND(B)) = M(IND(A)) \Leftrightarrow IND(B) = IND(A)$.
According to Definition 1, it is immediate. □

In Proposition 13, the equivalent formulation of attribute reduction is not convenient for us to compute the attribute reduction. We consider to use the girth function of matroids to compute it.

Definition 9. *(Girth function [29,30]) Let* $M = (U, \boldsymbol{I})$ *be a matroid. The girth* $g(M)$ *of* M *is defined as:*

$$g(M) = \begin{cases} min\{|C| : C \in \boldsymbol{C}(M)\}, & \boldsymbol{C}(M) \neq \emptyset; \\ \infty, & \boldsymbol{C}(M) = \emptyset. \end{cases}$$

For all $X \subseteq U$, *the girth function* g_M *is defined as* $g_M(X) = g(M|X)$. $g_M(X)$ *is called the girth of* X *in* M.

According to Definition 9, the girth function is related to circuits. Thus, the following lemma presents the family of all circuits of $M(R)$.

Lemma 6. *Ref. [38] Let R be an equivalence relation on U. Then,*

$$\mathbf{C}(M(R)) = \{\{x, y\} \subseteq U : x \neq y \wedge (x, y) \in R\}.$$

Example 7. *(Continued from Example 5)* $\mathbf{C}(M(R)) = \{\{a, b\}, \{c, d\}, \{c, e\}, \{d, e\}\}$.

Based on the characteristics of the matroid induced by an equivalence relation, a type of matroids is abstracted, which is called a 2-circuit matroid. M is called a 2-circuit matroid if $|C| = 2$ for all $C \in \mathbf{C}(M)$. Note that, if $\mathbf{C}(M) = \emptyset$, then M is also a 2-circuit matroid. In this section, we don't consider this case. The matroid $M(R)$ is a 2-circuit matroid.

Proposition 14. *Let R be an equivalence relation on U and $X \subseteq U$. Then,*

$$g(M(R)) = \begin{cases} 2, & \mathbf{C}(M(R)) \neq \emptyset; \\ \infty, & \mathbf{C}(M(R)) = \emptyset; \end{cases}$$

$$g_{M(R)}(X) = \begin{cases} 2, & \mathbf{C}(M(R)|X) \neq \emptyset; \\ \infty, & \mathbf{C}(M(R)|X) = \emptyset. \end{cases}$$

Proof. Since $M(R)$ is a 2-circuit matroid, $|C| = 2$ for all $C \in \mathbf{C}(M(R))$. According to Definition 9, it is immediate. □

Corollary 2. *Let R be an equivalence relation on U and $X \subseteq U$. Then,*

$$g(M(R)) = \begin{cases} 2, & \exists x \in U, s.t., |RN(x)| \geq 2; \\ \infty, & otherwise, \end{cases}$$

$$g_{M(R)}(X) = \begin{cases} 2, & \exists x \in X, s.t., |RN(x) \bigcap X| \geq 2; \\ \infty, & otherwise. \end{cases}$$

Proof. According to Lemma 6,

$$\begin{aligned} \mathbf{C}(M(R)) \neq \emptyset & \Leftrightarrow \exists x, y \subseteq U, s.t., x \neq y \wedge (x, y) \in R \\ & \Leftrightarrow \exists x \in U, s.t., |RN(x)| \geq 2. \end{aligned}$$

Hence,

$$g(M(R)) = \begin{cases} 2, & \exists x \in U, s.t., |RN(x)| \geq 2; \\ \infty, & otherwise. \end{cases}$$

Since $\mathbf{C}(M(R)|X) = \{C \subseteq X : C \in \mathbf{C}(M(R))\} = \{\{x, y\} \subseteq X : x \neq y \wedge (x, y) \in R\}$,

$$\begin{aligned} \mathbf{C}(M(R)|X) \neq \emptyset & \Leftrightarrow \exists x, y \subseteq X, s.t., x \neq y \wedge (x, y) \in R \\ & \Leftrightarrow \exists x \in U, s.t., |RN(x) \bigcap X| \geq 2. \end{aligned}$$

Hence,

$$g_{M(R)}(X) = \begin{cases} 2, & \exists x \in X, s.t., |RN(x) \bigcap X| \geq 2; \\ \infty, & otherwise. \end{cases}$$

□

Lemma 7. *Refs. [1,41] Let R_1 and R_2 be two equivalence relations on U, respectively. Then, for any $x \in U$,*

$$(R_1 \bigcap R_2)N(x) = R_1N(x) \bigcap R_2N(x).$$

According to Corollary 2, the girth function of the matroid induced by attribute subsets is presented in the following proposition.

Proposition 15. *Let $IS = (U, A)$ be an information system and $X \subseteq U$. Then, for all $B \subseteq A$,*

$$g(M(IND(B))) = \begin{cases} 2, & \exists x \in U, s.t., |\bigcap_{R_i \in B} R_iN(x)| \geq 2; \\ \infty, & otherwise, \end{cases}$$

$$g_{M(IND(B))}(X) = \begin{cases} 2, & \exists x \in X, s.t., |(\bigcap_{R_i \in B} R_iN(x)) \bigcap X| \geq 2; \\ \infty, & otherwise. \end{cases}$$

Proof. According to Lemma 7 and Corollary 2, it is immediate. □

Note that R_i in R_iN denotes the equivalence relation induced by attribute $R_i \in A$. According to the girth axiom, we know that a matroid is corresponding to one and only one girth function.

Proposition 16. *(Girth axiom [29,30]) Let $g : 2^U \to \mathbb{Z}^+ \bigcup \{0, \infty\}$ be a function. Then, there exists one and only one matroid M such that $g = g_M$ iff g satisfies the following three conditions:*
(G1) If $X \subseteq U$ and $g(X) < \infty$, then X has a subset Y such that $g(X) = g(Y) = |Y|$.
(G2) If $X \subseteq Y \subseteq U$, then $g(X) \geq g(Y)$.
(G3) If X and Y are distinct subsets of U with $g(X) = |X|$, $g(Y) = |Y|$, then $g((X \bigcup Y) - \{e\}) < \infty$ for any $e \in X \bigcap Y$.

Inspired by Propositions 13 and 16, we can use the girth function in matroids to compute attribute reduction.

Theorem 2. *Let $IS = (U, A)$ be an information system. For all $B \subseteq A$, B is a reduct of IS if and only if it satisfies the following two conditions:*

(1) For all $b \in B$, there exists $X \subseteq U$ such that $g_{M(IND(B))}(X) \neq g_{M(IND(B-b))}(X)$.
(2) For all $X \subseteq U$, $g_{M(IND(B))}(X) = g_{M(IND(A))}(X)$.

Proof. According to Propositions 13 and 16, it is immediate. □

4.2. The Process of the Matroidal Methodology

In this subsection, we give the process of the matroidal approach to compute attribute reduction in information systems according to the equivalent description in Section 4.1.

In order to obtain all results of an information system $IS = (U, A)$, we need to compute $g_{M(IND(B))}(X)$ for all $B \subseteq A$ and $X \subseteq U$ based on Theorem 2. According to Definition 1, we know a reduct of IS will not be $\emptyset$. Hence, we only consider $B \subseteq A$ and $B \neq \emptyset$. On the other hand, for all $X \subseteq U$ and $B \subseteq A$, if $|X| \leq 1$, then $g_{M(IND(B))}(X) = g_{M(IND(A))}(X)$. According to Theorem 2, we only consider X whose $|X| \geq 2$. Therefore, the process is shown as follows:

- **Input**: An information system $IS = (U, A)$, where $U = \{u_1, u_2, \cdots, u_n\}$ and $A = \{a_1, a_2, \cdots, a_m\}$.
- **Output**: All results of IS.
- **Step 1**: Suppose $B_i \subset A$ ($B_i \neq \emptyset$ and $i = 1, 2, \cdots, 2^m - 2$), we compute all $IND(B_i)$ and $IND(A)$.

- **Step 2:** For any $i = 1, 2, \cdots, 2^m - 2$, we compute $g_{M(IND(B_i))}(X)$ and $g_{M(IND(A))}(X)$ for any $X \subseteq U$ and $|X| \geq 2$.
- **Step 3:** Obtain all results of IS according to Theorem 2.

4.3. An Applied Example

Example 8. *Let us consider the following information system $IS = (U, A)$ as is shown in Table 2.*

Table 2. An information system.

	a_1	a_2	a_3
u_1	0	1	0
u_2	1	2	2
u_3	1	0	0
u_4	2	1	1
u_5	1	1	2

Let $B_1 = \{a_1\}$, $B_2 = \{a_2\}$, $B_3 = \{a_3\}$, $B_4 = \{a_1, a_2\}$, $B_5 = \{a_1, a_3\}$, $B_6 = \{a_2, a_3\}$, $A = \{a_1, a_1, a_3\}$. g_{B_i} denotes $g_{M(IND(B_i))}$ for $1 \leq i \leq 6$ and g_A denotes $g_{M(IND(A))}$. All girth functions induced by attribute subsets as is shown in Table 3.

Table 3. Girth functions induced by attribute subsets.

	g_{B_1}	g_{B_2}	g_{B_3}	g_{B_4}	g_{B_5}	g_{B_6}	g_A
u_1, u_2	∞	∞	∞	∞	∞	∞	∞
u_1, u_3	∞	∞	2	∞	∞	∞	∞
u_1, u_4	∞	2	∞	∞	∞	∞	∞
u_1, u_5	∞	2	∞	∞	∞	∞	∞
u_2, u_3	2	∞	∞	∞	∞	∞	∞
u_2, u_4	∞	∞	∞	∞	∞	∞	∞
u_2, u_5	∞	∞	2	∞	2	∞	∞
u_3, u_4	∞	∞	∞	∞	∞	∞	∞
u_3, u_5	∞	∞	∞	∞	∞	∞	∞
u_4, u_5	∞	2	∞	∞	∞	∞	∞
u_1, u_2, u_3	2	∞	2	∞	∞	∞	∞
u_1, u_2, u_4	∞	2	∞	∞	∞	∞	∞
u_1, u_2, u_5	∞	2	2	∞	2	∞	∞
u_1, u_3, u_4	∞	2	2	∞	∞	∞	∞
u_1, u_3, u_5	∞	2	2	∞	∞	∞	∞
u_1, u_4, u_5	∞	2	∞	∞	∞	∞	∞
u_2, u_3, u_4	2	∞	∞	∞	∞	∞	∞
u_2, u_3, u_5	2	∞	∞	∞	2	∞	∞
u_2, u_4, u_5	∞	2	2	∞	2	∞	∞
u_3, u_4, u_5	∞	2	∞	∞	∞	∞	∞
u_1, u_2, u_3, u_4	2	2	2	∞	∞	∞	∞
u_1, u_2, u_3, u_5	2	2	2	∞	2	∞	∞
u_1, u_2, u_4, u_5	∞	2	2	∞	2	∞	∞
u_1, u_3, u_4, u_5	∞	2	2	∞	∞	∞	∞
u_2, u_3, u_4, u_5	2	2	2	∞	2	∞	∞
u_1, u_2, u_3, u_4, u_5	2	2	2	∞	2	∞	∞

Accordingly, there are two reducts of IS: $B_4 = \{a_1, a_2\}$ and $B_6 = \{a_2, a_3\}$.

5. Conclusions

In this paper, we generalize four operators of rough sets to four operators of matroids through the interior axiom, the closure axiom, the exterior axiom and the boundary axiom, respectively. Moreover, we present a matroidal approach to compute attribute reduction in information systems. The main conclusions in this paper and the continuous work to do are listed as follows:

1. There are four matroids induced by these four operators of rough sets. In fact, these four matroids are the same one, which implies the relationship about operators between rough sets and matroids. In this work, we assume an equivalence relation. However, there are other structures have been used in rough set theory, among them, tolerance relations [43], similarity relations [44], and binary relations [15,45]. Hence, they can suggest as a future research, the possibility of extending their ideas to these types of settings.

2. The girth function of matroids is used to compute attribute reduction in information systems. This work can be viewed as a bridge linking matroids and information systems in the theoretical impact. In the practical impact, it is a novel method by which calculations will become algorithmic and can be implemented by a computer. Based on this work, we can use the girth function of matroids for attribute reduction in decision systems in the future.

3. In the future, we will further expand the research content of this paper based on some new studies on neutrosophic sets and related algebraic structures [46–50].

Author Contributions: This paper is written through contributions of all authors. The individual contributions and responsibilities of all authors can be described as follows: the idea of this whole thesis was put forward by X.Z., and he also completed the preparatory work of the paper. J.W. analyzed the existing work of rough sets and matroids, and wrote the paper.

References

1. Pawlak, Z. *Rough Sets: Theoretical Aspects of Reasoning about Data*; Kluwer Academic Publishers: Boston, MA, USA, 1991.
2. Pawlak, Z. Rough sets. *Int. J. Comput. Inf. Sci.* **1982**, *11*, 341–356. [CrossRef]
3. Du, Y.; Hu, Q.; Zhu, P.; Ma, P. Rule learning for classification based on neighborhood covering reduction. *Inf. Sci.* **2011**, *181*, 5457–5467. [CrossRef]
4. Wang, X.; Tsang, E.C.; Zhao, S.; Chen, D.; Yeung, D.S. Learning fuzzy rules from fuzzy samples based on rough set technique. *Inf. Sci.* **2007**, *177*, 4493–4514. [CrossRef]
5. Chen, J.; Lin, Y.; Lin, G.; Li, J.; Ma, Z. The relationship between attribute reducts in rough sets and minimal vertex covers of graphs. *Inf. Sci.* **2015**, *325*, 87–97. [CrossRef]
6. Dai, J.; Hu, Q.; Hu, H. Neighbor inconsistent pair selection for attribute reduction by rough set approach. *IEEE Trans. Fuzzy Syst.* **2018**, *26*, 937–950. [CrossRef]
7. Yang, Y.; Chen, D.; Wang, H. Active sample selection based incremental algorithm for attribute reduction with rough sets. *IEEE Trans. Fuzzy Syst.* **2017**, *25*, 825–838. [CrossRef]
8. Chen, Y.; Miao, D.; Wang, R.; Wu, K. A rough set approach to feature selection based on power set tree. *Knowl.-Based Syst.* **2011**, *24*, 275–281. [CrossRef]
9. Javidi, M.; Eskandari, S. Streamwise feature selection: A rough set method. *Int. J. Mach. Learn. Cybern.* **2018**, *9*, 667–676. [CrossRef]
10. Cai, Z.; Zhu, W. Multi-label feature selection via feature manifold learning and sparsity regularization. *Int. J. Mach. Learn. Cybern.* **2018**, 9, 1321–1334. [CrossRef]
11. Johnson, J.A.; Liu, M.; Chen, H. Unification of knowledge discovery and data mining using rough sets approach in a real-world application. *Rough Sets Curr. Trends Comput.* **2001**, *2005*, 330–337.
12. Alcantud, J.C.R. Revealed indifference and models of choice behavior. *J. Math. Psychol.* **2002**, *46*, 418–430. [CrossRef]
13. Luce, R.D. Semiorders and a theory of utility discrimination. *Econometrica* **1956**, *24*, 178–191. [CrossRef]
14. Wang, C.; He, Q.; Shao, M.; Xu, Y.; Hu, Q. A unified information measure for general binary relations. *Knowl.-Based Syst.* **2017**, *135*, 18–28. [CrossRef]

15. Zhu, W. Generalized rough sets based on relations. *Inf. Sci.* **12007**, *177*, 4997–5001. [CrossRef]
16. Zhu, W.; Wang, F. Reduction and axiomatization of covering generalized rough sets. *Inf. Sci.* **2003**, *152*, 217–230. [CrossRef]
17. Zhu, W. Relationship among basic concepts in covering-based rough sets. *Inf. Sci.* **2009**, *179*, 2478–2486. [CrossRef]
18. Yang, B.; Hu, B. On some types of fuzzy covering-based rough sets. *Fuzzy Sets Syst.* **2017**, 312, 36–65. [CrossRef]
19. Wang, C.; He, Q.; Wang, X.; Chen, D.; Qian, Y. Feature selection based on neighborhood discrimination index. *IEEE Trans. Neural Netw. Learn. Syst.* **2018**, *29*, 2986–2999. [CrossRef] [PubMed]
20. Zhao, Z. On some types of covering rough sets from topological points of view. *Int. J. Approx. Reason.* **2016**, *68*, 1–14. [CrossRef]
21. Gao, N.; Li, Q.; Han, H.; Li, Z. Axiomatic approaches to rough approximation operators via ideal on a complete completely distributive lattice. *Soft Comput.* **2018**, *22*, 2329–2339. [CrossRef]
22. Zhang, X.; Miao, D.; Liu, C.; Le, M. Constructive methods of rough approximation operators and multigranulation rough sets. *Knowl.-Based Syst.* **2016**, *91*, 114–125. [CrossRef]
23. Chiaselotti, G.; Ciucci, D.; Gentile, T. Rough set theory and digraphs. *Fundam. Inf.* **2017**, *153*, 291–325. [CrossRef]
24. Wang, J.; Zhu, W. Applications of bipartite graphs and their adjacency matrices to covering-based rough sets. *Fundam. Inf.* **2017**, *156*, 237–254. [CrossRef]
25. Liang, D.; Xu, Z.; Liu, D. Three-way decisions with intuitionistic fuzzy decision-theoretic rough sets based on point operators. *Inf. Sci.* **2017**, *375*, 183–201. [CrossRef]
26. Zhang, X.; Dai, J.; Yu, Y. On the union and intersection operations of rough sets based on various approximation spaces. *Inf. Sci.* **2015**, *292*, 214–229. [CrossRef]
27. Qian, Y.; Liang, J.; Pedrycz, W.; Dang, C. Positive approximation: An accelerator for attribute reduction in rough set theory. *Artif. Intell.* **2010**, *174*, 597–618. [CrossRef]
28. Edmonds, J. Matroids and the greedy algorithm. *Math. Program.* **1971**, *1*, 127–136. [CrossRef]
29. Lai, H. *Matroid Theory*; Higher Education Press: Beijing, China, 2001.
30. Oxley, J.G. *Matroid Theory*; Oxford University Press: Oxford, UK, 1993.
31. Liu, S.; Manganiello, F.; Kschischang, F.R. Matroidal structure of skew polynomial rings with application to network coding. *Finite Fields Their Appl.* **2017**, *46*, 236–246. [CrossRef]
32. Dougherty, R.; Freiling, C.; Zeger, K. Networks, matroids, and non-Shannon information inequalities. *IEEE Trans. Inf. Theory* **2007**, *53*, 1949–1969. [CrossRef]
33. Zhu, W.; Wang, S. Matroidal approaches to generalized rough sets based on relations. *Int. J. Mach. Learn. Cybern.* **2011**, *2*, 273–279. [CrossRef]
34. Liu, Y.; Zhu, W. Parametric matroid of rough set. *Int. J. Uncertain. Fuzziness Knowl.-Based Syst.* **2015**, *23*, 893–908. [CrossRef]
35. Li, X.; Liu, S. Matroidal approaches to rough set theory via closure operators. *Int. J. Approx. Reason.* **2012**, *53*, 513–527. [CrossRef]
36. Li, X.; Yi, H.; Liu, S. Rough sets and matroids from a lattice-theoretic viewpoint. *Inf. Sci.* **2016**, *342*, 37–52. [CrossRef]
37. Su, L.; Zhu, W. Closed-set lattice and modular matroid induced by covering-based rough sets. *Int. J. Mach. Learn. Cybern.* **2017**, *8*, 191–201. [CrossRef]
38. Wang, S.; Zhu, Q.; Zhu, W.; Min, F. Matroidal structure of rough sets and its characterization to attribute reduction. *Knowl.-Based Syst.* **2012**, 36, 155–161. [CrossRef]
39. Wang, J.; Zhu, W. Applications of matrices to a matroidal structure of rough sets. *J. Appl. Math.* **2013**, *2013*. [CrossRef]
40. Wang, J.; Zhu, W. Contraction to matroidal structure of rough sets. In *Rough Sets and Knowledge Technology*; Springer: Berlin/Heidelberg, Germany, 2013; Volume 8171, pp. 75–86.
41. Yao, Y. Constructive and algebraic methods of theory of rough sets. *Inf. Sci.* **1998**, *109*, 21–47. [CrossRef]
42. Guo, J.; Li, S. The property of operator in matroid. *J. Shaanxi Normal Univ.* **2007**, *35*, 13–16.
43. Skowron, A.; Stepaniuk, J. Tolerance approximation spaces. *Fundam. Inf.* **1996**, *27*, 245–253.
44. Slowinski, R.; Vanderpooten, D. A generalized definition of rough approximations based on similarity. *IEEE Trans. Knowl. Data Eng.* **2000**, *12*, 331–336. [CrossRef]

45. Liu, G.; Zhu, W. The algebraic structures of generalized rough set theory. *Inf. Sci.* **2008**, *178*, 4105–4113. [CrossRef]
46. Zhang, X.; Smarandache, F.; Liang, X. Neutrosophic duplet semi-group and cancellable neutrosophic triplet groups. *Symmetry* **2017**, *9*, 275. [CrossRef]
47. Zhang, X. Fuzzy anti-grouped filters and fuzzy normal filters in pseudo-BCI algebras. *J. Intell. Fuzzy Syst.* **2017**, *33*, 1767–1774. [CrossRef]
48. Zhang, X.; Bo, C.; Smarandache, F.; Park, C. New operations of totally dependent-neutrosophic sets and totally dependent-neutrosophic soft sets. *Symmetry* **2018**, *10*, 187. [CrossRef]
49. Zhang, X.; Bo, C.; Smarandache, F.; Dai, J. New inclusion relation of neutrosophic sets with applications and related lattice structure. *Int. J. Mach. Learn. Cybern.* **2018**, *9*, 1753–1763. [CrossRef]
50. Zhang, X.; Park, C.; Wu, S. Soft set theoretical approach to pseudo-BCI algebras. *J. Intell. Fuzzy Syst.* **2018**, *34*, 559–568. [CrossRef]

7

Bounds for the Generalized Distance Eigenvalues of a Graph

Abdollah Alhevaz [1,*], Maryam Baghipur [1], Hilal Ahmad Ganie [2] and Yilun Shang [3]

1 Faculty of Mathematical Sciences, Shahrood University of Technology, Shahrood P.O. Box: 316-3619995161, Iran; maryamb8989@gmail.com
2 Department of Mathematics, University of Kashmir, Srinagar 190006, India; hilahmad1119kt@gmail.com
3 Department of Computer and Information Sciences, Northumbria University, Newcastle NE1 8ST, UK; yilun.shang@northumbria.ac.uk
* Correspondence: a.alhevaz@gmail.com or a.alhevaz@shahroodut.ac.ir

Abstract: Let G be a simple undirected graph containing n vertices. Assume G is connected. Let $D(G)$ be the distance matrix, $D^L(G)$ be the distance Laplacian, $D^Q(G)$ be the distance signless Laplacian, and $Tr(G)$ be the diagonal matrix of the vertex transmissions, respectively. Furthermore, we denote by $D_\alpha(G)$ the generalized distance matrix, i.e., $D_\alpha(G) = \alpha Tr(G) + (1-\alpha)D(G)$, where $\alpha \in [0,1]$. In this paper, we establish some new sharp bounds for the generalized distance spectral radius of G, making use of some graph parameters like the order n, the diameter, the minimum degree, the second minimum degree, the transmission degree, the second transmission degree and the parameter α, improving some bounds recently given in the literature. We also characterize the extremal graphs attaining these bounds. As an special cases of our results, we will be able to cover some of the bounds recently given in the literature for the case of distance matrix and distance signless Laplacian matrix. We also obtain new bounds for the k-th generalized distance eigenvalue.

Keywords: distance matrix (spectrum); distance signlees Laplacian matrix (spectrum); (generalized) distance matrix; spectral radius; transmission regular graph

MSC: Primary: 05C50, 05C12; Secondary: 15A18

---

1. Introduction

We will consider simple finite graphs in this paper. A (simple) graph is denoted by $G = (V(G), E(G))$, where $V(G) = \{v_1, v_2, \ldots, v_n\}$ represents its vertex set and $E(G)$ represents its edge set. The *order* of G is the number of vertices represented by $n = |V(G)|$ and its *size* is the number of edges represented by $m = |E(G)|$. The *neighborhood* $N(v)$ of a vertex v consists of the set of vertices that are adjacent to it. The *degree* $d_G(v)$ or simply $d(v)$ is the number of vertices in $N(v)$. In a *regular* graph, all its vertices have the same degree. Let d_{uv} be the *distance* between two vertices $u, v \in V(G)$. It is defined as the length of a shortest path. $D(G) = (d_{uv})_{u,v \in V(G)}$ is called the *distance matrix* of G. $\overline{G}$ is the *complement* of the graph G. It has the same vertex set with G but its edge set consists of the edges not present in G. Moreover, the complete graph K_n, the complete bipartite graph $K_{s,t}$, the path P_n, and the cycle C_n are defined in the conventional way.

The *transmission* $Tr_G(v)$ of a vertex v is the sum of the distances from v to all other vertices in G, i.e., $Tr_G(v) = \sum\limits_{u \in V(G)} d_{uv}$. A graph G is said to be *k-transmission regular* if $Tr_G(v) = k$, for each $v \in V(G)$. The *transmission* (also called the *Wiener index*) of a graph G, denoted by $W(G)$, is the sum of distances between all unordered pairs of vertices in G. We have $W(G) = \frac{1}{2} \sum\limits_{v \in V(G)} Tr_G(v)$.

For a vertex $v_i \in V(G)$, $Tr_G(v_i)$ is also referred to as the *transmission degree*, or shortly Tr_i. The sequence of transmission degrees $\{Tr_1, Tr_2, \ldots, Tr_n\}$ is the *transmission degree sequence* of the graph. $T_i = \sum_{j=1}^{n} d_{ij} Tr_j$ is called the second transmission degree of v_i.

Distance matrix and its spectrum has been studied extensively in the literature, see e.g., [6]. Compared to adjacency matrix, distance matrix encapsulates more information such as a wide range of walk-related parameters, which can be applicable in thermodynamic calculations and have some biological applications in terms of molecular characterization. It is known that embedding theory and molecular stability have to do with graph distance matrix.

Almost all results obtained for the distance matrix of trees were extended to the case of weighted trees by Bapat [12] and Bapat et al. [13]. Not only different classes of graphs but the definition of distance matrix has been extended. Indeed, Bapat et al. [14] generalized the concept of the distance matrix to that of q-analogue of the distance matrix. Let $Tr(G) = diag(Tr_1, Tr_2, \ldots, Tr_n)$ be the diagonal matrix of vertex transmissions of G. The works [7–9] introduced the distance Laplacian and the distance signless Laplacian matrix for a connected graph G. The matrix $D^L(G) = Tr(G) - D(G)$ is referred to as the *distance Laplacian matrix* of G, while the matrix $D^Q(G) = Tr(G) + D(G)$ is the *distance signless Laplacian matrix* of G. Spectral properties of $D(G)$ and $D^Q(G)$ have been extensively studied since then.

Let A be the adjacency matrix and $\mathrm{Deg}(G) = \mathrm{diag}(d_1, d_2, \ldots, d_n)$ be the degree matrix G. $Q(G) = \mathrm{Deg}(G) + A$ is the *signless Laplacian matrix* of G. This matrix has been put forth by Cvetkovic in [16] and since then studied extensively by many researchers. For detailed coverage of this research see [17–20] and the references therein. To digging out the contribution of these summands in $Q(G)$, Nikiforov in [33] proposed to study the *α-adjacency matrix* $A_\alpha(G)$ of a graph G given by $A_\alpha(G) = \alpha\,\mathrm{Deg}(G) + (1-\alpha)A$, where $\alpha \in [0,1]$. We see that $A_\alpha(G)$ is a convex combination of the matrices A and $\mathrm{Deg}(G)$. Since $A_0(G) = A$ and $2A_{1/2}(G) = Q(G)$, the matrix $A_\alpha(G)$ can underpin a unified theory of A and $Q(G)$. Motivated by [33], Cui et al. [15] introduced the convex combinations $D_\alpha(G)$ of $Tr(G)$ and $D(G)$. The matrix $D_\alpha(G) = \alpha Tr(G) + (1-\alpha)D(G)$, $0 \le \alpha \le 1$, is called *generalized distance matrix* of G. Therefore the generalized distance matrix can be applied to the study of other less general constructions. This not only gives new results for several matrices simultaneously, but also serves the unification of known theorems.

Since the matrix $D_\alpha(G)$ is real and symmetric, its eigenvalues can be arranged as: $\partial_1 \ge \partial_2 \ge \cdots \ge \partial_n$, where ∂_1 is referred to as the *generalized distance spectral radius* of G. For simplicity, $\partial(G)$ is the shorthand for $\partial_1(G)$. By the Perron-Frobenius theorem, $\partial(G)$ is unique and it has a unique *generalized distance Perron vector*, X, which is positive. This is due to the fact that $D_\alpha(G)$ is non-negative and irreducible.

A column vector $X = (x_1, x_2, \ldots, x_n)^T \in \mathbb{R}^n$ is a function defined on $V(G)$. We have $X(v_i) = x_i$ for all i. Moreover,

$$X^T D_\alpha(G) X = \alpha \sum_{i=1}^{n} Tr(v_i) x_i^2 + 2(1-\alpha) \sum_{1 \le i < j \le n} d(v_i, v_j) x_i x_j,$$

and λ has an eigenvector X if and only if $X \neq \mathbf{0}$ and

$$\lambda x_v = \alpha Tr(v_i) x_i + (1-\alpha) \sum_{j=1}^{n} d(v_i, v_j) x_j.$$

They are often referred to as the *(λ, x)-eigenequations* of G. If $X \in \mathbb{R}^n$ has at least one non-negative element and it is normalized, then in the light of the Rayleigh's principle, it can be seen that

$$\partial(G) \ge X^T D_\alpha(G) X,$$

where the equality holds if and only if X becomes the generalized distance Perron vector of G.

Spectral graph theory has been an active research field for the past decades, in which for example distance signless Laplacian spectrum has been intensively explored. The work [41] identified the graphs with minimum distance signless Laplacian spectral radius among some special classes of graphs. The unique graphs with minimum and second-minimum distance signless Laplacian spectral radii among all bicyclic graphs of the same order are identified in [40]. In [24], the authors show some bounding inequalities for distance signless Laplacian spectral radius by utilizing vertex transmissions. In [26], chromatic number is used to derive a lower bound for distance signless Laplacian spectral radius. The distance signless Laplacian spectrum has varies connections with other interesting graph topics such as chromatic number [10]; domination and independence numbers [21], Estrada indices [4,5,22,23,34–36,38], cospectrality [11,42], multiplicity of the distance (signless) Laplacian eigenvalues [25,29,30] and many more, see e.g., [1–3,27,28,32].

The rest of the paper is organized as follows. In Section 2, we obtain some bounds for the generalized distance spectral radius of graphs using the diameter, the order, the minimum degree, the second minimum degree, the transmission degree, the second transmission degree and the parameter α. We then characterize the extremal graphs. In Section 3, we are devoted to derive new upper and lower bounds for the k-th generalized distance eigenvalue of the graph G using signless Laplacian eigenvalues and the α-adjacency eigenvalues.

2. Bounds on Generalized Distance Spectral Radius

In this section, we obtain bounds for the generalized distance spectral radius, in terms of the diameter, the order, the minimum degree, the second minimum degree, the transmission degree, the second transmission degree and the parameter α.

The following lemma can be found in [31].

Lemma 1. *If A is an $n \times n$ non-negative matrix with the spectral radius $\lambda(A)$ and row sums $r_1, r_2, \ldots, r_n$, then*

$$\min_{1 \le i \le n} r_i \le \lambda(A) \le \max_{1 \le i \le n} r_i.$$

Moreover, if A is irreducible, then both of the equalities holds if and only if the row sums of A are all equal.

The following gives an upper bound for $\partial(G)$, in terms of the order n, the diameter d and the minimum degree δ of the graph G.

Theorem 1. *Let G be a connected graph of order n having diameter d and minimum degree δ. Then*

$$\partial(G) \le dn - \frac{d(d-1)}{2} - 1 - \delta(d-1), \tag{1}$$

with equality if and only if G is a regular graph with diameter ≤ 2.

Proof. First, it is easily seen that,

$$\begin{aligned} Tr_p = \sum_{j=1}^{n} d_{jp} &\le d_p + 2 + 3 + \cdots + (d-1) + d(n-1-d_p-(d-2)) \\ &= dn - \frac{d(d-1)}{2} - 1 - d_p(d-1), \quad \text{for all } p = 1, 2, \ldots, n. \end{aligned} \tag{2}$$

Let $Tr_{max} = \max\{Tr_G(v_i) : 1 \leq i \leq n\}$. For a matrix A denote $\lambda(A)$ its largest eigenvalue. We have

$$\begin{aligned} \partial(G) &= \lambda\Big(\alpha(Tr(G)) + (1-\alpha)D(G)\Big) \\ &\leq \alpha\lambda(Tr(G)) + (1-\alpha)\lambda(D(G)) \\ &\leq \alpha Tr_{max} + (1-\alpha)Tr_{max} = Tr_{max}. \end{aligned}$$

Applying Equation (2), the inequality follows.

Suppose that G is regular graph with diameter less than or equal to two, then all coordinates of the generalized distance Perron vector of G are equal. If $d = 1$, then $G \cong K_n$ and $\partial = n-1$. Thus equality in (1) holds. If $d = 2$, we get $\partial(G) = d_i + 2(n-1-d_i) = 2n-2-d_i$, and the equality in (1) holds. Note that the equality in (1) holds if and only if all coordinates the generalized distance Perron vector are equal, and hence $D_\alpha(G)$ has equal row sums.

Conversely, suppose that equality in (1) holds. This will force inequalities above to become equations. Then we get $Tr_1 = Tr_2 = \cdots = Tr_n = Tr_{\max}$, hence all the transmissions of the vertices are equal and so G is a transmission regular graph. If $d \geq 3$, then from the above argument, for every vertex v_i, there is exactly one vertex v_j with $d_G(v_i, v_j) = 2$, and thus $d = 3$, and for a vertex v_s of eccentricity 2,

$$\partial(G)x_s = d_s x_s + 2(n-1-d_s)x_s = \left(3n - \frac{3(3-1)}{2} - 1 - d_s(3-1)\right) x_s,$$

implying that $d_s = n-2$, giving that $G = P_4$. But the $D_\alpha(P_4)$ is not transmission regular graph. Therefore, G turns out to be regular and its diameter can not be greater than 2. □

Taking $\alpha = \frac{1}{2}$ in Theorem 1, we immediately get the following bound for the distance signless Laplacian spectral radius $\rho_1^Q(G)$, which was proved recently in [27].

Corollary 1. ([27], Theorem 2.6) *Let G be a connected graph of order $n \geq 3$, with minimum degree δ_1, second minimum degree δ_2 and diameter d. Then*

$$\rho_1^Q(G) \leq 2dn - d(d-1) - 2 - (\delta_1 + \delta_2)(d-1),$$

with equality if and only if G is (transmission) regular graph of diameter $d \leq 2$.

Proof. As $2D_{\frac{1}{2}}(G) = D^Q(G)$, letting $\delta = \delta_1$ in Theorem 1, we have

$$\rho_1^Q(G) = 2\partial(G) \leq 2dn - d(d-1) - 2 - 2\delta_1(d-1) \leq 2dn - d(d-1) - 2 - (\delta_1+\delta_2)(d-1),$$

and the result follows. □

Next, the generalized distance spectral radius $\partial(G)$ of a connected graph and its complement is characterized in terms of a Nordhaus-Gaddum type inequality.

Corollary 2. *Let G be a graph of order n, such that both G and its complement $\overline{G}$ are connected. Let δ and Δ be the minimum degree and the maximum degree of G, respectively. Then*

$$\partial(G) + \partial(\overline{G}) \leq 2nk - (t-1)(t+n+\delta-\Delta-1) - 2,$$

where $k = \max\{d, \overline{d}\}, t = \min\{d, \overline{d}\}$ and $d, \overline{d}$ are the diameters of G and $\overline{G}$, respectively.

Proof. Let $\overline{\delta}$ denote the minimum degree of $\overline{G}$. Then $\overline{\delta} = n - 1 - \Delta$, and by Theorem 1, we have

$$\begin{aligned}
\partial(G) + \partial(\overline{G}) &\leq dn - \frac{d(d-1)}{2} - 1 - \delta(d-1) + \bar{d}n - \frac{\bar{d}(\bar{d}-1)}{2} - 1 - \bar{\delta}(\bar{d}-1) \\
&= n(d+\bar{d}) - \frac{1}{2}(d(d-1) + \bar{d}(\bar{d}-1)) - 2 - \delta(d-1) - (n-1-\Delta)(\bar{d}-1) \\
&\leq 2nk - (t-1)(t+n+\delta-\Delta-1) - 2.
\end{aligned}$$

□

The following gives an upper bound for $\partial(G)$, in terms of the order n, the minimum degree $\delta = \delta_1$ and the second minimum degree δ_2 of the graph G.

Theorem 2. *Let G be a connected graph of order n having minimum degree δ_1 and second minimum degree δ_2. Then for $s = \delta_1 + \delta_2$, we have*

$$\partial(G) \leq \frac{\alpha\Psi + \sqrt{\alpha^2\Psi^2 + 4(1-2\alpha)\Theta}}{2}, \tag{3}$$

where $\Theta = \left(dn - \frac{d(d-1)}{2} - 1 - \delta_1(d-1)\right)\left(dn - \frac{d(d-1)}{2} - 1 - \delta_2(d-1)\right)$ *and* $\Psi = 2dn - d(d-1) - 2 - s(d-1)$. *Also equality holds if and only if G is a regular graph with diameter at most two.*

Proof. Let $X = (x_1, x_2, \ldots, x_n)^T$ be the generalized distance Perron vector of graph G and let $x_i = \max\{x_k | k = 1, 2, \ldots, n\}$ and $x_j = \max_{k \neq i}\{x_k | k = 1, 2, \ldots, n\}$. From the ith equation of $D_\alpha(G)X = \partial(G)X$, we obtain

$$\partial x_i = \alpha Tr_i x_i + (1-\alpha) \sum_{k=1, k\neq i}^{n} d_{ik} x_k \leq \alpha Tr_i x_i + (1-\alpha) Tr_i x_j. \tag{4}$$

Similarly, from the jth equation of $D_\alpha(G)X = \partial(G)X$, we obtain

$$\partial x_j = \alpha Tr_j x_j + (1-\alpha) \sum_{k=1, k\neq j}^{n} d_{jk} x_k \leq \alpha Tr_j x_j + (1-\alpha) Tr_j x_i. \tag{5}$$

Now, by (2), we have,

$$\left(\partial - \alpha\left(dn - \frac{d(d-1)}{2} - 1 - d_i(d-1)\right)\right) x_i \leq (1-\alpha)\left(dn - \frac{d(d-1)}{2} - 1 - d_i(d-1)\right) x_j$$
$$\left(\partial - \alpha\left(dn - \frac{d(d-1)}{2} - 1 - d_j(d-1)\right)\right) x_j \leq (1-\alpha)\left(dn - \frac{d(d-1)}{2} - 1 - d_j(d-1)\right) x_i.$$

Multiplying the corresponding sides of these inequalities and using the fact that $x_k > 0$ for all k, we obtain

$$\partial^2 - \alpha(2dn - d(d-1) - 2 - (d-1)(d_i + d_j))\partial - (1-2\alpha)\xi_i\xi_j \leq 0,$$

where $\xi_l = dn - \frac{d(d-1)}{2} - 1 - d_l(d-1)$, $l = i, j$, which in turn gives

$$\partial(G) \leq \frac{\alpha(2dn - d(d-1) - 2 - s(d-1)) + \sqrt{\alpha^2(2dn - d(d-1) - 2 - s(d-1))^2 + 4(1-2\alpha)\Theta}}{2}.$$

Now, using $d_i + d_j \geq \delta_1 + \delta_2$, the result follows.

Suppose that equality occurs in (3), then equality occurs in each of the above inequalities. If equality occurs in (4) and (5), the we obtain $x_i = x_k$, for all $k = 1, 2, \ldots, n$ giving that G is a transmission regular graph. Also, equality in (2), similar to that of Theorem 1, gives that G is a graph of diameter at most two and equality in $d_i + d_j \geq \delta_1 + \delta_2$ gives that G is a regular graph. Combining all these it follows that equality occurs in (3) if G is a regular graph of diameter at most two.

Conversely, if G is a connected δ-regular graph of diameter at most two, then $\partial(G) = Tr_i = dn - \frac{d(d-1)}{2} - 1 - d_i(d-1)$. Also

$$\begin{aligned} & \frac{\alpha(2dn - d(d-1) - 2 - s(d-1)) + \sqrt{\alpha^2(2dn - d(d-1) - 2 - s(d-1))^2 + 4(1-2\alpha)\Theta}}{2} \\ = & \frac{\alpha(2dn - d(d-1) - 2 - s(d-1)) + (2dn - d(d-1) - 2 - s(d-1))(1-\alpha)}{2} \\ = & dn - \frac{d(d-1)}{2} - 1 - \delta(d-1) = \partial(G). \end{aligned}$$

That completes the proof. □

Remark 1. *For any connected graph G of order n having minimum degree δ, the upper bound given by Theorem 2 is better than the upper bound given by Theorem 1. As*

$$\begin{aligned} & \frac{\alpha(2dn - d(d-1) - 2 - s(d-1)) + \sqrt{\alpha^2(2dn - d(d-1) - 2 - s(d-1))^2 + 4(1-2\alpha)\Theta}}{2}, \\ \leq & \frac{\alpha(2dn - d(d-1) - 2 - 2\delta(d-1)) + \sqrt{\alpha^2(2dn - d(d-1) - 2 - 2\delta(d-1))^2 + 4(1-2\alpha)\Phi}}{2}, \\ = & \frac{\alpha(2dn - d(d-1) - 2 - 2\delta(d-1)) + (2dn - d(d-1) - 2 - 2\delta(d-1))(1-\alpha)}{2} \\ = & dn - \frac{d(d-1)}{2} - 1 - \delta(d-1), \end{aligned}$$

where $\Phi = (2dn - d(d-1) - 2 - 2\delta(d-1))^2$.

The following gives an upper bound for $\partial(G)$ by using quantities like transmission degrees as well as second transmission degrees.

Theorem 3. *If the transmission degree sequence and the second transmission degree sequence of G are $\{Tr_1, Tr_2, \dots, Tr_n\}$ and $\{T_1, T_2, \dots, T_n\}$, respectively, then*

$$\partial(G) \leq \max_{1 \leq i \leq n} \left\{ \frac{-\beta + \sqrt{\beta^2 + 4(\alpha Tr_i^2 + (1-\alpha)T_i + \beta Tr_i)}}{2} \right\}, \tag{6}$$

where $\beta \geq 0$ is an unknown parameter. Equality occurs if and only if G is a transmission regular graph.

Proof. Let $X = (x_1, \dots, x_n)$ be the generalized distance Perron vector of G and $x_i = \max\{x_j | j = 1, 2, \dots, n\}$. Since

$$\begin{aligned} \partial(G)^2 X &= (D_\alpha(G))^2 X = (\alpha Tr + (1-\alpha)D)^2 X \\ &= \alpha^2 Tr^2 X + \alpha(1-\alpha) Tr D X + \alpha(1-\alpha) D Tr X + (1-\alpha)^2 D^2 X, \end{aligned}$$

we have

$$\partial^2(G) x_i = \alpha^2 Tr_i^2 x_i + \alpha(1-\alpha) Tr_i \sum_{j=1}^n d_{ij} x_j + \alpha(1-\alpha) \sum_{j=1}^n d_{ij} Tr_j x_j + (1-\alpha)^2 \sum_{j=1}^n \sum_{k=1}^n d_{ij} d_{jk} x_k.$$

Now, we consider a simple quadratic function of $\partial(G)$:

$$\begin{aligned}(\partial^2(G)+\beta\partial(G))X &= (\alpha^2 Tr^2 X+\alpha(1-\alpha)TrDX+\alpha(1-\alpha)DTrX+(1-\alpha)^2 D^2 X) \\ &+ \beta(\alpha TrX+(1-\alpha)DX).\end{aligned}$$

Considering the ith equation, we have

$$\begin{aligned}(\partial^2(G)+\beta\partial(G))x_i &= \alpha^2 Tr_i^2 x_i+\alpha(1-\alpha)Tr_i\sum_{j=1}^{n} d_{ij}x_j+\alpha(1-\alpha)\sum_{j=1}^{n} d_{ij}Tr_j x_j \\ &+ (1-\alpha)^2\sum_{j=1}^{n}\sum_{k=1}^{n} d_{ij}d_{jk}x_k+\beta\left(\alpha Tr_i x_i+\alpha(1-\alpha)\sum_{j=1}^{n} d_{ij}x_j\right).\end{aligned}$$

It is easy to see that the inequalities below are true

$$\alpha(1-\alpha)Tr_i\sum_{j=1}^{n} d_{ij}x_j\leq\alpha(1-\alpha)Tr_i^2 x_i,\ \alpha(1-\alpha)\sum_{j=1}^{n} d_{ij}Tr_j x_j\leq\alpha(1-\alpha)T_i x_i,$$

$$(1-\alpha)^2\sum_{j=1}^{n}\sum_{k=1}^{n} d_{jk}d_{ij}x_k\leq(1-\alpha)^2 T_i x_i,\ (1-\alpha)\sum_{j=1}^{n} d_{ij}x_j\leq(1-\alpha)Tr_i x_i.$$

Hence, we have

$$\begin{aligned}&(\partial^2(G)+\beta\partial(G))x_i\leq\alpha Tr_i^2 x_i-\alpha T_i x_i+T_i x_i+\beta Tr_i x_i\\ &\Rightarrow\ \partial^2(G)+\beta\partial(G)-(\alpha Tr_i^2-(\alpha-1)T_i+\beta Tr_i)\leq 0\\ &\Rightarrow\ \partial(G)\leq\frac{-\beta+\sqrt{\beta^2+4(\alpha Tr_i^2-(\alpha-1)T_i+\beta Tr_i)}}{2}.\end{aligned}$$

From this the result follows.

Now, suppose that equality occurs in (6), then each of the above inequalities in the above argument occur as equalities. Since each of the inequalities

$$\alpha(1-\alpha)Tr_i\sum_{j=1}^{n} d_{ij}x_j\leq\alpha(1-\alpha)Tr_i^2 x_i,\ \alpha(1-\alpha)\sum_{j=1}^{n} d_{ij}Tr_j x_j\leq\alpha(1-\alpha)T_i x_i$$

and

$$(1-\alpha)^2\sum_{j=1}^{n}\sum_{k=1}^{n} d_{jk}d_{ij}x_k\leq(1-\alpha)^2 T_i x_i,\ (1-\alpha)\sum_{j=1}^{n} d_{ij}x_j\leq(1-\alpha)Tr_i x_i,$$

occur as equalities if and only if G is a transmission regular graph. It follows that equality occurs in (6) if and only if G is a transmission regular graph. That completes the proof. □

The following upper bound for the generalized distance spectral radius $\partial(G)$ was obtained in [15]:

$$\partial(G)\leq\max_{1\leq i\leq n}\left\{\sqrt{\alpha Tr_i^2+(1-\alpha)T_i}\right\},\tag{7}$$

with equality if and only if $\alpha Tr_i^2+(1-\alpha)T_i$ is same for i.

Remark 2. *For a connected graph G having transmission degree sequence* $\{Tr_1, Tr_2, \ldots, Tr_n\}$ *and the second transmission degree sequence* $\{T_1, T_2, \ldots, T_n\}$, *provided that* $T_i \leq Tr_i^2$ *for all i, we have*

$$\frac{-\beta + \sqrt{\beta^2 + 4\alpha Tr_i^2 + 4(1-\alpha)T_i + 4\beta Tr_i}}{2} \leq \sqrt{\alpha Tr_i^2 + (1-\alpha)T_i}.$$

Therefore, the upper bound given by Theorem 3 is better than the upper bound given by (7).

If, in particular we take the parameter β in Theorem 3 equal to the vertex covering number τ, the edge covering number, the clique number ω, the independence number, the domination number, the generalized distance rank, minimum transmission degree, maximum transmission degree, etc., then Theorem 3 gives an upper bound for $\partial(G)$, in terms of the vertex covering number τ, the edge covering number, the clique number ω, the independence number, the domination number, the generalized distance rank, minimum transmission degree, maximum transmission degree, etc.

Let $x_i = \min\{x_j | j = 1, 2, \ldots, n\}$ be the minimum among the entries of the generalized distance Perron vector $X = (x_1, \ldots, x_n)$ of the graph G. Proceeding similar to Theorem 3, we obtain the following lower bound for $\partial(G)$, in terms of the transmission degrees, the second transmission degrees and a parameter β.

Theorem 4. *If the transmission degree sequence and the second transmission degree sequence of G are* $\{Tr_1, Tr_2, \ldots, Tr_n\}$ *and* $\{T_1, T_2, \ldots, T_n\}$, *respectively, then*

$$\partial(G) \geq \min_{1 \leq i \leq n} \left\{ \frac{-\beta + \sqrt{\beta^2 + 4(\alpha Tr_i^2 + (1-\alpha)T_i + \beta Tr_i)}}{2} \right\},$$

where $\beta \geq 0$ *is an unknown parameter. Equality occurs if and only if G is a transmission regular graph.*

Proof. Similar to the proof of Theorem 3 and is omitted. □

The following lower bound for the generalized distance spectral radius was obtained in [15]:

$$\partial(G) \geq \min_{1 \leq i \leq n} \left\{ \sqrt{\alpha Tr_i^2 + (1-\alpha)T_i} \right\}, \tag{8}$$

with equality if and only if $\alpha Tr_i^2 + (1-\alpha)T_i$ is same for i.

Similar to Remark 2, it can be seen that the lower bound given by Theorem 4 is better than the lower bound given by (8) for all graphs G with $T_i \geq Tr_i^2$, for all i.

Again, if in particular we take the parameter β in Theorem 4 equal to the vertex covering number τ, the edge covering number, the clique number ω, the independence number, the domination number, the generalized distance rank, minimum transmission degree, maximum transmission degree, etc, then Theorem 4 gives a lower bound for $\partial(G)$, in terms of the vertex covering number τ, the edge covering number, the clique number ω, the independence number, the domination number, the generalized distance rank, minimum transmission degree, maximum transmission degree, etc.

$G_1 \nabla G_2$ is referred to as *join* of G_1 and G_2. It is defined by joining every vertex in G_1 to every vertex in G_2.

Example 1. (a) *Let* C_4 *be the cycle of order 4. One can easily see that* C_4 *is a 4-transmission regular graph and the generalized distance spectrum of* C_4 *is* $\{4, 4\alpha, 6\alpha - 2^{[2]}\}$. *Hence,* $\partial(C_4) = 4$. *Moreover, the transmission degree sequence and the second transmission degree sequence of* C_4 *are* $\{4, 4, 4, 4\}$ *and*

$\{16,16,16,16\}$, respectively. Now, putting $\beta = Tr_{max} = 4$ in the given bound of Theorem 3, we can see that the equality holds:

$$\partial(C_4) \leq \frac{-4+\sqrt{16+4(16\alpha+16(1-\alpha)+16)}}{2} = \frac{-4+\sqrt{144}}{2} = 4.$$

(b) *Let W_{n+1} be the wheel graph of order $n+1$. It is well known that $W_{n+1} = C_n \nabla K_1$. The distance signless Laplacian matrix of W_5 is*

$$D^Q(W_5) = \begin{pmatrix} 5 & 1 & 2 & 1 & 1 \\ 1 & 5 & 1 & 2 & 1 \\ 2 & 1 & 5 & 1 & 1 \\ 1 & 2 & 1 & 5 & 1 \\ 1 & 1 & 1 & 1 & 4 \end{pmatrix}.$$

Hence the distance signless Laplacian spectrum of W_5 is $spec(W_5) = \left\{\frac{13+\sqrt{41}}{4}, \frac{13-\sqrt{41}}{4}, \frac{5}{2}, \frac{3}{2}^{[2]}\right\}$, and then the distance signless Laplacian spectral radius is $\rho_1^Q(W_5) = \frac{13+\sqrt{41}}{4}$. Also, the transmission degree sequence and the second transmission degree sequence of W_5 are $\{5,5,5,5,4\}$ and $\{24,24,24,24,20\}$, respectively. As $D_{\frac{1}{2}}(G) = \frac{1}{2}D^Q(G)$, taking $\alpha = \frac{1}{2}$ and $\beta = Tr_{max} = 5$ in the given bound of Theorem 3, we immediately get the following upper bound for the distance signless Laplacian spectral radius $\rho_1^Q(W_5)$:

$$\frac{1}{2}\rho_1^Q(W_5) \leq \frac{-5+\sqrt{25+50+48+100}}{2} = \frac{-5+\sqrt{223}}{2},$$

which implies that

$$\rho_1^Q(W_5) \leq -5+\sqrt{223} \simeq 9.93.$$

3. Bounds for the k-th Generalized Distance Eigenvalue

In this section, we discuss the relationship between the generalized distance eigenvalues and the other graph parameters.

The following lemma can be found in [37].

Lemma 2. *Let X and Y be Hermitian matrices of order n such that $Z = X+Y$, and denote the eigenvalues of a matrix M by $\lambda_1 \geq \lambda_2 \geq \cdots \geq \lambda_n$.Then*

$$\lambda_k(Z) \leq \lambda_j(X) + \lambda_{k-j+1}(Y), \quad n \geq k \geq j \geq 1,$$
$$\lambda_k(Z) \geq \lambda_j(X) + \lambda_{k-j+n}(Y), \quad n \geq j \geq k \geq 1,$$

where $\lambda_i(M)$ is the ith largest eigenvalue of the matrix M. Any equality above holds if and only if a unit vector can be an eigenvector corresponding to each of the three eigenvalues.

The following gives a relation between the generalized distance eigenvalues of the graph G of diameter 2 and the signless Laplacain eigenvalues of the complement $\overline{G}$ of the graph G. It also gives a relation between generalized distance eigenvalues of the graph G of diameter greater than or equal to 3 with the α-adjacency eigenvalues of the complement $\overline{G}$ of the graph G.

Theorem 5. *Let G be a connected graph of order $n \geq 4$ having diameter d. Let $\overline{G}$ be the complement of G and let $\overline{q}_1 \geq \overline{q}_2 \geq \cdots \geq \overline{q}_n$ be the signless Laplacian eigenvalues of $\overline{G}$. If $d = 2$, then for all $k = 1,2,\ldots,n$, we have*

$$(3\alpha-1)n - 2\alpha + (1-2\alpha)d_k + (1-\alpha)\overline{q}_k \leq \partial_k(G) \leq (2n-2)\alpha + (1-2\alpha)d_k + (1-\alpha)\overline{q}_k.$$

Equality occurs on the right if and only if $k = 1$ and G is a transmission regular graph and on the left if and only if $k \neq 1$ and G is a transmission regular graph.

If $d \geq 3$, then for all $k = 1, 2, \ldots, n$, we have

$$\alpha n - 1 + \lambda_k(A_\alpha(\overline{G})) + \lambda_n(M') \leq \partial_k(G) \leq n - 1 + \lambda_k(A_\alpha(\overline{G})) + \lambda_1(M'),$$

where $A_\alpha(\overline{G}) = \alpha\,\mathrm{Deg}(\overline{G}) + (1-\alpha)\overline{A}$ is the α-adjacency matrix of $\overline{G}$ and $M' = \alpha Tr'(G) + (1-\alpha)M$ with $M = (m_{ij})$ a symmetric matrix of order n having $m_{ij} = \max\{0, d_{ij} - 2\}$, d_{ij} is the distance between the vertices v_i, v_j and $Tr'(G) = \mathrm{diag}(Tr'_1, Tr'_2, \ldots, Tr'_n)$, $Tr'_i = \sum_{d_{ij} \geq 3} (d_{ij} - 2)$.

Proof. Let G be a connected graph of order $n \geq 4$ having diameter d. Let $\mathrm{Deg}(\overline{G}) = diag(n-1-d_1, n-1-d_2, \ldots, n-1-d_n)$ be the diagonal matrix of vertex degrees of $\overline{G}$. Suppose that diameter d of G is two, then transmission degree $Tr_i = 2n - 2 - d_i$, for all i, then the distance matrix of G can be written as $D(G) = A + 2\overline{A}$, where A and $\overline{A}$ are the adjacency matrices of G and $\overline{G}$, respectively. We have

$$\begin{aligned} D_\alpha(G) &= \alpha Tr(G) + (1-\alpha)D(G) = \alpha(2n-2)I - \alpha\,\mathrm{Deg}(G) + (1-\alpha)(A + 2\overline{A}) \\ &= \alpha(2n-2)I - \alpha\,\mathrm{Deg}(G) + (1-\alpha)(A + \overline{A}) + (1-\alpha)\overline{A} \\ &= (3n\alpha - n - 2\alpha)I + (1-\alpha)J + (1-2\alpha)\,\mathrm{Deg}(G) + (1-\alpha)Q(\overline{G}), \end{aligned}$$

where I is the identity matrix and J is the all one matrix of order n. Taking $Y = (3n\alpha - n - 2\alpha)I + (1 - 2\alpha)\,\mathrm{Deg}(G) + (1-\alpha)Q(\overline{G})$, $X = (1-\alpha)J$, $j = 1$ in the first inequality of Lemma 2 and using the fact that $spec(J) = \{n, 0^{[n-1]}\}$, it follows that

$$\partial_k(G) \leq (2n-2)\alpha + (1-2\alpha)d_k + (1-\alpha)\overline{q}_k, \quad \text{for all} \quad k = 1, 2, \ldots, n. \tag{9}$$

Again, taking $Y = (3n\alpha - n - 2\alpha)I + (1-2\alpha)\,\mathrm{Deg}(G) + (1-\alpha)Q(\overline{G})$, $X = (1-\alpha)J$ and $j = n$ in the second inequality of Lemma 2, it follows that

$$\partial_k(G) \geq (3\alpha - 1)n - 2\alpha + (1-2\alpha)d_k + (1-\alpha)\overline{q}_k, \quad \text{for all} \quad k = 1, 2, \ldots, n. \tag{10}$$

Combining (9) and (10) the first inequality follows. Equality occurs in first inequality if and only if equality occurs in (9) and (10). Suppose that equality occurs in (9), then by Lemma 2, the eigenvalues ∂_k, $(3n-2)\alpha - n + (1-2\alpha)d_k + (1-\alpha)\overline{q}_k$ and $n(1-\alpha)$ of the matrices $D_\alpha(G)$, X and Y have the same unit eigenvector. Since $\mathbf{1} = \frac{1}{n}(1, 1, \ldots, 1)^T$ is the unit eigenvector of Y for the eigenvalue $n(1-\alpha)$, it follows that equality occurs in (9) if and only if $\mathbf{1}$ is the unit eigenvector for each of the matrices $D_\alpha(G)$, X and Y. This gives that G is a transmission regular graph and $\overline{G}$ is a regular graph. Since a graph of diameter 2 is regular if and only if it is transmission regular and complement of a regular graph is regular. Using the fact that for a connected graph G the unit vector $\mathbf{1}$ is an eigenvector for the eigenvalue ∂_1 if and only if G is transmission regular graph, it follows that equality occurs in first inequality if and only if $k = 1$ and G is a transmission regular graph.

Suppose that equality occurs in (10), then again by Lemma 2, the eigenvalues ∂_k, $(3n-2)\alpha - n + (1-2\alpha)d_k + (1-\alpha)\overline{q}_k$ and 0 of the matrices $D_\alpha(G)$, X and Y have the same unit eigenvector $\mathbf{x}$. Since $J\mathbf{x} = 0$, it follows that $\mathbf{x}^T\mathbf{1} = 0$. Using the fact that the matrix J is symmetric(so its normalized eigenvectors are orthogonal [43]), we conclude that the vector $\mathbf{1}$ belongs to the set of eigenvectors of the matrix J and so of the matrices $D_\alpha(G)$, X. Now, $\mathbf{1}$ is an eigenvector of the matrices $D_\alpha(G)$ and X, gives that G is a regular graph. Since for a regular graph of diameter 2 any eigenvector of $Q(\overline{G})$ and $D_\alpha(G)$ is orthogonal to $\mathbf{1}$, it follows that equality occurs in (10) if and only if $k \neq 1$ and G is a regular graph.

If $d \geq 3$, we define the matrix $M = (m_{ij})$ of order n, where $m_{ij} = \max\{0, d_{ij} - 2\}$, d_{ij} is the distance between the vertices v_i and v_j. The transmission of a vertex v_i can be written as $Tr_i = d_i + 2\overline{d}_i + Tr'_i$, where $Tr'_i = \sum_{d_{ij}\geq 3}(d_{ij}-2)$, is the contribution from the vertices which are at distance more than two from v_i. For $Tr'(G) = \text{diag}(Tr'_1, Tr'_2, \ldots, Tr'_n)$, we have

$$\begin{aligned} D_\alpha(G) &= \alpha Tr(G) + (1-\alpha)D(G) = \alpha\,\text{Deg}(G) + 2\alpha\,\text{Deg}(\overline{G}) + \alpha Tr'(G) + (1-\alpha)(A + 2\overline{A} + M) \\ &= \alpha(\text{Deg}(G) + \text{Deg}(\overline{G})) + (1-\alpha)(A+\overline{A}) + (\alpha\,\text{Deg}(\overline{G}) + (1-\alpha)\overline{A}) + (\alpha Tr'(G) + (1-\alpha)M) \\ &= D_\alpha(K_n) + A_\alpha(\overline{G}) + M', \end{aligned}$$

where $A_\alpha(\overline{G})$ is the α-adjacency matrix of $\overline{G}$ and $M' = \alpha Tr'(G) + (1-\alpha)M$. Taking $X = D_\alpha(K_n)$, $Y = A_\alpha(\overline{G}) + M'$ and $j = 1$ in the first inequality of Lemma 2 and using the fact that $spec(D_\alpha(K_n)) = \{n-1, \alpha n - 1^{[n-1]}\}$, it follows that

$$\partial_k(G) \leq n - 1 + \lambda_k(A_\alpha(\overline{G}) + M'), \quad \text{for all} \quad k = 1, 2, \ldots, n.$$

Again, taking $Y = A_\alpha(\overline{G})$, $X = M'$ and $j = 1$ in the first inequality of Lemma 2, we obtain

$$\partial_k(G) \leq n - 1 + \lambda_k(A_\alpha(\overline{G})) + \lambda_1(M'), \quad \text{for all} \quad k = 1, 2, \ldots, n. \tag{11}$$

Similarly, taking $X = D_\alpha(K_n)$, $Y = A_\alpha(\overline{G}) + M'$ and $j = n$ and then $Y = A_\alpha(\overline{G})$, $X = M'$ and $j = n$ in the second inequality of Lemma 2, we obtain

$$\partial_k(G) \geq \alpha n - 1 + \lambda_k(A_\alpha(\overline{G})) + \lambda_n(M'), \quad \text{for all} \quad k = 1, 2, \ldots, n. \tag{12}$$

From (11) and (12) the second inequality follows. That completes the proof. □

It can be seen that the matrix M' defined in Theorem 5 is positive semi-definite for all $\frac{1}{2} \leq \alpha \leq 1$. Therefore, we have the following observation from Theorem 5.

Corollary 3. *Let G be a connected graph of order $n \geq 4$ having diameter $d \geq 3$. If $\frac{1}{2} \leq \alpha \leq 1$, then*

$$\partial_k(G) \geq \alpha n - 1 + \lambda_k(A_\alpha(\overline{G})), \quad \textit{for all} \quad k = 1, 2, \ldots, n,$$

where $A_\alpha(\overline{G}) = \alpha\,\text{Deg}(\overline{G}) + (1-\alpha)\overline{A}$ is the α-adjacency matrix of $\overline{G}$.

It is clear from Corollary 3 that for $\frac{1}{2} \leq \alpha \leq 1$, any lower bound for the α-adjacency $\lambda_k(A_\alpha(\overline{G}))$ gives a lower bound for ∂_k and conversely any upper bound for ∂ gives an upper bound for $\lambda_k(A_\alpha(\overline{G}))$. We note that Theorem 5 generalizes one of the Theorems (namely Theorem 3.8) given in [8].

Example 2. (a) *Let C_n be a cycle of order n. It is well known (see [7]) that C_n is a k-transmission regular graph with $k = \frac{n^2}{4}$ if n is even and $k = \frac{n^2-1}{4}$ if n is odd. Let $n = 4$. It is clear that the distance spectrum of the graph C_4 is $\{4, 0, -2^{[2]}\}$. Also, since C_4 is a 4-transmission regular graph, then $Tr(C_4) = 4I_4$ and so $D_\alpha(C_4) = 4\alpha I_4 + (1-\alpha)D(C_4)$. Hence the generalized distance spectrum of C_4 is $\{4, 4\alpha, 6\alpha - 2^{[2]}\}$. Moreover, the signless Laplacian spectrum of $\overline{C_4}$ is $\{2^{[2]}, 0^{[2]}\}$. Since the diameter of C_4 is 2, hence, applying Theorem 5, for $k = 1$, we have,*

$$4\alpha = 4(3\alpha - 1) - 2\alpha + 2(1-2\alpha) + 2(1-\alpha) \leq \partial_1(C_4) = 4 \leq 6\alpha + 2(1-2\alpha) + 2(1-\alpha) = 4,$$

which shows that the equality occurs on right for $k = 1$ and transmission regular graph C_4.

Also, for $k = 2$, we have

$$4\alpha = 4(3\alpha - 1) - 2\alpha + 2(1-2\alpha) + 2(1-\alpha) \leq \partial_2(C_4) = 4\alpha \leq 6\alpha + 2(1-2\alpha) + 2(1-\alpha) = 4,$$

which shows that the equality occurs on left for $k = 2$ and transmission regular graph C_4.

(b) *Let C_6 be a cycle of order 6. It is clear that the distance spectrum of the graph C_6 is $\{9, 0^{[2]}, -1, -4^{[2]}\}$. Since C_6 is a 9-transmission regular graph, then $Tr(C_6) = 9I_6$ and so $D_\alpha(C_6) = 9\alpha I_6 + (1-\alpha)D(C_6)$. Hence, the generalized distance spectrum of C_6 is $\{9, 9\alpha^{[2]}, 10\alpha - 1, 13\alpha - 4^{[2]}\}$. Also, the α-adjacency spectrum of C_6 is $\{3, 2\alpha + 1, 3\alpha^{[2]}, 5\alpha - 2^{[2]}\}$. Let M' be the matrix defined by the Theorem 5, hence the spectrum of M' is $\{1^{[3]}, 2\alpha - 1^{[3]}\}$. Since diameter of the graph C_6 is 3, hence, applying Theorem 5, for $k = 1$, we have*

$$8\alpha + 1 = 6\alpha - 1 + 3 + 2\alpha - 1 \leq \partial_1(C_6) = 9 \leq 5 + 3 + 1 = 9.$$

Also for $k = 2$, we have

$$10\alpha - 1 = 6\alpha - 1 + 2\alpha + 1 + 2\alpha - 1 \leq \partial_2(C_6) = 9\alpha \leq 5 + 2\alpha + 1 + 1 = 2\alpha + 7.$$

We need the following lemma proved by Hoffman and Wielandt [39].

Lemma 3. *Suppose we have $C = A + B$. Here, all these matrices are symmetric and have order n. Suppose they have the eigenvalues α_i, β_i, and γ_i, where $1 \leq i \leq n$, respectively arranged in non-increasing order. Therefore, $\sum_{i=1}^{n}(\gamma_i - \alpha_i)^2 \leq \sum_{i=1}^{n}\beta_i^2$.*

The following gives relation between generalized distance spectrum and distance spectrum for a simple connected graph G. We use $[n]$ to denote the set of $\{1, 2, \ldots, n\}$. For each subset S of $[n]$, we use S^c to denote $[n] - S$.

Theorem 6. *Let G be a connected graph of order n and let $\mu_1, \ldots, \mu_n$ be the eigenvalues of the distance matrix of G. Then for each non-empty subset $S = \{r_1, r_2, \ldots, r_k\}$ of $[n]$, we have the following inequalities:*

$$\begin{aligned} & \frac{2k\alpha W(G) - \sqrt{k(n-k)\left(n\sum_{i=1}^{n}\alpha^2 Tr_i^2 - 4\alpha^2 W^2(G)\right)}}{n} \\ \leq & \sum_{i\in S}(\partial_i + (\alpha-1)\mu_i) \\ \leq & \frac{2k\alpha W(G) + \sqrt{k(n-k)\left(n\sum_{i=1}^{n}\alpha^2 Tr_i^2 - 4\alpha^2 W^2(G)\right)}}{n}. \end{aligned}$$

Proof. Since $D_\alpha(G) = \alpha Tr(G) + (1-\alpha)D(G)$, then by the fact that $2\alpha W(G) = \sum_{i=1}^{n}(\partial_i + (\alpha-1)\mu_i)$, we get $2\alpha W(G) - \sum_{i\in S}(\partial_i + (\alpha-1)\mu_i) = \sum_{i\in S^c}(\partial_i + (\alpha-1)\mu_i)$. By Cauchy-Schwarz inequality, we further have that

$$\left(2\alpha W(G) - \sum_{i\in S}(\partial_i + (\alpha-1)\mu_i)\right)^2 \leq \sum_{i\in S^C} 1^2 \sum_{i\in S^C}(\partial_i + (\alpha-1)\mu_i)^2.$$

Therefore

$$\begin{aligned} & \left(2\alpha W(G) - \sum_{i\in S}(\partial_i + (\alpha-1)\mu_i)\right)^2 \\ & \leq (n-k)\left(\sum_{i=1}^{n}(\partial_i + (\alpha-1)\mu_i)^2 - \sum_{i\in S}(\partial_i + (\alpha-1)\mu_i)^2\right). \end{aligned}$$

By Lemma 3, we have that

$$\left(2\alpha W(G) - \sum_{i\in S}(\partial_i + (\alpha-1)\mu_i)\right)^2 + (n-k)\sum_{i\in S}(\partial_i + (\alpha-1)\mu_i)^2$$
$$\leq (n-k)\sum_{i=1}^{n}(\partial_i + (\alpha-1)\mu_i)^2 \leq (n-k)\sum_{i=1}^{n}\alpha^2 Tr_i^2.$$

Again by Cauchy-Schwarz inequality, we have that

$$\left(\frac{n-k}{k}\right)\left(\sum_{i\in S}(\partial_i + (\alpha-1)\mu_i)\right)^2 = \left(\sum_{i\in S}\sqrt{\frac{n-k}{k}}(\partial_i + (\alpha-1)\mu_i)\right)^2$$
$$\leq \sum_{i\in S}\left(\frac{n-k}{k}\right)\sum_{i\in S}(\partial_i + (\alpha-1)\mu_i)^2 = (n-k)\sum_{i\in S}(\partial_i + (\alpha-1)\mu_i)^2.$$

Therefore, we have the following inequality

$$\left(2\alpha W(G) - \sum_{i\in S}(\partial_i + (\alpha-1)\mu_i)\right)^2 + \left(\frac{n-k}{k}\right)\left(\sum_{i\in S}(\partial_i + (\alpha-1)\mu_i)\right)^2$$
$$\leq (n-k)\sum_{i=1}^{n}\alpha^2 Tr_i^2.$$

Solving the quadratic inequality for $\sum_{i\in S}(\partial_i + (\alpha-1)\mu_i)$, so we complete the proof. □

Notice that $\sum_{i=1}^{n}(\partial_i - \alpha Tr_i) = 0$ and by Lemma 3, we also have $\sum_{i=1}^{n}(\partial_i - \alpha Tr_i)^2 \leq (1-\alpha)^2\sum_{i=1}^{n}\mu_i^2 = 2(1-\alpha)^2\sum_{1\leq i<j\leq n}^{n} d_{ij}^2$. We can similarly prove the following theorem.

Theorem 7. *Let G be a connected graph of order n. Then for each non-empty subset $S = \{r_1, r_2, \ldots, r_k\}$ of $[n]$, we have:*

$$\left|\sum_{i\in S}(\partial_i - \alpha Tr_i)\right| \leq \sqrt{\frac{2k(n-k)(1-\alpha)^2\sum_{1\leq i<j\leq n} d_{ij}^2}{n}}.$$

We conclude by giving the following bounds for the k-th largest generalized distance eigenvalue of a graph.

Theorem 8. *Assume G is connected and is of order n. Suppose it has diameter d and δ is its minimum degree. Let*

$$\begin{aligned}\varphi(G) &= \min\Bigg\{n^2(n-1)\left(\frac{\alpha^2 n^2(n-1)}{4} + (1-\alpha)^2 d^2\right) - 4\alpha^2 W^2(G),\\ &\quad n\left(\alpha^2\left(nd - \frac{d(d-1)}{2} - 1 - \delta(d-1)\right)^2 + (1-\alpha)^2 n(n-1)d^2\right) - 4\alpha^2 W^2(G)\Bigg\}.\end{aligned}$$

Then for $k = 1, \ldots, n$,

$$\frac{1}{n}\left\{2\alpha W(G) - \sqrt{\frac{k-1}{n-k+1}\varphi(G)}\right\} \leq \partial_k(G) \leq \frac{1}{n}\left\{2\alpha W(G) + \sqrt{\frac{n-k}{k}\varphi(G)}\right\}. \tag{13}$$

Proof. First we prove the upper bound. It is clear that

$$\text{trace}(D_\alpha^2(G)) = \sum_{i=1}^{k} \partial_i^2 + \sum_{i=k+1}^{n} \partial_i^2 \geq \frac{(\sum_{i=1}^{k} \partial_i)^2}{k} + \frac{(\sum_{i=k+1}^{n} \partial_i)^2}{n-k}.$$

Let $M_k = \sum_{i=1}^{k} \partial_i$. Then

$$\text{trace}(D_\alpha^2(G)) \geq \frac{M_k^2}{k} + \frac{(2\alpha W(G) - M_k)^2}{n-k},$$

which implies

$$\partial_k(G) \leq \frac{M_k}{k} \leq \frac{1}{n}\left\{2\alpha W(G) + \sqrt{\frac{n-k}{k}[n \cdot \text{trace}(D_\alpha^2(G)) - 4\alpha^2 W^2(G)]}\right\}.$$

We observe that

$$\begin{aligned} n \cdot \text{trace}(D_\alpha^2(G)) - 4\alpha^2 W^2(G) &= n\alpha^2 \sum_{i=1}^{n} Tr_i^2 + 2n(1-\alpha)^2 \sum_{1 \leq i < j \leq n} (d_{ij})^2 - 4\alpha^2 W^2(G) \\ &\leq n\alpha^2 \frac{n^3(n-1)^2}{4} + 2n(1-\alpha)^2 \frac{n(n-1)}{2} d^2 - 4\alpha^2 W^2(G) \\ &= n^2(n-1)\left(\frac{\alpha^2 n^2(n-1)}{4} + (1-\alpha)^2 d^2\right) - 4\alpha^2 W^2(G), \end{aligned}$$

since $Tr_i \leq \frac{n(n-1)}{2}$, and

$$\begin{aligned} & n \cdot \text{trace}(D_\alpha^2(G)) - 4\alpha^2 W^2(G) \\ &= n\alpha^2 \sum_{i=1}^{n} Tr_i^2 + 2n(1-\alpha)^2 \sum_{1 \leq i < j \leq n} (d_{ij})^2 - 4\alpha^2 W^2(G) \\ &\leq n\alpha^2 \left(nd - \frac{d(d-1)}{2} - 1 - \delta(d-1)\right)^2 + 2n(1-\alpha)^2 \frac{n(n-1)}{2} d^2 - 4\alpha^2 W^2(G) \\ &= n\left(\alpha^2 \left(nd - \frac{d(d-1)}{2} - 1 - \delta(d-1)\right)^2 + (1-\alpha)^2 n(n-1) d^2\right) - 4\alpha^2 W^2(G), \end{aligned}$$

since $Tr_i \leq nd - \frac{d(d-1)}{2} - 1 - d_i(d-1)$. Hence, we get the right-hand side of the inequality (13).

Now, we prove the lower bound. Let $N_k = \sum_{i=k}^{n} \partial_i$. Then we have

$$\begin{aligned} \text{trace}(D_\alpha^2(G)) = \sum_{i=1}^{k-1} \partial_i^2 + \sum_{i=k}^{n} \partial_i^2 &\geq \frac{\left(\sum_{i=1}^{k-1} \partial_i\right)^2}{k-1} + \frac{\left(\sum_{i=k}^{n} \partial_i\right)^2}{n-k+1} \\ &= \frac{(2\alpha W(G) - N_k)^2}{k-1} + \frac{N_k^2}{n-k+1}. \end{aligned}$$

Hence

$$\partial_k(G) \geq \frac{N_k}{n-k+1} \geq \frac{1}{n}\left\{2\alpha W(G) - \sqrt{\frac{k-1}{n-k+1}[n \cdot \text{trace}(D_\alpha^2(G)) - 4\alpha^2 W^2(G)]}\right\},$$

and we get the left-hand side of the inequality (13). □

By a *chemical tree*, we mean a tree which has all vertices of degree less than or equal to 4.

Example 3. *In Figure 1, we depicted a chemical tree of order* $n = 5$.

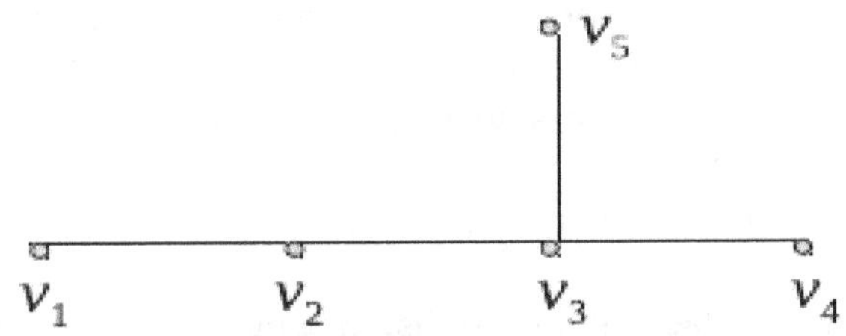

Figure 1. A chemical tree T.

The distance matrix of T is

$$D(T) = \begin{pmatrix} 0 & 1 & 2 & 3 & 3 \\ 1 & 0 & 1 & 2 & 2 \\ 2 & 1 & 0 & 1 & 1 \\ 3 & 2 & 1 & 0 & 2 \\ 3 & 2 & 1 & 2 & 0 \end{pmatrix}.$$

Let $\mu_1, \ldots, \mu_5$ *be the distance eigenvalues of the tree T. Then one can easily see that* $\mu_1 = 7.46$, $\mu_2 = -0.51$, $\mu_3 = -1.08$, $\mu_4 = -2$ *and* $\mu_5 = -3.86$. *Note that, as* $D_0(T) = D(T)$, *taking* $\alpha = 0$ *in Theorem 8, then for* $n = 5$ *we get* $-6\sqrt{\frac{k-1}{6-k}} \le \mu_k \le 6\sqrt{\frac{5-k}{k}}$, *for any* $1 \le k \le 5$. *For example,* $-6 \le \mu_1 \le 12$ *and* $-3 \le \mu_2 \le 7.3$.

4. Conclusions

Motivated by an article entitled "Merging the A- and Q-spectral theories" by V. Nikiforov [33], recently, Cui et al. [15] dealt with the integration of spectra of distance matrix and distance signless Laplacian through elegant convex combinations accommodating vertex transmissions as well as distance matrix. For $\alpha \in [0,1]$, the generalized distance matrix is known as $D_\alpha(G) = \alpha Tr(G) + (1-\alpha)D(G)$. Our results shed light on some properties of $D_\alpha(G)$ and contribute to establishing new inequalities (such as lower and upper bounds) connecting varied interesting graph invariants. We established some bounds for the generalized distance spectral radius for a connected graph using various identities like the number of vertices n, the diameter, the minimum degree, the second minimum degree, the transmission degree, the second transmission degree and the parameter α, improving some bounds recently given in the literature. We also characterized the extremal graphs attaining these bounds. Notice that the current work mainly focuses to determine some bounds for the spectral radius (largest eigenvalue) of the generalized distance matrix. It would be interesting to derive some bounds for other important eigenvalues such as the smallest eigenvalue as well as the second largest eigenvalue of this matrix.

Author Contributions: conceptualization, A.A., M.B. and H.A.G.; formal analysis, A.A., M.B., H.A.G. and Y.S.; writing—original draft preparation, A.A., M.B. and H.A.G.; writing—review and editing, A.A., M.B., H.A.G. and Y.S.; project administration, A.A.; funding acquisition, Y.S.

Acknowledgments: The authors would like to thank the academic editor and the four anonymous referees for their constructive comments that helped improve the quality of the paper.

References

1. Alhevaz, A.; Baghipur, M.; Ganie, H.A.; Pirzada, S. Brouwer type conjecture for the eigenvalues of distance signless Laplacian matrix of a graph. *Linear Multilinear Algebra* **2019**. [CrossRef]
2. Alhevaz, A.; Baghipur, M.; Hashemi, E. Further results on the distance signless Laplacian spectrum of graphs. *Asian-Eur. J. Math.* **2018**, *11*, 1850066. [CrossRef]

3. Alhevaz, A.; Baghipur, M.; Hashemi, E.; Ramane, H.S. On the distance signless Laplacian spectrum of graphs. *Bull. Malay. Math. Sci. Soc.* **2019**, *42*, 2603–2621. [CrossRef]
4. Alhevaz, A.; Baghipur, M.; Paul, S. On the distance signless Laplacian spectral radius and the distance signless Laplacian energy of graphs. *Discrete Math. Algorithms Appl.* **2018**, *10*, 1850035. [CrossRef]
5. Alhevaz, A.; Baghipur, M.; Pirzada, S. On distance signless Laplacian Estrada index and energy of graphs. *Kragujevac J. Math.* **2021**, *45*, 837–858.
6. Aouchiche, M.; Hansen, P. Distance spectra of graphs: A survey. *Linear Algebra Appl.* **2014**, *458*, 301–386. [CrossRef]
7. Aouchiche, M.; Hansen, P. Two Laplacians for the distance matrix of a graph. *Linear Algebra Appl.* **2013**, *439*, 21–33. [CrossRef]
8. Aouchiche, M.; Hansen, P. On the distance signless Laplacian of a graph. *Linear Multilinear Algebra* **2016**, *64*, 1113–1123. [CrossRef]
9. Aouchiche, M.; Hansen, P. Some properties of distance Laplacian spectra of a graph. *Czechoslovak Math. J.* **2014**, *64*, 751–761. [CrossRef]
10. Aouchiche, M.; Hansen, P. Distance Laplacian eigenvalues and chromatic number in graphs. *Filomat* **2017**, *31*, 2545–2555. [CrossRef]
11. Aouchiche, M.; Hansen, P. Cospectrality of graphs with respect to distance matrices. *Appl. Math. Comput.* **2018**, *325*, 309–321. [CrossRef]
12. Bapat, R.B. Determinant of the distance matrix of a tree with matrix weights. *Linear Algebra Appl.* **2006**, *416*, 2–7. [CrossRef]
13. Bapat, R.B.; Kirkland, S.J.; Neumann, M. On distance matrices and Laplacians. *Linear Algebra Appl.* **2005**, *401*, 193–209. [CrossRef]
14. Bapat, R.B.; Lal, A.K.; Pati, S. A q-analogue of the distance matrix of a tree. *Linear Algebra Appl.* **2006**, *416*, 799–814. [CrossRef]
15. Cui, S.Y.; He, J.X.; Tian, G.X. The generalized distance matrix. *Linear Algebra Appl.* **2019**, *563*, 1–23. [CrossRef]
16. Cvetković, D. Signless Laplacians and line graphs. *Bull. Acad. Serbe Sci. Arts Cl. Sci. Math. Natur. Sci. Math.* **2005**, *131*, 85–92. [CrossRef]
17. Cvetković, D. New theorems for signless Laplacians eigenvalues. *Bull. Acad. Serbe Sci. Arts Cl. Sci. Math. Natur. Sci. Math.* **2008**, *137*, 131–146.
18. Cvetković, D.; Simić, S.K. Towards a spectral theory of graphs based on the signless Laplacian I. *Publ. Inst. Math. (Beograd)* **2009**, *85*, 19–33. [CrossRef]
19. Cvetković, D.; Simić, S.K. Towards a spectral theory of graphs based on the signless Laplacian II. *Linear Algebra Appl.* **2010**, *432*, 2257–2272. [CrossRef]
20. Cvetković, D.; Simić, S.K. Towards a spectral theory of graphs based on the signless Laplacian III. *Appl. Anal. Discrete Math.* **2010**, *4*, 156–166.
21. Das, K.C.; Aouchiche, M.; Hansen, P. On distance Laplacian and distance signless Laplacian eigenvalues of graphs. *Linear Multilinear Algebra* **2019**, *67*, 2307–2324. [CrossRef]
22. Das, K.C.; Aouchiche, M.; Hansen, P. On (distance) Laplacian energy and (distance) signless Laplacian energy of graphs. *Discrete Appl. Math.* **2018**, *243*, 172–185. [CrossRef]
23. Diaz, R.C.; Rojo, O. Sharp upper bounds on the distance energies of a graph. *Linear Algebra Appl.* **2018**, *545*, 55–75. [CrossRef]
24. Duan, X.; Zhou, B. Sharp bounds on the spectral radius of a nonnegative matrix. *Linear Algebra Appl.* **2013**, *439*, 2961–2970. [CrossRef]
25. Fernandes, R.; de Freitas, M.A.A.; da Silva, C.M., Jr.; Del-Vecchio, R.R. Multiplicities of distance Laplacian eigenvalues and forbidden subgraphs. *Linear Algebra Appl.* **2018**, *541*, 81–93. [CrossRef]
26. Li, X.; Fan, Y.; Zha, S. A lower bound for the distance signless Laplacian spectral radius of graphs in terms of chromatic number. *J. Math. Res. Appl.* **2014**, *34*, 289–294.
27. Li, D.; Wang, G.; Meng, J. On the distance signless Laplacian spectral radius of graphs and digraphs. *Electr. J. Linear Algebra* **2017**, *32*, 438–446. [CrossRef]
28. Liu, S.; Shu, J. On the largest distance (signless Laplacian) eigenvalue of non-transmission-regular graphs. *Electr. J. Linear Algebra* **2018**, *34*, 459–471. [CrossRef]
29. Lu, L.; Huang, Q.; Huang, X. On graphs with distance Laplacian spectral radius of multiplicity $n-3$. *Linear Algebra Appl.* **2017**, *530*, 485–499. [CrossRef]

30. Lu, L.; Huang, Q.; Huang, X. On graphs whose smallest distance (signless Laplacian) eigenvalue has large multiplicity. *Linear Multilinear Algebra* **2018**, *66*, 2218–2231. [CrossRef]
31. Minć, H. *Nonnegative Matrices*; John Wiley & Sons: New York, NY, USA, 1988.
32. Lin, H.; Zhou, B. The distance spectral radius of trees. *Linear Multilinear Algebra* **2019**, *67*, 370–390. [CrossRef]
33. Nikiforov, V. Merging the *A*- and *Q*-spectral theories. *Appl. Anal. Discrete Math.* **2017**, *11*, 81–107. [CrossRef]
34. Shang, Y. Distance Estrada index of random graphs. *Linear Multilinear Algebra* **2015**, *63*, 466–471. [CrossRef]
35. Shang, Y. Estimating the distance Estrada index. *Kuwait J. Sci.* **2016**, *43*, 14–19.
36. Shang, Y. Bounds of distance Estrada index of graphs. *Ars Combin.* **2016**, *128*, 287–294.
37. So, W. Commutativity and spectra of Hermitian matrices. *Linear Algebra Appl.* **1994**, *212–213*, 121–129. [CrossRef]
38. Yang, J.; You, L.; Gutman, I. Bounds on the distance Laplacian energy of graphs. *Kragujevac J. Math.* **2013**, *37*, 245–255.
39. Wilkinson, J.H. *The Algebraic Eigenvalue Problem*; Oxford University Press: New York, NY, USA, 1965.
40. Xing, R.; Zhou, B. On the distance and distance signless Laplacian spectral radii of bicyclic graphs. *Linear Algebra Appl.* **2013**, *439*, 3955–3963. [CrossRef]
41. Xing, R.; Zhou, B.; Li, J. On the distance signless Laplacian spectral radius of graphs. *Linear Multilinear Algebra* **2014**, *62*, 1377–1387. [CrossRef]
42. Xue, J.; Liu, S.; Shu, J. The complements of path and cycle are determined by their distance (signless) Laplacian spectra. *Appl. Math. Comput.* **2018**, *328*, 137–143. [CrossRef]
43. Zhang, F. *Matrix Theory: Basic Results and Techniques*; Springer: New York, NY, USA, 1999.

8

Fuzzy Normed Rings

Aykut Emniyet [1] and Memet Şahin [2,*]

1 Department of Mathematics, Osmaniye Korkut Ata University, 8000 Osmaniye, Turkey; aykutemniyet@osmaniye.edu.tr
2 Department of Mathematics, Gaziantep University, 27310 Gaziantep, Turkey
* Correspondence: mesahin@gantep.edu.tr.

Abstract: In this paper, the concept of fuzzy normed ring is introduced and some basic properties related to it are established. Our definition of normed rings on fuzzy sets leads to a new structure, which we call a fuzzy normed ring. We define fuzzy normed ring homomorphism, fuzzy normed subring, fuzzy normed ideal, fuzzy normed prime ideal, and fuzzy normed maximal ideal of a normed ring, respectively. We show some algebraic properties of normed ring theory on fuzzy sets, prove theorems, and give relevant examples.

Keywords: Fuzzy sets; ring; normed space; fuzzy normed ring; fuzzy normed ideal

---

1. Introduction

Normed rings attracted attention of researchers after the studies by Naimark [1], a generalization of normed rings [2] and commutative normed rings [3]. Naimark defined normed rings in an algebraic fashion, while Gel'fand addressed them as complex Banach spaces and introduced the notion of commutative normed rings. In Reference [4], Jarden defined the ultrametric absolute value and studied the properties of normed rings in a more topological perspective. During his invaluable studies, Zadeh [5] presented fuzzy logic theory, changing the scientific history forever by making a modern definition of vagueness and using the sets without strict boundaries. As, in almost every aspect of computational science, fuzzy logic also became a convenient tool in classical algebra. Zimmermann [6] made significant contributions to the fuzzy set theory. Mordeson, Bhutani, and Rosenfeld [7] defined fuzzy subgroups, Liu [8], Mukherjee, and Bhattacharya [9] examined normal fuzzy subgroups. Liu [8] also discussed fuzzy subrings and fuzzy ideals. Wang, Ruan and Kerre [10] studied fuzzy subrings and fuzzy rings. Swamy and Swamy [11] defined and proved major theorems on fuzzy prime ideals of rings. Gupta and Qi [12] are concerned with T-norms, T-conorms and T-operators. In this study, we use the definitions of Kolmogorov, Silverman, and Formin [13] on linear spaces and norms. Uluçay, Şahin, and Olgun [14] worked out on normed Z-Modules and also on soft normed rings [15]. Şahin, Olgun, and Uluçay [16] defined normed quotient rings while Şahin and Kargın [17] presented neutrosophic triplet normed space. In Reference [18], Olgun and Şahin investigated fitting ideals of the universal module and while Olgun [19] found a method to solve a problem on universal modules. Şahin and Kargin proposed neutrosophic triplet inner product [20] and Florentin, Şahin, and Kargin introduced neutrosophic triplet G-module [21]. Şahin and et al defined isomorphism theorems for soft G-module in [22]. Fundamental homomorphism theorems for neutrosophic extended triplet groups [23] were introduced by Mehmet, Moges, and Olgun in 2018. In Reference [24], Bal, Moges, and Olgun introduced neutrosophic triplet cosets and quotient groups, and deal with its application areas in neutrosophic logic.

This paper anticipates a normed ring on R and fuzzy rings are defined in the previous studies. Now, we use that norm on fuzzy sets, hence a fuzzy norm is obtained and by defining our fuzzy norm on fuzzy rings, we get fuzzy normed rings in this study. The organization of this paper is as follows.

In Section 2, we give preliminaries and fuzzy normed rings. In Section 3, consists of further definitions and relevant theorems on fuzzy normed ideals of a normed ring. Fuzzy normed prime and fuzzy normed maximal ideals of a normed ring are introduced in Section 4. The conclusions are summarized in Section 5.

2. Preliminaries

In this section, definition of normed linear space, normed ring, Archimedean strict T-norm and concepts of fuzzy sets are outlined.

Definition 1. *[13] A functional $\|\|$ defined on a linear space L is said to be a norm (in L) if it has the following properties:*

N1: $\|x\| \geq 0$ for all $x \in L$, where $\|x\| = 0$ if and only if $x = 0$;
N2: $\|\alpha \cdot x\| = |\alpha|.\|x\|$; (and hence $\|x\| = \|-x\|$), for all $x \in L$ and for all α;
N3: Triangle inequality: $\|x + y\| \leq \|x\| + \|y\|$ for all $x, y \in L$.
A linear space L, equipped with a norm is called a normed linear space.

Definition 2. *[3] A ring A is said to be a normed ring if A possesses a norm $\|\|$, that is, a non-negative real-valued function $\|\| : A \to R$ such that for any $a, b \in A$,*

1. $\|a\| = 0 \Leftrightarrow a = 0$,
2. $\|a + b\| \leq \|a\| + \|b\|$,
3. $\|-a\| = \|a\|$, *(and hence $\|1_A\| = 1 = \|-1\|$ if identity exists), and*
4. $\|ab\| \leq \|a\| \|b\|$.

Definition 3. *[12] Let $* : [0,1] \times [0,1] \to [0,1]$. $*$ is an Archimedean strict T-norm iff for all $x, y, z \in [0,1]$:*

(1) *$*$ is commutative and associative, that is, $* (x,y) = * (y,x)$ and $* (x, * (y,z)) = * (* (x,y), z)$,*
(2) *$*$ is continuous,*
(3) *$* (x,1) = x$,*
(4) *$*$ is monotone, which means $* (x,y) \leq * (x,z)$ if $y \leq z$,*
(5) *$* (x,x) < x$ for $x \in (0,1)$, and*
(6) *when $x < z$ and $y < t$, $* (x,y) < * (z,t)$ for all $x, y, z, t \in (0,1)$.*

For convenience, we use the word t-norm shortly and show it as $x * y$ instead of $* (x,y)$. Some examples of t-norms are $x * y = \min\{x,y\}$, $x * y = \max\{x + y - 1, 0\}$ and $x * y = x.y$.

Definition 4. *[12] Let $\diamond : [0,1] \times [0,1] \to [0,1]$. $*$ is an Archimedean strict T-conorm iff for all $x, y, z \in [0,1]$:*

(1) *$\diamond$ is commutative and associative, that is, $\diamond (x,y) = \diamond (y,x)$ and $\diamond (x, \diamond (y,z)) = \diamond (\diamond (x,y), z)$,*
(2) *$\diamond$ is continuous,*
(3) *$\diamond (x,0) = x$,*
(4) *$\diamond$ is monotone, which means $\diamond (x,y) \leq \diamond (x,z)$ if $y \leq z$,*
(5) *$\diamond (x,x) > x$ for $x \in (0,1)$, and*
(6) *when $x < z$ and $y < t$, $\diamond (z,t) < \diamond (x,y)$ for all $x, y, z, t \in (0,1)$.*

For convenience, we use the word s-norm shortly and show it as $x \diamond y$ instead of $\diamond (x,y)$. Some examples of s-norms are $x \diamond y = \max\{x,y\}$, $x * y = \min\{x + y, 1\}$ and $x \diamond y = x + y - x.y$.

Definition 5. *[6] The fuzzy set B on a universal set X* is a set of ordered pairs

$$B = \{(x, \mu_B(x) : x \in X)\}$$

Here, $\mu_B(x)$ is the membership function or membership grade of x in B. For all $x \in X$, we have $0 \leq \mu_B(x) \leq 1$. If $x \notin B$, $\mu_B(x) = 0$, and if x is entirely contained in B, $\mu_B(x) = 1$. The membership grade of x in B is shown as $B(x)$ in the rest of this paper.

Definition 6. *[6] For the fuzzy sets A and B, the membership functions of the intersection, union and complement are defined pointwise as follows respectively:*

$$(A \cap B)(x) = \min\{A(x), B(x)\},$$

$$(A \cup B)(x) = \max\{A(x), B(x)\},$$

$$\overline{A}(x) = 1 - A(x).$$

Definition 7. *[10] Let $(R, +, .)$ be a ring and $F(R)$ be the set of all fuzzy subsets of R. As $A \in F(R)$, $\wedge$ is the fuzzy intersection and $\vee$ is the fuzzy union functions, for all $x, y \in R$, if A satisfies (1) $A(x - y) \geq A(x) \wedge A(y)$ and (2) $A(x.y) \geq A(x) \wedge A(y)$ then A is called a fuzzy subring of R. If A is a subring of R for all $a \in A$, then A is itself a fuzzy ring.*

Definition 8. *[11] A non-empty fuzzy subset A of R is said to be an ideal (in fact a fuzzy ideal) if and only if, for any $x, y \in R$, $A(x - y) \geq A(x) \wedge A(y)$ and $A(x.y) \geq A(x) \vee A(y)$.*

Note: *The fuzzy operations of the fuzzy subsets $A, B \in F(R)$ on the ring R can be extended to the operations below by t-norms and s-norms:*

For all $z \in R$,

$$(A + B)(z) = \mathop{\diamond}_{x+y=z} (A(x) * B(y));$$

$$(A - B)(z) = \mathop{\diamond}_{x-y=z} (A(x) * B(y));$$

$$(A.B)(z) = \mathop{\diamond}_{x.y=z} (A(x) * B(y)).$$

3. Fuzzy Normed Rings and Fuzzy Normed Ideals

In this section, there has been defined the fuzzy normed ring and some basic properties related to it. Throughout the rest of this paper, R is the set of real numbers, R will denote an associative ring with identity, NR is a normed ring and $F(X)$ is the set of all fuzzy subsets of the set X.

Definition 9. *Let $*$ be a continuous t-norm and $\diamond$ a continuous s-norm, NR a normed ring and let A be a fuzzy set. If the fuzzy set $A = \{(x, \mu_A(x)) : x \in NR\}$ over a fuzzy normed ring $F(NR)$ satisfy the following conditions then A is called a fuzzy normed subring of the normed ring $(NR, +, .)$:*

For all $x, y \in NR$,

(i) $A(x - y) \geq A(x) * A(y)$

(ii) $A(x.y) \geq A(x) * A(y)$.

Let 0 be the zero of the normed ring NR. For any fuzzy normed subring A and for all $x \in NR$, we have $A(x) \leq A(0)$, since $A(x - x) \geq A(x) * A(x) \Rightarrow A(0) \geq A(x)$.

Example 1. *Let A fuzzy set and $R = (Z, +, .)$ be the ring of all integers. Define a mapping $f : A \to F(NR(Z))$ where, for any $a \in A$ and $x \in Z$,*

$$A_f(x) = \begin{cases} 0 \text{ if } x \text{ is odd} \\ \frac{1}{a} \text{ if } x \text{ is even} \end{cases}$$

Corresponding t-norm $(*)$ *and t-conorm* $(\diamond)$ *are defined as* $a * b = min\{a,b\}$, $a \diamond b = max\{a,b\}$; *then, A is a fuzzy set as well as a fuzzy normed ring over* $[(Z,+,.),A]$.

Lemma 1. $A \in F(NR)$ *is a fuzzy normed subring of the normed ring NR if and only if* $A - A \subseteq A$ *and* $A.A \subseteq A$.

Proof. Let A be a fuzzy normed subring of NR. By [10], it is clear that A is a fuzzy group under addition and so $A - A \subseteq A$. Also for all $z \in NR$,

$$(A.A)(z) = \diamond_{x.y=z}(A(x) * A(y)) \leq \diamond_{x.y=z} A(xy) = A(z) \Rightarrow A.A \subseteq A$$

Now we suppose $A - A \subseteq A$ and $A.A \subseteq A$. For all $x,y \in NR$,

$$A(x-y) \geq (A-A)(x-y) = \diamond_{s-t=x-y}(A(s) * A(t)) \geq A(x) * A(y).$$

Similarly,

$$A(xy) \geq (A.A)(xy) = \diamond_{st=xy}(A(s) * A(t)) \geq A(x) * A(y).$$

Thus, A is a fuzzy normed subring of NR. □

Lemma 2.

i. *Let A be a fuzzy normed subring of the normed ring NR and let* $f : NR \to NR\prime$ *be a ring homomorphism. Then,* $f(A)$ *is a fuzzy normed subring of* $NR\prime$.

ii. *Let* $f : NR \to NR\prime$ *be a normed ring homomorphism. If B is a fuzzy normed subring of* $NR\prime$, *then* $f^{-1}(B)$ *is a fuzzy normed subring of* NR.

Proof. (i) Take $u, v \in NR\prime$. As f is onto, there exists $x, y \in NR$ such that $f(x) = u$ and $f(y) = v$. So,

$$\begin{aligned}
(f(A))(u) * (f(A))(v) &= \left(\diamond_{f(x)=u} A(x)\right) * \left(\diamond_{f(y)=v} A(y)\right) \\
&= \diamond_{f(x)=u, f(y)=v}(A(x) * A(y)) \\
&\leq \diamond_{f(x)=u, f(y)=v}(A(x-y)) \text{ (as } A \text{ is a fuzzy normed subring of } NR) \\
&\leq \diamond_{f(x)-f(y)=u-v}(A(x-y)) \\
&= \diamond_{f(x-y)=u-v}(A(x-y)) \text{ (since } f \text{ is a homomorphism)} \\
&= \diamond_{f(z)=u-v} A(z) \\
&= (f(A))(u-v).
\end{aligned}$$

Similarly, it is easy to see that

$$(f(A))(u.v) \geq (f(A)(u) * f(A)(v)).$$

Therefore, $f(A)$ is a fuzzy normed subring of $NR\prime$.
(ii) Proof is straightforward and similar to the proof of (i). □

Definition 10. *Let* A_1 *and* A_2 *be two fuzzy normed rings over the normed ring NR. Then* A_1 *is a fuzzy normed subring of* A_2 if

$$A_1(x) \leq A_2(x)$$

for all $x \in NR$.

Definition 11. *Let NR be a normed ring, $A \in F(NR)$ and let $A \neq \varnothing$. If for all $x, y \in NR$*

(i) $A(x - y) \geq A(x) * A(y)$ *and*
(ii) $A(x.y) \geq A(y)$ $(A(x.y) \geq A(x))$,

then A is called a fuzzy left (right) normed ideal of NR.

Definition 12. *If the fuzzy set A is both a fuzzy normed right and a fuzzy normed left ideal of NR, then A is called a fuzzy normed ideal of NR; i.e., if for all $x, y \in NR$*

(i) $A(x - y) \geq A(x) * A(y)$ *and*
(ii) $A(x.y) \geq A(x) \diamond A(y)$,

then $A \in F(NR)$ is a fuzzy normed ideal of NR.

Remark 1. *Let the multiplicative identity of NR (if exists) be 1_{NR}. As $A(x.y) \geq A(x) \diamond A(y)$ for all $x, y \in NR$, $A(x.1_{NR}) \geq A(x) \diamond A(1_{NR})$ and therefore for all $x \in NR$, $A(x) \geq A(1_{NR})$.*

Example 2. *Let A and B be two (fuzzy normed left, fuzzy normed right) ideals of a normed ring NR. Then, $A \cap B$ is also a (fuzzy normed left, fuzzy normed right) ideal of NR.*

Solution: Let $x, y \in NR$.

$$\begin{aligned}(A \cap B)(x - y) &= \min\{A(x - y), B(x - y)\} \\ &\geq \min\{A(x) * A(y), B(x) * B(y)\} \\ &\geq \min\{(A \cap B)(x), (A \cap B)(y)\}.\end{aligned}$$

On the other hand, as A and B are fuzzy normed left ideals, using $A(x.y) \geq A(y)$ and $B(x.y) \geq B(y)$ we have

$$(A \cap B)(x.y) = \min\{A(x.y), B(x.y)\} \geq \min\{A(y), B(y)\} = (A \cap B)(y).$$

So $A \cap B$ is a fuzzy normed left ideal. Similarly, it is easy to show that $A \cap B$ is a fuzzy normed right ideal. As a result $A \cap B$ is an fuzzy normed ideal of NR.

Example 3. *Let A be a fuzzy ideal of NR. The subring $A^0 = \{x : \mu_A(x) = \mu_A(0_{NR})\}$ is a fuzzy normed ideal of NR, since for all $x \in NR$, $A^0(x) \leq A^0(0)$.*

Theorem 1. *Let A be a fuzzy normed ideal of NR, $X = \{a_1, a_2, \ldots, a_m\} \subseteq NR$, $x, y \in NR$ and let $FN(X)$ be the fuzzy normed ideal generated by the set X in NR.* Then,

(i) $w \in FN(X) \Rightarrow A(w) \geq \underset{1 \leq i \leq m}{*}(A(a_i))$,
(ii) $x \in (y) \Rightarrow A(x) \geq A(y)$,
(iii) $A(0) \geq A(x)$ *and*
(iv) *if 1 is the multiplicative identity of NR, then* $A(x) \geq A(1)$.

Proof. (ii), (iii), and (iv) can be proved using (i). The set $FN(X)$ consists of the finite sums in the form $ra + as + uav + na$ where $a \in X$, $r, s, u, v \in NR$ and n is an integer. Let $w \in FN(X)$. So there exists an integer n and $r, s, u, v \in NR$ such that $w = ra_i + a_is + ua_iv + na_i$ where $1 \leq i \leq m$. As A is a fuzzy normed ideal,

$$A(ra_i + a_is + ua_iv + na_i) \geq A(ra_i) * A(a_is) * A(ua_iv) * A(na_i) \geq A(a_i).$$

Therefore

$$A(w) \geq \underset{1 \leq i \leq m}{*}(A(a_i)).$$

□

4. Fuzzy Normed Prime Ideal and Fuzzy Normed Maximal Ideal

In this section, fuzzy normed prime ideal and fuzzy normed maximal ideal are outlined.

Definition 13. *Let A and B be two fuzzy subsets of the normed ring NR. We define the operation $A \circ B$* as follows:

$$A \circ B(x) = \begin{cases} \underset{x=yz}{\diamond}(A(y) * B(z)) & \text{, if } x \text{ can be defined as } x = yz \\ 0 & \text{, otherwise .} \end{cases}$$

If the normed ring NR has a multiplicative inverse, namely if $NR.NR = NR$, then the second case does not occur.

Lemma 3. *If A and B are a fuzzy normed right and a fuzzy normed left ideal of a normed ring NR, respectively, $A \circ B \subseteq A \cap B$ and hence $(A \circ B)(x) \leq (A \cap B)(x)$ for all $x \in NR$.*

Proof. It is shown in Example 2 that if A and B are fuzzy normed left ideals of NR, then $A \cap B$ is also a fuzzy normed left ideal. Now, let A and B be a fuzzy normed right and a fuzzy normed left ideal of NR, respectively. If $A \circ B(x) = 0$, the proof is trivial.

Let

$$(A \circ B)(x) = \underset{x=yz}{\diamond}(A(y) * B(z)).$$

As A is a fuzzy normed right ideal and B is a fuzzy normed left ideal, we have

$$A(y) \leq A(yz) = A(x)$$

and

$$B(y) \leq B(yz) = B(x)$$

Thus,

$$\begin{aligned} (A \circ B)(x) &= \underset{x=yz}{\diamond}(A(y) * B(z)) \\ &\leq \min(A(x), B(x)) \\ &= (A \cap B)(x) \end{aligned}$$

□

Definition 14. *Let A and B be fuzzy normed ideals of a normed ring NR and let FNP be a non-constant function, which is not an ideal of NR. If*

$$A \circ B \subseteq FNP \Rightarrow A \subseteq FNP \text{ or } B \subseteq FNP,$$

then FNP is called a fuzzy normed prime ideal of NR.

Example 4. *Show that if the fuzzy normed ideal I $(I \neq NR)$ is a fuzzy normed prime ideal of NR, then the characteristic function λ_I* is also a fuzzy normed prime ideal.

Solution: As $I \neq NR$, λ_I is a non-constant function on NR. Let A and B be two fuzzy normed ideals on NR such that $A \circ B \subseteq \lambda_I$, but $A\lambda_I$ and $B\lambda_I$. There exists $x, y \in NR$ such that $A(x) \leq \lambda_I(x)$ and $B(y) \leq \lambda_I(y)$. In this case, $A(x) \neq 0$ and $B(y) \neq 0$, but $\lambda_I(x) = 0$ and $\lambda_I(y) = 0$. Therefore

$x \notin I, y \notin I$. As I is a fuzzy normed prime ideal, there exists an $r \in NR$, such that $xry \notin I$. This is obvious, because if I is fuzzy normed prime, $A \circ B(xry) \subseteq I \Rightarrow A(x) \subseteq I$ or $B(ry) \subseteq I$ and therefore as $(NRxNR)(\underline{NRry}NR) = (NRxNR)(NRyNR) \subseteq I$, we have either $NRxNR \subseteq I$ or $NRyNR \subseteq I$. Assume $NRxNR \subseteq I$. Then $xxx = (x)^3 \in I \Rightarrow x \subseteq I$, but this contradicts with the fact that $\lambda_I(x) = 0$. Now let $a = xry$. $\lambda_I(a) = 0$. Thus, $A \circ B(a) = 0$. On the other hand,

$$\begin{aligned} A \circ B(a) &= \underset{a=cd}{\diamond}(A(c) * B(d)) \\ &\geq A(x) * B(ry) \\ &\geq A(x) * B(y) \\ &\geq 0 \text{ (as } A(x) \neq 0 \text{ and } B(y) \neq 0). \end{aligned}$$

This is a contradiction, since $A \circ B(a) = 0$. Therefore if A and B are fuzzy normed ideals of a normed ring NR, then $A \circ B \subseteq \lambda_I \Rightarrow A \subseteq \lambda_I$ or $B \subseteq \lambda_I$. As a result, the characteristic function λ_I is a fuzzy normed prime ideal.

Theorem 2. *Let FNP be a fuzzy normed prime ideal of a normed ring NR. The ideal defined by $FNP^0 = \{x : x \in NR, FNP(x) = FNP(0)\}$ and is also a fuzzy normed prime ideal of NR.*

Proof: Let $x, y \in FNP^0$. As FNP is an fuzzy normed ideal, $FNP(x-y) \geq FNP(x) * FNP(y) = FNP(0)$. On the other hand, by Theorem 1, we have $FNP(0) \geq FNP(x-y)$. So, $FNP(x-y) = FNP(0)$ and $x - y \in FNP^0$. Now, let $x \in FNP^0$ and $r \in NR$. In this case, $FNP(rx) \geq FNP(x) = FNP(0)$ and thus $FNP(rx) = FNP(0)$. Similarly, $FNP(xr) = FNP(0)$. Now, for all $x \in FNP^0$ and $r \in NR$, $rx, xr \in FNP^0$. Therefore, FNP^0 is a fuzzy normed ideal of NR. Let I and J be two ideals of NR, such that $IJ \subseteq FNP^0$. Now, we define fuzzy normed ideals $A = FNP^0\lambda_I$ and $B = FNP^0\lambda_J$. We will show that $(A \circ B)(x) \leq FNP(x)$ for all $x \in NR$. Assume $(A \circ B)(x) \neq 0$. Recall $A \circ B = \underset{x=yz}{\diamond}(A(y) * B(z))$, so we only need to take the cases of $A(y) * B(z) \neq 0$ under consideration. However, in all these cases, $A(y) = FNP(0)$ or $A(y) = 0$ and similarly $B(z) = FNP(0)$ or $B(z) = 0$ and hence $A(y) = B(z) = FNP(0)$. Now, $\lambda_I(y) = 1$ and $\lambda_J(z) = 1$ implies $y \in I$, $z \in J$ and $x \in IJ \subseteq FNP^0$. Thus, $FNP(x) = FNP(0)$ and for all $x \in NR$, we get $(A \circ B)(x) \leq FNP(x)$. As FNP is a fuzzy normed prime ideal and A and B are fuzzy normed ideals, either $A \subseteq FNP$ or $B \subseteq FNP$. Assume $A = FNP^0\lambda_I \subseteq FNP$. We need to show that $I \subseteq FNP^0$. Let $IFNP$. Then, there exists an $a \in I$, such that $a \notin FNP^0$; i.e., $FNP(a) \neq FNP(0)$. It is evident that $FNP(0) \geq FNP(a)$. Thus, $FNP(a) < FNP(0)$. However, $A(a) = FNP^0\lambda_I(a) = FNP(0) > FNP(a)$ and this is a contradiction to the assumption $A \subseteq FNP$. So, $I \subseteq FNP^0$. Similarly, one can show that $B \subseteq FNP$ and $J \subseteq FNP^0$. Thus, FNP^0 is a fuzzy normed prime ideal. □

Definition 15. *Let A be a fuzzy normed ideal of a normed ring NR. If A is non-constant and for all fuzzy normed ideals B of NR, $A \subseteq B$ implies $A^0 = B^0$ or $B = \lambda_{NR}$, A is called a fuzzy normed maximal ideal of the normed ring NR.* Fuzzy normed maximal left(right) ideals are defined similarly.

Example 5. *Let A be a fuzzy normed maximal left (right) ideal of a normed ring NR. Then, $A^0 = \{x \in NR : A(x) = A(0)\}$ is a fuzzy normed maximal left (right) ideal of NR.*

Theorem 3. *If A is a fuzzy normed left(right) maximal ideal of a normed ring NR, then $A(0) = 1$.*

Proof. Assume $A(0) \neq 1$. Let $A(0) < t < 1$ and let B be a fuzzy subset of NR such that $B(x) = t$ for all $x \in NR$. B is trivially an ideal of NR. Also it is easy to verify that $A \subset B$, $B \neq \lambda_{NR}$ and $B^0 = \{x \in NR : B(x) = B(0)\} = NR$. But, despite the fact that $A \subset B$, $A^0 \neq B^0$ and $B \neq \lambda_{NR}$ is a contradiction to the fuzzy normed maximality of A. Thus, $A(0) = 1$. □

5. Conclusions

In this paper, we defined a fuzzy normed ring. Here we examine the algebraic properties of fuzzy sets in ring structures. Some related notions, e.g., the fuzzy normed ring homomorphism, fuzzy normed subring, fuzzy normed ideal, fuzzy normed prime ideal and fuzzy normed maximal ideal are proposed. We hope that this new concept will bring a new opportunity in research and development of fuzzy set theory. To extend our work, further research can be done to study the properties of fuzzy normed rings in other algebraic structures such as fuzzy rings and fuzzy fields.

Author Contributions: All authors contributed equally.

Acknowledgments: We thank Vakkas Uluçay for the arrangement.

References

1. Naĭmark, M.A. *Normed Rings*; Noordhoff: Groningen, The Netherlands, 1964.
2. Arens, R. A generalization of normed rings. *Pac. J. Math.* **1952**, *2*, 455–471. [CrossRef]
3. Gel'fand, I.; Raikov, M.D.A.; Shilov, G.E. *Commutative Normed Rings*; Chelsea Publishing Company: New York, NY, USA, 1964.
4. Jarden, M. *Normed Rings-Algebraic Patching*; Springer: Berlin/Heidelberg, Germany, 2011.
5. Zadeh, L.A. Fuzzy Sets. *Inf. Control* **1965**, *8*, 338–353. [CrossRef]
6. Zimmermann, H.J. *Fuzzy Set Theory and Its Application*, 2nd ed.; Kluwer Academic Publishers: Norwell, MA, USA, 1991.
7. Mordeson, J.N.; Bhutani, K.R.; Rosenfeld, A. *Fuzzy Group Theory*; Springer: Heidelberg, Germany, 2005.
8. Liu, W. Fuzzy Invariant Subgroups and Fuzzy Ideals. *Fuzzy Sets Syst.* **1982**, *8*, 133–139. [CrossRef]
9. Mukherjee, N.P.; Bhattacharya, P. Fuzzy Normal Subgroups and Fuzzy Cosets. *Inf. Sci.* **1984**, *34*, 225–239. [CrossRef]
10. Wang, X.; Ruan, D.; Kerre, E.E. *Mathematics of Fuzziness—Basic Issues*; Springer: Berlin/Heidelberg, Germany, 2009.
11. Swamy, U.M.; Swamy, K.L.N. Fuzzy Prime Ideals of Rings. *J. Math. Anal. Appl.* **1988**, *134*, 94–103. [CrossRef]
12. Gupta, M.M.; Qi, J. Theory of T-norms and Fuzzy Inference Methods. *Fuzzy Sets Syst.* **1991**, *40*, 431–450. [CrossRef]
13. Kolmogorov, A.N.; Silverman, R.A.; Fomin, S.V. *Introductory Real Analysis*; Dover Publications, Inc.: New York, NY, USA, 1970.
14. Uluçay, V.; Şahin, M.; Olgun, N. Normed Z-Modules. *Int. J. Pure Appl. Math.* **2017**, *112*, 425–435.
15. Şahin, M.; Olgun, N.; Uluçay, V. Soft Normed Rings. *Springer Plus* **2016**, *5*, 1950–1956.
16. Şahin, M.; Olgun, N.; Uluçay, V. Normed Quotient Rings. *New Trends Math. Sci.* **2018**, *6*, 52–58.
17. Şahin, M.; Kargın, A. Neutrosohic Triplet Normed Space. *Open Phys.* **2017**, *15*, 697–704. [CrossRef]
18. Olgun, N.; Şahin, M. Fitting ideals of universal modules. In Proceedings of the International Conference on Natural Science and Engineering (ICNASE'16), Kilis, Turkey, 19–20 March 2016; pp. 1–8.
19. Olgun, N. A Problem on Universal Modules. *Commun. Algebra* **2015**, *43*, 4350–4358. [CrossRef]
20. Şahin, M.; Kargın, A. Neutrosophic Triplet Inner Product. *Neutrosophic Oper. Res.* **2017**, *2*, 193–205.
21. Smarandache, F.; Şahin, M.; Kargın, A. Neutrosophic Triplet G-Module. *Mathematics* **2018**, *6*, 53. [CrossRef]
22. Şahin, M.; Olgun, N.; Kargın, A.; Uluçay, V. Isomorphism theorems for soft G-modules. *Afr. Mat.* **2018**, 1–8. [CrossRef]
23. Çelik, M.; Shalla, M.; Olgun, N. Fundamental Homomorphism Theorems for Neutrosophic Extended Triplet Groups. *Symmetry* **2018**, *10*, 321–335.
24. Bal, M.; Shalla, M.M.; Olgun, N. Neutrosophic Triplet Cosets and Quotient Groups. *Symmetry* **2018**, *10*, 126. [CrossRef]

9

Sufficient Conditions for Triangular Norms Preserving ⊗−Convexity

Lifeng Li [1,2,*] and Qinjun Luo [3]

1 School of Science, Xi'an University of Posts and Telecommunications, Xi'an 710121, China
2 Shaanxi Key Laboratory of Network Data Analysis and Intelligent Processing, Xi'an University of Posts and Telecommunications, Xi'an 710121, China
3 School of Statistics, Xi'an University of Finance and Economics, Xi'an 710100, China; tjxy@mail.xaufe.edu.cn
* Correspondence: lilifeng@xupt.edu.cn.

Abstract: The convexity in triangular norm (for short, $\otimes-$convexity) is a generalization of Zadeh's quasiconvexity. The aggregation of two $\otimes-$convex sets is under the aggregation operator $\otimes$ is also $\otimes-$convex, but the aggregation operator $\otimes$ is not unique. To solve it in complexity, in the present paper, we give some sufficient conditions for aggregation operators preserve $\otimes-$convexity. In particular, when aggregation operators are triangular norms, we have that several results such as arbitrary triangular norm preserve $\otimes_D-$convexity and $\otimes_a-$convexity on bounded lattices, $\otimes_M$ preserves $\otimes_H-$convexity in the real unite interval $[0,1]$.

Keywords: aggregation operator; triangular norm; $\otimes-$convex set

---

1. Introduction

Fuzzy set theory introduced by Zadeh in 1965, as an mathematical tool to deal with uncertainty in information system and knowledge base, has been widely used in various fields of science and technology. By applying fuzzy set theory, Zadeh in [1] proposed the concept of quasiconvex fuzzy set, and has attracted wide attention of researchers and practitioners from many different areas such as fuzzy mathematics, optimization and engineering. Subsequently, Zadeh's quasiconvex fuzzy set was generalized with a lattice L instead of the interval $[0,1]$. A fuzzy set $\mu : \mathbb{R}^n \to L$ is quasiconvex if for any $x, y \in \mathbb{R}^n$ and all $\lambda \in [0,1]$ the inequality

$$\mu(\lambda x + (1-\lambda)y) \geq \mu(x) \wedge \mu(y) \tag{1}$$

holds.

A quasiconvex fuzzy set has an important property: intersection of quasiconvex fuzzy sets is a quasiconvex fuzzy set, i.e., let $X \subseteq \mathbb{R}^n$, for any fuzzy sets μ and ν,

$$\mu \textit{ and } \nu \textit{ are quasiconvex} \Rightarrow \min\{\mu, \nu\} \textit{ is quasiconvex}. \tag{2}$$

The above condition is called intersection preserving quasiconvexity. This property is also true for lattice valued fuzzy sets.

The theory of aggregation operators [2], has been successfully used in mathematics, complex networks and decision making etc (e.g., see [3–6]). The arithmetic mean, the ordered weighted averaging operator and the probabilistic aggregation are widely used examples. In reference [7] Janiš, Král and Renčová pointed that the intersection of fuzzy sets is not the only operator preserving quasiconvexity in general, and they gave someconditions in order that an aggregation operator preserves quasiconvexity.

Triangular norms are kinds of binary aggregation operations that become an essential tool in fuzzy logic, information science and computer sciences. By using triangular norms, properties of fuzzy convexity and various generalizations of fuzzy convexity were considered by many authors (for example, see [8–11]). Suppose $\otimes : [0,1]^2 \to [0,1]$ is a triangular norm, Nourouzi [10] given the concept of $\otimes-$convex set which generalized Zadeh's quasiconvex fuzzy set. A $\otimes-$convex set as defined in [10] can also be generalized as being lattice-valued in the following sense. Let L be a lattice and let $\otimes : L^2 \to L$ be a triangular norm. A fuzzy set $\mu : \mathbb{R}^n \to L$ is called $\otimes-$convex if for any $x, y \in \mathbb{R}^n$ and all $\lambda \in [0,1]$ the inequality

$$\mu(\lambda x + (1-\lambda)y) \geq \mu(x) \otimes \mu(y) \tag{3}$$

holds.

Following [7,10], in the present paper, we continue to study sufficient conditions for aggregation operators and triangular norms that preserve $\otimes-$convexity on a bounded lattice. In Section 3, we give some sufficient conditions for aggregation operator preserving $\otimes-$convexity, those results are generalizations of Propositions 2 and 3 (in [7]). Triangular norm is a kind of important aggregation operator, we give some sufficient conditions for triangular norm preserving $\otimes-$convexity in Section 4. And Section 5 is conclusion.

2. Preliminaries

We first give the basic definitions and results from the existing literature. In following, we use L denote a bounded lattice $(L \leq, 0_L, 1_L)$.

Definition 1. *[2] An aggregation operation is a function $A : L^n \to L$ which satisfies*

(i) $A(a_1, a_2, \ldots, a_n) \leq A(a'_1, a'_2, \ldots, a'_n)$ *whenever* $a_i \leq a'_i$ *for* $1 \leq i \leq n$.
(ii) $A(0_L, 0_L, \ldots, 0_L) = 0_L$ *and* $A(1_L, 1_L, \ldots, 1_L) = 1_L$.

A binary aggregation operation is said to be symmetric if for any $a_1, a_2 \in L$, $A(a_1, a_2) = A(a_2, a_1)$. A special aggregation function is a triangular norm defined as following.

Definition 2. *[12] A map $\otimes : L^2 \to L$ is called a triangular norm if*

(T1) $a \otimes b = b \otimes a$.
(T2) $a_1 \otimes b \leq a_2 \otimes b$ *if* $a_1 \leq a_2$.
(T3) $a \otimes (b \otimes c) = (a \otimes b) \otimes c$.
(T4) $a \otimes 1_L = a$.

Example 1. *The two basic triangular norms $\otimes_M$ and $\otimes_D$ defined as the following are the strongest and the weakest triangular norms on L, respectively.*

$$a \otimes_M b = a \wedge b,$$

$$a \otimes_D b = \begin{cases} a \wedge b, & a, b \in \{1_L\}, \\ 0, & otherwise. \end{cases}$$

Example 2. *Suppose $H = (0, \lambda) \subseteq [0,1)$ and let $* : H^2 \to H$ be an operation on H which satisfies (T1)–(T3) and*

$$a * b \leq \min\{a, b\},$$

$$a \otimes_H b = \begin{cases} a * b, & (a,b) \in H^2; \\ \min\{a,b\} & otherwise. \end{cases}$$

Then $\otimes_H$ is a kind of triangular norms on $[0,1]$ follows from Proposition 3.60 in [13].

3. Sufficient Conditions for an Aggregation Operator Preserving $\otimes-$Convexity

In this Section, we generalize Propositions 2 and 3 (in [7]), and give some sufficient conditions for an aggregation operator which preserves $\otimes-$convexity.

Theorem 1. *Let $A : L^2 \to L$ be an aggregation operator on L, let $\mu, \nu : \mathbb{R}^n \to L$ be arbitrarily $\otimes-$convex fuzzy sets. If $A(a \otimes b, c \otimes d) = A(a,c) \otimes A(b,d)$ for each $a,b,c,d \in L$, then $A(\mu,\nu)$ is $\otimes-$convex.*

Proof. Let $\mu, \nu : \mathbb{R}^n \to L$ be arbitrarily $\otimes-$convex fuzzy sets, and $x, y \in \mathbb{R}^n$. Then we see

$$\begin{aligned} & A(\mu,\nu)(\lambda x + (1-\lambda)y) \\ = \; & A(\mu(\lambda x + (1-\lambda)y), \nu(\lambda x + (1-\lambda)y)) \\ \geq \; & A(\mu(x) \otimes \mu(y), \nu(x) \otimes \nu(y)) \\ = \; & A(\mu(x), \nu(x)) \otimes A(\mu(y), \nu(y)) \\ = \; & A(\mu,\nu)(x) \otimes A(\mu,\nu)(y). \end{aligned}$$

Thus, $A(\mu,\nu)$ is $\otimes-$convex. □

The converse of Theorem 1, however, is in general not true. For example,

Example 3. *Consider a lattice $L = (0_L, a, b, 1_L)$, where $0_L \leq a \leq 1_L$, $0_L \leq b \leq 1_L$, and a, b are incomparable elements and the aggregation operator defined in Table 1. Let $\mu, \nu : \mathbb{R}^n \to L$ be arbitrarily $\otimes_D-$convex fuzzy sets. For any $x, y \in \mathbb{R}^n$ and all $\lambda \in [0,1]$*

$$\begin{aligned} & A(\mu,\nu)(\lambda x + (1-\lambda)y) \\ = \; & A(\mu(\lambda x + (1-\lambda)y), \nu(\lambda x + (1-\lambda)y)) \\ \geq \; & A(\mu(x) \otimes_D \mu(y), \nu(x) \otimes_D \nu(y)) \\ = \; & \begin{cases} A(\mu(y),\nu(y)), & \mu(x) = \nu(x) = 1_L, \\ A(\mu(y),\nu(x)), & \mu(x) = \nu(y) = 1_L, \\ A(\mu(x),\nu(x)), & \mu(y) = \nu(y) = 1_L, \\ A(\mu(x),\nu(y)), & \mu(y) = \nu(x) = 1_L, \\ 0_L, & otherwise, \end{cases} \end{aligned}$$

we have

$$\begin{aligned} & A(\mu,\nu)(x) \otimes_D A(\mu,\nu)(y) \\ = \; & \begin{cases} A(\mu(y),\nu(y)), & A(\mu,\nu)(x) = 1_L, \\ A(\mu(x),\nu(x)), & A(\mu,\nu)(y) = 1_L, \\ 0_L, & otherwise, \end{cases} \\ = \; & \begin{cases} A(\mu(y),\nu(y)), & \mu(x) = \nu(x) = 1_L, \\ A(\mu(x),\nu(x)), & \mu(y) = \nu(y) = 1_L, \\ 0_L, & otherwise. \end{cases} . \end{aligned}$$

Hence, $A(\mu,\nu)$ is $\otimes_D-$convex. And $A(1_L \otimes_D b, a \otimes_D 1_L) = A(b,a) = a$, $A(1_L, a) \otimes_D A(b, 1_L) = a \otimes_D b = 0_L$.

Table 1. Aggregation operator A.

A	0_L	a	b	1_L
0_L	0_L	0_L	0_L	0_L
a	0_L	0	b	b
b	0_L	a	b	b
1_L	0_L	a	b	1_L

Theorem 2. *Let $A : L^2 \to L$ be an aggregation operator on L, let $\mu, \nu : \mathbb{R}^n \to L$ be arbitrary $\otimes-$convex fuzzy sets. If $A(\mu,\nu)$ is $\otimes-$convex, then $A(a\otimes b, c\otimes d) \geq A(a,c)\otimes A(b,d)$ for each $a,b,c,d \in L$. Moreover if the triangular norm $\otimes$ is idempotent, then $A(a\otimes b, c\otimes d) = A(a,c)\otimes A(b,d)$ for each $a,b,c,d \in L$.*

Proof. Suppose that $A(\mu,\nu)$ is $\otimes-$convex. Let a,b,c,d be arbitrary elements of L. For $x,y \in \mathbb{R}^n$ and $z = \lambda x + (1-\lambda)y$, define

$$\mu(t) = \begin{cases} a, & t = z+\theta(y-z), \theta < 0; \\ a\otimes b, & t = z; \\ b, & t = z+\theta(y-z), \theta > 0; \\ 0_L, & otherwise, \end{cases} \qquad \nu(t) = \begin{cases} c, & t = z+\theta(y-z), \theta < 0; \\ c\otimes d, & t = z; \\ d, & t = z+\theta(y-z), \theta > 0; \\ 0_L, & otherwise. \end{cases}$$

Clearly μ,ν are $\otimes-$convex. And

$$A(\mu,\nu)(t) = \begin{cases} A(a,c), & t = z+\theta(y-z), \theta < 0; \\ A(a\otimes b, c\otimes d), & t = z; \\ A(b,d), & t = z+\theta(y-z), \theta > 0; \\ 0_L, & otherwise. \end{cases}$$

As $A(\mu,\nu)$ has to be a $\otimes-$convex fuzzy set, we have

$$A(a\otimes b, c\otimes d) \geq A(a,c)\otimes A(b,d).$$

From the monotonicity of A it follows that $A(a\otimes b, c\otimes d) \leq A(a,c)$ and $A(a\otimes b, c\otimes d) \leq A(b,d)$. Hence

$$A(a\otimes b, c\otimes d)\otimes A(a\otimes b, c\otimes d) \leq A(a,c)\otimes A(b,d).$$

Therefore, since the operator $\otimes$ is idempotent it follows that

$$A(a\otimes b, c\otimes d) \leq A(a,c)\otimes A(b,d).$$

□

Since the triangular norm $a\otimes_M b = a\wedge b$ is idempotent, Proposition 2 (in [7]) follows from Theorems 1 and 2.

Theorem 3. *Let $A : L^2 \to L$ be an aggregation operator on L, and let $\mu,\nu : \mathbb{R}^n \to L$ be arbitrary $\otimes-$convex fuzzy sets. If $A(a,b) = A(a,a)\otimes A(b,b) = A(a\otimes b, a\otimes b)$ for each $a,b \in L$, then $A(\mu,\nu)$ is $\otimes-$convex.*

Proof. Let $\mu, \nu : R^n \to L$ be arbitrary $\otimes-$convex fuzzy sets. For any $x, y \in \mathbb{R}^n$ and all $\lambda \in [0,1]$

$$\begin{aligned}
& A(\mu,\nu)(\lambda x + (1-\lambda)y) \\
= \; & A(\mu(\lambda x + (1-\lambda)y), \nu(\lambda x + (1-\lambda)y)) \\
= \; & A(\mu(\lambda x + (1-\lambda)y), \mu(\lambda x + (1-\lambda)y)) \otimes A(\nu(\lambda x + (1-\lambda)y), \nu(\lambda x + (1-\lambda)y)) \\
\geq \; & A(\mu(x) \otimes \mu(y), \mu(x) \otimes \mu(y)) \otimes A(\nu(x) \otimes \nu(y), \nu(x) \otimes \nu(y)) \\
= \; & A(\mu(x), \mu(y)) \otimes A(\nu(x), \nu(y)) \\
= \; & (A(\mu(x), \mu(x)) \otimes A(\mu(y), \mu(y))) \otimes (A(\nu(x), \nu(x)) \otimes A(\nu(y), \nu(y))) \\
= \; & (A(\mu(x), \mu(x)) \otimes A(\nu(x), \nu(x))) \otimes (A(\mu(y), \mu(y)) \otimes A(\nu(y), \nu(y))) \\
= \; & A(\mu,\nu)(x) \otimes A(\mu,\nu)(y).
\end{aligned}$$

Thus, $A(\mu, \nu)$ is $\otimes-$convex. □

The following shows that the converse of Theorem 3 is in general not true.

Example 4. *Consider a lattice $L = (0_L, a, b, 1_L)$, where $0_L \leq a \leq 1_L$, $0_L \leq b \leq 1_L$, and a, b are incomparable elements and the binary symmetric aggregation operator A defined in Table 2. Let $\mu, \nu : \mathbb{R}^n \to L$ be arbitrary $\otimes_D-$convex fuzzy sets. For any $x, y \in \mathbb{R}^n$ and all $\lambda \in [0,1]$, can prove that $A(\mu, \nu)$ is $\otimes_D-$convex. And $A(b, a) = a$, $A(b, b) \otimes_D A(a, a) = a \otimes_D a = 0_L$, and $A(b \otimes_D b, a \otimes_D a) = A(0_L, 0_L) = 0_L$.*

Table 2. Aggregation operator A.

A	0_L	a	b	1_L
0_L	0_L	0_L	0_L	0_L
a	0_L	a	a	a
b	0_L	a	a	b
1_L	0_L	a	b	1_L

Theorem 4. *Let $A : L^2 \to L$ be an symmetric aggregation operator on L, let $\mu, \nu : \mathbb{R}^n \to L$ be arbitrary $\otimes-$convex fuzzy sets. If $A(\mu, \nu)$ is $\otimes-$convex, then $A(a, b) \geq A(a, a) \otimes A(b, b)$ for each $a, b \in L$. Moreover if the triangular norm $\otimes$ is idempotent, then $A(a, b) = A(a, a) \otimes A(b, b) = A(a \otimes b, a \otimes b)$ for each $a, b \in L$.*

Proof. Suppose that $A(\mu, \nu)$ is $\otimes-$convex. Let a, b be arbitrary elements of L, and put, for $x, y \in \mathbb{R}^n$ and $0 < \lambda < 1$, $z = \lambda x + (1-\lambda)y$. We define

$$\mu(t) = \begin{cases} a, & t = z + \theta(y-z), \theta \leq 0; \\ b, & t = z + \theta(y-z), \theta > 0; \\ 0_L, & \textit{otherwise}, \end{cases} \qquad \nu(t) = \begin{cases} a, & t = z + \theta(y-z), \theta < 0; \\ b, & t = z + \theta(y-z), \theta \geq 0; \\ 0_L, & \textit{otherwise}. \end{cases}$$

Clearly μ, ν are $\otimes-$convex and as A preserves $\otimes-$convexity, then we have

$$A(a, b) \geq A(a, a) \otimes A(b, b).$$

Suppose that the triangular norm $\otimes$ is idempotent. Let $x, y \in \mathbb{R}^n$ and $z = \lambda x + (1-\lambda)y$, define

$$\mu(t) = \begin{cases} a, & t = z + \theta(y-z), \theta \leq 0; \\ 1_L, & t = z + \theta(y-z), \theta > 0; \\ 0_L, & \textit{otherwise}, \end{cases} \qquad \nu(t) = \begin{cases} a, & t = z + \theta(y-z), \theta < 0; \\ 1_L, & t = z + \theta(y-z), \theta \geq 0; \\ 0_L, & \textit{otherwise}. \end{cases}$$

Clearly μ, ν are $\otimes-$convex. Since, in addition, A preserves $\otimes-$convexity this can be combined with the fact that the triangular norm $\otimes$ is idempotent, we deduce

$$A(a, a) \geq A(a, 1_L) \otimes A(1_L, a) = A(1_L, a) \otimes A(1_L, a) = A(1_L, a).$$

From the monotony of A it follows that $A(a,a) \leq A(1_L,a)$. Hence

$$A(a,a) = A(1_L,a).$$

Therefore

$$A(a,b) \leq A(1_L,b) = A(b,b), A(a,b) \leq A(1_L,a) = A(a,a).$$

Hence

$$A(a,b) = A(a,b) \otimes A(a,b) \leq A(a,a) \otimes A(b,b).$$

Thus

$$A(a,b) = A(a,a) \otimes A(b,b).$$

Let $c = a, d = b$, from Theorem 2 we have

$$A(a,b) = A(a \otimes b, a \otimes b).$$

□

Then Proposition 3 (in [7]) follows from Theorems 3 and 4 due to $a \otimes_M b = a \wedge b$ is idempotent.

Since the triangular norm $a \otimes_M b = \min\{a,b\}$ is the strongest triangular norm on $[0,1]$, from the definition of $\otimes-$convexity we can prove the following theorem.

Theorem 5. *If $f_1, f_2 : [0,1] \to [0,1]$ are both nondecreasing, $\min\{f_1(0), f_2(0)\} = 0, f_1(1) = f_2(1) = 1$. Let $A : [0,1]^2 \to [0,1]$ defined by $A(a,b) = \min\{f_1(a), f_2(b)\}$, then $A(\mu,\nu)$ preserves $\otimes-$convexity for any triangular norm on $[0,1]$. But the converse statement is in general not true.*

Example 5. *Suppose $L = [0,1]$, $A(a,b) = \frac{1}{2}(a+b)$. Then $A(\mu,\nu)(\lambda x + (1-\lambda)y) \geq A(\mu,\nu)(x) \otimes_D A(\mu,\nu)(y)$. i.e., $A(\mu,\nu)$ is $\otimes_D-$convex. And $A(a,b) = \frac{1}{2}(a+b) \neq \min\{f_1(a), f_2(b)\}$.*

4. Sufficient Conditions for Triangular Norm Preserving ⊗−Convexity

In this section we give some sufficient conditions which guarantee that a triangular norm preserves $\otimes-$convexity. The following theorem is obvious.

Theorem 6. *Let $\otimes : L^2 \to L$ be a triangular norm on L. If $\mu, \nu : \mathbb{R}^n \to L$ are arbitrary $\otimes-$convex fuzzy sets, then $\mu \otimes \nu$ is $\otimes-$convex.*

Theorem 7. *Let $\otimes : L^2 \to L$ be a triangular norm on L. If $\mu, \nu : \mathbb{R}^n \to L$ are arbitrary $\otimes_D-$convex fuzzy sets, then $\mu \otimes \nu$ is $\otimes_D-$convex.*

Proof. Let $\mu, \nu : \mathbb{R}^n \to L$ be arbitrary $\otimes_D$−convex fuzzy sets. For any $x, y \in \mathbb{R}^n$ and all $\lambda \in [0,1]$

$$
\begin{aligned}
& (\mu \otimes \nu)(\lambda x + (1-\lambda)y) \\
= \; & \mu(\lambda x + (1-\lambda)y) \otimes \nu(\lambda x + (1-\lambda)y) \\
\geq \; & (\mu(x) \otimes_D \mu(y)) \otimes (\nu(x) \otimes_D \nu(y)) \\
= \; & \begin{cases} \mu(x) \otimes \nu(x), & \mu(y) = \nu(y) = 1_L, \\ \mu(y) \otimes \nu(y), & \mu(x) = \nu(x) = 1_L, \\ \mu(x) \otimes \nu(y), & \mu(y) = \nu(x) = 1_L, \\ \mu(y) \otimes \nu(x), & \mu(x) = \nu(y) = 1_L, \\ 0_L, & otherwise. \end{cases}
\end{aligned}
$$

Then we see

$$
\begin{aligned}
& (\mu \otimes \nu)(x) \otimes_D (\mu \otimes \nu)(y) \\
= \; & (\mu(x) \otimes \nu(x)) \otimes_D (\mu(y) \otimes \nu(y)) \\
= \; & \begin{cases} \mu(x) \otimes \nu(x), & \mu(y) \otimes \nu(y) = 1_L, \\ \mu(y) \otimes \nu(y), & \mu(x) \otimes \nu(x) = 1_L, \\ 0_L, & otherwise, \end{cases} \\
= \; & \begin{cases} \mu(x) \otimes \nu(x), & \mu(y) = \nu(y) = 1_L, \\ \mu(y) \otimes \nu(y), & \mu(x) = \nu(x) = 1_L, \\ 0_L, & otherwise. \end{cases}
\end{aligned}
$$

Hence

$$(\mu \otimes \nu)(\lambda x + (1-\lambda)y) \geq (\mu \otimes \nu)(x) \otimes_D (\mu \otimes \nu)(y).$$

Thus, $\mu \otimes \nu$ is $\otimes_D$−convex.

□

Let $\otimes$ be a triangular norm on L. Li in [14] given a family triangular norms $(\otimes_a)_{a \in L}$ as follows

$$x \otimes_a y = \begin{cases} 0_L, & x \otimes y \leq a \ \ and \ \ x, y \neq 1_L; \\ x \otimes y, & otherwise. \end{cases}$$

Theorem 8. *Let $\otimes : L^2 \to L$ be a triangular norm on L, and $a \in L$. If $\mu, \nu : \mathbb{R}^n \to L$ are arbitrary $\otimes_a$−convex fuzzy sets, then $\mu \otimes \nu$ is $\otimes_a$−convex.*

Proof. Let $\mu, \nu : \mathbb{R}^n \to L$ be arbitrary $\otimes_a$−convex fuzzy sets. For any $x, y \in \mathbb{R}^n$ and all $\lambda \in [0,1]$

$$
\begin{aligned}
& (\mu \otimes \nu)(\lambda x + (1-\lambda)y) \\
= \; & \mu(\lambda x + (1-\lambda)y) \otimes \nu(\lambda x + (1-\lambda)y) \\
\geq \; & (\mu(x) \otimes_a \mu(y)) \otimes (\nu(x) \otimes_a \nu(y)) \\
= \; & \begin{cases} 0_L, & \mu(x) \otimes \mu(y) \leq a \ \ or \ \ \nu(x) \otimes \nu(y) \leq a, \\ \mu(x) \otimes \mu(y) \otimes \nu(x) \otimes \nu(y), & otherwise. \end{cases}
\end{aligned}
$$

Then we have

$$\begin{aligned} & (\mu \otimes \nu)(x) \otimes_a (\mu \otimes \nu)(y) \\ = \;& (\mu(x) \otimes \nu(x)) \otimes_a (\mu(y) \otimes \nu(y)) \\ = \;& \begin{cases} 0_L, & \mu(x) \otimes \nu(x) \otimes \mu(y) \otimes \nu(y) \leq a, \\ \mu(x) \otimes \mu(y) \otimes \nu(x) \otimes \nu(y), & \textit{otherwise}. \end{cases} \end{aligned}$$

Since $\mu(x) \otimes \mu(y) \leq a$ or $\nu(x) \otimes \nu(y) \leq a$ implies $\mu(x) \otimes \nu(x) \otimes \mu(y) \otimes \nu(y) \leq a$, we have

$$(\mu \otimes \nu)(\lambda x + (1-\lambda)y) \geq (\mu \otimes \nu)(x) \otimes_a (\mu \otimes \nu)(y).$$

Thus, $\mu \otimes \nu$ is $\otimes_a$−convex.
□

Example 6. *Consider the lattice $(L = \{0_L, a, b, c, d, 1_L\}, \leq, 0, 1)$ given in Figure 1. Consider the function $\otimes_b$ on L defined by*

$$\alpha \otimes_b \beta = \begin{cases} 0_L, & \alpha \wedge \beta \leq b \textit{ and } \alpha, \beta \neq 1_L; \\ \alpha \wedge \beta, & \textit{otherwise}, \end{cases}$$

then $\otimes_b$ is a triangular norm and $\otimes_b$ is described in Table 3.

Hence, for any $\otimes_b$-convex sets $\mu, \nu : \mathbb{R}^n \to L$, $\mu \otimes_M \nu = \mu \wedge \nu$ is also a $\otimes_b$-convex set.

Table 3. Triangular norm $\otimes_b$.

T_b	0_L	a	b	c	d	1_L
0_L	0_L	0_L	0_L	0_L	0_L	0_L
a	0_L	0_L	0_L	0_L	0_L	a
b	0_L	0_L	0_L	0_L	0_L	b
c	0_L	0_L	0_L	c	0_L	c
d	0_L	0_L	0_L	0_L	d	d
1_L	0_L	a	b	c	d	1_L

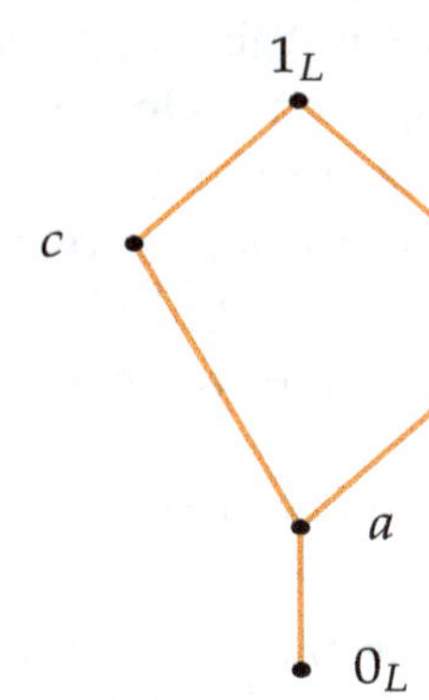

Figure 1. The order ≤ on L.

Theorem 9. *Let $\mu, \nu : \mathbb{R}^n \to [0,1]$ be arbitrary $\otimes_H$−convex fuzzy sets. Then $\min\{\mu, \nu\}$ is a $\otimes_H$−convex fuzzy set.*

Proof. Let $\mu, \nu : \mathbb{R}^n \to L$ be arbitrary $\otimes_H-$convex fuzzy sets. For any $x, y \in \mathbb{R}^n$ and all $\lambda \in [0,1]$

$$\begin{aligned}
& \min\{\mu,\nu\}(\lambda x + (1-\lambda)y) \\
= & \min\{\mu(\lambda x + (1-\lambda)y), \nu(\lambda x + (1-\lambda)y)\} \\
\geq & \min\{\mu(x) \otimes_H \mu(y), \nu(x) \otimes_H \nu(y)\} \\
= & \begin{cases} \min\{\mu(x) * \mu(y), \nu(x) * \nu(y)\}, & (\mu(x),\mu(y)) \in H^2 \ \textit{and}\ (\nu(x),\nu(y)) \in H^2, \\ \min\{\mu(x) * \mu(y), \nu(x), \nu(y)\}, & (\mu(x),\mu(y)) \in H^2 \ \textit{and}\ (\nu(x),\nu(y)) \notin H^2, \\ \min\{\mu(x), \mu(y), \nu(x) * \nu(y)\}, & (\mu(x),\mu(y)) \notin H^2 \ \textit{and}\ (\nu(x),\nu(y)) \in H^2, \\ \min\{\mu(x), \mu(y), \nu(x), \nu(y)\}, & \textit{otherwise}. \end{cases}
\end{aligned}$$

Then we deduce

$$\begin{aligned}
& \min\{\mu,\nu\}(x) \otimes_H \min\{\mu,\nu\}(y) \\
= & \min\{\mu(x),\nu(x)\} \otimes_H \min\{\mu(y),\nu(y)\} \\
= & \begin{cases} \min\{\mu(x),\nu(x)\} * \min\{\mu(y),\nu(y)\}, & (\min\{\mu(x),\nu(x)\}, \min\{\mu(y),\nu(y)\}) \in H^2, \\ \min\{\mu(x),\mu(y),\nu(x),\nu(y)\}, & \textit{otherwise}. \end{cases}
\end{aligned}$$

Since $\min\{\mu(x),\mu(y)\} \geq \mu(x) * \mu(y) \geq \min\{\mu(x),\nu(x)\} * \min\{\mu(y),\nu(y)\}, \min\{\nu(x),\nu(y)\} \geq \nu(x) * \nu(y) \geq \min\{\mu(x),\nu(x)\} * \min\{\mu(y)$ we have

$$\min\{\mu,\nu\}(\lambda x + (1-\lambda)y) \geq \min\{\mu,\nu\}(x) \otimes_H \min\{\mu,\nu\}(y).$$

Thus, $\min\{\mu,\nu\}$ is a $\otimes_H-$convex fuzzy set.
□

Example 7. *Suppose* $H = (0, \frac{1}{2})$ *and the triangular norm* $\otimes_H$ *is*

$$a \otimes_H b = \begin{cases} \frac{ab}{2}, & (a,b) \in (0,\frac{1}{2})^2; \\ \min\{a,b\} & \textit{otherwise}, \end{cases}$$

then, $\min\{\mu,\nu\}$ *is a* $\otimes_H-$*convex fuzzy set.*

5. Conclusions

The authors of the paper [7] discuss properties which are preserved under aggregation for arbitrary lattices and arbitrary pairs of mappings Results in this paper are also discussed under aggregation for an arbitrary lattice and an arbitrary pair of mappings. However, this does not mean that even without these conditions the aggregation of SOME quasiconvex ($\otimes-$convex) mappings to SOME lattices need not be quasiconvex ($\otimes-$convex). Which are the properties of a lattice L and an aggregation A (weaker than those from the paper by Janis, Kral and Rencova in [7]), such that A preserves quasiconvexity ($\otimes-$convex) for mappings into L? We hope to solve this problem in future work.

Author Contributions: Conceptualization, L.L.; methodology, L.L. and Q.L.

Acknowledgments: This work was supported by the National Natural Science Foundation of China (Grant Nos. 11701446, 61702389) and Shaanxi key disciplines of special funds projects.

References

1. Zadeh, L.A. Fuzzy sets. *Inf. Control* **1965**, *8*, 338–353. [CrossRef]
2. Lizasoain, I.; Moreno, C. OWA operators defined on complete lattices. *Fuzzy Sets Syst.* **2013**, *224*, 36–52. [CrossRef]

3. Aggarwal, M. Discriminative aggregation operators for multi criteria decision making. *Appl. Soft Comput.* **2017**, *52*, 1058–1069. [CrossRef]
4. Jiang, W.; Wei, B.; Zhan, J.; Xie, C.; Zhou, D. A Visibility Graph Power Averaging Aggregation Operator: A Methodology Based on Network Analysis. *Comput. Ind. Eng.* **2016**, *101*, 260–268. [CrossRef]
5. Liu, P.; Chen, S.M. Group Decision Making Based on Heronian Aggregation Operators of Intuitionistic Fuzzy Numbers. *IEEE Trans. Cybern.* **2017**, *99*, 2514–2530. [CrossRef] [PubMed]
6. Scellato, S.; Fortuna, L.; Frasca, M.; Gómez-Gardenes, J.; Latora, V. Traffic optimization in transport networks based on local routing. *Eur. Phys. J. B* **2010**, *73*, 303–308. [CrossRef]
7. Janiš, V.; Král, P.; Renčová, M. Aggregation operators preserving quasiconvexity. *Inf. Sci.* **2013**, *228*, 37–44. [CrossRef]
8. Hua, X.J.; Xin, X.L.; Zhu, X. Generalized (convex) fuzzy sublattices. *Comput. Math. Appl.* **2011**, *62*, 699–708. [CrossRef]
9. Pan, X.D.; Meng, D. Triangular norm based graded convex fuzzy sets. *Fuzzy Sets Syst.* **2012**, *209*, 1–13. [CrossRef]
10. Nourouzi, K.; Aghajani, A. Convexity in triangular norm of fuzzy sets. *Chaos Solitons Fractals* **2008**, *36*, 883–889. [CrossRef]
11. Tahayori, H.; Tettamanzi, G.B.; Antoni, G.D.; Visconti, A. On the calculation of extended max and min operations between convex fuzzy sets of the real line. *Fuzzy Sets Syst.* **2009**, *160*, 3103–3114. [CrossRef]
12. De Baets, B.; Mesiar, R. Triangular norms on product lattices. *Fuzzy Sets Syst.* **1999**, *104*, 61–75. [CrossRef]
13. Klement, E.P.; Mesiar, R.; Pap, E. *Triangular Norms*; Kluwer Academic Publishers: Dordrecht, The Netherlands, 2000.
14. Li, L.; Zhang, J.; Zhou, C. Sufficient conditions for a T-partial order obtained from triangular norms to be a lattice. *Kybernetika* **2018**, submitted.

10

Edge Even Graceful Labeling of Cylinder Grid Graph

Ahmed A. Elsonbaty [1,2] and Salama Nagy Daoud [1,3,*]

1 Department of Mathematics, Faculty of Science, Taibah University, Al-Madinah 41411, Saudi Arabia; ahmad_elsonbaty@hotmail.com
2 Department of Mathematics, Faculty of Science, Ain Shams University, Cairo 11566, Egypt
3 Department of Mathematics and Compuer Science, Faculty of Science, Menoufia University, Shebin El Kom 32511, Egypt
* Correspondence: sdaoud@taibahu.edu.sa

Abstract: Edge even graceful labeling (e.e.g., l.) of graphs is a modular technique of edge labeling of graphs, introduced in 2017. An e.e.g., l. of simple finite undirected graph $G = (V(G), E(G))$ of order $P = |(V(G)|$ and size $q = |E(G)|$ is a bijection $f : E(G) \to \{2, 4, \ldots, 2q\}$, such that when each vertex $v \in V(G)$ is assigned the modular sum of the labels (images of f) of the edges incident to v, the resulting vertex labels are distinct mod$2r$, where $r = \max(p, q)$. In this work, the family of cylinder grid graphs are studied. Explicit formulas of e.e.g., l. for all of the cases of each member of this family have been proven.

Keywords: graceful labeling; edge even graceful labeling; cylinder grid graph

---

1. Introduction

The field of graph theory plays an important role in various areas of pure and applied sciences. One of the important areas in graph theory is graph labeling of a graph G which is an assignment of integers either to the vertices or edges or both subject to certain conditions. Graph labeling began nearly 50 years ago. Over these decades, more than 200 methods of labeling techniques were invented, and more than 2500 papers were published. In spite of this huge literature, just few general results were discovered. Nowadays, graph labeling has much attention from different brilliant researchers in graph theory, which has rigorous applications in many disciplines, e.g., communication networks, coding theory, X-ray crystallography, radar, astronomy, circuit design, communication network addressing, database management, and graph decomposition problems. More interesting applications of graph labeling can be found in References [1–11]. A function f is called a graceful labeling of a graph G if $f : V(G) \to \{0, 1, 2, \ldots, q\}$ is injective and the induced function $f^* : E(G) \to \{1, 2, \ldots, q\}$, defined as $f^*(e = uv) = |f(u) - f(v)|$, is bijective. This type of graph labeling was first introduced by Rosa in 1967 [12] as a $\beta-$ valuation, and later, Solomon W. Golomb [13] termed it as graceful labeling. A function f is called an odd graceful labeling of a graph G if $f : V(G) \to \{0, 1, 2, \ldots, 2q-1\}$ is injective and the induced function $f^* : E(G) \to \{1, 3, \ldots, 2q-1\}$, defined as $f^*(e = uv) = |f(u) - f(v)|$, is bijective. This type of graph labeling first introduced by Gnanajothi in 1991 [14]. For more results on this type of labeling, see References [15,16]. A function f is called an edge graceful labeling of a graph G if $f : E(G) \to \{1, 2, \ldots, q\}$ is bijective and the induced function $f^* : V(G) \to \{0, 1, 2, \ldots, p-1\}$, defined as $f^*(u) = \sum_{e=uv \in E(G)} f(e)(\mathrm{mod}p)$, is bijective. This type of graph labeling was first introduced by Lo in 1985 [17]. For more results on this labeling see [18,19]. A function f is called an edge odd graceful labeling of a graph G if $f : E(G) \to \{1, 3, \ldots, 2q-1\}$ is bijective and the induced function $f^* : V(G) \to \{0, 1, 2, \ldots, 2q-1\}$ defined as $f^*(u) = \sum_{e=uv \in E(G)} f(e)(\mathrm{mod}2q)$ is injective. This type of graph labeling was first introduced by Solairaju and Chithra in 2009 [20]. For more results on this labeling, see References [21–23]. A function f is called an edge even graceful labeling of a graph G if $f : E(G) \to \{2, 4, \ldots, 2q-2\}$ is bijective and the induced function $f^* : V(G) \to \{0, 2, 4, \ldots, 2q-2\}$, defined as $f^*(u) = \sum_{e=uv \in E(G)} f(e) (\text{mod} 2r)$ where $r = \max\{p, q\}$, is injective. This type of graph labeling was first introduced by Elsonbaty and Daoud in 2017 [24,25]. For a summary of the results on these five types of graceful labels as well as all known labeling techniques, see Reference [26].

2. Cylinder Grid Graph

The Cartesian product $G_1 \times G_2$ of two graphs G_1 and G_2, is the graph with vertex set $V(G_1) \times V(G_2)$, and any two vertices (u_1, v_1) and (u_2, v_2) are adjacent in $G_1 \times G_2$ whenever $u_1 = u_2$ and $v_1 v_2 \in E(G_2)$ or $v_1 = v_2$ and $u_1 u_2 \in E(G_1)$. The cylinder grid graph $C_{m,n}$ is the graph formed from the Cartesian product $P_m \times C_n$ of the path graph P_m and the cycle graph C_n. That is, the cylinder grid graph consists of m copies of C_n represented by circles, and will be numbered from the innermost circle to the outer circle as $C_n^{(1)}, C_n^{(2)}, C_n^{(3)}, \ldots, C_n^{(m-1)}, C_n^{(m)}$ and we call them simply circles; n copies of P_m represented by paths transverse the m circles and will be numbered clockwise as $P_m^{(1)}, P_m^{(2)}, P_m^{(3)}, \ldots, P_m^{(n-1)}, P_m^{(n)}$ and we call them paths (see Figure 1).

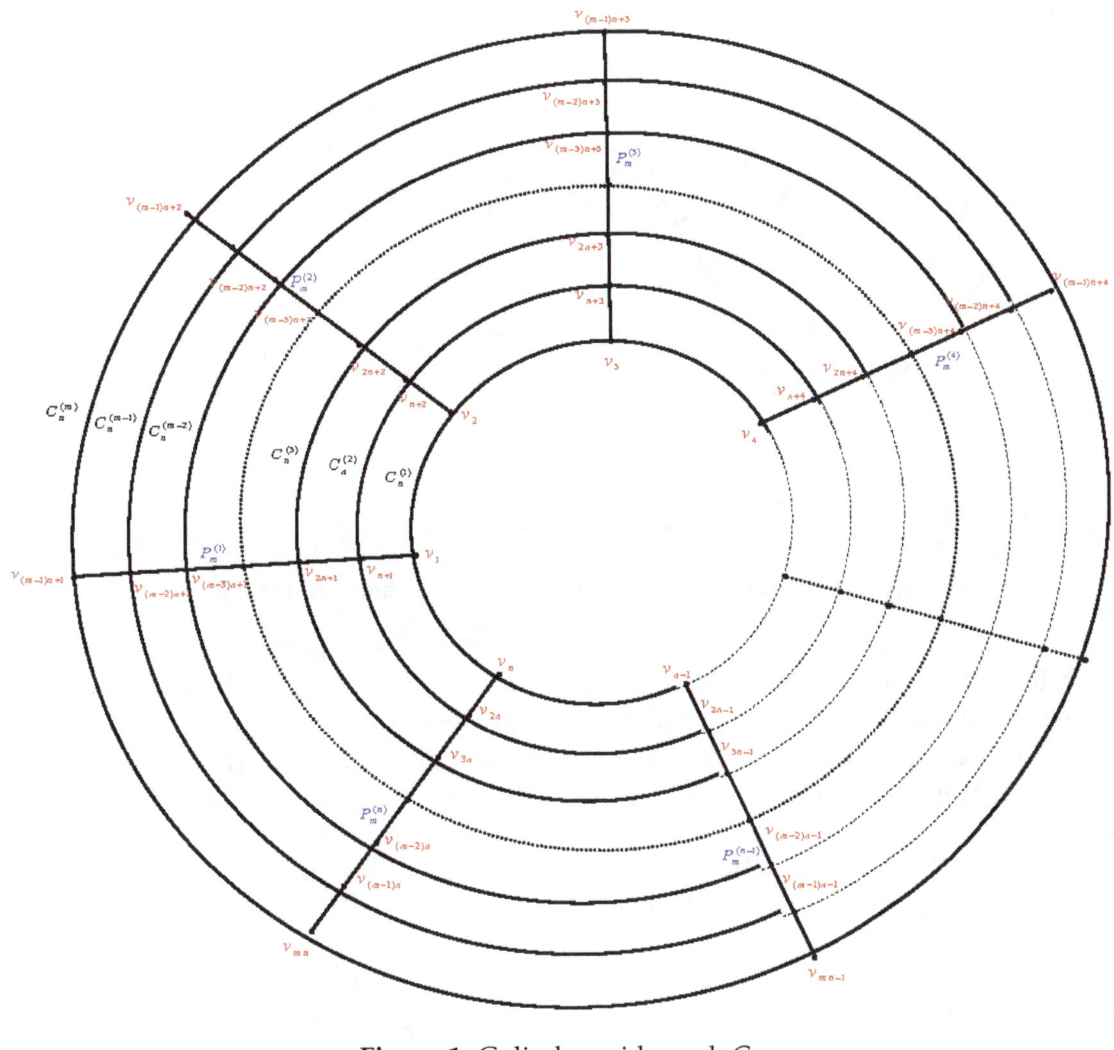

Figure 1. Cylinder grid graph $C_{m,n}$.

Theorem 1. *If m is an even positive integer greater than or equal 2 and $n \geq 2$, then the cylinder grid graph $C_{m,n}$, is an edge even graceful graph.*

Proof. Using standard notation $p = |V(C_{m,n})| = mn$, $q = |E(C_{m,n})| = 2mn - n$ and $r = \max(p,q) = 2mn - n$ and $f : E(C_{m,n}) \rightarrow \{2, 4, 6, \ldots, 4mn - 2n - 2\}$. Let the cylinder grid graph $C_{m,n}$ be as in Figure 2. □

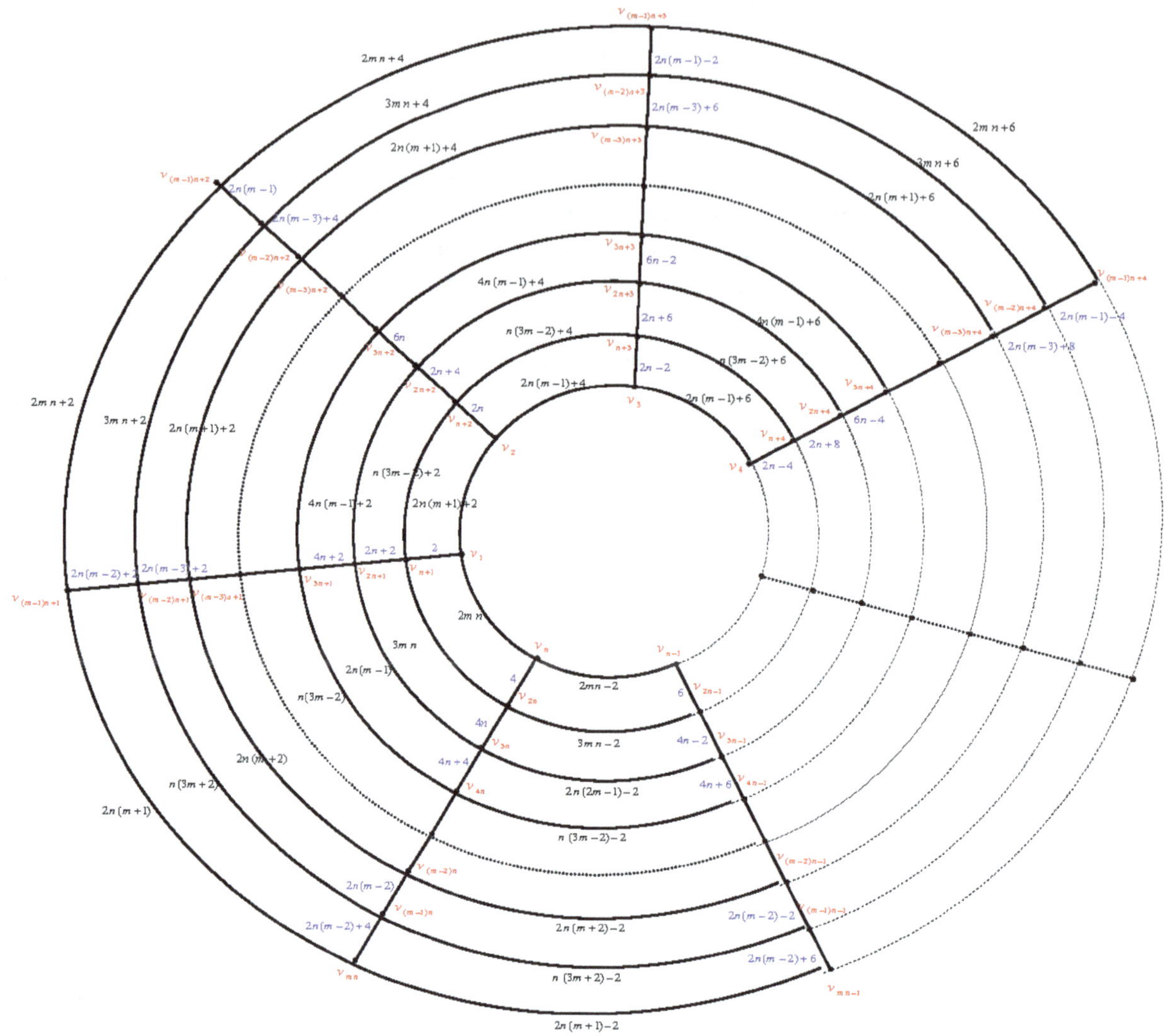

Figure 2. The cylinder grid graph $C_{m,n}$, m is even and $n \geq 2$.

First, we label the edges of the paths $P_m^{(k)}, 1 \leq k \leq n$ beginning with the edges of the path $P_m^{(1)}$ as follows: Move anticlockwise to label the edges $v_1v_{n+1}, v_nv_{2n}, v_{n-1}v_{2n-1}, \ldots, v_3v_{n+3}, v_2v_{n+2}$ by $2, 4, 6, \ldots, 2n-2, 2n$, then move clockwise to label the edges $v_{n+1}v_{2n+1}, v_{n+2}v_{2n+2}, v_{n+3}v_{2n+3}, \ldots,$ $v_{2n-1}v_{3n-1}, v_{2n}v_{3n}$ by $2n+2, 2n+4, 2n+6, \ldots, 4n-2, 4n$, then move anticlockwise to label the edges $v_{2n+1}v_{3n+1}, v_{3n}v_{4n}, v_{3n-1}v_{4n-1}, \ldots, v_{2n+3}v_{3n+3}, v_{2n+2}v_{3n+2}$ by $4n+2, 4n+4, 4n+6, \ldots, 6n-2, 6n$ and so on. Finally, move anticlockwise to label the edges $v_{(m-2)n+1}v_{(m-1)n+1}, v_{(m-1)n}v_{mn}, v_{(m-1)n-1}v_{mn-1}, \ldots,$ $v_{(m-2)n+3}v_{(n-1)m+3}, v_{(m-2)n+2}v_{(m-1)n+2}$ by $2n(m-1)+2, 2n(m-2)+4, 2n(m-2)+6, 2n(m-2)+8, \ldots, 2n(m-1)-2, 2n(m-1)$.

Secondly, we label the edges of the circles $C_n^{(k)}, 1 \leq k \leq m$ beginning with the edges of the innermost circle $C_n^{(1)}$ then the edges of outer circle $C_n^{(m)}$, then the edges of the circles $C_n^{(m-2)}, C_n^{(m-4)}, \ldots, C_n^{(2)}$.

Finally, we label the edges of the circles $C_n^{(m-1)}, C_n^{(m-3)}, \ldots, C_n^{(3)}$ as follows: $f(v_iv_{i+1}) = 2n(m-1) + 2i, 1 \leq i \leq n-1, f(v_nv_1) = 2mn; f(v_{(m-1)n+i}v_{(m-1)n+i+1}) = 2mn + 2i, 1 \leq i \leq n-1, f(v_{mn}v_{(m-1)n+1}) =$

$2n(m+1)$; $f(v_{(k-1)n+i}v_{(k-1)n+i+1}) = n(3m-k)+2i, 1 \leq i \leq n-1, f(v_{kn}v_{(k-1)n+1}) = n(3m-k+2), 2 \leq k \leq m-2$; $f(v_{(k-1)n+i}v_{(k-1)n+i+1}) = n(4m-k-1)+2i, 1 \leq i \leq n-1, f(v_{kn}v_{(k-1)n+1}) = n(4m-k+1)$, $3 \leq k \leq m-1, k$ is odd.

Thus, the labels of corresponding vertices $\bmod(4mn-2n)$ will be: $f^*(v_i) \equiv 2i+2$; $f^*(v_{n+i}) \equiv 2mn+2n+4i+2$; $f^*(v_{2n+i}) \equiv 4n+4i+2$; $f^*(v_{3n+i}) \equiv 2mn+6n+4i+2; \ldots; f^*(v_{(m-3)n+i}) \equiv 4mn-6n+4i+2$; $f^*(v_{(m-2)n+i}) \equiv 2mn-4n+4i+2$; $f^*(v_{(m-1)n+i}) \equiv 2mn+2i+2, 1 \leq i \leq n$.

Illustration: An e.e.g., l, of the cylinder grid graphs $C_{8,11}$ and $C_{8,12}$ are shown in Figure 3.

Theorem 2. *If $m = 3$ and n is an odd positive integer greater than 3, then the cylinder grid graph $C_{3,n}$, is an edge even graceful graph.*

Proof. Using standard notation $p = |V(C_{3,n})| = 3n$, $q = |E(C_{3,n})| = 5n$, $r = \max(p,q) = 5n$, and $f: E(C_{3,n}) \to \{2,4,6,\ldots,10n-2\}$. There are three cases:

Case (1): If $n \equiv 1 \bmod 6$, let the cylinder grid graph $C_{3,n}$ be as in Figure 4.

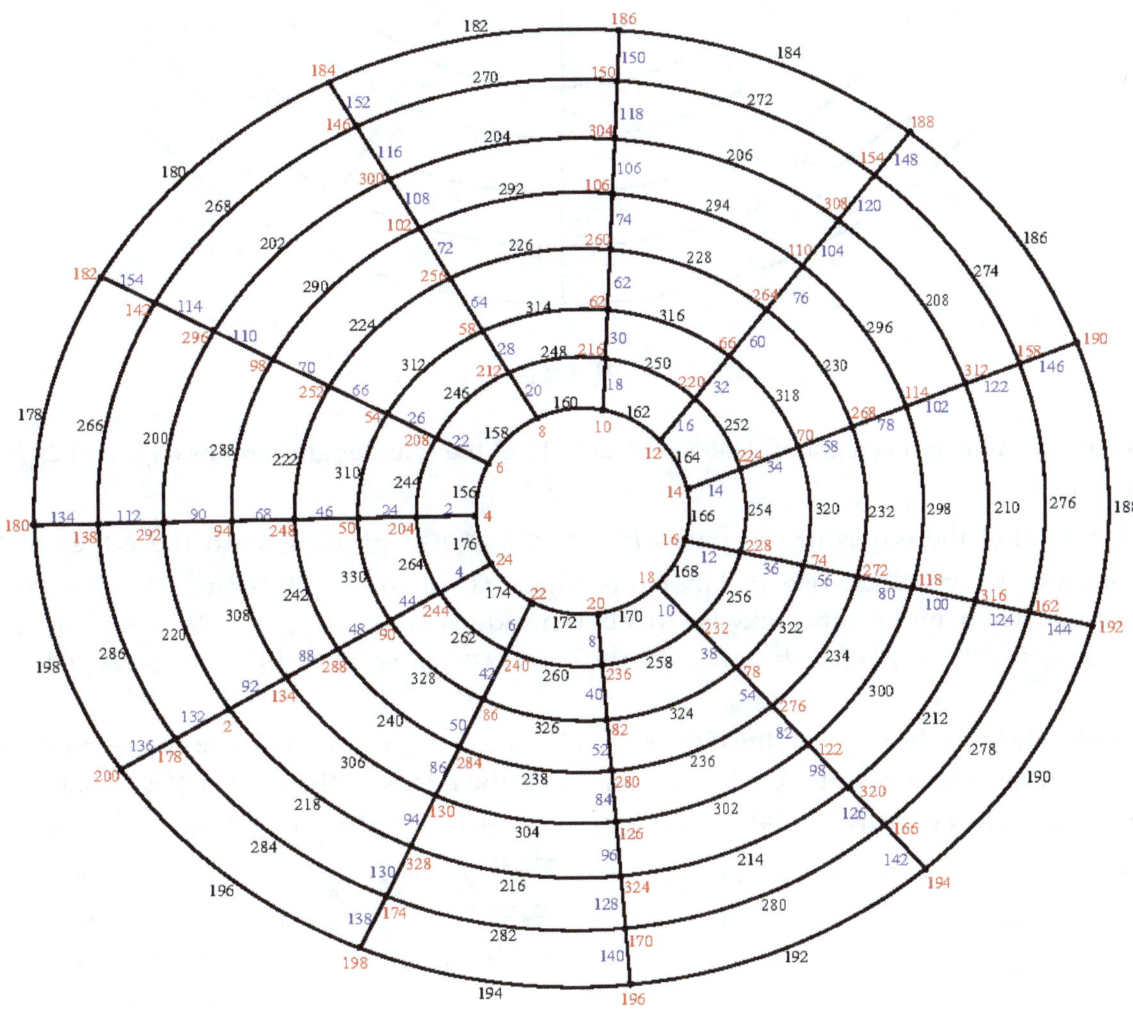

(a) $C_{8,11}$

Figure 3. *Cont.*

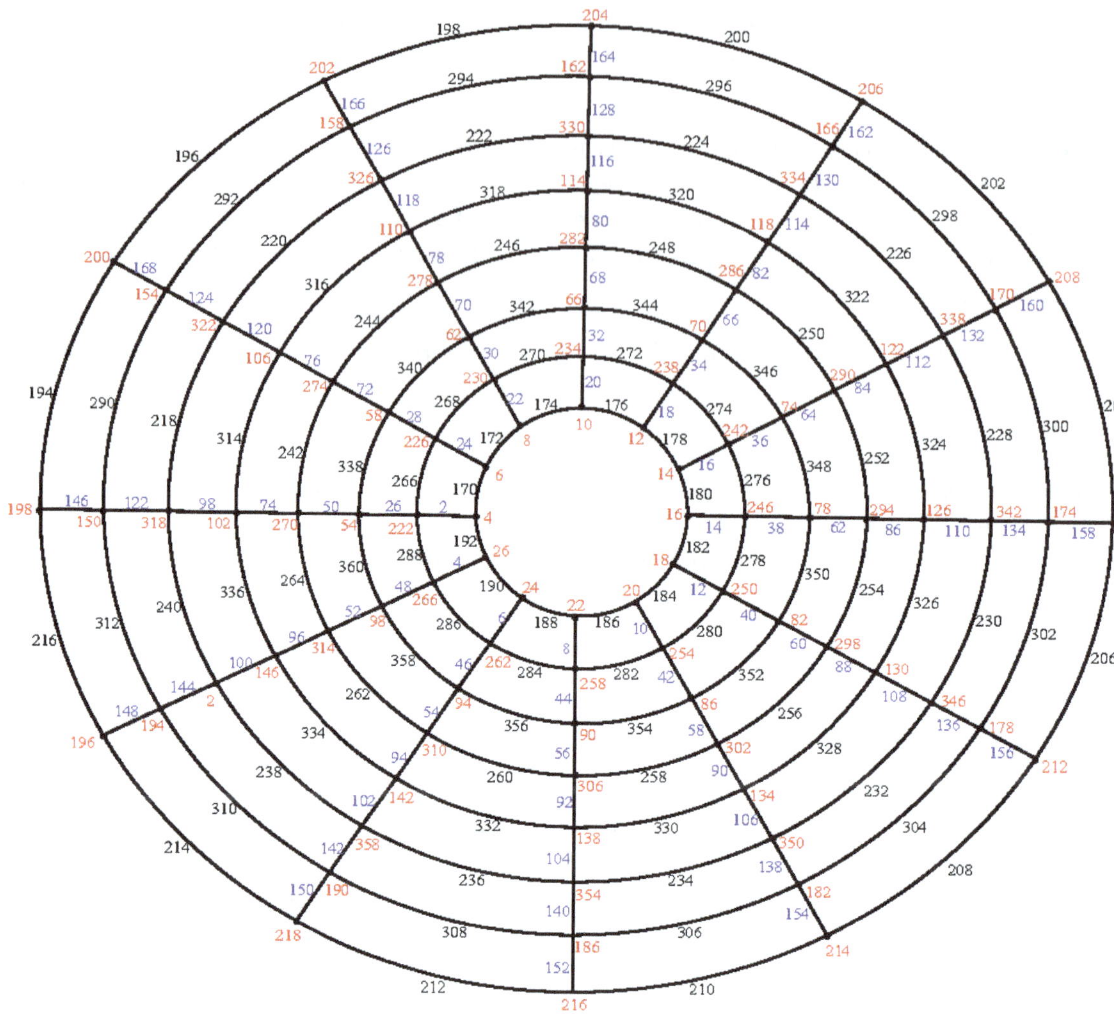

(b) $C_{8,12}$

Figure 3. An edge even graceful labeling (e.g., l.) of the cylinder grid graphs $C_{8,11}$ and $C_{8,12}$.

First, we label the edges of the paths $P_3^{(k)}, 1 \le k \le n$ beginning with the edges of the path $P_3^{(1)}$ as follows: Move clockwise to label the edges $f(v_1v_{n+1}) = 2, f(v_2v_{n+2}) = 6, f(v_iv_{n+i}) = 2i+2, 3 \le i \le n$. Then, move anticlockwise to label the edges $f(v_{n+1}v_{2n+1}) = 2n+4, f(v_{2n}v_{3n}) = 2n+6, f(v_{2n-1}v_{3n-1}) = 2n+8, f(v_{2n-2}v_{3n-2}) = 2n+10, \ldots, f(v_{n+3}v_{2n+3}) = 4n, f(v_{n+2}v_{2n+2}) = 4n+2$.

Secondly, we label the edges of the circles $C_n^{(k)}, 1 \le k \le 3$ beginning with the edges of the innermost circle $C_n^{(1)}$, then the edges of outer circle $C_n^{(3)}$, and then the edges of the circle $C_n^{(2)}$. Label the edges of the circle $C_n^{(1)}$ as follows: $f(v_1v_2) = 4n+4, f(v_2v_3) = 4n+6, \ldots, f(v_{\frac{n-1}{3}}v_{\frac{n+2}{3}}) = \frac{14n+10}{3}, f(v_{\frac{n+2}{3}}v_{\frac{n+5}{3}}) = \frac{14n+4}{3}, f(v_{\frac{n+5}{3}}v_{\frac{n+8}{3}}) = \frac{14n+16}{3}, f(v_{\frac{n+8}{3}}v_{\frac{n+11}{3}}) = \frac{14n+22}{3}, f(v_{\frac{n+11}{3}}v_{\frac{n+14}{3}}) = \frac{14n+34}{3}, f(v_{\frac{n+14}{3}}v_{\frac{n+17}{3}}) = \frac{14n+28}{3}, f(v_{\frac{n+17}{3}}v_{\frac{n+20}{3}}) = \frac{14n+40}{3}, f(v_{\frac{n+20}{3}}v_{\frac{n+23}{3}}) = \frac{14n+46}{3}, f(v_{\frac{n+23}{3}}v_{\frac{n+26}{3}}) = \frac{14n+58}{3}, f(v_{\frac{n+26}{3}}v_{\frac{n+29}{3}}) = \frac{14n+52}{3}, f(v_{\frac{n+29}{3}}v_{\frac{n+32}{3}}) = \frac{14n+64}{3}, f(v_{\frac{n+32}{3}}v_{\frac{n+35}{3}}) = \frac{14n+70}{3}, f(v_{\frac{n+35}{3}}v_{\frac{n+38}{3}}) = \frac{14n+82}{3}, f(v_{\frac{n+38}{3}}v_{\frac{n+41}{3}}) = \frac{14n+76}{3}, f(v_{\frac{n+41}{3}}v_{\frac{n+44}{3}}) = \frac{14n+88}{3}, f(v_{\frac{n+44}{3}}v_{\frac{n+47}{3}}) = \frac{14n+94}{3}, \ldots, f(v_{n-13}v_{n-12}) = 6n-22, f(v_{n-12}v_{n-11}) = 6n-24, f(v_{n-11}v_{n-10}) = 6n-20, f(v_{n-10}v_{n-9}) = 6n-18, f(v_{n-9}v_{n-8}) = 6n-14, f(v_{n-8}v_{n-7}) = 6n-16, f(v_{n-7}v_{n-6}) = 6n-12, f(v_{n-6}v_{n-5}) = 6n-10, f(v_{n-5}v_{n-4}) = 6n-6, f(v_{n-4}v_{n-3}) = 6n-8, f(v_{n-3}v_{n-2}) = 6n-4, f(v_{n-2}v_{n-1}) = 6n-2, f(v_{n-1}v_n) = 6n+2, f(v_nv_1) = 6n$. □

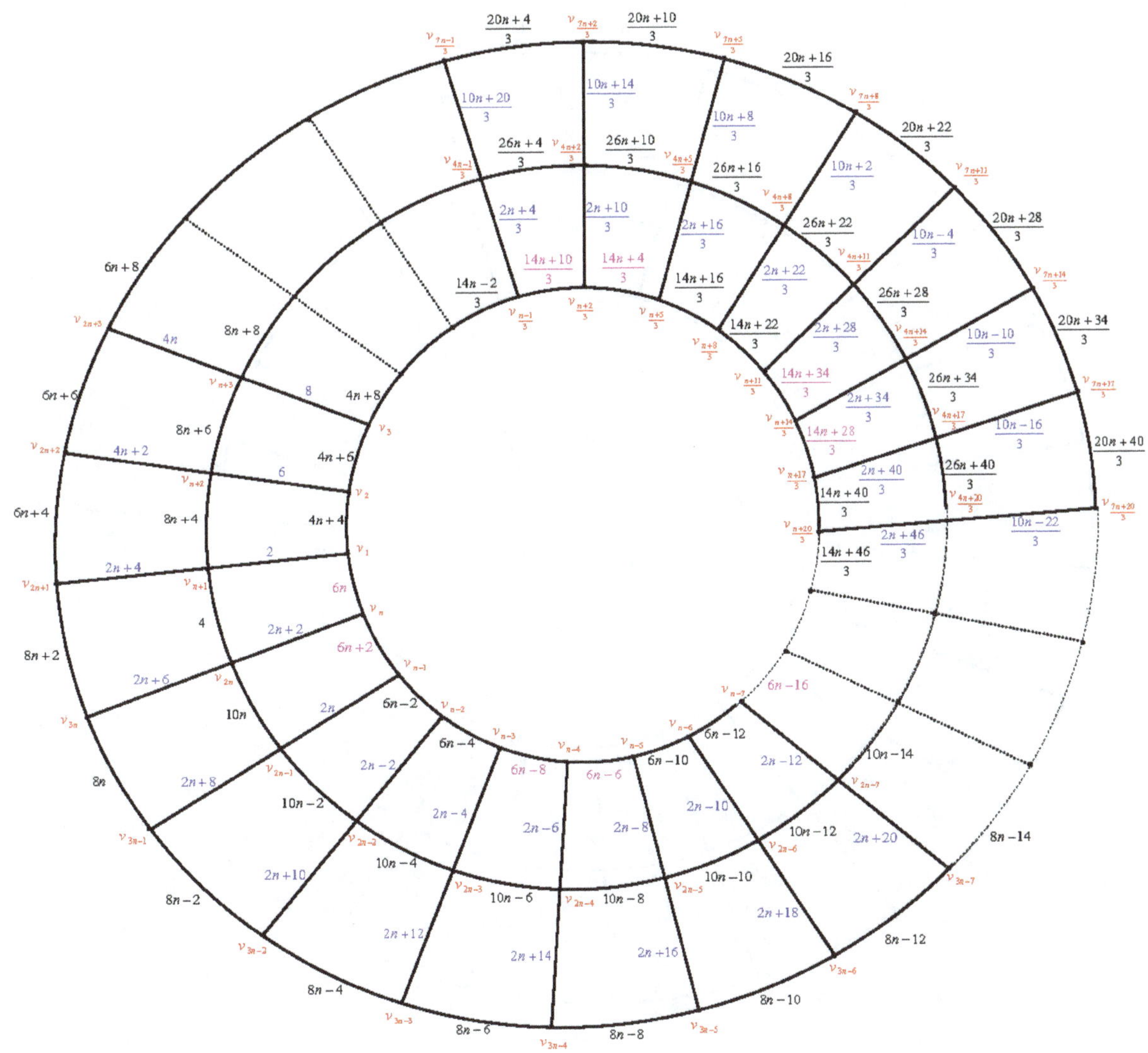

Figure 4. The cylinder grid graph $C_{3,n}$, $n \equiv 1 \bmod 6$.

Label the edges of the circle $C_n^{(2)}$ as follows: $f(v_{n+i}v_{n+i+1}) = 8n + 2i + 2, 1 \le i \le n-1, f(v_{2n}v_{n+1}) = 4$.

Label the edges of the circle $C_n^{(3)}$ as follows: $f(v_{2n+i}v_{2n+i+1}) = 6n + 2i + 2, 1 \le i \le n$.

The labels of corresponding of vertices mod$10n$ are as follows:

The labels of vertices of the circle $C_n^{(1)}$ are as follows: $f^*(v_1) \equiv 6, f^*(v_2) \equiv 8n+16, f^*(v_3) \equiv 8n+22, \ldots, f^*(v_{\frac{n-1}{3}}) \equiv 4, f^*(v_{\frac{n+2}{3}}) \equiv 8, f^*(v_{\frac{n+5}{3}}) \equiv 12, f^*(v_{\frac{n+8}{3}}) \equiv 20, f^*(v_{\frac{n+11}{3}}) \equiv 28, f^*(v_{\frac{n+14}{3}}) \equiv 32, f^*(v_{\frac{n+17}{3}}) \equiv 36, f^*(v_{\frac{n+20}{3}}) \equiv 44, f^*(v_{\frac{n+23}{3}}) \equiv 52, f^*(v_{\frac{n+26}{3}}) \equiv 56, f^*(v_{\frac{n+29}{3}}) \equiv 60, f^*(v_{\frac{n+32}{3}}) \equiv 68, f^*(v_{\frac{n+35}{3}}) \equiv 76, f^*(v_{\frac{n+38}{3}}) \equiv 80, f^*(v_{\frac{n+41}{3}}) \equiv 84, f^*(v_{\frac{n+44}{3}}) \equiv 92, f^*(v_{\frac{n+47}{3}}) \equiv 100, \ldots, f^*(v_{n-12}) \equiv 4n-68, f^*(v_{n-11}) \equiv 4n-64, f^*(v_{n-10}) \equiv 4n-56, f^*(v_{n-9}) \equiv 4n-48, f^*(v_{n-8}) \equiv 4n-44, f^*(v_{n-7}) \equiv 4n-40, f^*(v_{n-6}) \equiv 4n-32, f^*(v_{n-5}) \equiv 4n-24, f^*(v_{n-4}) \equiv 4n-20, f^*(v_{n-3}) \equiv 4n-16, f^*(v_{n-2}) \equiv 4n-8, f^*(v_{n-1}) \equiv 4n, f^*(v_n) \equiv 4n+4$.

The labels of vertices of the circle $C_n^{(2)}$ are $f^*(v_{i+1}) = 4i + 10, 1 \le i \le n-1, f^*(v_{2n}) = 4n + 12$.

The labels of vertices of the circle $C_n^{(3)}$ are $f^*(v_{2i+1}) = 6n + 2i + 8, 1 \le i \le n$.

Case (2): If $n \equiv 3 \bmod 6$, let the cylinder grid graph $C_{3,n}$ be as in Figure 5.

First, we label the edges of the paths $P_3^{(k)}, 1 \le k \le n$ beginning with the edges of the path $P_3^{(1)}$ as the same in case (1).

Secondly, we label the edges of the circles $C_n^{(k)}, 1 \le k \le 3$ beginning with the edges of the innermost circle $C_n^{(1)}$, then the edges of outer circle $C_n^{(3)}$, and then the edges of the circle $C_n^{(2)}$.

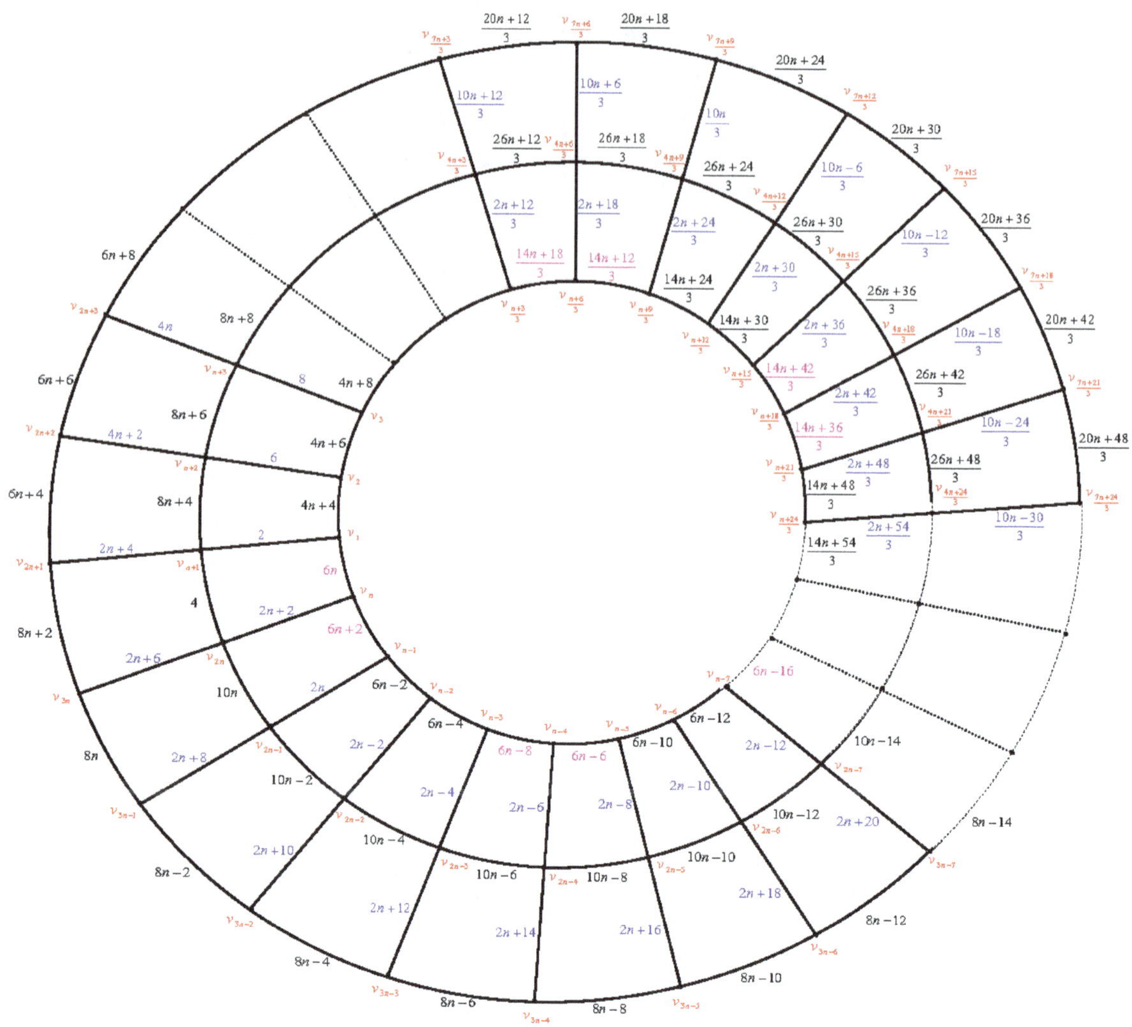

Figure 5. The cylinder grid graph $C_{3,n}$, $n \equiv 3\text{mod}6$.

Label the edges of the circle $C_n^{(1)}$ as follows: $f(v_1v_2) = 4n+4, f(v_2v_3) = 4n+6, \ldots, f(v_{\frac{n+3}{3}}v_{\frac{n+6}{3}}) = \frac{14n+18}{3}, f(v_{\frac{n+6}{3}}v_{\frac{n+9}{3}}) = \frac{14n+12}{3}, f(v_{\frac{n+9}{3}}v_{\frac{n+12}{3}}) = \frac{14n+24}{3}, f(v_{\frac{n+12}{3}}v_{\frac{n+15}{3}}) = \frac{14n+30}{3}, f(v_{\frac{n+15}{3}}v_{\frac{n+18}{3}}) = \frac{14n+42}{3}, f(v_{\frac{n+18}{3}}v_{\frac{n+21}{3}}) = \frac{14n+36}{3}, f(v_{\frac{n+21}{3}}v_{\frac{n+24}{3}}) = \frac{14n+48}{3}, f(v_{\frac{n+24}{3}}v_{\frac{n+27}{3}}) = \frac{14n+54}{3}, f(v_{\frac{n+27}{3}}v_{\frac{n+30}{3}}) = \frac{14n+66}{3}, f(v_{\frac{n+30}{3}}v_{\frac{n+33}{3}}) = \frac{14n+60}{3}, f(v_{\frac{n+33}{3}}v_{\frac{n+36}{3}}) = \frac{14n+72}{3}, f(v_{\frac{n+36}{3}}v_{\frac{n+39}{3}}) = \frac{14n+78}{3}, f(v_{\frac{n+39}{3}}v_{\frac{n+42}{3}}) = \frac{14n+90}{3}, f(v_{\frac{n+42}{3}}v_{\frac{n+45}{3}}) = \frac{14n+84}{3}, f(v_{\frac{n+45}{3}}v_{\frac{n+48}{3}}) = \frac{14n+96}{3}, f(v_{\frac{n+48}{3}}v_{\frac{n+51}{3}}) = \frac{14n+102}{3}, \ldots, f(v_{n-13}v_{n-12}) = 6n-22, f(v_{n-12}v_{n-11}) = 6n-24, f(v_{n-11}v_{n-10})f(v_{n-9}v_{n-8}) = 6n-14, f(v_{n-8}v_{n-7}) = 6n-16, f(v_{n-7}v_{n-6}) = 6n-12, f(v_{n-6}v_{n-5}) = 6n-10, f(v_{n-5}v_{n-4}) = 6n-6, f(v_{n-4}v_{n-3}) = 6n-8, f(v_{n-3}v_{n-2}) = 6n-4, f(v_{n-2}v_{n-1}) = 6n-2, f(v_{n-1}v_n) = 6n+2, f(v_nv_1) = 6n.$

The labels of corresponding vertices mod$10n$ are as follows: The label of vertices of the circle $C_n^{(1)}$ are $f^*(v_1) \equiv 6, f^*(v_2) \equiv 8n+16, f^*(v_3) \equiv 8n+22, \ldots, f^*(v_{\frac{n}{3}-1}) \equiv 10n-2, f^*(v_{\frac{n}{3}}) \equiv 4, f^*(v_{\frac{n}{3}+1}) \equiv 12, f^*(v_{\frac{n}{3}+2}) \equiv 16, f^*(v_{\frac{n}{3}+3}) \equiv 20, f^*(v_{\frac{n}{3}+4}) \equiv 28, f^*(v_{\frac{n}{3}+5}) \equiv 36, f^*(v_{\frac{n}{3}+6}) \equiv 40, f^*(v_{\frac{n}{3}+7}) \equiv 44, f^*(v_{\frac{n}{3}+8}) \equiv 52, f^*(v_{\frac{n}{3}+9}) \equiv 60, f^*(v_{\frac{n}{3}+10}) \equiv 64, f^*(v_{\frac{n}{3}+11}) \equiv 68, f^*(v_{\frac{n}{3}+12}) \equiv 76, f^*(v_{\frac{n}{3}+13}) \equiv 84, f^*(v_{\frac{n}{3}+14}) \equiv 88, f^*(v_{\frac{n}{3}+15}) \equiv 92, f^*(v_{\frac{n}{3}+16}) \equiv 100, \ldots, f^*(v_{n-12}) \equiv 4n-68, f^*(v_{n-11}) \equiv 4n-64, f^*(v_{n-10}) \equiv 4n-56, f^*(v_{n-9}) \equiv 4n-48, f^*(v_{n-8}) \equiv 4n-44, f^*(v_{n-7}) \equiv 4n-40, f^*(v_{n-6}) \equiv 4n-32, f^*(v_{n-5}) \equiv 4n-24, f^*(v_{n-4}) \equiv 4n-20, f^*(v_{n-3}) \equiv 4n-16, f^*(v_{n-2}) \equiv 4n-8, f^*(v_{n-1}) \equiv 4n, f^*(v_n) \equiv 4n+4.$

The labels of vertices of the circles $C_n^{(2)}$ and $C_n^{(3)}$ are the same as in case (1).

Case (3): If $n \equiv 5\text{mod}6$, let the cylinder grid graph $C_{3,n}$ be as in Figure 6.

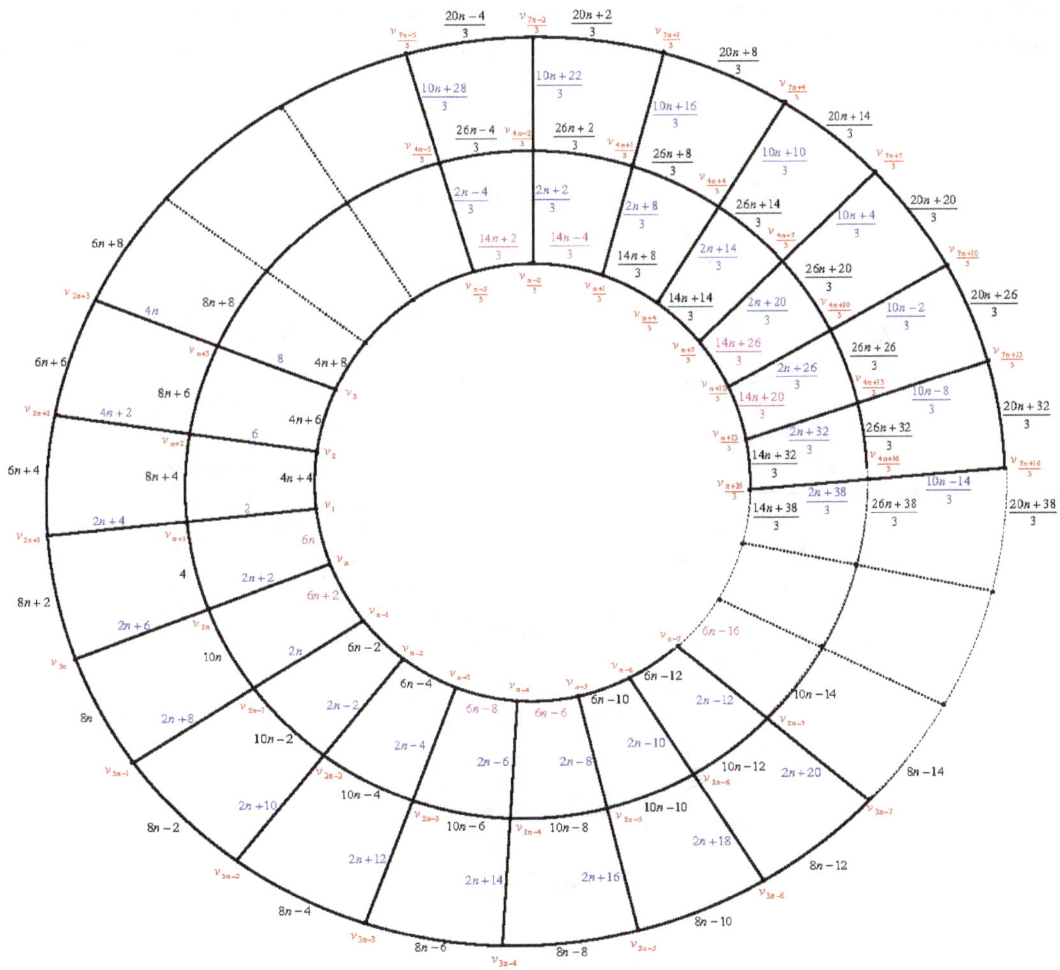

Figure 6. The cylinder grid graph $C_{3,n}$, $n \equiv 5 \mod 6$.

First, we label the edges of the paths $P_3^{(k)}, 1 \le k \le 2$ beginning with the edges of the path $P_3^{(1)}$ as the same in case (1). Second, we label the edges of the circles $C_n^{(k)}, 1 \le k \le 3$ beginning with the edges of the innermost circle $C_n^{(1)}$, then the edges of outer circle $C_n^{(3)}$, and then the edges of the circle $C_n^{(2)}$.

Label the edges of the circle $C_n^{(1)}$ as follows: $f(v_1v_2) = 4n+4, f(v_2v_3) = 4n+6, \ldots, f(v_{\frac{n-5}{3}}v_{\frac{n-2}{3}}) = \frac{14n+2}{3}, f(v_{\frac{n-2}{3}}v_{\frac{n+1}{3}}) = \frac{14n-4}{3}, f(v_{\frac{n+1}{3}}v_{\frac{n+4}{3}}) = \frac{14n+8}{3}, f(v_{\frac{n+4}{3}}v_{\frac{n+7}{3}}) = \frac{14n+14}{3}, f(v_{\frac{n+7}{3}}v_{\frac{n+10}{3}}) = \frac{14n+26}{3}, f(v_{\frac{n+10}{3}}v_{\frac{n+13}{3}}) = \frac{14n+20}{3}, f(v_{\frac{n+13}{3}}v_{\frac{n+16}{3}}) = \frac{14n+32}{3}, f(v_{\frac{n+16}{3}}v_{\frac{n+19}{3}}) = \frac{14n+38}{3}, f(v_{\frac{n+19}{3}}v_{\frac{n+22}{3}}) = \frac{14n+50}{3}, f(v_{\frac{n+22}{3}}v_{\frac{n+25}{3}}) = \frac{14n+44}{3}, f(v_{\frac{n+25}{3}}v_{\frac{n+28}{3}}) = \frac{14n+56}{3}, f(v_{\frac{n+28}{3}}v_{\frac{n+31}{3}}) = \frac{14n+62}{3}, f(v_{\frac{n+31}{3}}v_{\frac{n+34}{3}}) = \frac{14n+74}{3}, f(v_{\frac{n+34}{3}}v_{\frac{n+37}{3}}) = \frac{14n+68}{3}, f(v_{\frac{n+37}{3}}v_{\frac{n+40}{3}}) = \frac{14n+80}{3}, f(v_{\frac{n+40}{3}}v_{\frac{n+43}{3}}) = \frac{14n+86}{3}, \ldots, f(v_{n-13}v_{n-12}) = 6n-22, f(v_{n-12}v_{n-11}) = 6n-24, f(v_{n-11}v_{n-10}) = 6n-20, f(v_{n-10}v_{n-9}) = 6n-18, f(v_{n-9}v_{n-8}) = 6n-14, f(v_{n-8}v_{n-7}) = 6n-16, f(v_{n-7}v_{n-6}) = 6n-12, f(v_{n-6}v_{n-5}) = 6n-10, f(v_{n-5}v_{n-4}) = 6n-6, f(v_{n-4}v_{n-3}) = 6n-8, f(v_{n-3}v_{n-2}) = 6n-4, f(v_{n-2}v_{n-1}) = 6n-2, f(v_{n-1}v_n) = 6n+2, f(v_nv_1) = 6n.$

The labels of corresponding vertices mod$10n$ are as follows: The labels of vertices of the circle $C_n^{(1)}$: $f^*(v_1) \equiv 6, f^*(v_2) \equiv 8n+16, f^*(v_3) \equiv 8n+22, \ldots, f^*(v_{\frac{n}{3}-5}) \equiv 10n-4, f^*(v_{\frac{n-2}{3}}) \equiv 0, f^*(v_{\frac{n+1}{3}}) \equiv 4, f^*(v_{\frac{n+4}{3}}) \equiv 12, f^*(v_{\frac{n+7}{3}}) \equiv 20, f^*(v_{\frac{n+10}{3}}) \equiv 24, f^*(v_{\frac{n+13}{3}}) \equiv 28, f^*(v_{\frac{n+16}{3}}) \equiv 36, f^*(v_{\frac{n}{3}+7}) \equiv 44, f^*(v_{\frac{n+19}{3}}) \equiv 44, f^*(v_{\frac{n+22}{3}}) \equiv 48, f^*(v_{\frac{n+25}{3}}) \equiv 52, f^*(v_{\frac{n+28}{3}}) \equiv 60, f^*(v_{\frac{n+31}{3}}) \equiv 68, f^*(v_{\frac{n+34}{3}}) \equiv 72, f^*(v_{\frac{n+37}{3}}) \equiv 76, f^*(v_{\frac{n+40}{3}}) \equiv 84, f^*(v_{\frac{n+43}{3}}) \equiv 92, f^*(v_{\frac{n+46}{3}}) \equiv 96, f^*(v_{\frac{n+49}{3}}) \equiv 100, \ldots, f^*(v_{n-12}) \equiv 4n-68, f^*(v_{n-11}) \equiv 4n-64, f^*(v_{n-10}) \equiv 4n-56, f^*(v_{n-9}) \equiv 4n-48, f^*(v_{n-8}) \equiv 4n-44, f^*(v_{n-7}) \equiv 4n-40, f^*(v_{n-6}) \equiv 4n-32, f^*(v_{n-5}) \equiv 4n-24, f^*(v_{n-4}) \equiv 4n-20, f^*(v_{n-3}) \equiv 4n-16, f^*(v_{n-2}) \equiv 4n-8, f^*(v_{n-1}) \equiv 4n, f^*(v_n) \equiv 4n+4.$

The labels of vertices of the circles $C_n^{(2)}$ and $C_n^{(3)}$ are the same as in case (1).

Illustration: An e.e.g., l. of the cylinder grid graphs $C_{3,25}, C_{3,27}$ and $C_{3,29}$ are shown in Figure 7.

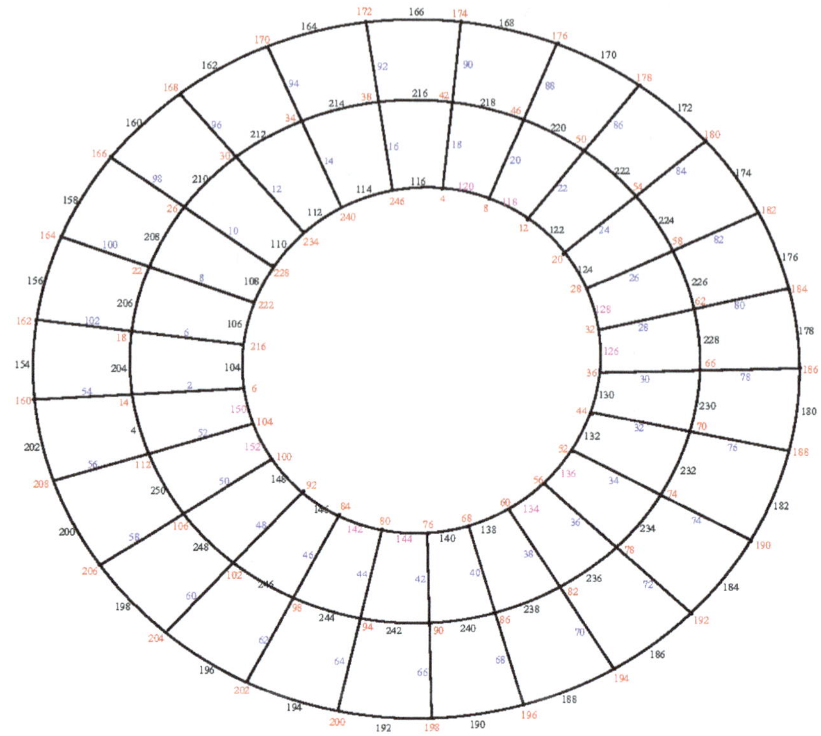

(a) $C_{3,25}$

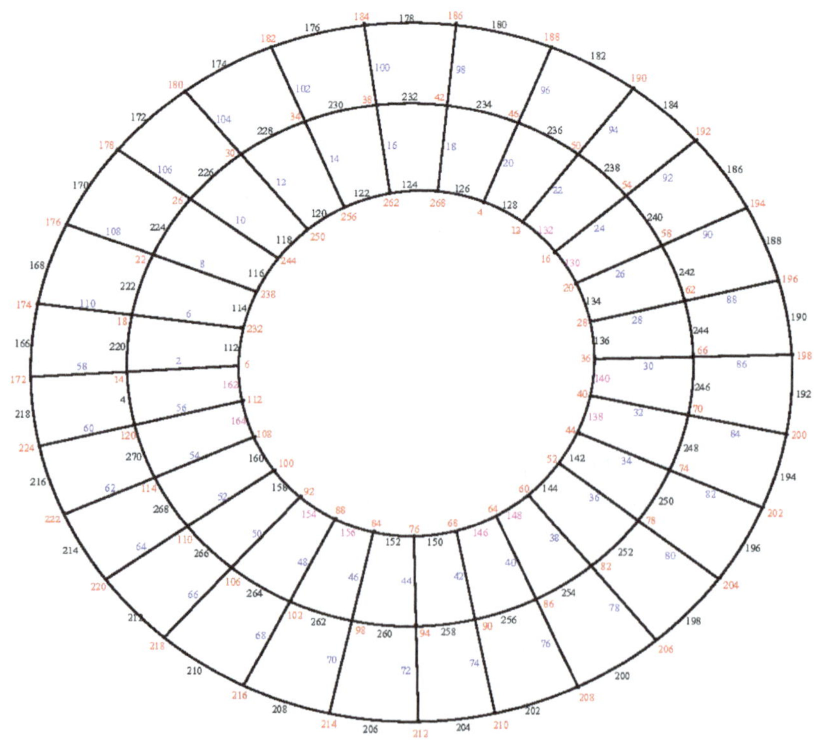

(b) $C_{3,27}$

Figure 7. *Cont.*

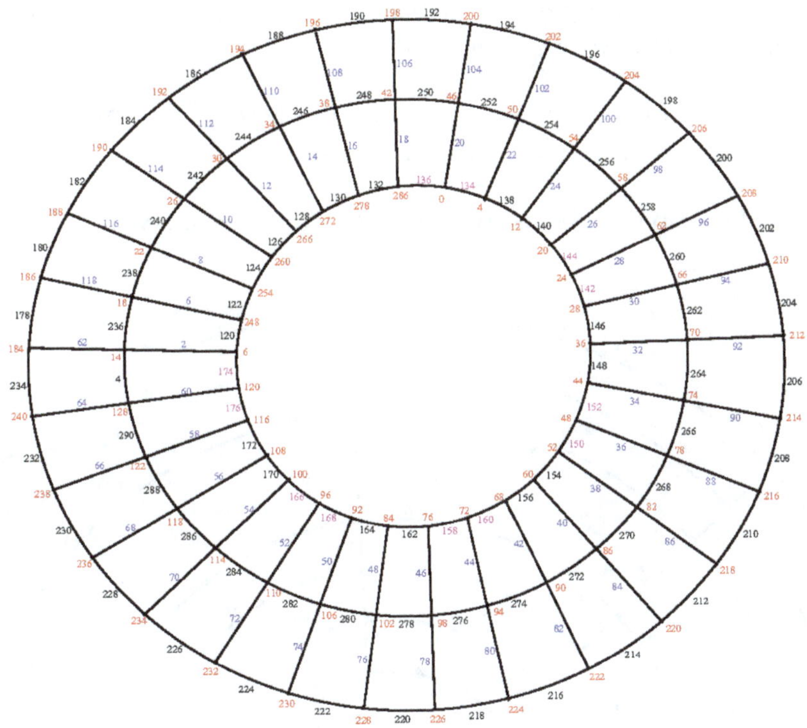

(c) $C_{3,29}$

Figure 7. An e.e.g., l. of the cylinder grid graphs $C_{3,25}$, $C_{3,27}$ and $C_{3,29}$.

Remark 1. *Note that $C_{3,5}$ is an edge even graceful graph but it does not follow the pervious rule (see Figure 8).*

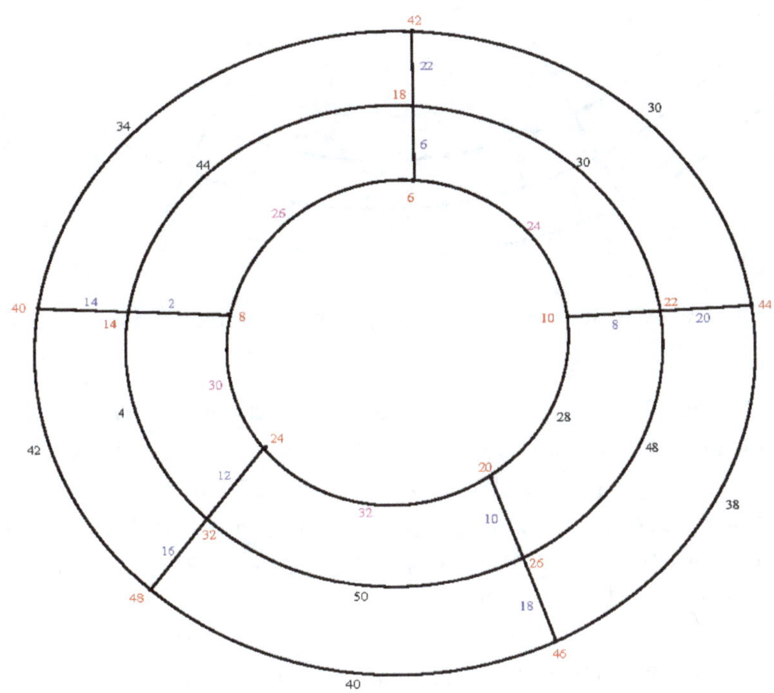

Figure 8. An e.e.g., l. of the cylinder grid graph $C_{3,5}$.

Theorem 3. *If m is an odd positive integer greater than 3 and n is an even positive integer, $n \geq 2$, then the cylinder grid graph $C_{m,n}$, is an edge even graceful graph.*

Proof. Using standard notation $p = |V(C_{m,n})| = mn$, $q = |E(C_{m,n})| = 2mn - n$ and $r = \max(p,q) = 2mn - n$ and $f : E(C_{m,n}) \rightarrow \{2, 4, 6, \ldots, 4mn - 2n - 2\}$. □

Let the cylinder grid graph $C_{m,n}$ be as in Figure 9. There are six cases:

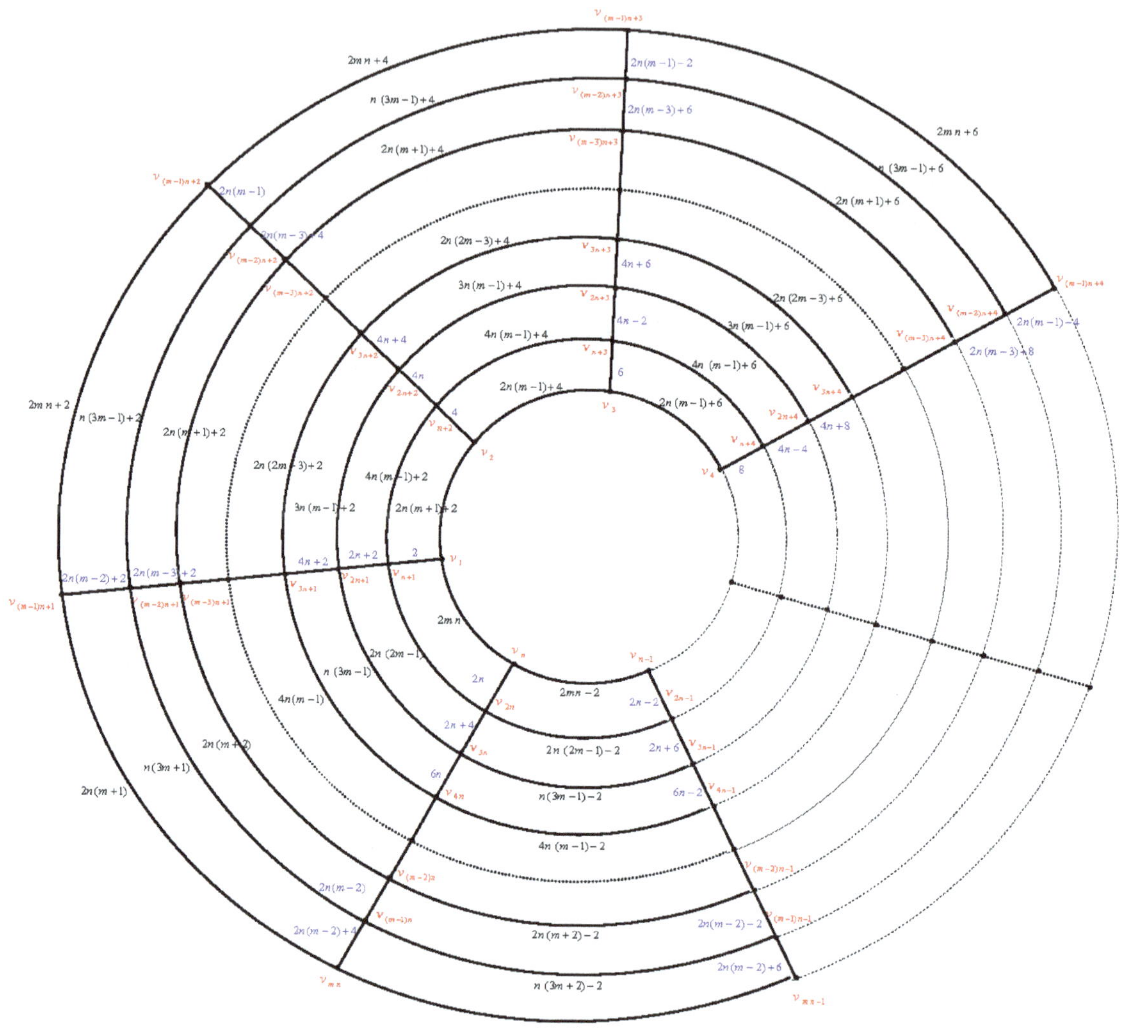

Figure 9. The cylinder grid graph $C_{m,n}$,m is odd greater than 3 and $n \geq 2$.

Case (1): $n \equiv 0 \mod 12$. First, we label the edges of the paths $P_m^{(k)}, 1 \leq k \leq n$ beginning with the edges of the path $P_m^{(1)}$ as follows: Move clockwise to label the edges $v_1v_{n+1}, v_2v_{n+2}, v_3v_{n+3}, \ldots, v_{n-1}v_{2n-1}, v_nv_{2n}$ by $2, 4, 6, \ldots, 2n-2, 2n$, then move anticlockwise to label the edges $v_{n+1}v_{n+2}, v_{2n}v_{3n}, v_{2n-1}v_{3n-1}, \ldots, v_{n+3}v_{2n+3}, v_{n+2}v_{2n+2}$ by $2n+2, 2n+4, 2n+6, \ldots, 4n-2, 4n$, then move clockwise to label the edges $v_{2n+1}v_{3n+1}, v_{2n+2}v_{3n+2}, v_{2n+3}v_{3n+3}, \ldots, v_{3n-1}v_{4n-1}, v_{3n}v_{4n}$ by $4n+2, 4n+4, 4n+6, \ldots, 6n-2, 6n$ and so on.

Finally, move anticlockwise to label the edges $v_{(m-2)n+1}v_{(m-1)n+1}, v_{(m-1)n}v_{mn}, v_{(m-1)n-1}v_{mn-1}, \ldots, v_{(m-2)n+3}v_{(m-1)n+3}, v_{(m-2)n+2}v_{m(n-1)+2}$ by $2n(m-2)+2, 2n(m-2)+4, 2n(m-2)+6, \ldots, 2n(m-1)-2, 2n(m-1)$.

Secondly, we label the edges of the circles $C_n^{(k)}, 1 \leq k \leq m$ beginning with the edges of the inner most circle $C_n^{(1)}$, then the edges of outer circle $C_n^{(m)}$, then the edges of the circles $C_m^{(m-1)}, C_m^{(m-3)}, \ldots, C_m^{(2)}$. Finally, we label the edges of the circles $C_m^{(m-1)}, C_m^{(m-3)}, \ldots, C_m^{(2)}$.

Label the edges of the circle $C_n^{(1)}$ as follows: $f(v_1v_2) = 2n(m-1)+2, f(v_2v_3) = 2n(m-1)+6, f(v_3v_4) = 2n(m-1)+4, f(v_4v_5) = 2n(m-1)+8, f(v_5v_6) = 2n(m-1)+10, f(v_6v_7) = 2n(m-1)+14, f(v_7v_8) = 2n(m-1)+12, f(v_8v_9) = 2n(m-1)+16, f(v_9v_{10}) = 2n(m-1)+18, f(v_{10}v_{11}) = 2n(m-1)+22, f(v_{11}v_{12}) = 2n(m-1)+20, f(v_{12}v_{13}) = 2n(m-1)+24, \ldots, f(v_{n-7}v_{n-6}) = 2mn-14, f(v_{n-6}v_{n-5}) = 2mn-10, f(v_{n-5}v_{n-4}) = 2mn-12, f(v_{n-4}v_{n-3}) = 2mn-8, f(v_{n-3}v_{n-2}) = 2mn-2, f(v_{n-2}v_{n-1}) = 2mn-6, f(v_{n-1}v_n) = 2mn, f(v_nv_1) = 2mn-4$.

Label the edges of the circle $C_n^{(m)}$ as follows: $f(v_{(m-1)n+i}v_{(m-1)n+i+1}) = 2mn+2i, f(v_{mn}v_{(m-1)n+1}) = 2n(m+1), 1 \leq i \leq n-1$.

Label the edges of the circle $C_n^{(m-2)}$ as follows: $f(v_{(m-3)n+i}v_{(m-3)n+i+1}) = 2n(m+1)+2i, f(v_{(m-2)n}v_{(m-3)n+1}) = 2n(m+2), 1 \leq i \leq n-1$.

Label the edges of the circle $C_n^{(m-4)}$ as follows: $f(v_{(m-5)n+i}v_{(m-5)n+i+1}) = 2n(m+2)+2i, f(v_{(m-4)n}v_{(m-5)n+1}) = 2n(m+3), 1 \leq i \leq n-1$, and so on.

Label the edges of the circle $C_n^{(3)}$ as follows: $f(v_{2n+i}v_{2n+i+1}) = 3n(m-1)+2i, f(v_{3n}v_{2n+1}) = n(3m-1), 1 \leq i \leq n-1$,

Label the edges of the circle $C_n^{(m-1)}$ as follows: $f(v_{(m-2)n+i}v_{(m-2)n+i+1}) = n(3m-1)+2i, f(v_{(m-1)n}v_{(m-1)n+1}) = n(3m+1)-1, 1 \leq i \leq n-1$,

Label the edges of the circle $C_n^{(m-3)}$ as follows: $f(v_{(m-4)n+i}v_{(m-4)n+i+1}) = n(3m+2)+2i, f(v_{(m-3)n}v_{(m-4)n+1}) = 3n(m+1), 1 \leq i \leq n-1, \ldots$, and so on.

Label the edges of the circle $C_n^{(4)}$ as follows: $f(v_{3n+i}v_{3n+i+1}) = 2n(2m-3)+2i, f(v_{4n}v_{3n+1}) = 4n(m-1), 1 \leq i \leq n-1$,

Label the edges of $C_n^{(2)}$ as follows: $f(v_{n+i}v_{n+i+1}) = 4n(m-1)+2i, f(v_{2n}v_{2n+1}) = 2n(m-1), 1 \leq i \leq n-1$,

Thus, the labels of corresponding vertices $\mathrm{mod}(4mn-2n)$ will be:

The label the vertices of $C_n^{(1)}$ are: $f^*(v_1) \equiv 0; f^*(v_2) \equiv 4mn-4n+12; f^*(v_3) \equiv 4mn-4n+16; f^*(v_4) \equiv 4mn-4n+20; f^*(v_5) \equiv 4mn-4n+28; f^*(v_6) \equiv 4mn-4n+36; f^*(v_7) \equiv 4mn-4n+40; f^*(v_8) \equiv 4mn-4n+44; f^*(v_9) \equiv 4mn-4n+52; f^*(v_{10}) \equiv 4mn-4n+60; f^*(v_{11}) \equiv 4mn-4n+64; f^*(v_{12}) \equiv 4mn-4n+68; \ldots; f^*(v_{n-6}) \equiv 4n-36; f^*(v_{n-5}) \equiv 4n-32; f^*(v_{n-4}) \equiv 4n-28; f^*(v_{n-3}) \equiv 4n-16; f^*(v_{n-2}) \equiv 4n-12; f^*(v_{n-1}) \equiv 4n-8; f^*(v_n) \equiv 4n-4$.

The label the vertices of $C_n^{(2)}, C_n^{(3)}, C_n^{(4)}, \ldots, C_n^{(m-2)}, C_n^{(m-1)}, C_n^{(m)}$ respectively are: $f^*(v_{n+i}) \equiv 4i+2; f^*(v_{2n+i}) \equiv 2mn+4n+4i+2; f^*(v_{3n+i}) \equiv 4n+4i+2; \ldots; f^*(v_{(m-3)n+i}) \equiv 4mn-6n+4i+2;$ $f^*(v_{(m-2)n+i}) \equiv 2mn-6n+4i+2; f^*(v_{(m-1)n+i}) \equiv 2mn+2i+2, 1 \leq i \leq n$.

Case (2): $n \equiv 2 \mathrm{mod} 12, n \neq 2$.

First, we label the edges of the paths $P_m^{(k)}, 1 \leq k \leq n$ begin with the edges of the path $P_m^{(1)}$ as the same in case (1).

Secondly, we label the edges of the circles $C_n^{(k)}, 1 \leq k \leq m$ begin with the edges of the inner most circle $C_n^{(1)}$, then the edges of outer circle $C_n^{(m)}$, then the edges of the circles $C_n^{(m-2)}, C_n^{(m-4)}, \ldots, C_n^{(3)}$.

Finally, we label the edges of the circles $C_m^{(m-1)}, C_m^{(m-3)}, \ldots, C_m^{(2)}$.

Label the edges of the circle $C_n^{(1)}$ as follows: $f(v_1v_2) = 2n(m-1)+2, f(v_2v_3) = 2n(m-1)+6, f(v_3v_4) = 2n(m-1)+4, f(v_4v_5) = 2n(m-1)+8, f(v_5v_6) = 2n(m-1)+10, f(v_6v_7) = 2n(m-1)+14, f(v_7v_8) = 2n(m-1)+12, f(v_8v_9) = 2n(m-1)+16, f(v_9v_{10}) = 2n(m-1)+18, f(v_{10}v_{11}) = 2n(m-1)+22, f(v_{11}v_{12}) = 2n(m-1)+20, f(v_{12}v_{13}) = 2n(m-1)+24, \ldots, f(v_{n-9}v_{n-8}) = 2mn-18, f(v_{n-8}v_{n-7}) = 2mn-14, f(v_{n-7}v_{n-6}) = 2mn-16, f(v_{n-6}v_{n-5}) = 2mn-12, f(v_{n-5}v_{n-4}) = 2mn-10, f(v_{n-4}v_{n-3}) = 2mn-6, f(v_{n-3}v_{n-2}) = 2mn-8, f(v_{n-2}v_{n-1}) = 2mn-4, f(v_{n-1}v_n) = 2mn-2, f(v_nv_1) = 2mn$.

Label the edges of the circle $C_n^{(2)}$ as follows: $f(v_{n+1}v_{n+2}) = 4n(m-1)+4, f(v_{n+2}v_{n+3}) = 4n(m-1)+2, f(v_{n+3}v_{n+4}) = 4n(m-1)+8, f(v_{n+4}v_{n+5}) = 4n(m-1)+6,\ f(v_{n+i}v_{n+i+1}) = 4n(m-1)+2i, 6 \le i \le n-2, f(v_{2n-1}v_{2n}) = 2n(2m-1), f(v_{2n}v_{n+1}) = 2n(2m-1)-2$. Label the edges of $C_n^{(m)}, C_n^{(m-2)}, C_n^{(m-4)}, \ldots, C_n^{(3)}$ and $C_n^{(m-1)}, C_n^{(m-3)}, C_n^{(m-5)}, \ldots, C_n^{(4)}$ as in case (1).

Thus, the labels of corresponding vertices mod$(4mn-2n)$ will be:

The label the vertices of $C_n^{(1)}$ are: $f^*(v_1) \equiv 4, f^*(v_2) \equiv 4mn-4n+12, f^*(v_3) \equiv 4mn-4n+16, f^*(v_4) \equiv 4mn-4n+20, f^*(v_5) \equiv 4mn-4n+28,\ f^*(v_6) \equiv 4mn-4n+36, f^*(v_7) \equiv 4mn-4n+40, f^*(v_8) \equiv 4mn-4n+44, f^*(v_9) \equiv 4mn-4n+52,\ f^*(v_{10}) \equiv 4mn-4n+60, f^*(v_{11}) \equiv 4mn-4n+64, f^*(v_{12}) \equiv 4mn-4n+68, f^*(v_{13}) \equiv 4mn-4n+76, \ldots,\ f^*(v_{n-8}) \equiv 4n-48, f^*(v_{n-7}) \equiv 4n-44, f^*(v_{n-6}) \equiv 4n-40, f^*(v_{n-5}) \equiv 4n-32, f^*(v_{n-4}) \equiv 4n-24, f^*(v_{n-3}) \equiv 4n-20,\ f^*(v_{n-2}) \equiv 4n-16, f^*(v_{n-1}) \equiv 4n-8, f^*(v_n) \equiv 4n-2$.

The label the vertices of the circle $C_n^{(2)}$ are: $f^*(v_{n+1}) \equiv 6, f^*(v_{n+2}) \equiv 10, f^*(v_{n+3}) \equiv 14, f^*(v_{n+4}) \equiv 18, f^*(v_{n+5}) \equiv 20, f^*(v_{n+i}) \equiv 4i+2, 6 \le i \le n-2,\ f^*(v_{2n-1}) \equiv 4n, f^*(v_{2n}) \equiv 4n+2$.

The label the vertices of $C_n^{(3)}, C_n^{(4)}, \ldots, C_n^{(m-2)}, C_n^{(m-1)}, C_n^{(m)}$ respectively are as the same as in case (1).

Remark 2. *In case $n=2$. Let the edges of the cylinder grid graph $C_{m,2}$ are labeled as shown in Figure 10. The corresponding labels of vertices mod$(8m-4)$ are as follows:* $f^*(v_1) \equiv 8, f^*(v_{2i+1}) \equiv 4m+8i+4, 1 \le i \le \frac{m-3}{2}, f^*(v_{2i}) \equiv 8i+6, 1 \le i \le \frac{m-1}{2};\ f^*(v'_1) \equiv 12, f^*(v'_2) \equiv 20, f^*(v'_{2i+1}) \equiv 4m+8i+18, 1 \le i \le \frac{m-3}{2}, f^*(v'_{2i}) \equiv 8i+10, 2 \le i \le \frac{m-1}{2}$.

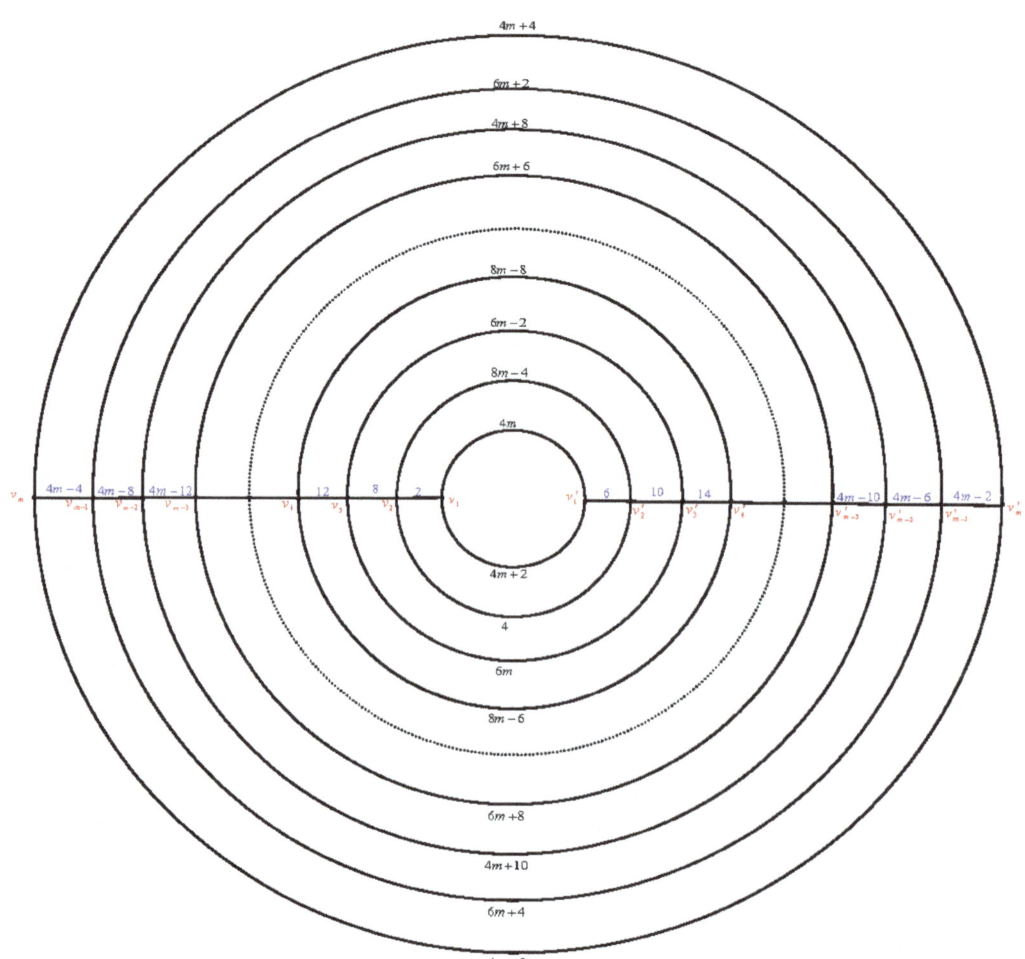

Figure 10. The cylinder grid graph $C_{m,2}$.

Case (3): $n \equiv 4 \bmod 12$.

First we label the edges of the paths $P_m^{(k)}, 1 \le k \le n$ begin with the edges of the path $P_m^{(1)}$ as the same in case (1).

Second we label the edges of the circles $C_n^{(k)}, 1 \le k \le m$ begin with the edges of the inner most circle $C_n^{(1)}$, then the edges of outer circle $C_n^{(m)}$, then the edges of the circles $C_n^{(m-2)}, C_n^{(m-4)}, \ldots, C_n^{(3)}$.

Finally we label the edges of the circles $C_m^{(m-1)}, C_m^{(m-3)}, \ldots, C_m^{(2)}$.

Label the edges of the circle $C_n^{(1)}$ as follows: $f(v_1v_2) = 2n(m-1)+2, f(v_2v_3) = 2n(m-1)+6, f(v_3v_4) = 2n(m-1)+4, f(v_4v_5) = 2n(m-1)+8, f(v_5v_6) = 2n(m-1)+10, f(v_6v_7) = 2n(m-1)+14, f(v_7v_8) = 2n(m-1)+12, f(v_8v_9) = 2n(m-1)+16, f(v_9v_{10}) = 2n(m-1)+18, f(v_{10}v_{11}) = 2n(m-1)+22, f(v_{11}v_{12}) = 2n(m-1)+20, f(v_{12}v_{13}) = 2n(m-1)+24, \ldots, f(v_{n-8}v_{n-7}) = 2mn-16, f(v_{n-7}v_{n-6}) = 2mn-14, f(v_{n-6}v_{n-5}) = 2mn-10, f(v_{n-5}v_{n-4}) = 2mn-12, f(v_{n-4}v_{n-3}) = 2mn-8, f(v_{n-3}v_{n-2}) = 2mn-2, f(v_{n-2}v_{n-1}) = 2mn-6, f(v_{n-1}v_n) = 2mn$, and $f(v_nv_1) = 2mn-4$.

Label the edges of $C_n^{(m)}, C_n^{(m-2)}, C_n^{(m-4)}, \ldots, C_n^{(3)}$ and $C_n^{(m-1)}, C_n^{(m-3)}, C_n^{(m-5)}, \ldots, C_n^{(4)}, C_n^{(2)}$ as in case (1).

Thus we have the labels of corresponding vertices of the circle $C_n^{(1)}$ $\bmod(4mn-2n)$ will be: $f^*(v_1) \equiv 0, f^*(v_2) \equiv 4mn-4n+12, f^*(v_3) \equiv 4mn-4n+16, f^*(v_4) \equiv 4mn-4n+20, f^*(v_5) \equiv 4mn-4n+28, f^*(v_6) \equiv 4mn-4n+36, f^*(v_7) \equiv 4mn-4n+40, f^*(v_8) \equiv 4mn-4n+44, f^*(v_9) \equiv 4mn-4n+52, f^*(v_{10}) \equiv 4mn-4n+60, f^*(v_{11}) \equiv 4mn-4n+64, f^*(v_{12}) \equiv 4mn-4n+68, f^*(v_{13}) \equiv 4mn-4n+76, \ldots, f^*(v_{n-7}) \equiv 4n-44, f^*(v_{n-6}) \equiv 4n-36, f^*(v_{n-5}) \equiv 4n-32, f^*(v_{n-4}) \equiv 4n-28, f^*(v_{n-3}) \equiv 4n-16, f^*(v_{n-2}) \equiv 4n-12, f^*(v_{n-1}) \equiv 4n-8, f^*(v_n) \equiv 4n-4$.

The label the vertices of $C_n^{(2)}, C_n^{(3)}, C_n^{(4)}, \ldots, C_n^{(m-2)}, C_n^{(m-1)}, C_n^{(m)}$ respectively are as same in case (1)

Remark 3. *In case $n = 4$. Let the the edges of the cylinder grid graph $C_{m,4}$ are labeled as shown in Figure 11. The corresponding labels of vertices mod$(16m-8)$ are as follows: $f^*(v_1) \equiv 6, f^*(v_2) \equiv 8, f^*(v_3) \equiv 16, f^*(v_4) \equiv 20; f^*(v_{4i+1}) \equiv 4i+10, 1 \le i \le 3, f^*(v_8) \equiv 28; f^*(v_{8k+i}) \equiv 8m+4i+16k-10,\ 1 \le i \le 4, 1 \le k \le \frac{m-5}{2}; f^*(v_{4m-11}) \equiv 0, f^*(v_{4m-10}) \equiv 2, f^*(v_{4m-9}) \equiv 4, f^*(v_{4m-8}) \equiv 10, f^*(v_{8k+4+i}) \equiv 4i+16k+10, 1 \le i \le 4, 1 \le k \le \frac{m-3}{2}$.*

Case (4): $n \equiv 6 \bmod 12$.

First, we label the edges of the paths $P_m^{(k)}, 1 \le k \le n$ begin with the edges of the path $P_m^{(1)}$ as the same in case (1).

Secondly, we label the edges of the circles $C_n^{(k)}, 1 \le k \le m$ begin with the edges of the inner most circle $C_n^{(1)}$, then the edges of outer circle $C_n^{(m)}$, then the edges of the circles $C_n^{(m-2)}, C_n^{(m-4)}, \ldots, C_n^{(3)}$.

Finally, we label the edges of the circles $C_m^{(m-1)}, C_m^{(m-3)}, \ldots, C_m^{(2)}$.

Label the edges of the circle $C_n^{(1)}$ as follows: $f(v_1v_2) = 2n(m-1)+2, f(v_2v_3) = 2n(m-1)+6, f(v_3v_4) = 2n(m-1)+4, f(v_4v_5) = 2n(m-1)+8, f(v_5v_6) = 2n(m-1)+10, f(v_6v_7) = 2n(m-1)+14, f(v_7v_8) = 2n(m-1)+12, f(v_8v_9) = 2n(m-1)+16, f(v_9v_{10}) = 2n(m-1)+18, f(v_{10}v_{11}) = 2n(m-1)+22, f(v_{11}v_{12}) = 2n(m-1)+20, f(v_{12}v_{13}) = 2n(m-1)+24, \ldots, f(v_{n-9}v_{n-8}) = 2mn-18, f(v_{n-8}v_{n-7}) = 2mn-14, f(v_{n-7}v_{n-6}) = 2mn-16, f(v_{n-6}v_{n-5}) = 2mn-12, f(v_{n-5}v_{n-4}) = 2mn-10, f(v_{n-4}v_{n-3}) = 2mn-6, f(v_{n-3}v_{n-2}) = 2mn-8, f(v_{n-2}v_{n-1}) = 2mn-4, f(v_{n-1}v_n) = 2mn+2, f(v_nv_1) = 2mn-2$.

Label the edges of $C_n^{(m-4)}, \ldots, C_n^{(3)}$ and $C_n^{(m-1)}, C_n^{(m-3)}, C_n^{(m-5)}, \ldots, C_n^{(4)}, C_n^{(2)}$ as in case (1).

Label the edges of the circle $C_n^{(m-2)}$ as follows: $f(v_{(m-3)n+1}v_{(m-3)n+2}) = 2n(m+2), f(v_{(m-3)n+i}v_{(m-3)n+i+1}) = 2n(m+1)+2i, 2 \le i \le n-1, f(v_{(m-2)n}v_{(m-3)n+1}) = 2n(m+2)+2$.

Label the edges of the circle $C_n^{(m)}$ as follows: $f(v_{(m-1)n+1}v_{(m-1)n+2}) = 2mn, f(v_{(m-1)n+i}v_{(m-1)n+i+1}) = 2mn+2i, 2 \le i \le n-1, f(v_{(m-2)n}v_{(m-3)n+1}) = 2n(m+2)$.

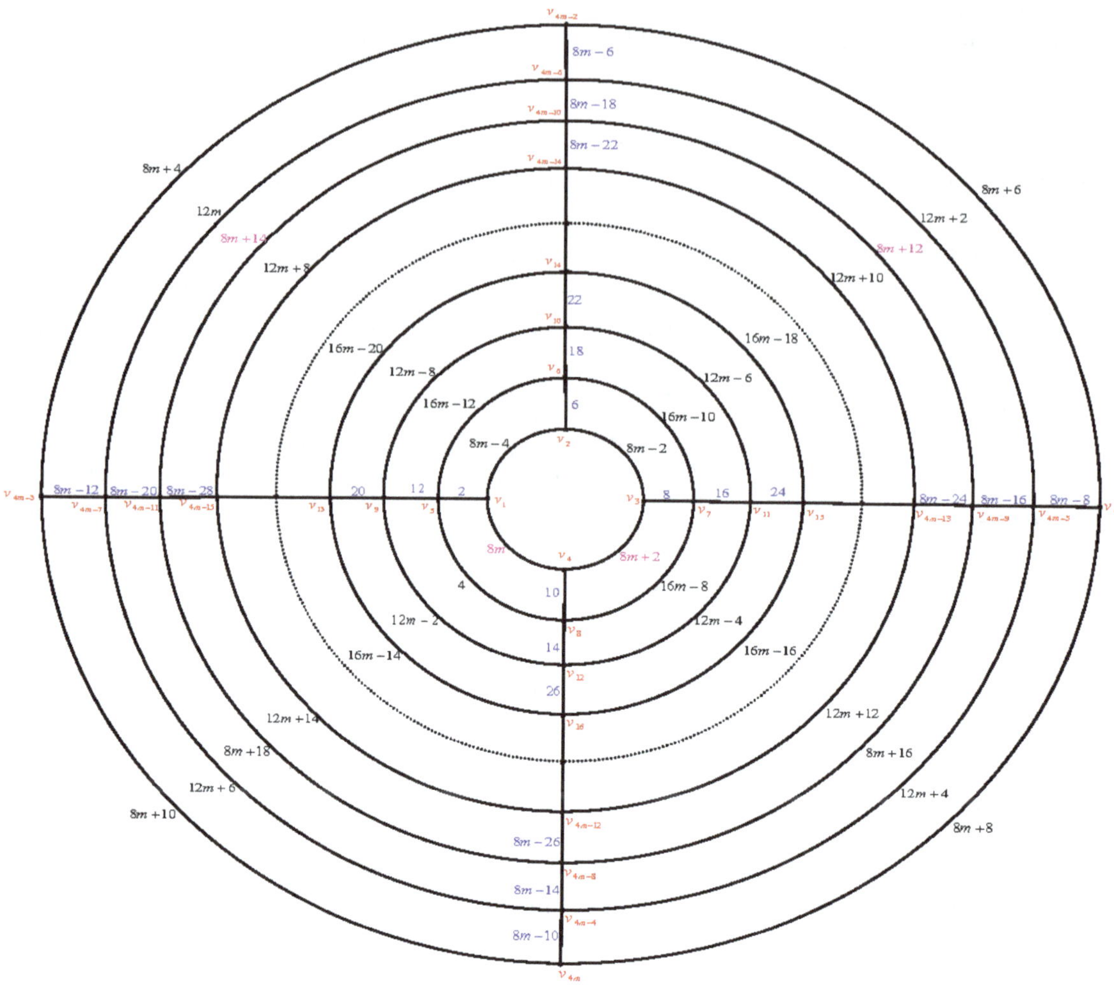

Figure 11. An e.e.g., l. of the cylinder grid graph $C_{m,4}$.

Thus we have the labels of corresponding vertices mod$(4mn - 2n)$ will be:

The labels the vertices of the circle $C_n^{(1)}$ are: $f^*(v_1) \equiv 2, f^*(v_2) \equiv 4mn - 4n + 12, f^*(v_3) \equiv 4mn - 4n + 16, f^*(v_4) \equiv 4mn - 4n + 20, f^*(v_5) \equiv 4mn - 4n + 28, f^*(v_6) \equiv 4mn - 4n + 36, f^*(v_7) \equiv 4mn - 4n + 40, f^*(v_8) \equiv 4mn - 4n + 44, f^*(v_9) \equiv 4mn - 4n + 52, f^*(v_{10}) \equiv 4mn - 4n + 60, f^*(v_{11}) \equiv 4mn - 4n + 64, f^*(v_{12}) \equiv 4mn - 4n + 68, \ldots, f^*(v_{n-8}) \equiv 4n - 48, f^*(v_{n-7}) \equiv 4n - 44, f^*(v_{n-6}) \equiv 4n - 40, f^*(v_{n-5}) \equiv 4n - 32, f^*(v_{n-4}) \equiv 4n - 24, f^*(v_{n-3}) \equiv 4n - 20, f^*(v_{n-2}) \equiv 4n - 16, f^*(v_{n-1}) \equiv 4n - 4; f^*(v_n) \equiv 4n.$

The labels the vertices of the circle $C_n^{(m-2)}$ are: $f^*(v_{(m-3)n+1}) \equiv 4mn - 6n + 6; f^*(v_{(m-3)n+2}) \equiv 4mn - 6n + 8; f^*(v_{(m-3)n+i}) \equiv 4mn - 6n + 4i + 2, 3 \leq i \leq n - 1, f^*(v_{(m-2)n}) \equiv 4mn - 4n + 4.$

The labels the vertices of $C_n^{(2)}, C_n^{(3)}, C_n^{(4)}, \ldots, C_n^{(m-1)}, C_n^{(m-3)}$ respectively are the same as in case (1).

The labels the vertices of $C_n^{(m)}$ are: $f^*(v_{(m-1)n+1}) \equiv 2mn + 2, f^*(v_{(m-1)n+2}) \equiv 2mn + 4, f^*(v_{(m-1)n+i}) \equiv 2mn + 2i + 2, 3 \leq i \leq n$. Case (5): $n \equiv 8 \bmod 12$.

First, we label the edges of the paths $P_m^{(k)}, 1 \leq k \leq n$ begin with the edges of the path $P_m^{(1)}$ as the same in case (1).

Secondly, we label the edges of the circles $C_n^{(k)}, 1 \leq k \leq m$ begin with the edges of the inner most circle $C_n^{(1)}$, then the edges of outer circle $C_n^{(m)}$, then the edges of the circles $C_n^{(m-2)}, C_n^{(m-4)}, \ldots, C_n^{(3)}$.

Finally we label the edges of the circles $C_m^{(m-1)}, C_m^{(m-3)}, \ldots, C_m^{(2)}$

Label the edges of the circle $C_n^{(1)}$ as follows: $f(v_1v_2) = 2n(m-1) + 2, f(v_2v_3) = 2n(m-1) + 6, f(v_3v_4) = 2n(m-1) + 4, f(v_4v_5) = 2n(m-1) + 8, f(v_5v_6) = 2n(m-1) + 10, f(v_6v_7) = 2n(m-1) + 14, f(v_7v_8) = 2n(m-1) + 12, f(v_8v_9) = 2n(m-1) + 16, f(v_9v_{10}) = 2n(m-1) + 18, f(v_{10}v_{11}) =$

$2n(m-1)+22, f(v_{11}v_{12}) = 2n(m-1)+20, f(v_{12}v_{13}) = 2n(m-1)+24,\ f(v_{13}v_{14}) = 2n(m-1)+26, \ldots, f(v_{n-9}v_{n-8}) = 2mn-20, f(v_{n-8}v_{n-7}) = 2mn-16, f(v_{n-7}v_{n-6}) = 2mn-14,\ f(v_{n-6}v_{n-5}) = 2mn-10, f(v_{n-5}v_{n-4}) = 2mn-12, f(v_{n-4}v_{n-3}) = 2mn-8, f(v_{n-3}v_{n-2}) = 2mn-6,\ f(v_{n-2}v_{n-1}) = 2mn-2, f(v_{n-1}v_n) = 2mn-4, f(v_nv_1) = 2mn+4$.

Label the edges of the circle $C_n^{(m)}$ as follows $f(v_{(m-1)n+1}v_{(m-1)n+2}) = 2mn+2, f(v_{(m-1)n+2}v_{(m-1)n+3}) = 2mn+6, f(v_{(m-1)n+i}v_{(m-1)n+i+1}) = 2mn+2i+2,\ 3 \le i \le n-1, f(v_{mn}v_{(m-1)n+1}) = 2mn$.

Label the edges of $C_n^{(m-2)}, C_n^{(m-4)}, \ldots, C_n^{(3)}$ and $C_n^{(m-1)}, C_n^{(m-3)}, C_n^{(m-5)}, \ldots, C_n^{(4)}, C_n^{(2)}$ as the same in case (1).

Thus we labels of corresponding vertices of the circle $C_n^{(1)}$ mod$(4mn-2n)$ will be: $f^*(v_1) \equiv 8, f^*(v_2) \equiv 4mn-4n+12, f^*(v_3) \equiv 4mn-4n+16, f^*(v_4) \equiv 4mn-4n+20, f^*(v_5) \equiv 4mn-4n+28, f^*(v_6) \equiv 4mn-4n+36, f^*(v_7) \equiv 4mn-4n+40, f^*(v_8) \equiv 4mn-4n+44, f^*(v_9) \equiv 4mn-4n+52, f^*(v_{10}) \equiv 4mn-4n+60, f^*(v_{11}) \equiv 4mn-4n+64, f^*(v_{12}) \equiv 4mn-4n+68, f^*(v_{13}) \equiv 4mn-4n+76, \ldots, f^*(v_{n-8}) \equiv 4n-52, f^*(v_{n-7}) \equiv 4n-44, f^*(v_{n-6}) \equiv 4n-36, f^*(v_{n-5}) \equiv 4n-32, f^*(v_{n-4}) \equiv 4n-28, f^*(v_{n-3}) \equiv 4n-20, f^*(v_{n-2}) \equiv 4n-12, f^*(v_{n-1}) \equiv 4n-8, f^*(v_n) \equiv 4n$.

The labels the vertices of the circle $C_n^{(m)}$ are: $f^*(v_{(m-1)n+1}) \equiv 2mn-4n+14, f^*(v_{(m-1)n+2}) \equiv 2mn+8, f^*(v_{(m-1)n+i}) \equiv 2mn+2i+6, 3 \le i \le n-1, f^*(v_{mn}) \equiv 2mn+4$. The labels the vertices of $C_n^{(2)}, C_n^{(3)}, C_n^{(4)}, \ldots, C_n^{(m-2)}, C_n^{(m-1)}$ respectively are as the same in case (1).

Case (6): $n \equiv 10 \bmod 12$. First we label the edges of the paths $P_m^{(k)}, 1 \le k \le n$ begin with the edges of the path $P_m^{(1)}$ as the same as in case (1).

Second we label the edges of the circles $C_n^{(k)}, 1 \le k \le m$ begin with the edges of the inner most circle $C_n^{(1)}$, then the edges of outer circle $C_n^{(m)}$, then the edges of the circles $C_n^{(m-2)}, C_n^{(m-4)}, \ldots, C_n^{(3)}$.

Finally we label the edges of the circles $C_m^{(m-1)}, C_m^{(m-3)}, \ldots, C_m^{(2)}$.

Label the edges of the circle $C_n^{(1)}$ as follows: $f(v_1v_2) = 2n(m-1)+2, f(v_2v_3) = 2n(m-1)+6, f(v_3v_4) = 2n(m-1)+4, f(v_4v_5) = 2n(m-1)+8,\ f(v_5v_6) = 2n(m-1)+10, f(v_6v_7) = 2n(m-1)+14, f(v_7v_8) = 2n(m-1)+12, f(v_8v_9) = 2n(m-1)+16,\ f(v_9v_{10}) = 2n(m-1)+18, f(v_{10}v_{11}) = 2n(m-1)+22, f(v_{11}v_{12}) = 2n(m-1)+20, f(v_{12}v_{13}) = 2n(m-1)+24,\ f(v_{13}v_{14}) = 2n(m-1)+26, \ldots, f(v_{n-9}v_{n-8}) = 2mn-18, f(v_{n-8}v_{n-7}) = 2mn-14, f(v_{n-7}v_{n-6}) = 2mn-16,\ f(v_{n-6}v_{n-5}) = 2mn-12, f(v_{n-5}v_{n-4}) = 2mn-10, f(v_{n-4}v_{n-3}) = 2mn-6, f(v_{n-3}v_{n-2}) = 2mn-8,\ f(v_{n-2}v_{n-1}) = 2mn-4, f(v_{n-1}v_n) = 2mn-2, f(v_nv_1) = 2mn$.

Label the edges of the circle $C_n^{(2)}$ as follows: $f(v_{n+1}v_{n+2}) = 4n(m-1)+4, f(v_{n+2}v_{n+3}) = 4n(m-1)+2, f(v_{n+i}v_{n+i+1}) = 4n(m-1)+2i, 3 \le i \le n-2,\ f(v_{2n-1}v_{2n}) = 2n(2m-1), f(v_{2n}v_{2n+1}) = 2n(2m-1)-2$.

Label the edges of $C_n^{(m)}, C_n^{(m-2)}, C_n^{(m-4)}, \ldots, C_n^{(3)}$ and $C_n^{(m-1)}, C_n^{(m-3)}, C_n^{(m-5)}, \ldots, C_n^{(4)}$ as in case (1).

Thus we have the labels of corresponding vertices mod$(4mn-2n)$ will be:

The labels the vertices of the circle $C_n^{(1)}$ are as follows: $f^*(v_1) \equiv 4, f^*(v_2) \equiv 4mn-4n+12, f^*(v_3) \equiv 4mn-4n+16, f^*(v_4) \equiv 4mn-4n+20, f^*(v_5) \equiv 4mn-4n+28,\ f^*(v_6) \equiv 4mn-4n+36, f^*(v_7) \equiv 4mn-4n+40, f^*(v_8) \equiv 4mn-4n+44, f^*(v_9) \equiv 4mn-4n+52, f^*(v_{10}) \equiv 4mn-4n+60, f^*(v_{11}) \equiv 4mn-4n+64, f^*(v_{12}) \equiv 4mn-4n+68, f^*(v_{13}) \equiv 4mn-4n+76,\ f^*(v_{14}) \equiv 4mn-4n+84, \ldots, f^*(v_{n-8}) \equiv 4n-48, f^*(v_{n-7}) \equiv 4n-44, f^*(v_{n-6}) \equiv 4n-40, f^*(v_{n-5}) \equiv 4n-32,\ f^*(v_{n-4}) \equiv 4n-24, f^*(v_{n-3}) \equiv 4n-20, f^*(v_{n-2}) \equiv 4n-16, f^*(v_{n-1}) \equiv 4n-8, f^*(v_n) \equiv 4n-2$.

The labels the vertices of the circle $C_n^{(2)}$ are as follows: $f^*(v_{n+1}) \equiv 6, f^*(v_{n+2}) \equiv 10, f^*(v_{n+3}) \equiv 12, f^*(v_{n+i}) \equiv 4i+2, 4 \le i \le n-2, f^*(v_{2n-1}) \equiv 4n, f^*(v_{2n}) \equiv 4n+2$. Label the vertices of $C_n^{(3)}, C_n^{(4)}, \ldots, C_n^{(m-2)}, C_n^{(m-1)}, C_n^{(m)}$ respectively are as the same as in case (1).

Illustration: The edge even graceful labeling of the cylinder grid graphs $C_{9,2}, C_{9,4}, C_{7,10}, C_{7,12}, C_{7,14}C_{7,16}C_{7,18}$ and $C_{7,20}$ are shown in Figure 12.

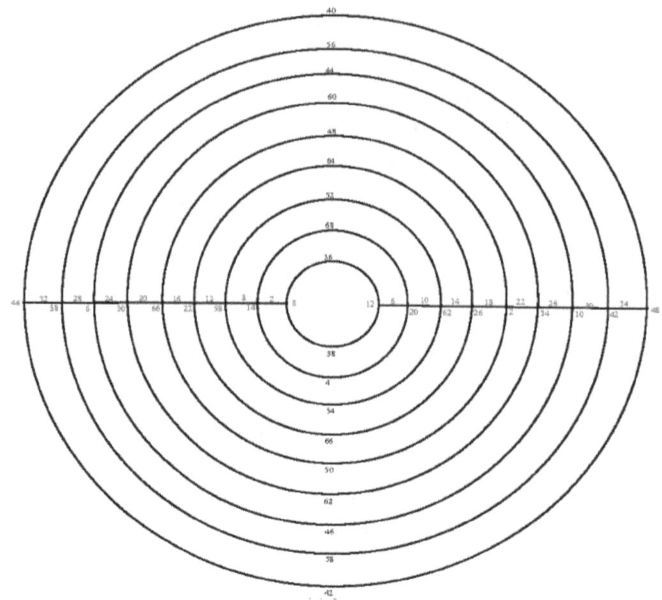

(a) $C_{9,2}$

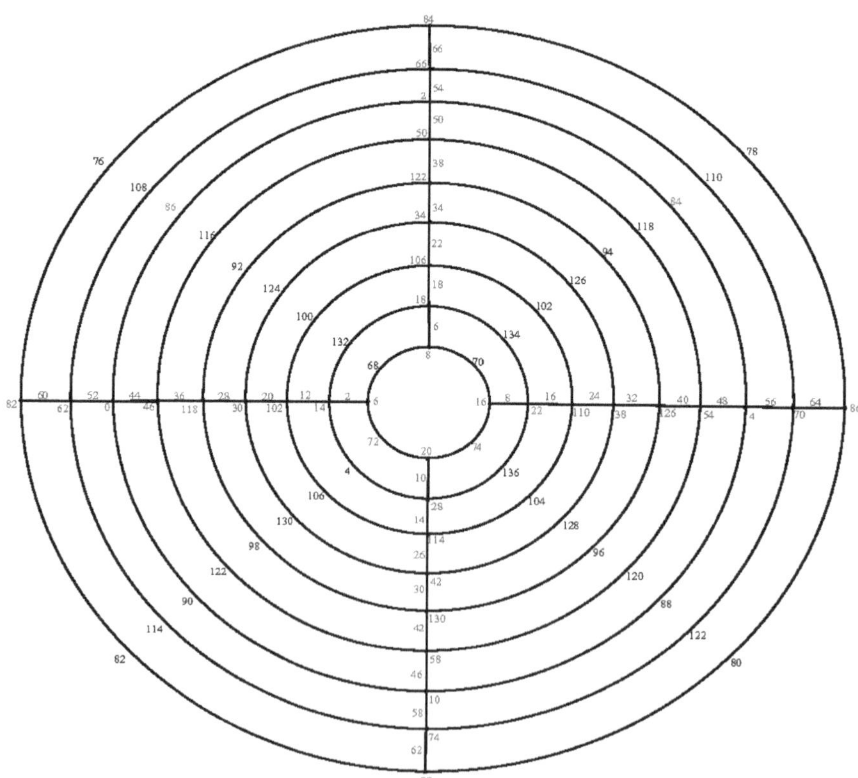

(b) $C_{9,4}$

Figure 12. *Cont.*

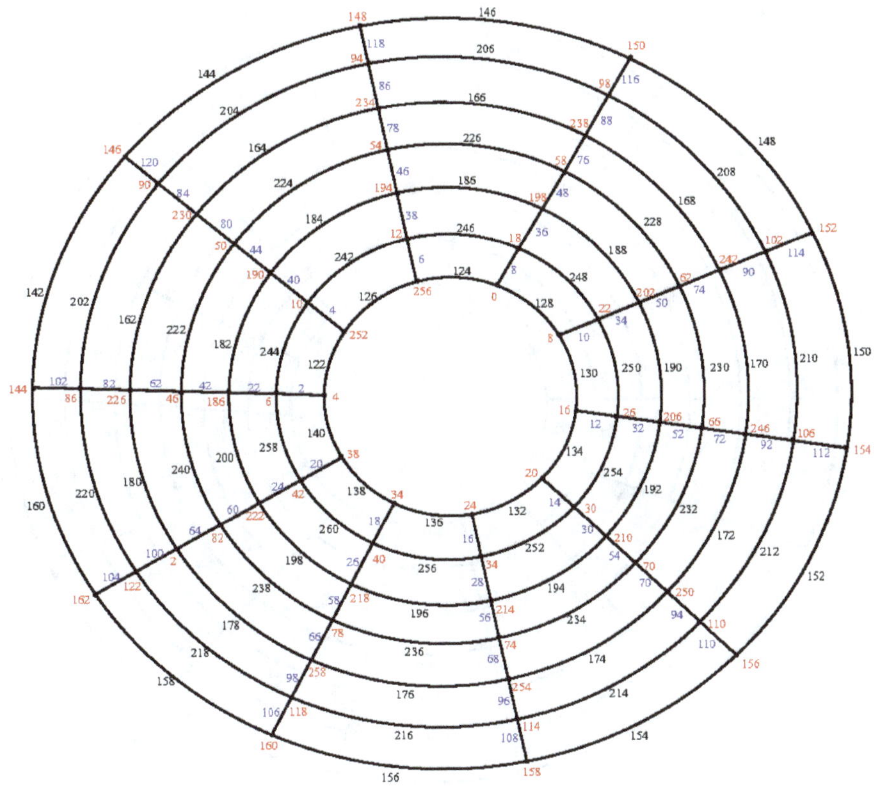

(c) $C_{7,10}$

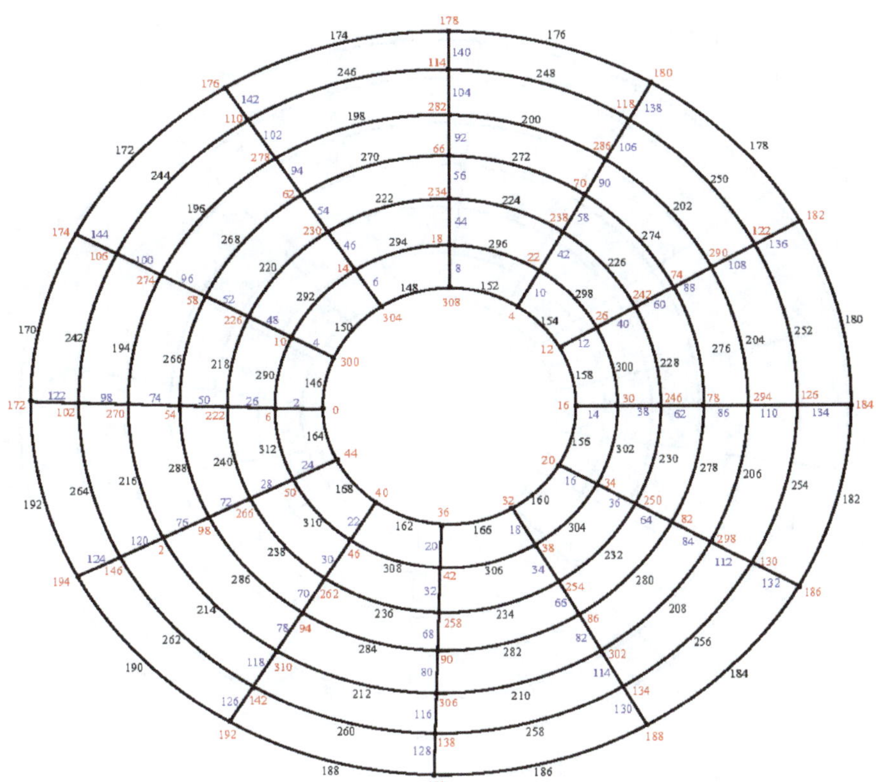

(d) $C_{7,12}$

Figure 12. *Cont.*

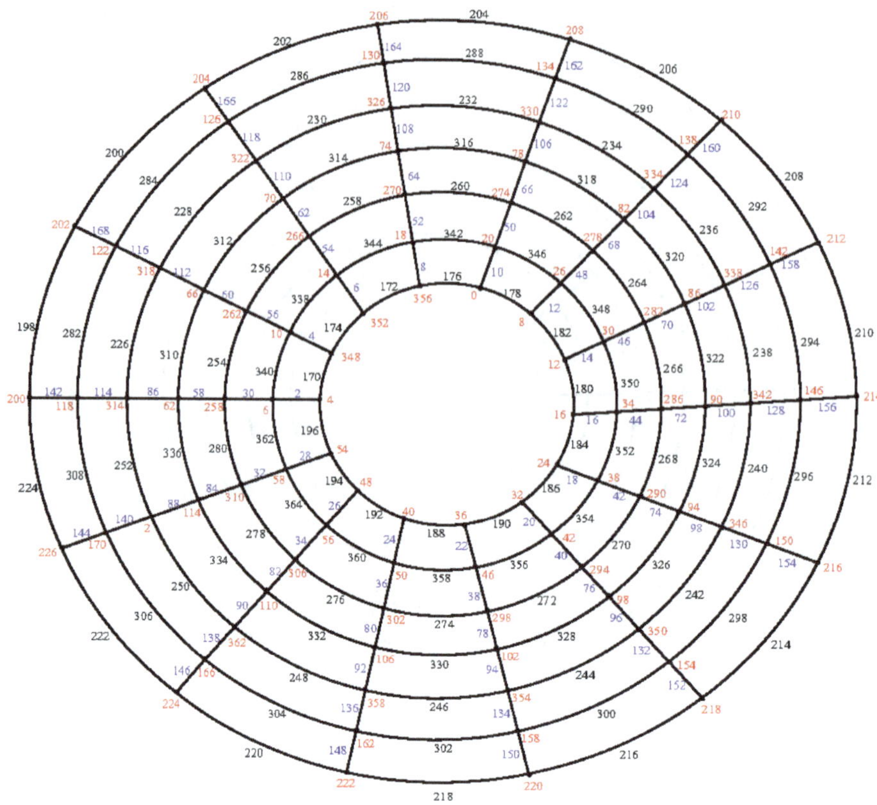

(e) $C_{7,14}$

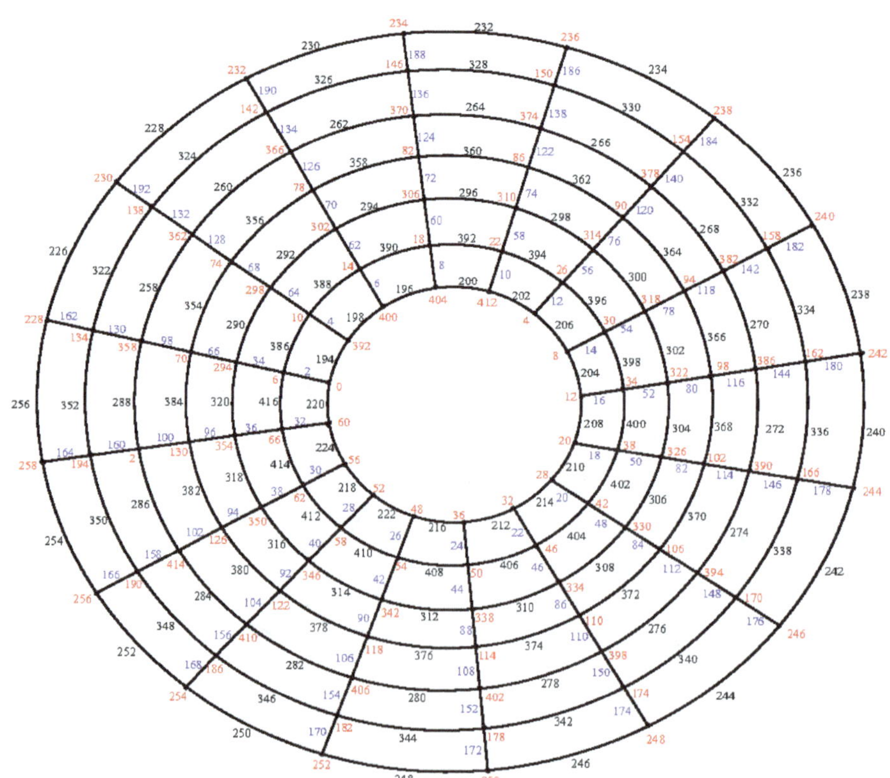

(f) $C_{7,16}$

Figure 12. *Cont.*

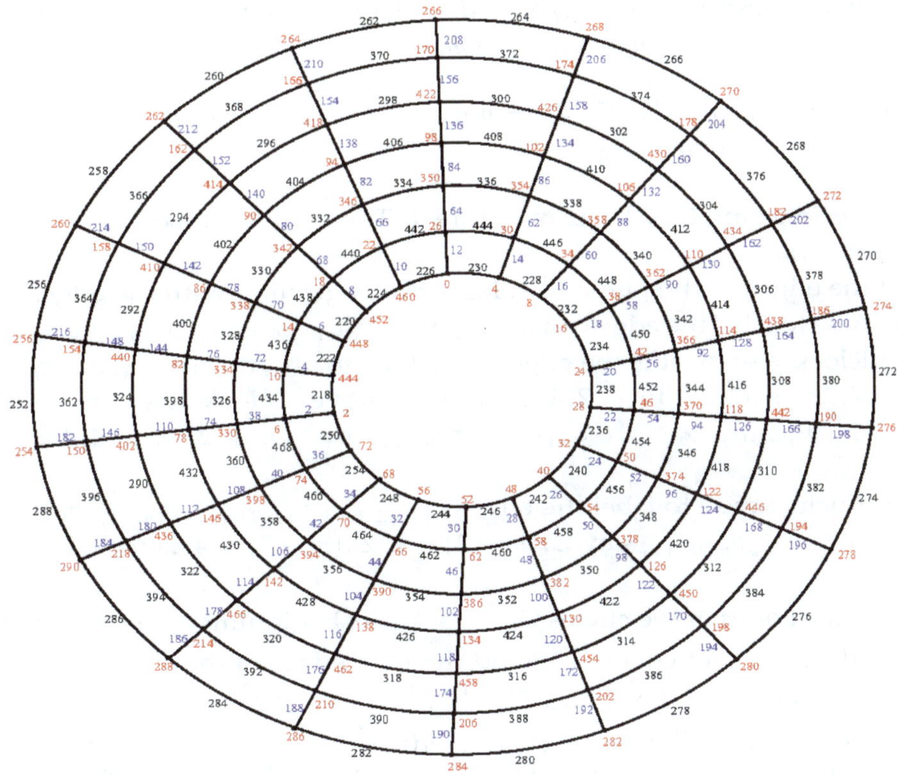

(g) $C_{7,18}$

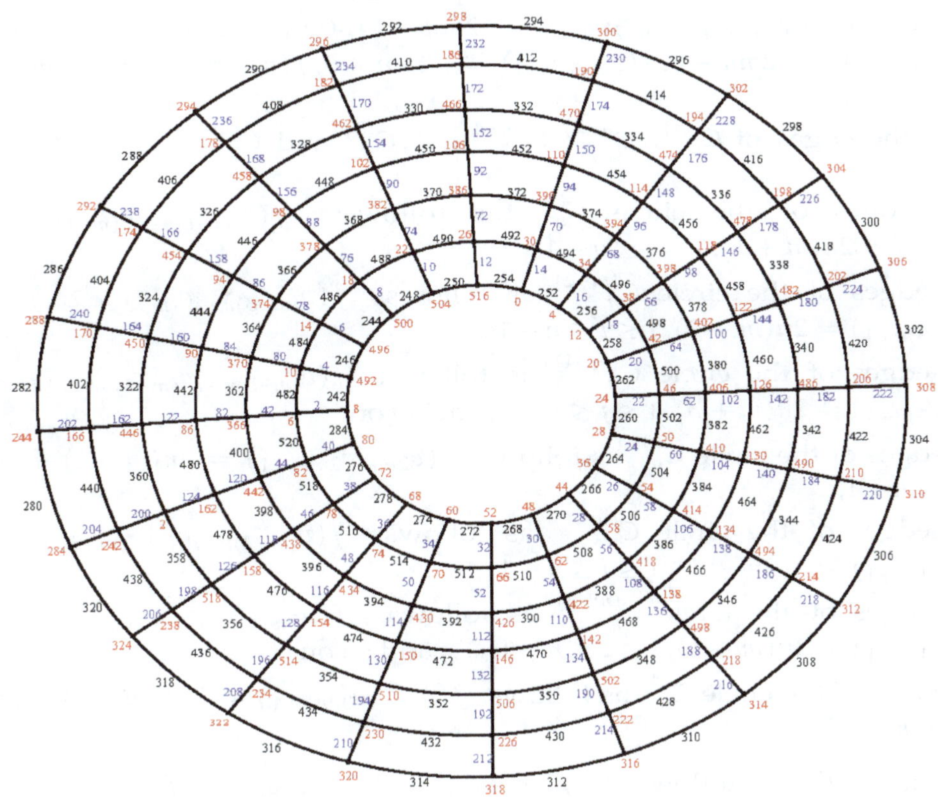

(h) $C_{7,20}$

Figure 12. An e.e.g., l. of the cylinder grid graphs $C_{9,2}, C_{9,4}, C_{7,10}, C_{7,12}, C_{7,14}, C_{7,16}, C_{7,18}$, and $C_{7,20}$.

Theorem 4. *If m is an odd positive integer greater than 3 and n is an odd positive integer, $n \geq 3$, then the cylinder grid graph $C_{m,n}$, is an edge even graceful graph.*

Proof. Using standard notation $p = |V(C_{m,n})| = mn, q = |E(C_{m,n})| = 2mn - n, r = \max(p, q) = 2mn - n$, and $f : E(C_{m,n}) \to \{2, 4, 6, \ldots, 4mn - 2n - 2\}$. □

Let the cylinder grid graph $C_{m,n}$ be as in Figure 9. There are six cases:

Case (1): $n \equiv 1 \bmod 12$.

First, we label the edges of the paths $P_m^{(k)}, 1 \leq k \leq n$ beginning with the edges of the path $P_m^{(1)}$ as follows: Move clockwise to label the edges $v_1v_{n+1}, v_2v_{n+2}, v_3v_{n+3}, \ldots, v_{n-1}v_{2n-1}, v_nv_{2n}$ by $2, 4, 6, \ldots, 2n-2, 2n$, then move anticlockwise to label the edges $v_{n+1}v_{n+2}, v_{2n}v_{3n}, v_{2n-1}v_{3n-1}, \ldots, v_{n+3}v_{2n+3}, v_{n+2}v_{2n+2}$ by $2n+2, 2n+4, 2n+6, \ldots, 4n-2, 4n$, then move clockwise to label the edges $v_{2n+1}v_{3n+1}, v_{2n+2}v_{3n+2}, v_{2n+3}v_{3n+3}, \ldots, v_{3n-1}v_{4n-1}, v_{3n}v_{4n}$ by $4n+2, 4n+4, 4n+6, \ldots, 6n-2, 6n$, and so on.

Finally, move anticlockwise to label the edges $v_{(m-2)n+1}v_{(m-1)n+1}, v_{(m-1)n}v_{mn}, v_{(m-1)n-1}v_{mn-1}, \ldots, v_{(m-2)n+3}v_{(m-1)n+3}, v_{(m-2)n+2}v_{m(n-1)+2}$ by $2n(m-2)+2, 2n(m-2)+4, 2n(m-2)+6, \ldots, 2n(m-1)-2, 2n(m-1)$.

Second, we label the edges of the circles $C_n^{(k)}, 1 \leq k \leq m$ beginning with the edges of the innermost circle $C_n^{(1)}$, then the edges of outer circle $C_n^{(m)}$, and then the edges of the circles $C_n^{(m-2)}, C_n^{(m-4)}, \ldots, C_n^{(3)}$.

Finally, we label the edges of the circles $C_m^{(m-1)}, C_m^{(m-3)}, \ldots, C_m^{(2)}$.

Label the edges of $C_n^{(1)}$ as follows: $f(v_1v_2) = 2n(m-1)+2, f(v_2v_3) = 2n(m-1)+4, f(v_3v_4) = 2n(m-1)+8, f(v_4v_5) = 2n(m-1)+6, f(v_5v_6) = 2n(m-1)+10, f(v_6v_7) = 2n(m-1)+12, f(v_7v_8) = 2n(m-1)+16, f(v_8v_9) = 2n(m-1)+14, f(v_9v_{10}) = 2n(m-1)+18, f(v_{10}v_{11}) = 2n(m-1)+20, f(v_{11}v_{12}) = 2n(m-1)+24, f(v_{12}v_{13}) = 2n(m-1)+22, f(v_{13}v_{14}) = 2n(m-1)+26, f(v_{14}v_{15}) = 2n(m-1)+28, \ldots, f(v_{n-7}v_{n-6}) = 2mn-14, f(v_{n-6}v_{n-5}) = 2mn-10, f(v_{n-5}v_{n-4}) = 2mn-12, f(v_{n-4}v_{n-3}) = 2mn-8, f(v_{n-3}v_{n-2}) = 2mn-6, f(v_{n-2}v_{n-1}) = 2mn-2, f(v_{n-1}v_n) = 2mn-4, f(v_nv_1) = 2mn$.

Then, label the edges of $C_n^{(m)}, C_n^{(m-2)}, C_n^{(m-4)}, \ldots, C_n^{(3)}$ and $C_n^{(m-1)}, C_n^{(m-3)}, C_n^{(m-5)}, \ldots, C_n^{(4)}, C_n^{(2)}$ as follows:

Label the edges of the circle $C_n^{(m)}$ as follows: $f(v_{(m-1)n+i}v_{(m-1)n+i+1}) = 2mn + 2i, f(v_{mn}v_{(m-1)n+1}) = 2n(m+1), 1 \leq i \leq n-1$.

Label the edges of the circle $C_n^{(m-2)}$ as follows: $f(v_{(m-3)n+i}v_{(m-3)n+i+1}) = 2n(m+1) + 2i, f(v_{(m-2)n}v_{(m-3)n+1}) = 2n(m+2), 1 \leq i \leq n-1$.

Label the edges of the circle $C_n^{(m-4)}$ as follows: $f(v_{(m-5)n+i}v_{(m-5)n+i+1}) = 2n(m+2) + 2i, f(v_{(m-4)n}v_{(m-5)n+1}) = 2n(m+3), 1 \leq i \leq n-1$, and so on.

Label the edges of the circle $C_n^{(3)}$ as follows: $f(v_{2n+i}v_{2n+i+1}) = 3n(m-1)+2i, f(v_{3n}v_{2n+1}) = n(3m-1), 1 \leq i \leq n-1$,

Label the edges of the circle $C_n^{(m-1)}$ as follows: $f(v_{(m-2)n+i}v_{(m-2)n+i+1}) = n(3m-1) + 2i, f(v_{(m-1)n}v_{(m-1)n+1}) = n(3m+1), 1 \leq i \leq n-1$,

Label the edges of the circle $C_n^{(m-3)}$ as follows: $f(v_{(m-4)n+i}v_{(m-4)n+i+1}) = n(3m+2) + 2i, f(v_{(m-3)n}v_{(m-4)n+1}) = 3n(m+1), 1 \leq i \leq n-1, \ldots$, and so on.

Label the edges of the circle $C_n^{(4)}$ as follows: $f(v_{3n+i}v_{3n+i+1}) = 2n(2m-3)+2i, f(v_{4n}v_{3n+1}) = 4n(m-1), 1 \leq i \leq n-1$,

Label the edges of $C_n^{(2)}$ as follows: $f(v_{n+i}v_{n+i+1}) = 4n(m-1)+2i, f(v_{2n}v_{2n+1}) = 2n(m-1), 1 \leq i \leq n-1$,

Thus, the labels of corresponding vertices of the circle $C_n^{(1)}$ mod$(4mn-2n)$ will be: $f^*(v_1) \equiv 4, f^*(v_2) \equiv 4mn-4n+10, f^*(v_3) \equiv 4mn-4n+18, f^*(v_4) \equiv 4mn-4n+22, f^*(v_5) \equiv 4mn-4n+26, f^*(v_6) \equiv 4mn-4n+34, f^*(v_7) \equiv 4mn-4n+42, f^*(v_8) \equiv 4mn-4n+46, f^*(v_9) \equiv 4mn-4n+50, f^*(v_{10}) \equiv 4mn-$

$4n + 58, f^*(v_{11}) \equiv 4mn - 4n + 66, f^*(v_{12}) \equiv 4mn - 4n + 70, f^*(v_{13}) \equiv 4mn - 4n + 74, f^*(v_{14}) \equiv 4mn - 4n + 82, \ldots, f^*(v_{n-7}) \equiv 4n - 44, f^*(v_{n-6}) \equiv 4n - 36, f^*(v_{n-5}) \equiv 4n - 32, f^*(v_{n-4}) \equiv 4n - 28, f^*(v_{n-3}) \equiv 4n - 20, f^*(v_{n-2}) \equiv 4n - 12, f^*(v_{n-1}) \equiv 4n - 8, f^*(v_n) \equiv 4n - 4$.

The labels of the vertices of $C_n^{(2)}, C_n^{(3)}, C_n^{(4)}, \ldots, C_n^{(m-2)}, C_n^{(m-1)}, C_n^{(m)}$, respectively, are as follows: $f^*(v_{n+i}) \equiv 4i + 2; f^*(v_{2n+i}) \equiv 2mn + 4n + 4i + 2; f^*(v_{3n+i}) \equiv 4n + 4i + 2; \ldots; f^*(v_{(m-3)n+i}) \equiv 4mn - 6n + 4i + 2; f^*(v_{(m-2)n+i}) \equiv 2mn - 6n + 4i + 2; f^*(v_{(m-1)n+i}) \equiv 2mn + 2i + 2, 1 \le i \le n$.

Case (2): $n \equiv 3 \bmod 12$.

First, we label the edges of the paths $P_m^{(k)}, 1 \le k \le n$ beginning with the edges of the path $P_m^{(1)}$ as the same in case (1).

Second, we label the edges of the circles $C_n^{(k)}, 1 \le k \le m$ beginning with the edges of the innermost circle $C_n^{(1)}$, then the edges of outer circle $C_n^{(m)}$, and then the edges of the circles $C_n^{(m-2)}, C_n^{(m-4)}, \ldots, C_n^{(3)}$.

Finally, we label the edges of the circles $C_m^{(m-1)}, C_m^{(m-3)}, \ldots, C_m^{(2)}$.

Label the edges of the circle $C_n^{(1)}$ as follows: $f(v_1v_2) = 2n(m-1) + 2, f(v_2v_3) = 2n(m-1) + 4, f(v_3v_4) = 2n(m-1) + 8, f(v_4v_5) = 2n(m-1) + 6, f(v_5v_6) = 2n(m-1) + 10, f(v_6v_7) = 2n(m-1) + 12, f(v_7v_8) = 2n(m-1) + 16, f(v_8v_9) = 2n(m-1) + 14, f(v_9v_{10}) = 2n(m-1) + 18, f(v_{10}v_{11}) = 2n(m-1) + 20, f(v_{11}v_{12}) = 2n(m-1) + 24, f(v_{12}v_{13}) = 2n(m-1) + 22, f(v_{13}v_{14}) = 2n(m-1) + 26, \ldots, f(v_{n-9}v_{n-8}) = 2mn - 18, f(v_{n-8}v_{n-7}) = 2mn - 14, f(v_{n-7}v_{n-6}) = 2mn - 16, f(v_{n-6}v_{n-5}) = 2mn - 12, f(v_{n-5}v_{n-4}) = 2mn - 10, f(v_{n-4}v_{n-3}) = 2mn - 6, f(v_{n-3}v_{n-2}) = 2mn - 8, f(v_{n-2}v_{n-1}) = 2mn, f(v_{n-1}v_n) = 2mn - 2, f(v_nv_1) = 2mn - 4$.

Label the edges of the circle $C_n^{(2)}$ as follows: $f(v_{n+i}v_{n+i+1}) = 4n(m-1) + 2i, 1 \le i \le n - 9, f(v_{2n-9}v_{2n-8}) = 2n(2m-1) - 18, f(v_{2n-8}v_{2n-7}) = 2n(2m-1) - 14, f(v_{2n-7}v_{2n-6}) = 2n(2m-1) - 16, f(v_{2n-6}v_{2n-5}) = 2n(2m-1) - 10, f(v_{2n-5}v_{2n-4}) = 2n(2m-1) - 12, f(v_{2n-4}v_{2n-3}) = 2n(2m-1) - 6, f(v_{2n-3}v_{2n-2}) = 2n(2m-1) - 8, f(v_{2n-2}v_{2n-1}) = 2n(2m-1) - 4, f(v_{2n-1}v_{2n}) = 2n(2m-1) - 2, f(v_{2n}v_{n+1}) = 2n(2m-1)$.

Label the edges of $C_n^{(m)}, C_n^{(m-2)}, C_n^{(m-4)}, \ldots, C_n^{(3)}$ and $C_n^{(m-1)}, C_n^{(m-3)}, C_n^{(m-5)}, \ldots, C_n^{(4)}$ as in case (1).

Thus, the labels of corresponding vertices $\bmod(4mn - 2n)$ will be:

The labels of the vertices of $C_n^{(1)}$ are as follows: $f^*(v_1) \equiv 0, f^*(v_2) \equiv 4mn - 4n + 10, f^*(v_3) \equiv 4mn - 4n + 18, f^*(v_4) \equiv 4mn - 4n + 22, f^*(v_5) \equiv 4mn - 4n + 26, f^*(v_6) \equiv 4mn - 4n + 34, f^*(v_7) \equiv 4mn - 4n + 42, f^*(v_8) \equiv 4mn - 4n + 46, f^*(v_9) \equiv 4mn - 4n + 50, f^*(v_{10}) \equiv 4mn - 4n + 58, f^*(v_{11}) \equiv 4mn - 4n + 66, f^*(v_{12}) \equiv 4mn - 4n + 70, f^*(v_{13}) \equiv 4mn - 4n + 74, f^*(v_{14}) \equiv 4mn - 4n + 82, \ldots, f^*(v_{n-8}) \equiv 4n - 48, f^*(v_{n-7}) \equiv 4n - 44, f^*(v_{n-6}) \equiv 4n - 40, f^*(v_{n-5}) \equiv 4n - 32, f^*(v_{n-4}) \equiv 4n - 24, f^*(v_{n-3}) \equiv 4n - 20, f^*(v_{n-2}) \equiv 4n - 12, f^*(v_{n-1}) \equiv 4n - 4, f^*(v_n) \equiv 4n - 6$.

The labels of the vertices of the circle $C_n^{(2)}$ are as follows: $f^*(v_{n+i}) \equiv 4i + 2, 1 \le i \le n - 9, f^*(v_{2n-8}) \equiv 4n - 28, f^*(v_{2n-7}) \equiv 4n - 26, f^*(v_{2n-6}) \equiv 4n - 22, f^*(v_{2n-5}) \equiv 4n - 18, f^*(v_{2n-4}) \equiv 4n - 14, f^*(v_{2n-3}) \equiv 4n - 10, f^*(v_{2n-2}) \equiv 4n - 8, f^*(v_{2n-1}) \equiv 4n - 2, f^*(v_{2n}) \equiv 4n + 2$.

The labels of the vertices of $C_n^{(3)}, C_n^{(4)}, \ldots, C_n^{(m-2)}, C_n^{(m-1)}, C_n^{(m)}$, respectively, are the same as in case (1).

Remark 4. *In case $n = 3$ and m is odd, $m \ge 3$.*

Let the label of edges of the cylinder grid graph $C_{m,3}$ be as in Figure 13.

Thus, the labels of corresponding vertices $\bmod(12m - 6)$ are as follows:

The labels of the vertices of the circle $C_3^{(1)}$ are $f^*(v_1) \equiv 8, f^*(v_2) \equiv 12, f^*(v_3) \equiv 16$.

The labels of the vertices of the circle $C_3^{(3)}$ are $f^*(v_{3m-2}) \equiv 6m + 10, f^*(v_{3m-1}) \equiv 6m + 12, f^*(v_{3m}) \equiv 6m + 14$.

The labels of the vertices of the circles $C_3^{(2)}, C_3^{(4)}, \ldots, C_3^{(m-1)}$ are $f^*(v_{3k+i}) \equiv 4i + 6k + 4, 1 \le i \le 3, 1 \le k \le m - 2, k$ is odd.

The labels of the vertices of the circles $C_3^{(3)}, C_3^{(5)}, \ldots, C_3^{(m-2)}$ are $f^*(v_{3k+i}) \equiv 6m + 4i + 6k + 10, 1 \le i \le 3, 2 \le k \le m - 3, k$ is even.

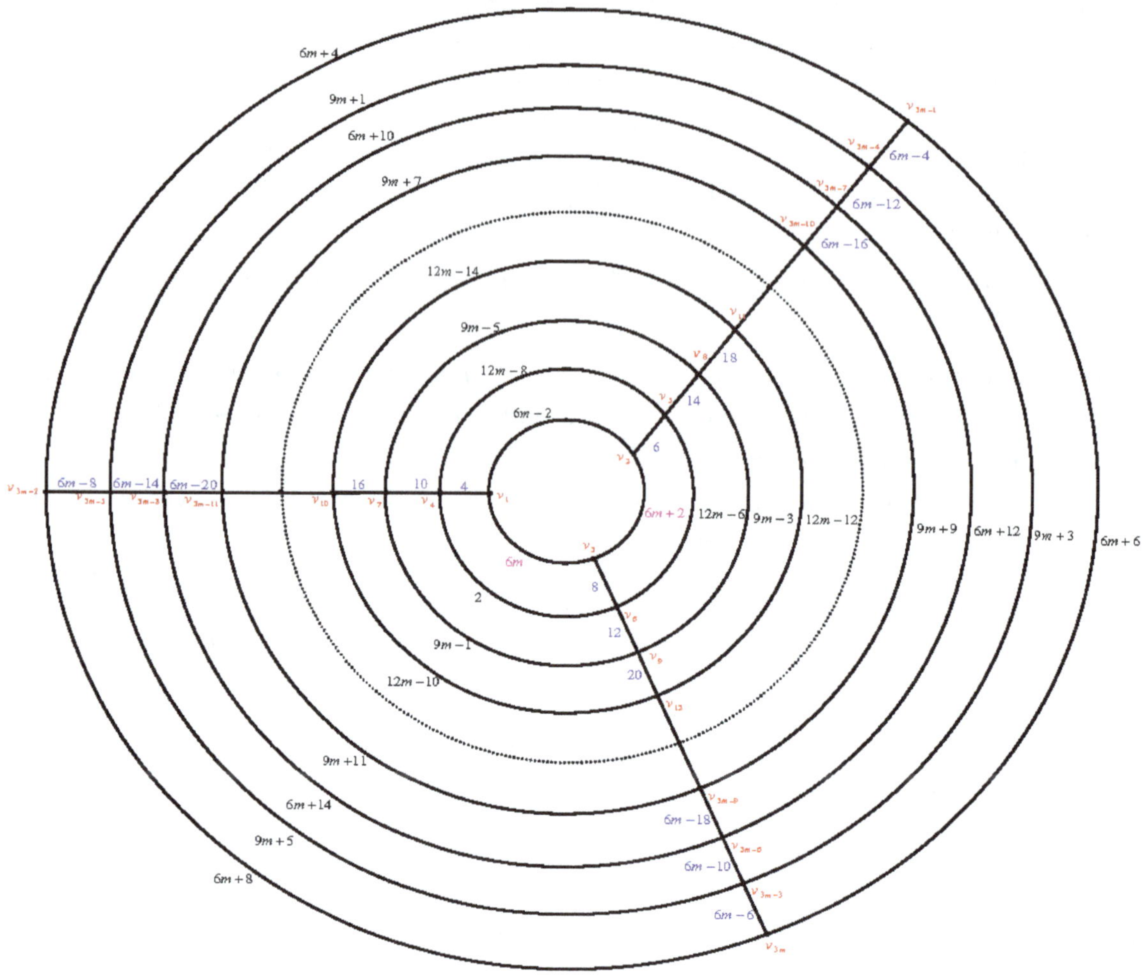

Figure 13. The cylinder grid graph $C_{m,3}$, m is odd, $m \geq 3$.

Case (3): $n \equiv 5\text{mod}12$.

First, we label the edges of the paths $P_m^{(k)}, 1 \leq k \leq n$ beginning with the edges of the path $P_m^{(1)}$ the same as in case (1).

Second, we label the edges of the circles $C_n^{(k)}, 1 \leq k \leq m$ beginning with the edges of the innermost circle $C_n^{(1)}$, then the edges of outer circle $C_n^{(m)}$, and then the edges of the circles $C_n^{(m-2)}, C_n^{(m-4)}, \ldots, C_n^{(3)}$.

Finally, we label the edges of the circles $C_m^{(m-1)}, C_m^{(m-3)}, \ldots, C_m^{(2)}$.

Label the edges of the circle $C_n^{(1)}$ as follows: $f(v_1v_2) = 2n(m-1)+2, f(v_2v_3) = 2n(m-1)+4, f(v_3v_4) = 2n(m-1)+8, f(v_4v_5) = 2n(m-1)+6,\ f(v_5v_6) = 2n(m-1)+10, f(v_6v_7) = 2n(m-1)+12, f(v_7v_8) = 2n(m-1)+16, f(v_8v_9) = 2n(m-1)+14,\ f(v_9v_{10}) = 2n(m-1)+18, f(v_{10}v_{11}) = 2n(m-1)+20, f(v_{11}v_{12}) = 2n(m-1)+24, f(v_{12}v_{13}) = 2n(m-1)+22,\ f(v_{13}v_{14}) = 2n(m-1)+26, \ldots, f(v_{n-8}v_{n-7}) = 2mn-16, f(v_{n-7}v_{n-6}) = 2mn-14, f(v_{n-6}v_{n-5}) = 2mn-10,\ f(v_{n-5}v_{n-4}) = 2mn-12, f(v_{n-4}v_{n-3}) = 2mn-8, f(v_{n-3}v_{n-2}) = 2mn-6, f(v_{n-2}v_{n-1}) = 2mn-2, f(v_{n-1}v_n) = 2mn-4, f(v_nv_1) = 2mn$. Label the edges of $C_n^{(m)}, C_n^{(m-2)}, C_n^{(m-4)}, \ldots, C_n^{(3)}$ and $C_n^{(m-1)}, C_n^{(m-3)}, C_n^{(m-5)}, \ldots, C_n^{(4)}, C_n^{(2)}$ as in case (1).

Thus, the labels of corresponding vertices of the circle $C_n^{(1)}$ $\text{mod}(4mn-2n)$ will be: $f^*(v_1) \equiv 4, f^*(v_2) \equiv 4mn-4n+10, f^*(v_3) \equiv 4mn-4n+18, f^*(v_4) \equiv 4mn-4n+22, f^*(v_5) \equiv 4mn-4n+26, f^*(v_6) \equiv 4mn-4n+34, f^*(v_7) \equiv 4mn-4n+42, f^*(v_8) \equiv 4mn-4n+46, f^*(v_9) \equiv 4mn-4n+50, f^*(v_{10}) \equiv 4mn-4n+58, f^*(v_{11}) \equiv 4mn-4n+66, f^*(v_{12}) \equiv 4mn-4n+70, f^*(v_{13}) \equiv 4mn-4n+74, f^*(v_{14}) \equiv 4mn-4n+82, \ldots, f^*(v_{n-7}) \equiv 4n-44, f^*(v_{n-6}) \equiv 4n-36, f^*(v_{n-5}) \equiv 4n-32, f^*(v_{n-4}) \equiv 4n-28, f^*(v_{n-3}) \equiv 4n-20,\ f^*(v_{n-2}) \equiv 4n-12, f^*(v_{n-1}) \equiv 4n-8, f^*(v_n) \equiv 4n-4$.

The labels of the vertices of $C_n^{(2)}, C_n^{(3)}, C_n^{(4)}, \ldots, C_n^{(m-2)}, C_n^{(m-1)}, C_n^{(m)}$, respectively, are the same as in case (1).

Case (4): $n \equiv 7 \mathrm{mod} 12$.

First, we label the edges of the paths $P_m^{(k)}, 1 \le k \le n$ beginning with the edges of the path $P_m^{(1)}$ the same as in case (1).

Second, we label the edges of the circles $C_n^{(k)}, 1 \le k \le m$ beginning with the edges of the innermost circle $C_n^{(1)}$, then the edges of outer circle $C_n^{(m)}$, and then the edges of the circles $C_n^{(m-2)}, C_n^{(m-4)}, \ldots, C_n^{(3)}$.

Finally, we label the edges of the circles $C_m^{(m-1)}, C_m^{(m-3)}, \ldots, C_m^{(2)}$.

Label the edges of the circle $C_n^{(1)}$ as follows: $f(v_1v_2) = 2n(m-1)+2, f(v_2v_3) = 2n(m-1)+4, f(v_3v_4) = 2n(m-1)+8, f(v_4v_5) = 2n(m-1)+6,\ f(v_5v_6) = 2n(m-1)+10, f(v_6v_7) = 2n(m-1)+12, f(v_7v_8) = 2n(m-1)+16, f(v_8v_9) = 2n(m-1)+14,\ f(v_9v_{10}) = 2n(m-1)+18, f(v_{10}v_{11}) = 2n(m-1)+20, f(v_{11}v_{12}) = 2n(m-1)+24, f(v_{12}v_{13}) = 2n(m-1)+22,\ f(v_{13}v_{14}) = 2n(m-1)+26, \ldots, f(v_{n-10}v_{n-9}) = 2mn-20, f(v_{n-9}v_{n-8}) = 2mn-18, f(v_{n-8}v_{n-7}) = 2mn-14,\ f(v_{n-7}v_{n-6}) = 2mn-16, f(v_{n-6}v_{n-5}) = 2mn-12, f(v_{n-5}v_{n-4}) = 2mn-10, f(v_{n-4}v_{n-3}) = 2mn-6, f(v_{n-3}v_{n-2}) = 2mn-8,\ f(v_{n-2}v_{n-1}) = 2mn, f(v_{n-1}v_n) = 2mn-2, f(v_nv_1) = 2mn-4$.

Label the edges of the circle $C_n^{(2)}$ as follows: $f(v_{n+i}v_{n+i+1}) = 4n(m-1)+2i, 1 \le i \le n-9, f(v_{2n-9}v_{2n-8}) = 2n(2m-1)-18, f(v_{2n-8}v_{2n-7}) = 2n(2m-1)-14,\ f(v_{2n-7}v_{2n-6}) = 2n(2m-1)-16, f(v_{2n-6}v_{2n-5}) = 2n(2m-1)-10, f(v_{2n-5}v_{2n-4}) = 2n(2m-1)-12,\ f(v_{2n-4}v_{2n-3}) = 2n(2m-1)-6, f(v_{2n-3}v_{2n-2}) = 2n(2m-1)-8, f(v_{2n-2}v_{2n-1}) = 2n(2m-1)-4,\ f(v_{2n-1}v_{2n}) = 2n(2m-1)-2, f(v_{2n}v_{n+1}) = 2n(2m-1)$.

Label the edges of $C_n^{(m)}, C_n^{(m-2)}, C_n^{(m-4)}, \ldots, C_n^{(3)}$ and $C_n^{(m-1)}, C_n^{(m-3)}, C_n^{(m-5)}, \ldots, C_n^{(4)}$ as in case (1).

Thus, the labels of corresponding vertices $\mathrm{mod}(4mn-2n)$ will be:

The labels of the vertices of the circle $C_n^{(1)}$ are as follows: $f^*(v_1) \equiv 0, f^*(v_2) \equiv 4mn-4n+10, f^*(v_3) \equiv 4mn-4n+18, f^*(v_4) \equiv 4mn-4n+22, f^*(v_5) \equiv 4mn-4n+26,\ f^*(v_6) \equiv 4mn-4n+34, f^*(v_7) \equiv 4mn-4n+42, f^*(v_8) \equiv 4mn-4n+46, f^*(v_9) \equiv 4mn-4n+50,\ f^*(v_{10}) \equiv 4mn-4n+58, f^*(v_{11}) \equiv 4mn-4n+66, f^*(v_{12}) \equiv 4mn-4n+70, f^*(v_{13}) \equiv 4mn-4n+74,\ f^*(v_{14}) \equiv 4mn-4n+82, \ldots, f^*(v_{n-9}) \equiv 4n-56, f^*(v_{n-8}) \equiv 4n-48, f^*(v_{n-7}) \equiv 4n-44, f^*(v_{n-6}) \equiv 4n-40,\ f^*(v_{n-5}) \equiv 4n-32, f^*(v_{n-4}) \equiv 4n-24, f^*(v_{n-3}) \equiv 4n-20, f^*(v_{n-2}) \equiv 4n-12, f^*(v_{n-1}) \equiv 4n-4, f^*(v_n) \equiv 4n-6$.

The labels of the vertices of the circle $C_n^{(2)}$ are as follows: $f^*(v_{n+i}) \equiv 4i+2, 1 \le i \le n-9, f^*(v_{2n-8}) \equiv 4n-28, f^*(v_{2n-7}) \equiv 4n-26, f^*(v_{2n-6}) \equiv 4n-22, f^*(v_{2n-5}) \equiv 4n-18,\ f^*(v_{2n-4}) \equiv 4n-14, f^*(v_{2n-3}) \equiv 4n-10, f^*(v_{2n-2}) \equiv 4n-8, f^*(v_{2n-1}) \equiv 4n-2, f^*(v_{2n}) \equiv 4n+2$.

The labels of the vertices of $C_n^{(3)}, C_n^{(4)}, \ldots, C_n^{(m-2)}, C_n^{(m-1)}, C_n^{(m)}$, respectively, are as in case (1).

Case (5): $n \equiv 9 \mathrm{mod} 12$.

First, we label the edges of the paths $P_m^{(k)}, 1 \le k \le n$ beginning with the edges of the path $P_m^{(1)}$ the same as in case (1).

Second, we label the edges of the circles $C_n^{(k)}, 1 \le k \le m$ beginning with the edges of the innermost circle $C_n^{(1)}$, then the edges of outer circle $C_n^{(m)}$, and then the edges of the circles $C_n^{(m-2)}, C_n^{(m-4)}, \ldots, C_n^{(3)}$.

Finally, we label the edges of the circles $C_m^{(m-1)}, C_m^{(m-3)}, \ldots, C_m^{(2)}$.

Label the edges of the circle $C_n^{(1)}$ as follows: $f(v_1v_2) = 2n(m-1)+2, f(v_2v_3) = 2n(m-1)+4, f(v_3v_4) = 2n(m-1)+8, f(v_4v_5) = 2n(m-1)+6,\ f(v_5v_6) = 2n(m-1)+10, f(v_6v_7) = 2n(m-1)+12, f(v_7v_8) = 2n(m-1)+16, f(v_8v_9) = 2n(m-1)+14,\ f(v_9v_{10}) = 2n(m-1)+18, f(v_{10}v_{11}) = 2n(m-1)+20, f(v_{11}v_{12}) = 2n(m-1)+24, f(v_{12}v_{13}) = 2n(m-1)+22,\ f(v_{13}v_{14}) = 2n(m-1)+26, \ldots, f(v_{n-11}v_{n-10}) = 2mn-22, f(v_{n-10}v_{n-9}) = 2mn-18, f(v_{n-9}v_{n-8}) = 2mn-20,\ f(v_{n-8}v_{n-7}) = 2mn-16, f(v_{n-7}v_{n-6}) = 2mn-14, f(v_{n-6}v_{n-5}) = 2mn-10, f(v_{n-5}v_{n-4}) = 2mn-12,\ f(v_{n-4}v_{n-3}) = 2mn-8, f(v_{n-3}v_{n-2}) = 2mn-6, f(v_{n-2}v_{n-1}) = 2mn, f(v_{n-1}v_n) = 2mn-2, f(v_nv_1) = 2mn-4$.

Label the edges of the circle $C_n^{(2)}$ as follows: $f(v_{n+i}v_{n+i+1}) = 4n(m-1)+2i, 1 \le i \le n-8, f(v_{2n-7}v_{2n-6}) = 2n(2m-1)-12, f(v_{2n-6}v_{2n-5}) = 2n(2m-1)-14,\ f(v_{2n-5}v_{2n-4}) = 2n(2m-1)-6, f(v_{2n-4}v_{2n-3}) = 2n(2m-1)-10, f(v_{2n-3}v_{2n-2}) = 2n(2m-1)-8,\ f(v_{2n-2}v_{2n-1}) = 2n(2m-1)-4, f(v_{2n}v_{n+1}) = 2n(2m-1)$.

Label the edges of $C_n^{(m)}, C_n^{(m-2)}, C_n^{(m-4)}, \ldots, C_n^{(3)}$ and $C_n^{(m-1)}, C_n^{(m-3)}, C_n^{(m-5)}, \ldots, C_n^{(4)}$ as in case (1).
Thus, the labels of corresponding vertices $\bmod(4mn-2n)$ will be:

The labels of the vertices of the circle $C_n^{(1)}$ are as follows: $f^*(v_1) \equiv 0, f^*(v_2) \equiv 4mn-4n+10, f^*(v_3) \equiv 4mn-4n+18, f^*(v_4) \equiv 4mn-4n+22, f^*(v_5) \equiv 4mn-4n+26, f^*(v_6) \equiv 4mn-4n+34, f^*(v_7) \equiv 4mn-4n+42, f^*(v_8) \equiv 4mn-4n+46, f^*(v_9) \equiv 4mn-4n+50, f^*(v_{10}) \equiv 4mn-4n+58, f^*(v_{11}) \equiv 4mn-4n+66, f^*(v_{12}) \equiv 4mn-4n+70, f^*(v_{13}) \equiv 4mn-4n+74, f^*(v_{14}) \equiv 4mn-4n+82, \ldots, f^*(v_{n-10}) \equiv 4n-60, f^*(v_{n-9}) \equiv 4n-56, f^*(v_{n-8}) \equiv 4n-52, f^*(v_{n-7}) \equiv 4n-44, f^*(v_{n-6}) \equiv 4n-36, f^*(v_{n-5}) \equiv 4n-32, f^*(v_{n-4}) \equiv 4n-28, f^*(v_{n-3}) \equiv 4n-20, f^*(v_{n-2}) \equiv 4n-10, f^*(v_{n-1}) \equiv 4n-4, f^*(v_n) \equiv 4n-6$.

The labels of the vertices of the circle $C_n^{(2)}$ are as follows: $f^*(v_{n+i}) \equiv 4i+2, 1 \le i \le n-8, f^*(v_{2n-7}) \equiv 4n-24, f^*(v_{2n-6}) \equiv 4n-22, f^*(v_{2n-5}) \equiv 4n-16, f^*(v_{2n-4}) \equiv 4n-12, f^*(v_{2n-3}) \equiv 4n-14, f^*(v_{2n-2}) \equiv 4n-8, f^*(v_{2n-1}) \equiv 4n-2, f^*(v_{2n}) \equiv 4n+2$.
The labels of the vertices of $C_n^{(3)}, C_n^{(4)}, \ldots, C_n^{(m-2)}, C_n^{(m-1)}, C_n^{(m)}$, respectively, are as in case (1).
Case (6): $n \equiv 11 \bmod 12$.

First, we label the edges of the paths $P_m^{(k)}, 1 \le k \le n$ beginning with the edges of the path $P_m^{(1)}$ the same as in case (1).

Second, we label the edges of the circles $C_n^{(k)}, 1 \le k \le m$ beginning with the edges of the innermost circle $C_n^{(1)}$, then the edges of outer circle $C_n^{(m)}$, and then the edges of the circles $C_n^{(m-2)}, C_n^{(m-4)}, \ldots, C_n^{(3)}$.

Finally, we label the edges of the circles $C_m^{(m-1)}, C_m^{(m-3)}, \ldots, C_m^{(2)}$.

Label the edges of the circle $C_n^{(1)}$ as follows: $f(v_1v_2) = 2n(m-1)+4, f(v_2v_3) = 2n(m-1)+2, f(v_3v_4) = 2n(m-1)+6, f(v_4v_5) = 2n(m-1)+8, f(v_5v_6) = 2n(m-1)+12, f(v_6v_7) = 2n(m-1)+10, f(v_7v_8) = 2n(m-1)+14, f(v_8v_9) = 2n(m-1)+16, f(v_9v_{10}) = 2n(m-1)+20, f(v_{10}v_{11}) = 2n(m-1)+18, f(v_{11}v_{12}) = 2n(m-1)+22, f(v_{12}v_{13}) = 2n(m-1)+24, f(v_{13}v_{14}) = 2n(m-1)+28, \ldots, f(v_{n-8}v_{n-7}) = 2mn-16, f(v_{n-7}v_{n-6}) = 2mn-14, f(v_{n-6}v_{n-5}) = 2mn-10, f(v_{n-5}v_{n-4}) = 2mn-12, f(v_{n-4}v_{n-3}) = 2mn-8, f(v_{n-3}v_{n-2}) = 2mn-6, f(v_{n-2}v_{n-1}) = 2mn-2, f(v_{n-1}v_n) = 2mn-4, f(v_nv_1) = 2mn$.

Label the edges of the circle $C_n^{(2)}$ as follows: $f(v_{n+i}v_{n+i+1}) = 4n(m-1)+2i, 1 \le i \le n-2, f(v_{2n-1}v_{2n}) = 4mn, f(v_{2n}v_{n+1}) = 4mn-2$.
Label the edges of $C_n^{(m)}, C_n^{(m-2)}, C_n^{(m-4)}, \ldots, C_n^{(3)}$ and $C_n^{(m-1)}, C_n^{(m-3)}, C_n^{(m-5)}, \ldots, C_n^{(4)}$ as in case (1).
Thus, the labels of corresponding vertices $\bmod(4mn-2n)$ will be:

The labels of the vertices of the circle $C_n^{(1)}$ are as follows: $f^*(v_1) \equiv 6, f^*(v_2) \equiv 4mn-4n+10, f^*(v_3) \equiv 4mn-4n+14, f^*(v_4) \equiv 4mn-4n+22, f^*(v_5) \equiv 4mn-4n+30, f^*(v_6) \equiv 4mn-4n+34, f^*(v_7) \equiv 4mn-4n+38, f^*(v_8) \equiv 4mn-4n+46, f^*(v_9) \equiv 4mn-4n+54, f^*(v_{10}) \equiv 4mn-4n+58, f^*(v_{11}) \equiv 4mn-4n+62, f^*(v_{12}) \equiv 4mn-4n+70, \ldots, f^*(v_{n-7}) \equiv 4n-44, f^*(v_{n-6}) \equiv 4n-36, f^*(v_{n-5}) \equiv 4n-32, f^*(v_{n-4}) \equiv 4n-28, f^*(v_{n-3}) \equiv 4n-20, f^*(v_{n-2}) \equiv 4n-12, f^*(v_{n-1}) \equiv 4n-8, f^*(v_n) \equiv 4n-4$.
The labels of the vertices of the circle $C_n^{(2)}$ are as follows: $f^*(v_{n+1}) \equiv 4, f^*(v_{n+i}) \equiv 4i+2, 2 \le i \le n-2, f^*(v_{2n-1}) \equiv 4n, f^*(v_{2n}) \equiv 4n+2$.

The labels of the vertices of $C_n^{(3)}, C_n^{(4)}, \ldots, C_n^{(m-2)}, C_n^{(m-1)}, C_n^{(m)}$, respectively, are as the same as in case (1).

Illustration: An e.e.g.l. of the cylinder grid graphs $C_{9,3}, C_{7,9}, C_{7,11}, C_{7,13}, C_{7,15}, C_{7,17}$ and $C_{7,19}$ is shown in Figure 14.

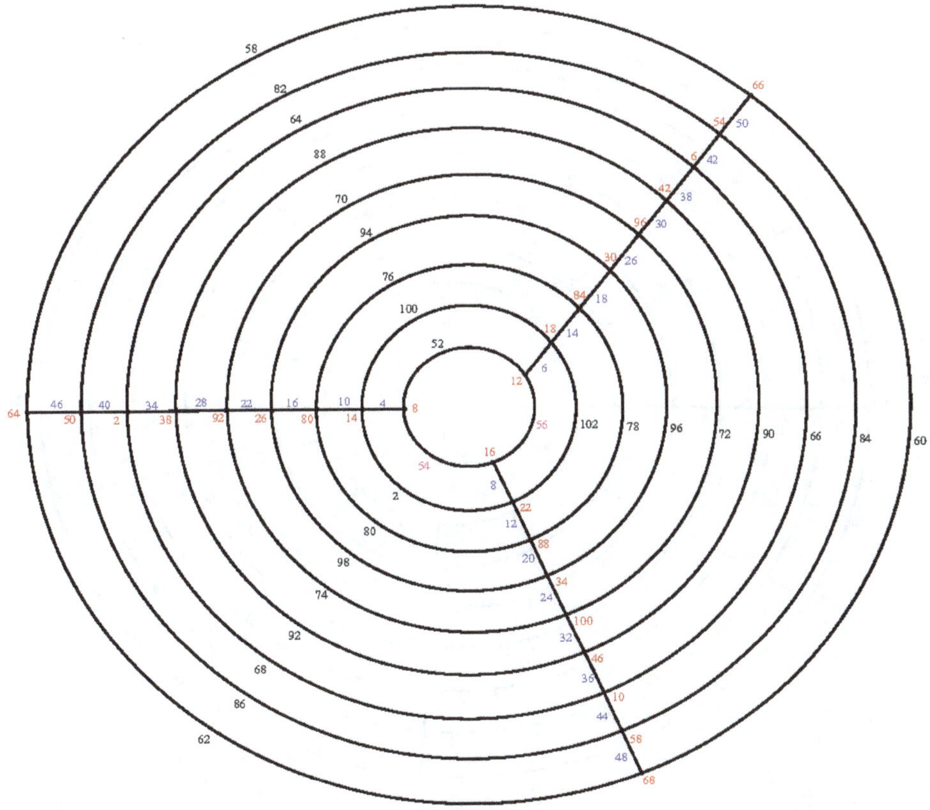

(a) $C_{9,3}$

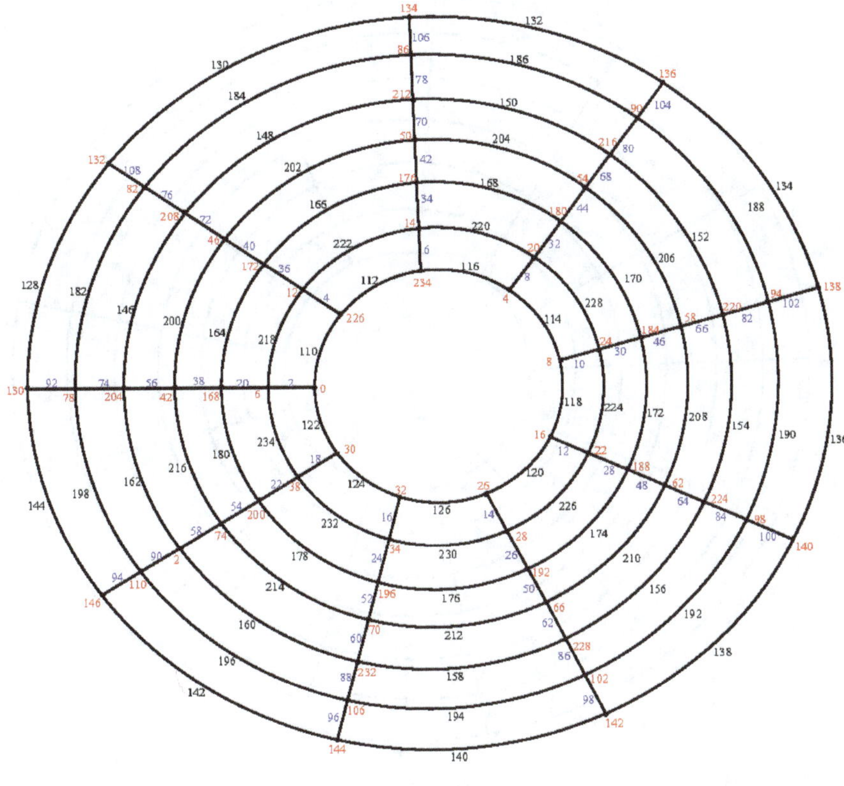

(b) $C_{7,9}$

Figure 14. *Cont.*

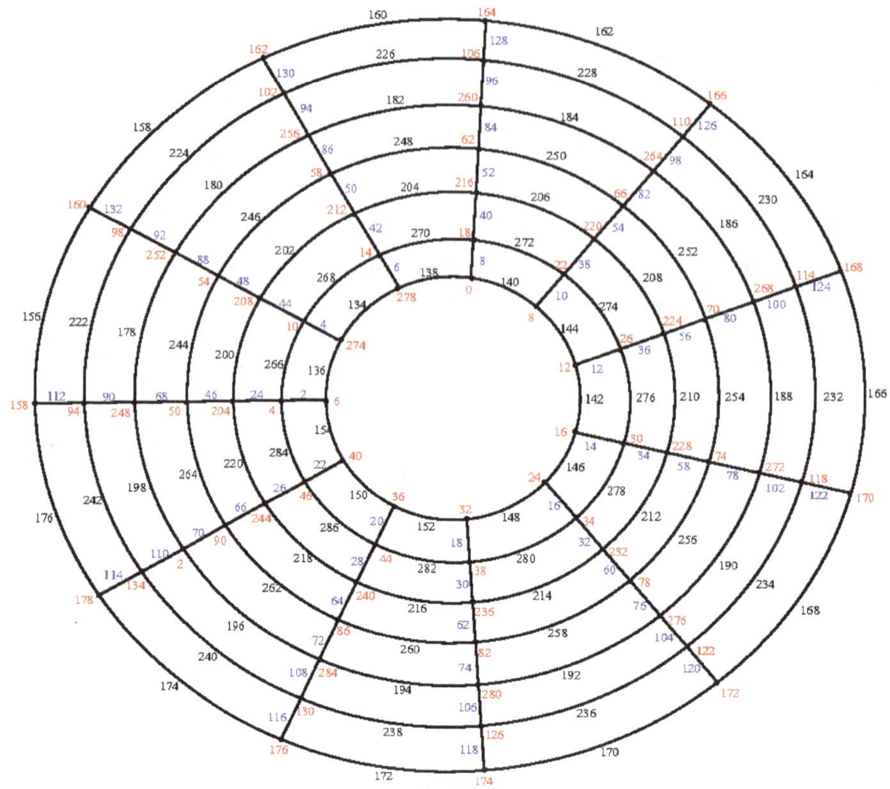

(c) $C_{7,11}$

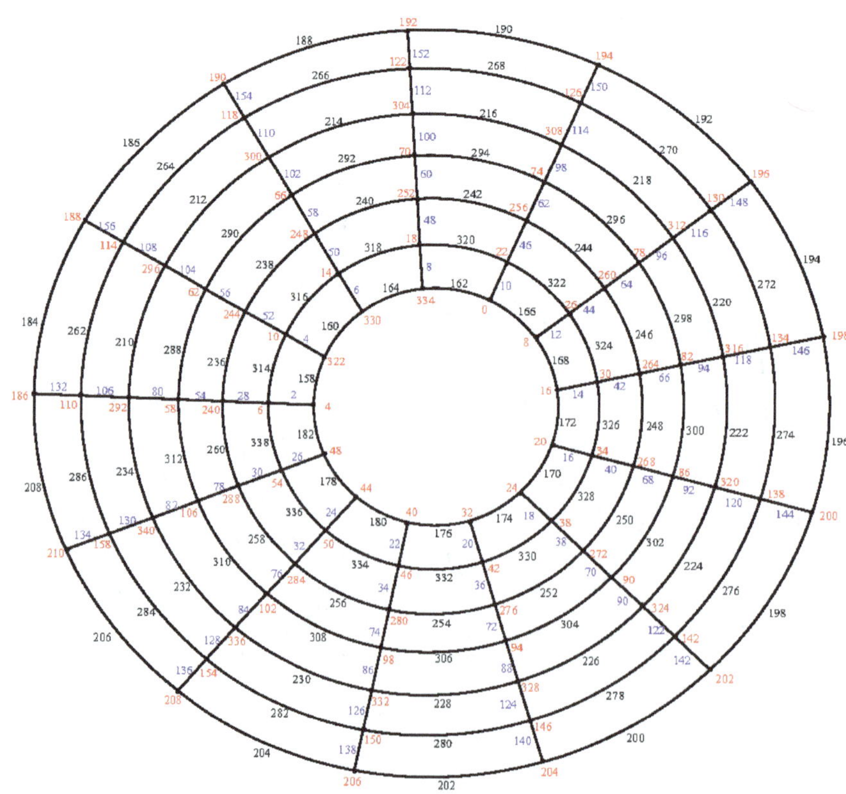

(d) $C_{7,13}$

Figure 14. *Cont.*

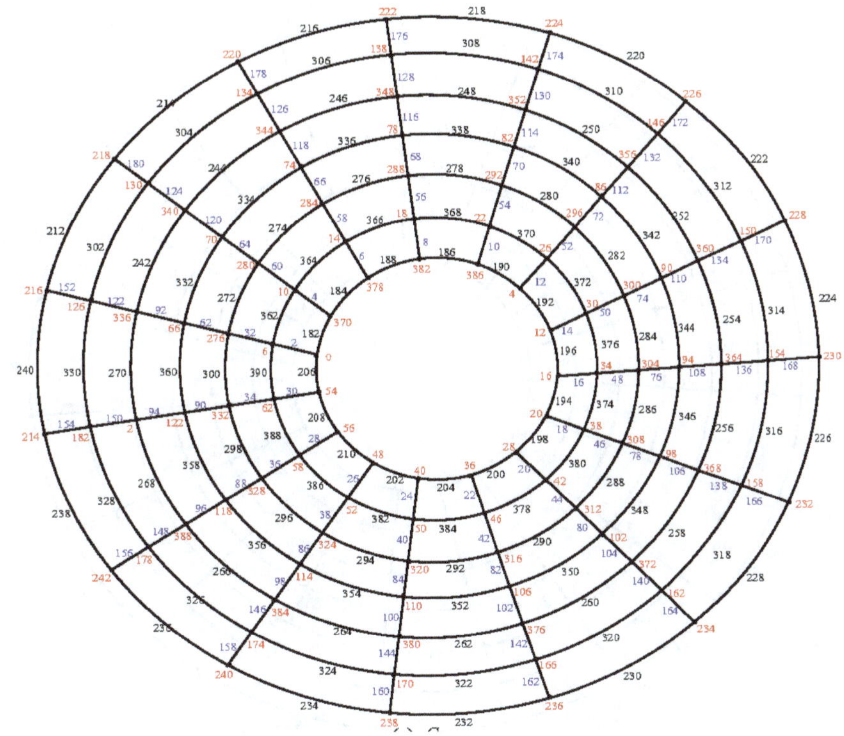

(e) $C_{7,15}$

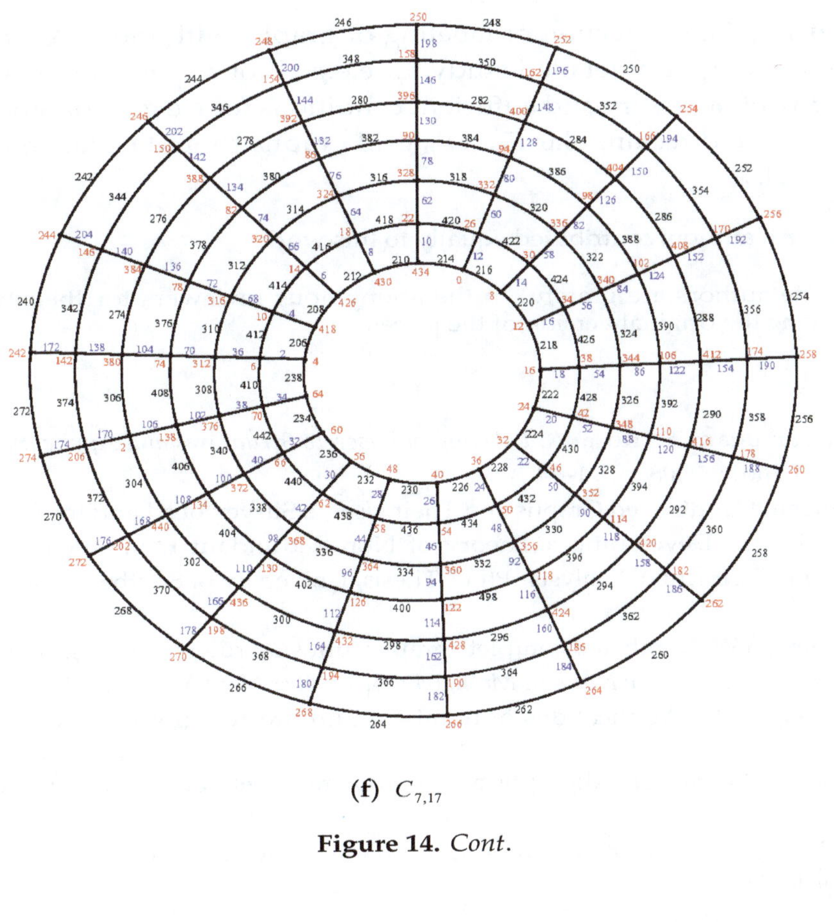

(f) $C_{7,17}$

Figure 14. *Cont.*

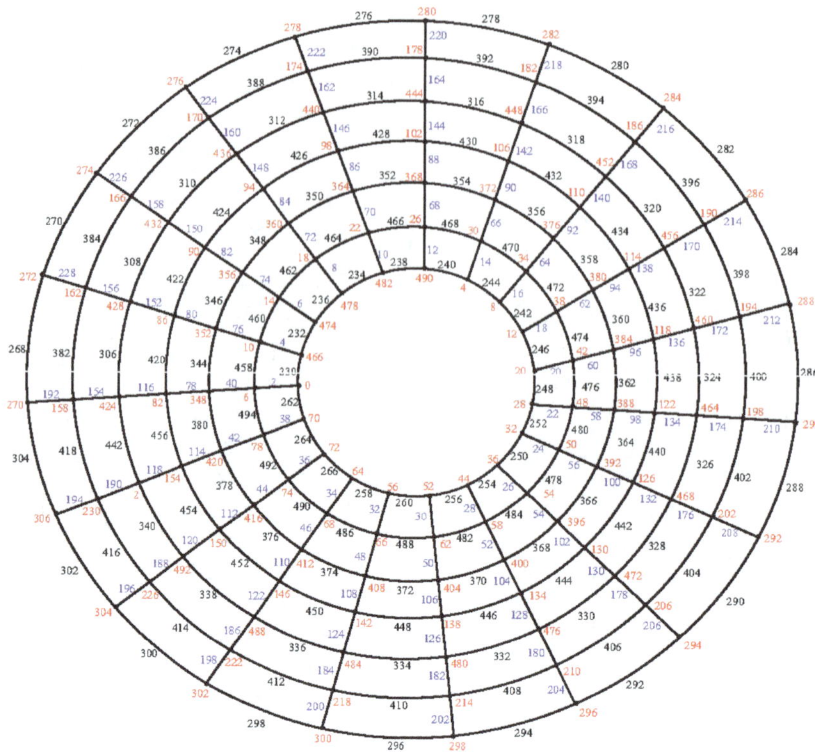

(g) $C_{7,19}$

Figure 14. An e.e.g.l. of the cylinder grid graphs $C_{9,3}, C_{7,9}, C_{7,11}, C_{7,13}, C_{7,15}, C_{7,17}$ and $C_{7,19}$.

3. Conclusions

In this paper, using the connection of labeling of graphs with modular arithmetic and theory of numbers in general, we give a detailed study for e.e.g., l. of all cases of members of the cylinder grid graphs. The study of necessary and sufficient conditions for e.e.g., l. of other important families including torus $C_m \times C_n$ and rectangular $P_m \times P_n$ grid graphs should be taken into consideration in future studies of e.e.g., l.

Author Contributions: All authors contributed equally to this work.

Acknowledgments: The authors are grateful to the anonymous reviewers for their helpful comments and suggestions for improving the original version of the paper.

References

1. Acharya, B.D.; Arumugam, S.; Rosa, A. *Labeling of Discrete Structures and Applications*; Narosa Publishing House: New Delhi, India, 2008; pp. 1–14.
2. Bloom, G.S. Numbered Undirected Graphs and Their Uses, a Survey of a Unifying Scientific and Engineering Concept and Its Use in Developing a Theory of Non-Redundant Homometric Sets Relating to Some Ambiguities in X-ray Diffraction Analysis. Ph.D. Thesis, University of Southern California, Los Angeles, CA, USA, 1975.
3. Bloom, G.S.; Golomb, S.W. Numbered complete graphs, unusual rulers, and assorted applications. In *Theory and Applications of Graphs, Lecture Notes in Math, 642*; Springer: New York, NY, USA, 1978; pp. 53–65.
4. Bloom, G.S.; Golomb, S.W. Applications of numbered undirected graphs. *Proc. IEEE* **1977**, *65*, 562–570. [CrossRef]
5. Bloom, G.S.; Hsu, D.F. On graceful digraphs and a problem in network addressing. *Congr. Numer.* **1982**, *35*, 91–103.
6. Graham, R.L.; Pollak, H.O. On the addressing problem for loop switching. *Bell Syst. Tech. J.* **1971**, *50*, 2495–2519. [CrossRef]

7. Sutton, M.; Labellings, S.G. Summable Graphs Labellings and Their Applications. Ph.D. Thesis, the University of Newcastle, New South Wales, Australia, 2001.
8. Shang, Y. More on the normalized Laplacian Estrada index. *Appl. Anal. Discret. Math.* **2014**, *8*, 346–357. [CrossRef]
9. Shang, Y. Geometric assortative growth model for small-world networks. *Sci. World J.* **2014**, *2014*, 1–8. [CrossRef] [PubMed]
10. Shang, Y. Deffuant model of opinion formation in one-dimensional multiplex networks. *J. Phys. A Math. Theor.* **2015**, *48*, 395101. [CrossRef]
11. Gross, J.; Yellen, J. *Graph Theory and Its Applications*; CRC Press: Boca Raton, FL, USA, 1999.
12. Rosa, A. On certain valuations of the vertices of a graph. In *Theory of Graphs, Proceedings of the International Symposium, Rome, Italy, July 1966*; Gordan and Breach, Dunod: New York, NY, USA, 1967; pp. 349–355.
13. Golomb, S.W. How to Number a Graph. In *Graph Theory and Computing*; Read, R.C., Ed.; Cademic Press: New York, NY, USA, 1972; pp. 23–37.
14. Gnanajothi, R.B. Topics in Graph Theory. Ph.D. Thesis, Madurai Kamaraj University, Tamil Nadu, India, 1991.
15. Seoud, M.A.; Abdel-Aal, M.E. On odd graceful graphs. *Ars Comb.* **2013**, *108*, 161–185.
16. Gao, Z. Odd graceful labelings of some union graphs. *J. Nat. Sci. Heilongjiang Univ.* **2007**, *24*, 35–39.
17. Lo, S.P. On edge-gracefull1abelings of graphs. *Congr. Numer.* **1985**, *50*, 231–241.
18. Kuan, Q.; Lee, S.; Mitchem, J.; Wang, A. On edge-graceful unicyclic graphs. *Congr. Numer.* **1988**, *61*, 65–74.
19. Lee, L.; Lee, S.; Murty, G. On edge-graceful labelings of complete graphs: Solutions of Lo's conjecture. *Congr. Numer.* **1988**, *62*, 225–233.
20. Solairaju, A.; Chithra, K. Edge-odd graceful graphs. *Electron. Notes Discret. Math.* **2009**, *33*, 15–20. [CrossRef]
21. Daoud, S.N. Edge odd graceful labeling of some path and cycle related graphs. *AKCE Int. J. Graphs Comb.* **2017**, *14*, 178–203. [CrossRef]
22. Daoud, S.N. Edge odd graceful labeling cylinder grid and torus grid graphs. *IEEE Access* **2019**, *7*, 10568–10592. [CrossRef]
23. Daoud, S.N. Vertex odd graceful labeling. *Ars Comb.* **2019**, *142*, 65–87.
24. Elsonbaty, A.; Daoud, S.N. Edge even graceful labeling of some path and cycle related graphs. *Ars Comb.* **2017**, *130*, 79–96.
25. Daoud, S.N. Edge even graceful labeling polar grid graph. *Symmetry* **2019**, *11*, 38. [CrossRef]
26. Gallian, J.A. A dynamic survey of graph labeling. *Electron. J. Comb.* **2017**, *22*, #DS6.

11

Binary Icosahedral Group and 600-Cell

Jihyun Choi and Jae-Hyouk Lee *

Department of Mathematics, Ewha Womans University 52, Ewhayeodae-gil, Seodaemun-gu, Seoul 03760, Korea; katherinejh.choi@gmail.com

* Correspondence: jaehyoukl@ewha.ac.kr.

Abstract: In this article, we have an explicit description of the binary isosahedral group as a 600-cell. We introduce a method to construct binary polyhedral groups as a subset of quaternions $\mathbb{H}$ via spin map of $SO(3)$. In addition, we show that the binary icosahedral group in $\mathbb{H}$ is the set of vertices of a 600-cell by applying the Coxeter–Dynkin diagram of H_4.

Keywords: binary polyhedral group; icosahedron; dodecahedron; 600-cell

MSC: 52B10; 52B11; 52B15

---

1. Introduction

The classification of finite subgroups in $SL_n(\mathbb{C})$ derives attention from various research areas in mathematics. Especially when $n = 2$, it is related to McKay correspondence and ADE singularity theory [1].

The list of finite subgroups of $SL_2(\mathbb{C})$ consists of cyclic groups ($\mathbb{Z}_n$), binary dihedral groups corresponded to the symmetry group of regular $2n$-gons, and binary polyhedral groups related to regular polyhedra. These are related to the classification of regular polyhedrons known as Platonic solids. There are five platonic solids (tetrahedron, cubic, octahedron, dodecahedron, icosahedron), but, as a regular polyhedron and its dual polyhedron are associated with the same symmetry groups, there are only three binary polyhedral groups (binary tetrahedral group $2T$, binary octahedral group $2O$, binary icosahedral group $2I$) related to regular polyhedrons. Moreover, it is a well-known fact that there is a correspondence between binary polyhedral groups and vertices of 4-polytopes as follows:

$$\begin{array}{lcl} 2T & \leftrightarrow & \text{vertices of 24-cell,} \\ 2O & \leftrightarrow & \text{vertices of dual compound of 24-cell,} \\ 2I & \leftrightarrow & \text{vertices of 600-cell,} \end{array}$$

where the dual compound of 24-cell means by the compound polytopes obtained from 24-cell and its dual polytope, which is also a 24-cell [2–5].

As the symmetries of polyhedrons are isometries, the related finite subgroups are also considered as the subgroups of $SU(2)$. As $SU(2) = Sp(1)$ is a spin group of $SO(3)$, we can regard $2T$, $2O$, and $2I$ as subgroups of quaternions $\mathbb{H}$. From this point of view, it is also well known that the vertices of 24 cell correspond to roots of D_4, and the set of vertices of the dual compound of 24-cell, which is the union of a 24-cell and a dual 24-cell forms a roots of F_4. The 600-cell is a complicated case of a reflection group of H_4-type [3,6].

The aim of this article is to provide explicit description of a binary icosahedron group $2I$ as a 600-cell. By applying spin covering map from $Sp(1)$ to $SO(3)$, we introduce a method to construct the binary polyhedral groups in terms of quaternions from the symmetries of regular polyhedrons. Then, by applying the theory of reflection groups along the Coxeter–Dynkin diagram, we show that the subgroup $2I$ in $\mathbb{H}$ is indeed the set of vertices of a 600-cell. We also discuss $2T$ related to 24-cell, but, because the dual compound of a 24-cell is not regular, its relation to $2O$ will be discussed in another article.

2. Binary Polyhedral Groups in Quaternions

Every finite subgroup of $SL_2(\mathbb{C})$ is conjugate to a finite subgroup of $SU(2)$ so that the classification of the finite subgroup of $SL_2(\mathbb{C})$ including binary polyhedral groups corresponds to the classification of the finite subgroup of $SU(2)$. As $SU(2) \simeq Sp(1)$, we can identify the binary polyhedral groups as certain subsets in quaternions $\mathbb{H}$. In fact, $Sp(1)$ is not only a unit sphere in $\mathbb{H}$ but also the spin group $Spin(3)$, which is a 2-covering map of $SO(3)$. In this section, we explain how an element in $SO(3)$ lifts to quaternions in $Sp(1)$.

The algebra of quaternions $\mathbb{H}$ is the four-dimensional vector space over $\mathbb{R}$ defined by

$$\mathbb{H} := \{a + bi + cj + dk \mid a, b, c, d \in \mathbb{R}\}$$

satisfying relations $i^2 = j^2 = k^2 = ijk = -1$. The quaternionic conjugate of $q = a + bi + cj + dk$ is defined by

$$\bar{q} := a - bi - cj - dk$$

and the corresponding norm $|q|$ is also defined by $|q| := \sqrt{q\bar{q}} = \sqrt{a^2 + b^2 + c^2 + d^2}$. Along this norm, quaternions satisfy $|pg| = |p|\,|g|$, which implies that it is one of the normed algebras whose classification consists of real numbers $\mathbb{R}$, complex numbers $\mathbb{C}$, quaternions $\mathbb{H}$, and octonions $\mathbb{O}$. A quaternion q is called *real* if $\bar{q} = q$ and is called *imaginary* if $\bar{q} = -q$. According to these facts, we can divide $\mathbb{H}$ into a real part and an imaginary part:

$$\mathbb{H} \simeq \mathbb{R}^4 \simeq Re(\mathbb{H}) \oplus Im(\mathbb{H}) = \mathbb{R} \oplus \mathbb{R}^3.$$

It is well known that the set of unit sphere $S^3 = \{q \in \mathbb{H} \mid |q| = 1\}$ in $\mathbb{H}$ is a Lie group $Sp(1)$, which is also isomorphic to $SU(2)$ as follows:

$$\begin{array}{ccc} Sp(1) & \simeq & SU(2) = \left\{ \begin{pmatrix} a & b \\ -\bar{b} & \bar{a} \end{pmatrix} \middle| \; a, b \in \mathbb{C}, |a|^2 + |b|^2 = 1 \right\}, \\ q = a + bj & \longleftrightarrow & \begin{pmatrix} a & b \\ -\bar{b} & \bar{a} \end{pmatrix}. \end{array}$$

Below, we use the identification between $\mathbb{R}^3$ and $Im(\mathbb{H})$. Along this, a vector $v = (v_1, v_2, v_3)$ in $\mathbb{R}^3$ (resp. a quaternion $q = a_1 i + a_2 j + a_3 k$ in $Im(\mathbb{H})$) is corresponded to a quaternion $(v)^{\#} = v_1 i + v_2 j + v_3 k$ in $Im(\mathbb{H})$ (resp. a vector $\vec{q} = (a_1, a_2, a_3)$ in $\mathbb{R}^3$).

Now, we define a map Φ,which is given by an action of $Sp(1)$ on $\mathrm{Im}(\mathbb{H}) \cong \mathbb{R}^3$

$$\begin{array}{rcccccc} \Phi: & Sp(1) & \rightarrow & SO(3), & & & \\ & x & \mapsto & \Phi(x) := \rho_x & : \mathrm{Im}(\mathbb{H}) & \rightarrow & \mathrm{Im}(\mathbb{H}), \\ & & & & v & \mapsto & xv\bar{x}. \end{array}$$

As a matter of fact, the map ρ_x must be defined as

$$\begin{array}{rccc} \rho_x & : \mathbb{R}^3 & \rightarrow & \mathbb{R}^3 \\ & v & \mapsto & \overrightarrow{(xv^{\#}\bar{x})} \end{array}$$

so that ρ_x is in $SO(3)$. However, we use a simpler definition instead. It is well known that Φ is a 2-covering map, which is also a group homomorphism.

In the next section, we will consider the preimage of Φ to define the lifting of symmetry groups of polyhedrons in $\mathbb{R}^3$, which are subgroups of $SO(3)$. For this purpose, we consider ρ_x further in below.

We observe that the multiplication of two pure quaternions p, q in $Im(\mathbb{H})$ can be written by the cross product $\times$ and the standard inner product $\cdot$ on $\mathbb{R}^3$:

$$pq = -\vec{p} \cdot \vec{q} + (\vec{p} \times \vec{q})^{\#}.$$

After we denote $x = x_0 + x_+$ ($x_0 \in \mathrm{Re}(\mathbb{H})$, $x_+ \in \mathrm{Im}(\mathbb{H})$), $\rho_x(v)$ can be written as

$$\begin{aligned}\rho_x(v) &= xv\bar{x} = (x_0 + x_+)v(x_0 - x_+)\\ &= \left(x_0 v - \vec{v} \cdot \overrightarrow{x_+} + (\overrightarrow{x_+} \times \vec{v})^{\#}\right)(x_0 - x_+)\\ &= (x_0^2 - |\overrightarrow{x_+}|^2)v + 2\left(\overrightarrow{x_+} \cdot \overrightarrow{v}\right)x_+ + 2x_0\left(\overrightarrow{x_+} \times \vec{v}\right)^{\#}.\end{aligned}$$

Here, since $|x|^2 = x_0^2 + |\overrightarrow{x_+}|^2$, we denote $x = \cos\theta + \sin\theta \frac{x_+}{|x_+|}$ where $\cos\theta = x_0$ and $\sin\theta = |\overrightarrow{x_+}|^2$ for some $\theta \in [0, \pi)$.

Now, to understand the meaning of $\rho_x(v)$, we consider two cases for $\overrightarrow{v}$ case (1) $\overrightarrow{v} \perp \overrightarrow{x_+}$ and case (2) $\overrightarrow{v} // \overrightarrow{x_+}$.

(1) (Case $\overrightarrow{v} \perp \overrightarrow{x_+}$) Since $\overrightarrow{v} \cdot \overrightarrow{x_+} = 0$, we have

$$\begin{aligned}\rho_x(v) &= (x_0^2 - |\overrightarrow{x_+}|^2)v + 2x_0\left(\overrightarrow{x_+} \times \vec{v}\right)^{\#}\\ &= (x_0^2 - |\overrightarrow{x_+}|^2)v + 2x_0|\overrightarrow{x_+}|\left(\frac{\overrightarrow{x_+}}{|\overrightarrow{x_+}|} \times \overrightarrow{v}\right)^{\#}\\ &= \cos 2\theta\ v\ + \sin 2\theta\ \left(\frac{\overrightarrow{x_+}}{|\overrightarrow{x_+}|} \times \overrightarrow{v}\right)^{\#}.\end{aligned}$$

(2) (Case $\overrightarrow{v} // \overrightarrow{x_+}$) After we denote $\overrightarrow{v} = t\,\overrightarrow{x_+}$ for some $t \in \mathbb{R}$,

$$\begin{aligned}\rho_x(v) &= (x_0^2 - |\overrightarrow{x_+}|^2)tx_+ + 2\left(\overrightarrow{x_+} \cdot t\,\overrightarrow{x_+}\right)x_+\\ &= \left(t\left(x_0^2 - |\overrightarrow{x_+}|^2\right) + 2\left(\overrightarrow{x_+} \cdot t\,\overrightarrow{x_+}\right)\right)x_+\\ &= t\left(x_0^2 + |\overrightarrow{x_+}|^2\right)x_+ = tx_+ = v.\end{aligned}$$

By the two cases above, we conclude $\rho_x(v)$ presents the rotation of vector $\overrightarrow{v}$ in $\mathbb{R}^3$ with respect to the axis $\overrightarrow{x_+}$ by $2\cos^{-1}\ x_0 \in [0, 2\pi)$.

By applying the above, we have the following lemma.

Lemma 1. *For each element A in $SO(3)$ presenting a rotation with respect to a unit vector $\vec{a}$ for angle $\alpha \in [0, 2\pi)$, the preimage of $\Phi : Sp(1) \to SO(3)$ is given as*

$$\Phi^{-1}(A) = \left\{\pm\left(\cos\frac{\alpha}{2} + \sin\frac{\alpha}{2}\left(\vec{a}\right)^{\#}\right)\right\} \subset \mathbb{H}.$$

Note if we choose unit vector $-\vec{a}$ instead $\vec{a}$ in the lemma, then the rotation performed for angle $-\alpha$. Therefore, the related lifting is given by $\cos\left(-\frac{\alpha}{2}\right) + \sin\left(-\frac{\alpha}{2}\right)\left(-\vec{a}\right)^{\#} = \cos\frac{\alpha}{2} + \sin\frac{\alpha}{2}\left(\vec{a}\right)^{\#}$. Hence $\Phi^{-1}(A)$ is well defined.

By applying Lemma 1, we can consider the preimage of any subset G in $SO(3)$. We call the preimage $\Phi^{-1}(G)$ the *lift* of G in $Sp(1) \subset \mathbb{H}$. When G is one of the symmetry groups of regular polyhedrons, the lift $\Phi^{-1}(G)$ is called *binary polyhedral group*. In particular, we consider *binary tetrahedral group 2T*, *binary octahedral group 2O*, and *binary icosahedral group 2I* , which are lifts of symmetry groups of tetrahedron, octahedron and icosahedron with order $24, 48, 120$, respectively.

2.1. *Symmetry Groups of Regular Polyhedrons*

A polyhedron considered here is convex and regular. According to convention, we denote a regular polyhedron by $\{p, q\}$, which means that the polyhedron has only one type of face which is a p-gon, and each vertex is contained in q faces. It is well known that there are only five regular polyhedrons, which are also called *Platonic solids*. Up to duality, we consider three classes of regular polyhedrons as follows:

$$\begin{array}{ccc} \text{Tetrahedron} & \leftrightarrow & \text{Tetrahedron(self-dual)}, \\ \text{Octahedron} & \leftrightarrow & \text{Cube}, \\ \text{Icosahedron} & \leftrightarrow & \text{Dodecahedron}. \end{array}$$

As the special linear group $SO(3)$ is generated by rotations on $\mathbb{R}^3$, we consider rotations of $\mathbb{R}^3$ preserving a regular polyhedron to study the symmetry group of it. When the axis of the rotation crosses a vertex (the barycenter of an edge, the barycenter or a face resp.), we call the rotation *vertex symmetry* (*edge symmetry*, *face symmetry* resp.). For instance, the tetrahedron has two different types of axes of rotations. One is the line passing through a vertex and the barycenter of the opposite face, and the other is the line connecting barycenters of the edges at the opposite position. We also say that a symmetry has *order* n if the order of the corresponding rotation is n. Note that the order of each edge symmetry is 2. One can figure all the possible orders of each type of symmetry for regular polyhedrons, as shown in Table 1.

Table 1. Order of symmetries.

Polyhedron	**Tetrahedron**	**Octahedron**	**Icosahedron**
Point Symmetry	3	4, 2	5
Edge Symmetry	2	2	2
Face Symmetry	3	3	3

Construction of binary polyhedral groups

Now, we will introduce a construction which provides a way to find the elements of binary polyhedral groups related to regular polyhedrons when the set of vertices of regular polyhedrons are given.

Assume we have a regular polyhedron $\{p, q\}$ whose barycenter is the origin of $\mathbb{R}^3$ and let $\{P_i\}$ be the set of vertices of the regular polyhedron:

(1) Find the barycenters of vertices, edges and faces (The barycenter of each vertex is itself).
(2) For each barycenter, derive all the related symmetries in $SO(3)$ by identifying corresponding axis of rotations and its order.
(3) For each symmetry obtained from step 2, we get related lifts in $\mathbb{H}$ by applying Lemma 1. It is useful to observe that we obtain the axis of rotation and its order instead of related angle where there can be more than one related angle.
(4) The union of lifts is a subset of binary polyhedral groups. In fact, its union with $\{\pm 1\}$ is the binary polyhedral group by counting elements.

Note: From the above, it is clear that two regular polyhedrons which are dual to each other are associated with the same binary polyhedral groups.

For example, let B be a barycenter of order 3. Then, there are two related angles $\frac{\pi}{3}$ and $\frac{2\pi}{3}$. Thus, the corresponding lift is

$$\left\{\pm\left(\cos\frac{\pi}{3} + \sin\frac{\pi}{3}\frac{B}{|B|}\right), \pm\left(\cos\frac{2\pi}{3} + \sin\frac{2\pi}{3}\frac{B}{|B|}\right)\right\}.$$

Here, we also observe that, if we begin with $-B$ instead of B, the corresponding lifts are still the same.

Binary tetrahedral group

We consider a tetrahedron consisting of vertices $\{P_1, P_2, P_3, P_4\}$.

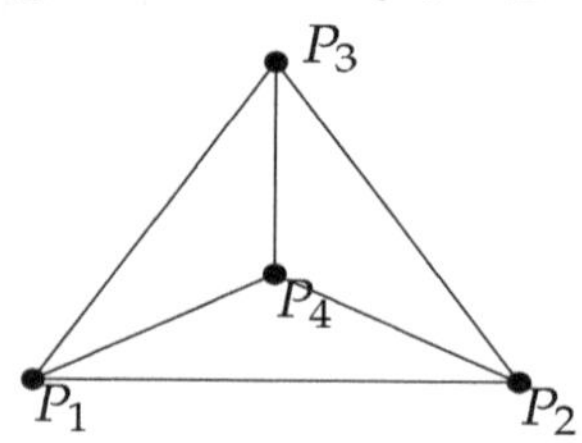

Tetrahedron

(1) Since vertex symmetry has order 3, each vertex symmetry has two angles $\frac{\pi}{3}$ and $\frac{2\pi}{3}$ and the union of lifts of vertex symmetries is

$$V_T := \bigcup_i \left\{ \pm \left(\cos\frac{\pi}{3} + \sin\frac{\pi}{3}\frac{P_i}{|P_i|} \right), \pm \left(\cos\frac{2\pi}{3} + \sin\frac{2\pi}{3}\frac{P_i}{|P_i|} \right) \right\}.$$

Thus, we have

$$|V_T| = |\text{vertices of a tetrahedron}| \times 4 = 16.$$

(2) As the edge symmetry has order 2, each edge symmetry has only one angle $\frac{\pi}{2}$ so that the related lift is

$$\left\{ \pm \left(\cos\frac{\pi}{2} + \sin\frac{\pi}{2} \frac{\frac{P_i + P_j}{2}}{\left|\frac{P_i + P_j}{2}\right|} \right) \right\}$$

and the union is given as

$$E_T := \bigcup_{i \neq j} \left\{ \pm \left(\frac{P_i + P_j}{|P_i + P_j|} \right) \right\}.$$

Since $P_1 + P_2 + P_3 + P_4 = 0$, two barycenters $\frac{1}{2}(P_1 + P_2)$ and $\frac{1}{2}(P_3 + P_4)$ of edges have the same lifts of edge symmetries. Similarly, the pairs of edges produce the same lifts of edge symmetries, and we get

$$|E_T| = \frac{|\text{edges of a tetrahedron}|}{2} \times 2 = 6.$$

(3) For a barycenter of face consisting of $\{P_1, P_2, P_3\}$, we have a relation

$$\frac{P_1 + P_2 + P_3}{3} = -\frac{P_4}{3}$$

since $P_1 + P_2 + P_3 + P_4 = 0$. Thus, the related lift of face symmetry is the same as the lift of vertex lift for a vertex P_4. Similarly, each lift of face symmetry corresponds to the lift of vertex symmetry.

Finally, the union $V_T \cup E_T$ of lifts of symmetries of a tetrahedron is a subset binary tetrahedral group $2T$ in $Sp(1)$. Since

$$|V_T \cup E_T \cup \{\pm 1\}| = 16 + 6 + 2 = 24 = |2T|,$$

the union $V_T \cup E_T \cup \{\pm 1\}$ is a binary tetrahedral group, namely

$$2T = V_T \cup E_T \cup \{\pm 1\}.$$

If we choose vertices $\{P_1, P_2, P_3, P_4\}$ of a tetrahedron as

$$\left\{P_1 = \frac{i+j+k}{2}, P_2 = \frac{i-j-k}{2}, P_3 = \frac{-i+j-k}{2}, P_4 = \frac{-i-j+k}{2}\right\},$$

the corresponding binary tetrahedral group is obtained as

$$2\hat{T} =: \left\{\pm 1,\ \pm i,\ \pm j,\ \pm k,\ \frac{1}{2}(\pm 1 \pm i \pm j \pm k)\right\}.$$

Remark: The subset $2\hat{T}$ is the unit integral quaternions which is also known as *Hurwitz integral quaternions*. ([7,8])

Binary Octahedral Group

We consider an octahedron consisting of vertices $\{P_i\ i = 1, ..., 8\}$ as below.

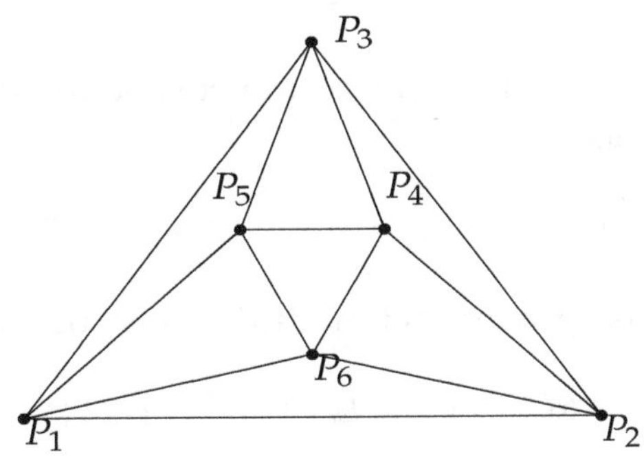

Octahedron

(1) The possible orders of vertex symmetry are 2 and 4. The vertex symmetry with order 4 has two angles $\frac{\pi}{4}$ and $\frac{3\pi}{4}$ and one with order 2 has one angle $\frac{\pi}{2}$. Thus, the union of lifts of vertex symmetries is

$$V_O := \bigcup_i \left\{\pm\left(\cos\frac{\pi}{4} + \sin\frac{\pi}{4}\frac{P_i}{|P_i|}\right),\ \pm\left(\cos\frac{3\pi}{4} + \sin\frac{3\pi}{4}\frac{P_i}{|P_i|}\right),\ \pm\left(\cos\frac{\pi}{2} + \sin\frac{\pi}{2}\frac{P_i}{|P_i|}\right)\right\}.$$

Since two antipodal vertices produce the same lifts of vertex symmetries, we obtain that

$$|V_O| = \frac{|\text{vertices of an octahedron}|}{2} \times 6 = 18.$$

(2) As the edge symmetry has order 2, each edge symmetry has only one angle $\frac{\pi}{2}$ so that the union of the related lift is

$$E_O := \bigcup \left\{\pm\left(\frac{P_i + P_j}{|P_i + P_j|}\right)\right\},$$

where the union is performed for all the pairs of P_i and P_j form an edge. For the barycenter of an edge given by $\frac{1}{2}(P_i + P_j)$, there is exactly one edge whose barycenter is antipodal to $\frac{1}{2}(P_i + P_j)$. Moreover, the pair of edges produce the same lifts of edge symmetries. Therefore, we get

$$|E_O| = \frac{|\text{edges of an octahedron}|}{2} \times 2 = 12.$$

(3) For a barycenter of face given as $\frac{P_i+P_j+P_k}{3}$, the face symmetry has order 3 and it is related to two angles $\frac{\pi}{3}$ and $\frac{2\pi}{3}$. Thus, the lifts of a face symmetry is

$$\left\{\pm\left(\cos\frac{\pi}{3}+\sin\frac{\pi}{3}\frac{\frac{P_i+P_j+P_k}{3}}{\left|\frac{P_i+P_j+P_k}{3}\right|}\right), \pm\left(\cos\frac{2\pi}{3}+\sin\frac{2\pi}{3}\frac{\frac{P_i+P_j+P_k}{3}}{\left|\frac{P_i+P_j+P_k}{3}\right|}\right)\right\}$$

and the union of lifting of face symmetries is

$$F_O := \bigcup\left\{\pm\left(\cos\frac{\pi}{3}+\sin\frac{\pi}{3}\frac{P_i+P_j+P_k}{|P_i+P_j+P_k|}\right), \pm\left(\cos\frac{2\pi}{3}+\sin\frac{2\pi}{3}\frac{P_i+P_j+P_k}{|P_i+P_j+P_k|}\right)\right\}.$$

Since the octahedron is symmetric for origin, for the barycenter of a face given by $\frac{1}{3}(P_i+P_j+P_k)$, there is exactly one face whose barycenter is antipodal to $\frac{1}{3}(P_i+P_j+P_k)$, and the pair of faces produce the same lifts of face symmetries. Therefore, we deduce

$$|F_O| = \frac{|\text{faces of an octahedron}|}{2} \times 4 = 16.$$

Finally, the union $V_O \cup E_O \cup F_O$ of lifts of symmetries of an octahedron is a subset of the binary octahedral group $2O$ in $Sp(1)$. Since

$$|V_O \cup E_O \cup F_O \cup \{\pm 1\}| = 18+12+16+2 = 48 = |2O|,$$

the union $V_O \cup E_O \cup F_O \cup \{\pm 1\}$ is a binary octahedral group, namely

$$2O = V_O \cup E_O \cup F_O \cup \{\pm 1\}.$$

One can take P_is as follows:

$$P_1 = i, P_2 = j, P_3 = k, P_4 = -i, P_5 = -j, P_6 = -k$$

so as to obtain

$$2\hat{O} := \left\{\begin{array}{c} \pm 1,\ \pm i,\ \pm j,\ \pm k,\ \frac{1}{2}(\pm 1 \pm i \pm j \pm k), \\ \frac{1}{\sqrt{2}}(\pm 1 \pm i), \frac{1}{\sqrt{2}}(\pm 1 \pm j), \frac{1}{\sqrt{2}}(\pm 1 \pm k), \frac{1}{\sqrt{2}}(\pm i \pm j), \frac{1}{\sqrt{2}}(\pm i \pm k), \frac{1}{\sqrt{2}}(\pm j \pm k) \end{array}\right\}.$$

Binary Icosahedral Group

Since both the regular icosahedron and its dual regular dodecahedron produce the binary icosahedral group, we consider a regular dodecahedron in $\mathbb{R}^3$ instead of a regular icosahedron. Moreover, for the sake of convenience, one can choose specific coordinates of vertices of a dodecahedron in $\mathbb{R}^3$ such as

$$\left\{(\pm 1, \pm 1, \pm 1),\ (\pm\tau, \pm\frac{1}{\tau}, 0),\ (0, \pm\tau, \pm\frac{1}{\tau}),\ (\pm\frac{1}{\tau}, 0, \pm\tau)\right\},$$

where $\tau = \frac{\sqrt{5}+1}{2} = 2\cos\frac{\pi}{5}$ and $\frac{1}{\tau} = \frac{\sqrt{5}-1}{2} = -2\cos\frac{2\pi}{5}$. It is also useful to know $\sin\frac{\pi}{5} = \frac{\sqrt{10-2\sqrt{5}}}{4}$ and $\sin\frac{2\pi}{5} = \frac{\sqrt{10+2\sqrt{5}}}{4}$.

In the following diagram, we consider the given set of vertices as a subset in $\operatorname{Im}\mathbb{H} = \mathbb{R}^3$ and depict the configuration among the vertices.

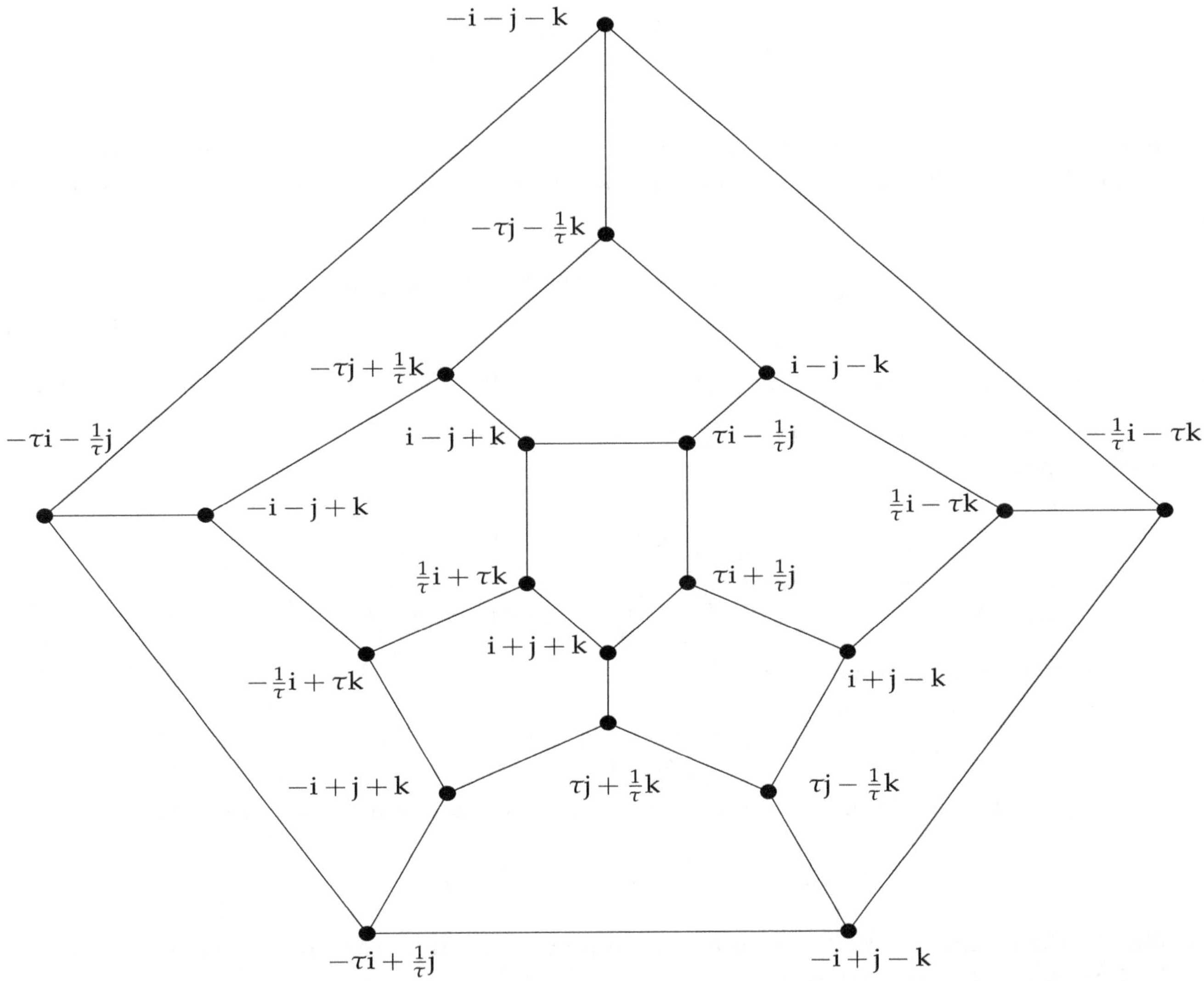

dodecahedron

For the above dodecahedron, we denote the set of vertices as $\{P_i \; i = 1, \dots, 20\}$ without a specific choice of order.

(1) Since the possible order of each vertex symmetry is 3, the vertex symmetry has two angles $\frac{\pi}{3}$ and $\frac{2\pi}{3}$. Thus, the union of lifts of vertex symmetries is

$$V_I := \bigcup_i \left\{ \pm\left(\cos\frac{\pi}{3} + \sin\frac{\pi}{3}\frac{P_i}{|P_i|}\right), \pm\left(\cos\frac{2\pi}{3} + \sin\frac{2\pi}{3}\frac{P_i}{|P_i|}\right) \right\}.$$

Since the dodecahedron is symmetric for origin, each vertex and its antipodal vertex produce the same lifts of vertex symmetries. Thus, we obtain that

$$|V_I| = \frac{|\text{vertices of a dodecahedron}|}{2} \times 4 = 40.$$

(2) As before, the edge symmetry has order 2, and each edge symmetry has only one angle $\frac{\pi}{2}$ so that the union of the related lifts is

$$E_I := \bigcup \left\{ \pm \left(\frac{P_i + P_j}{|P_i + P_j|} \right) \right\},$$

where the union is performed for all the pairs of P_i and P_j form an edge. Just like the lifts of edge symmetries for an octahedron, the pair of antipodal edges produce the same lifts of edge symmetries. Therefore, we get

$$|E_I| = \frac{|\text{edges of a dodecahedron}|}{2} \times 2 = 30.$$

(3) For a barycenter of face given as $\frac{1}{5}\left(P_{i_1} + P_{i_2} + P_{i_3} + P_{i_4} + P_{i_5}\right)$ where P_{i_l} $(l = 1,2,3,4,5)$ forms a face, the face symmetry has order 5 and it is related to four angles $\frac{a\pi}{5}$ $(a = 1,2,3,4)$. Thus, the lifts of a face symmetry are

$$\left\{ \pm \left(\cos\frac{a\pi}{5} + \sin\frac{a\pi}{5} \frac{\frac{1}{5}\left(P_{i_1} + P_{i_2} + P_{i_3} + P_{i_4} + P_{i_5}\right)}{\left|\frac{1}{5}\left(P_{i_1} + P_{i_2} + P_{i_3} + P_{i_4} + P_{i_5}\right)\right|} \right) (a = 1,2,3,4) \right\}$$

and the union of lifts of face symmetries is

$$F_I := \bigcup \left\{ \pm \left(\cos\frac{a\pi}{5} + \sin\frac{a\pi}{5} \frac{P_{i_1} + P_{i_2} + P_{i_3} + P_{i_4} + P_{i_5}}{|P_{i_1} + P_{i_2} + P_{i_3} + P_{i_4} + P_{i_5}|} \right) (a = 1,2,3,4) \right\}.$$

Since a pair of antipodal faces produce the same lifts of face symmetries, we deduce

$$|F_I| = \frac{|\text{faces of a dodecahedron}|}{2} \times 8 = 48.$$

Finally, the union $V_I \cup E_I \cup F_I$ of lifts of symmetries of a dodecahedron is a subset of a binary icosahedral group $2I$ in $Sp(1)$. Since

$$|V_I \cup E_I \ \cup \ F_I \cup \{\pm 1\}| = 40 + 30 + 48 + 2 = 120 = |2I|,$$

the union $V_I \cup E_I \cup F_I \cup \{\pm 1\}$ is a binary icosahedral group, namely

$$2I = V_I \cup E_I \ \cup \ F_I \cup \{\pm 1\}.$$

For the given vertices we have, we can obtain

$$\begin{aligned} V_I &= \left\{ \tfrac{1}{2}(\pm 1 \pm i \pm j \pm k), \tfrac{1}{2}\left(\pm 1 \pm \tau j \pm \tfrac{1}{\tau}k\right), \tfrac{1}{2}\left(\pm 1 \pm \tfrac{1}{\tau}i \pm \tau k\right), \tfrac{1}{2}\left(\pm 1 \pm \tau i \pm \tfrac{1}{\tau}j\right) \right\} \\ E_I &= \left\{ \pm i, \pm j, \pm k, \tfrac{1}{2}\left(\pm i \pm \tfrac{1}{\tau}j \pm \tau k\right), \tfrac{1}{2}\left(\pm \tfrac{1}{\tau}i \pm \tau j \pm k\right), \tfrac{1}{2}\left(\pm \tau i \pm j \pm \tfrac{1}{\tau}k\right) \right\} \\ F_I &= \left\{ \begin{array}{l} \tfrac{1}{2}\left(\pm \tfrac{1}{\tau} \pm \tau i \pm k\right), \tfrac{1}{2}\left(\pm \tau \pm \tfrac{1}{\tau}i \pm j\right), \tfrac{1}{2}\left(\pm \tau \pm \tfrac{1}{\tau}j \pm k\right), \\ \tfrac{1}{2}\left(\pm \tau \pm i \pm \tfrac{1}{\tau}k\right), \tfrac{1}{2}\left(\pm \tfrac{1}{\tau} \pm i \pm \tau j\right), \tfrac{1}{2}\left(\pm \tfrac{1}{\tau} \pm j \pm \tau k\right) \end{array} \right\}. \end{aligned}$$

As a result, we can identify all the elements of the binary icosahedral group as

$$2\hat{I} := \left\{ \begin{array}{c} \pm 1,\ \pm i,\ \pm j,\ \pm k,\ \frac{1}{2}(\pm 1 \pm i \pm j \pm k) \\ \frac{1}{2}\left(\pm i \pm \frac{1}{\tau} j \pm \tau k\right),\ \frac{1}{2}\left(\pm 1 \pm \tau j \pm \frac{1}{\tau} k\right),\ \frac{1}{2}\left(\pm \frac{1}{\tau} \pm \tau i \pm k\right),\ \frac{1}{2}\left(\pm \tau \pm \frac{1}{\tau} i \pm j\right), \\ \frac{1}{2}\left(\pm \frac{1}{\tau} i \pm \tau j \pm k\right),\ \frac{1}{2}\left(\pm \tau \pm \frac{1}{\tau} j \pm k\right),\ \frac{1}{2}\left(\pm \tau \pm i \pm \frac{1}{\tau} k\right),\ \frac{1}{2}\left(\pm \frac{1}{\tau} \pm i \pm \tau j\right), \\ \frac{1}{2}\left(\pm \tau i \pm j \pm \frac{1}{\tau} k\right),\ \frac{1}{2}\left(\pm \frac{1}{\tau} \pm j \pm \tau k\right),\ \frac{1}{2}\left(\pm 1 \pm \frac{1}{\tau} i \pm \tau k\right),\ \frac{1}{2}\left(\pm 1 \pm \tau i \pm \frac{1}{\tau} j\right) \end{array} \right\}.$$

Theorem 1. *The finite subsets $2\hat{T}$, $2\hat{O}$ and $2\hat{I}$ in $\mathbb{H}$ defined as above are a binary tetrahedral group, a binary octahedral group, and a binary icosahedral group, respectively.*

Note that it is well known that the a subset $\left\{\pm 1, \pm i, \pm j, \pm k, \frac{1}{2}(\pm 1 \pm i \pm j \pm k)\right\}$ in $2I$ is the vertices of 24-cell and the complementary subset in $2I$ is the vertices of a snub 24-cell.

3. 600-Cell

The Coxeter–Dynkin diagrams are the way of describing the group generated by reflections. For each graph, *node* represents a mirror (or a reflection hypersurface) and the label m attached to a branch between nodes marks the dihedral angle $\frac{\pi}{m}$ between two mirrors. By convention, no label is attached to a branch if the corresponding dihedral angle is $\frac{\pi}{3}$. When all the dihedral angles are $\frac{\pi}{3}$, the diagram is called *simply laced*. Ringed nodes present so called active mirrors where there is a point P not to sit in the hyperplanes of reflections corresponded to the mirrors. By successive applying the reflections in the diagram to the point P, we obtain a polytope whose symmetry group is the Weyl group generated by the Coxeter–Dynkin diagram. Moreover, the combinatorics of subpolytopes can also be decoded by the Coxeter–Dynkin diagram when it is simply laced with one ringed node (see [7,9,10]). In fact, a similar method can be applied for the diagram, which is not simply laced or has more than one ringed node.

The Coxeter–Dynkin diagram of 24-cell is an example of simply laced with one ringed node.

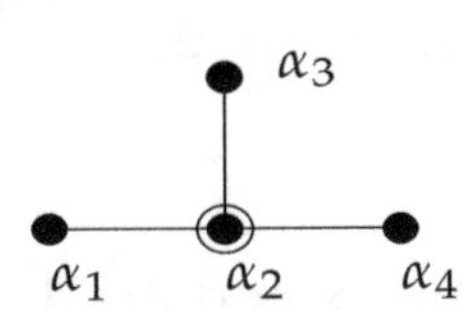

Coxeter–Dynkin diagram of 24-cell

The Weyl group associated with this diagram is D_4-type. In [7], the subpolytopes of 24-cell as shown in Table 2 are described by using the Coxeter–Dynkin diagram.

Table 2. Subpolytopes of 24-cell.

Subpolytope	Vertices	Edges	Faces	Cells
total number	24	96	96 ({3})	24 ({3,3})

The Coxeter–Dynkin diagram of 600-cell is given by

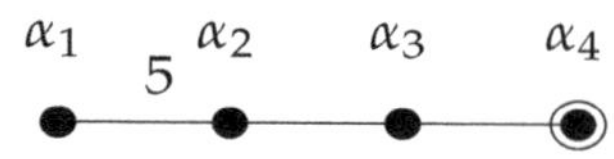

Coxeter–Dynkin diagram of 600-cell

whose Weyl group is H_4-type. Thus, the diagram is not simply laced and has one ringed node.

(1) Vertices

By removing a ringed node, we obtain the isotropy subgroup in the Weyl group of H_4 which fixed a vertex in the 600-cell. Here, the corresponding isotropy group is H_3 and we can compute the total number of vertices as

$$\frac{|H_4|}{|H_3|} = \frac{14400}{120} = 120.$$

For the remaining diagram above, we ring a node connected to the removed node. Then, we obtain the Coxeter–Dynkin diagram of an Icosahedron, which implies that the vertex figure of 600 cell is an icosahedron.

(2) Edges

For edges, we consider the ringed node that performs one reflection corresponding to an edge.

For the isotropy subgroup of the edge, we remove the unringed node connected to the ringed node. In addition, the remaining diagram generates the isotropy subgroup $H_2 \times A_1$. Thus, we compute the total number of edges as

$$\frac{|H_4|}{|H_2||A_1|} = \frac{14400}{10 \cdot 2} = 720.$$

(3) Faces

For faces, we consider the ringed node and extend the diagram to unringed nodes so as to obtain a subdiagram of A_2-type. The subdiagram of A_2 with one ringed node generate $\{3\}$, namely a triangle. Thus, the faces of 600-cell are all triangles.

For the isotropy subgroup of a face, we remove any unringed node connected to the subdiagram of a face. The remaining subdiagram generates the isotropy subgroup of a face, which is $A_1 \times A_2$. Thus, we compute the total number of faces as

$$\frac{|H_4|}{|A_1||A_2|} = \frac{14400}{2!3!} = 1200.$$

(4) Cells

To obtain a cell in a 600-cell, we consider an extended diagram from the ringed nodes to unringed nodes so as to obtain a subdiagram of A_3. The diagram of type A_3 with one ringed node on one side represents a tetrahedron.

For the isotropy subgroup of a cell, we consider that any unringed node connected to the subdiagram of A_3, and the subdiagram given by removing the node generates the isotropy subgroup, which is A_3. Thus, we compute the total number of cells, Table 3 shows the subpolytopes of 600:

$$\frac{|H_4|}{|A_3|} = \frac{14400}{4!} = 600.$$

Table 3. Subpolytopes of 600-cell.

Subpolytopes	Vertices	Edges	Faces	Cells
total number	120	720	1200 ({3})	600 ({3,3})

4. Binary Polyhedral Groups as Polytopes

In this section, we show that the binary icosahedral group $2I$ in $\mathbb{H}$ is the set of vertices of a 600-cell. Thus, the convex hull of $2I$ in $\mathbb{H}$ is a 600-cell.

For each α in $\mathbb{H}$ with $|\alpha| = 1$, we define a reflection on $\mathbb{H}$ as

$$\begin{array}{rccl} \sigma_\alpha : & \mathbb{H} & \longrightarrow & \mathbb{H}, \\ & x & \mapsto & \sigma_\alpha(x) := x - 2\,(\vec{x}\cdot\vec{\alpha})\,\alpha. \end{array}$$

Since $\mathbb{H}$ is a normed division algebra, $\sigma_\alpha(x)$ is also written as $\sigma_a(x) = -a\,\bar{x}\,a$ via quaternionic multiplication (see Ref. [7]). Since σ_a is a reflection for a vector α, σ_a has eigenvalues ± 1 where α is an eigenvector of -1 and the hyperplane perpendicular to α is the eigenspace of 1. Moreover, it is not an element in $SO(3)$.

Binary tetrahedral group $2\hat{T}$ in $\mathbb{H}$ and 24-cell

For 24-cell, we consider the Coxeter–Dynkin diagram of type D_4 given in Section 3, where

$$\alpha_1 = i, \alpha_2 = \frac{1}{2}(1 + i + j + k), \alpha_3 = j, \alpha_4 = k.$$

In Ref. [7], the Weyl group generated by the Coxeter–Dynkin diagram acts on the binary tetrahedral group $2\hat{T}$. Moreover, it is shown that $2\hat{T}$ is the set of vertices of a 24-cell. In fact, it is also the subset of unit integral quaternions.

Binary icosahedral group $2\hat{I}$ in $\mathbb{H}$ and 600-cell

Similarly, for 600-cell, we consider the Coxeter–Dynkin diagram of Type H_4 given in Section 3, where

$$\alpha_1 = i, \alpha_2 = \frac{1}{2}\left(\tau i + j - \frac{1}{\tau}k\right), \alpha_3 = j, \alpha_4 = \frac{1}{2}\left(-\frac{1}{\tau} + j + \tau k\right).$$

The Weyl group generated by the reflections $\{\sigma_{\alpha_i}\ i = 1, 2, 3, 4\}$ is denoted by W_H. In below, we want to show that **(1)** the Weyl group W_H acts on the binary icosahedral group $2\hat{I}$, and **(2)** $2\hat{I}$ is a single orbit, where it corresponds to the set of vertices of 600-cell.

Lemma 2. *The Weyl group W_H acts on the binary icosahedral group $2\hat{I}$ in $\mathbb{H}$.*

Proof. Since the Weyl group W_H is generated by the reflections σ_{α_i} $(i = 1, 2, 3, 4)$, we show that each σ_{α_i} acts on $2I$. For an arbitrary element $a + bi + cj + dk \in \mathbb{H}$, the reflections are written as follows:

$$\begin{aligned}
\sigma_{\alpha_1}(a + bi + cj + dk) &= a - bi + cj + dk,\\
\sigma_{\alpha_2}(a + bi + cj + dk) &= a + \left(-\frac{1}{2\tau}b - \frac{\tau}{2}c + \frac{1}{2}d\right) i\\
&\quad + \left(-\frac{\tau}{2}b + \frac{1}{2}c + \frac{1}{2\tau}d\right) j + \left(\frac{1}{2}b + \frac{1}{2\tau}c + \frac{\tau}{2}d\right) k,\\
\sigma_{\alpha_3}(a + bi + cj + dk) &= a + bi - cj + dk,\\
\sigma_{\alpha_4}(a + bi + cj + dk) &= \left(\frac{\tau}{2}a + \frac{1}{2\tau}c + \frac{1}{2}d\right) + bi\\
&\quad + \left(\frac{1}{2\tau}a + \frac{1}{2}c - \frac{\tau}{2}d\right) j + \left(\frac{1}{2}a - \frac{\tau}{2}c - \frac{1}{2\tau}d\right) k.
\end{aligned}$$

It is easy to see that σ_{α_1} and σ_{α_3} act on $2\hat{I}$. By choosing $\{1, i, j, k\}$ as an ordered orthonormal basis of $\mathbb{H}$, σ_{α_2} and σ_{α_4} can be written as

$$\sigma_{\alpha_2} = \begin{pmatrix} 1 & 0 & 0 & 0 \\ 0 & -\frac{1}{2\tau} & -\frac{\tau}{2} & \frac{1}{2} \\ 0 & -\frac{\tau}{2} & \frac{1}{2} & \frac{1}{2\tau} \\ 0 & \frac{1}{2} & \frac{1}{2\tau} & \frac{\tau}{2} \end{pmatrix} \text{ and } \sigma_{\alpha_4} = \begin{pmatrix} \frac{\tau}{2} & 0 & \frac{1}{2\tau} & \frac{1}{2} \\ 0 & 1 & 0 & 0 \\ \frac{1}{2\tau} & 0 & \frac{1}{2} & -\frac{\tau}{2} \\ \frac{1}{2} & 0 & -\frac{\tau}{2} & -\frac{1}{2\tau} \end{pmatrix}.$$

In addition, these are similar because $\sigma_{\alpha_4} = S^t\sigma_{\alpha_2}S$, where S is an orthogonal matrix

$$\begin{pmatrix} 0 & 1 & 0 & 0 \\ 0 & 0 & 0 & 1 \\ 0 & 0 & 1 & 0 \\ 1 & 0 & 0 & 0 \end{pmatrix}.$$

In fact, S is an element in $SO(4)$ defined by

$$1 \to k,\ i \to 1,\ j \to j,\ k \to i$$

and one can check that S acts on $2\hat{I}$ by simple calculation. Thus, it suffices to show that σ_{α_2} acts on $2I$ to check σ_{α_2} and σ_{α_4} act on $2\hat{I}$. For σ_{α_2}, we consider 3×3 submatrix A of σ_{α_2} defined as

$$A := \begin{pmatrix} -\frac{1}{2\tau} & -\frac{\tau}{2} & \frac{1}{2} \\ -\frac{\tau}{2} & \frac{1}{2} & \frac{1}{2\tau} \\ \frac{1}{2} & \frac{1}{2\tau} & \frac{\tau}{2} \end{pmatrix}.$$

This is a automorphism of $\operatorname{Im}\mathbb{H}$ which satisfies $A^tA = Id$ and $\det A = -1$. Moreover, A also acts on

$$\left\{(\pm 1, \pm 1, \pm 1),\ (\pm\tau, \pm\frac{1}{\tau}, 0),\ (0, \pm\tau, \pm\frac{1}{\tau}),\ (\pm\frac{1}{\tau}, 0, \pm\tau)\right\},$$

which is our choice of the vertices of a dodecahedron. Since A is a reflection, it is also a symmetry of the dodecahedron so that it also acts on the set of edges and the set of faces. According to the construction of binary icosahedral group $2\hat{I}$ in Section 2.1, the action of A on the icosahedron induces the action of σ_{α_2} on $2\hat{I}$. For example, an edge symmetry given by an edge $P_i + P_j$ is sent to another given by $AP_i + AP_j$ because

$$\sigma_{\alpha_2}\left(\frac{P_i + P_j}{|P_i + P_j|}\right) = \frac{AP_i + AP_j}{|P_i + P_j|} = \frac{AP_i + AP_j}{|AP_i + AP_j|}.$$

Similarly, we conclude that σ_{α_2} acts on $2\hat{I}$. □

By applying the above lemma, we obtain the following theorem.

Theorem 2. *The set $2\hat{I}$ of a binary icosahedral group is an orbit of the Weyl group W_H, and it is the set of vertices of a* 600*-cell.*

Proof. By Lemma 2, the Weyl group W_H acts on $2I$. Now, we consider an element 1 in $2\hat{I}$ and its orbit $W_H\{1\} \subset 2\hat{I}$. Since 1 is perpendicular to $\alpha_1 = i, \alpha_2 = \frac{1}{2}\left(\tau i + j - \frac{1}{\tau}k\right)$, $\alpha_3 = j$ and $1 \cdot \alpha_4 = (1,0,0,0) \cdot \left(-\frac{1}{2\tau}, 0, \frac{1}{2}, \frac{\tau}{2}\right) \neq 0$, the orbit $W_H\{1\}$ is given by the following Coxeter–Dynkin diagram 3 of 600-cell. By Section 3, we have $|W_H\{1\}| = 120 = |2\hat{I}|$. Therefore, we conclude $W_H\{1\} = 2\hat{I}$ and this gives the theorem. □

Author Contributions: This article is written by both authors.

Acknowledgments: The authors gratefully thank the Referees for the constructive comments and recommendations, which definitely helped to improve the readability and quality of the paper.

References

1. Van Hoboken, J. Platonic Solids, Binary Polyhedral Groups Kleinian Singularities and Lie Algebras of Type ADE. Master's Thesis, University of Amsterdam, Amsterdam, The Netherlands, March 2002.
2. Coxeter, H.S.M. The binary polyhedral groups, and other generalizations of the quaternion group. *Duke Math. J.* **1940**, *7*, 367–379. [CrossRef]
3. Koca, M.; Al-Ajmi, M.; Koç, R. Group theoretical analysis of 600-cell and 120-cell 4D polytopes with quaternions. *J. Phys. A* **2007**, *40*, 7633–7642. [CrossRef]
4. Koca, M.; Al-Ajmi, M.; Koç, R. Quaternionic representation of the Coxeter group $W(H_4)$ and the polyhedra. *J. Phys. A* **2006**, *39*, 14047–14054. [CrossRef]
5. Koca, M.; Al-Ajmi, M.; Koca, N.O. Quaternionic representation of snub 24-cell and its dual polytope derived from E_8 root system. *Linear Algebra Appl.* **2011**, *434*, 977–989. [CrossRef]
6. Dechant, P.-P. Clifford algebra unveils a surprising geometric significance of quaternionic root systems of Coxeter groups. *Adv. Appl. Clifford Algebra* **2013**, *23*, 301–321. [CrossRef]
7. Chang, W.N.; Lee, J.H.; Lee, S.H.; Lee, Y.J. Gosset polytopes in integral octonions. *Czechoslov. Math. J.* **2014**, *64*, 683–702. [CrossRef]
8. Conway, J.H.; Smith, D.A. *On Quaternions and Octonions: Their Geometry, Arithmetic, and Symmetry*; AK Peters: Natick, MA, USA, 2003.
9. Coxeter, H.S.M. *Regular Polytopes*, 3rd ed.; Dover Publications, Inc.: New York, NY, USA, 1973.
10. Lee, J.-H. Gosset polytopes in Picard groups of del Pezzo surfaces. *Can. J. Math.* **2012**, *64*, 123–150. [CrossRef]

12

The Structure Theorems of Pseudo-BCI Algebras in which Every Element is Quasi-Maximal

Xiaoying Wu [1] **and Xiaohong Zhang** [1,2,*]

1 Department of Mathematics, Shaanxi University of Science & Technology, Xi'an 710021, China; 46018@sust.edu.cn

2 Department of Mathematics, Shanghai Maritime University, Shanghai 201306, China

* Correspondence: zhangxh@shmtu.edu.cn or zhangxiaohong@sust.edu.cn

Abstract: For mathematical fuzzy logic systems, the study of corresponding algebraic structures plays an important role. Pseudo-BCI algebra is a class of non-classical logic algebras, which is closely related to various non-commutative fuzzy logic systems. The aim of this paper is focus on the structure of a special class of pseudo-BCI algebras in which every element is quasi-maximal (call it QM-pseudo-BCI algebras in this paper). First, the new notions of quasi-maximal element and quasi-left unit element in pseudo-BCK algebras and pseudo-BCI algebras are proposed and some properties are discussed. Second, the following structure theorem of QM-pseudo-BCI algebra is proved: every QM-pseudo-BCI algebra is a KG-union of a quasi-alternating BCK-algebra and an anti-group pseudo-BCI algebra. Third, the new notion of weak associative pseudo-BCI algebra (WA-pseudo-BCI algebra) is introduced and the following result is proved: every WA-pseudo-BCI algebra is a KG-union of a quasi-alternating BCK-algebra and an Abel group.

Keywords: fuzzy logic; pseudo-BCI algebra; quasi-maximal element; KG-union; quasi-alternating BCK-algebra

---

1. Introduction

In the study of t-norm based fuzzy logic systems [1–9], algebraic systems (such as residuated lattices, BL-algebras, MTL-algebras, pseudo-BL algebras, pseudo-MTL algebras, et al.) play an important role. In this paper, we discuss pseudo-BCI/BCK algebras which are connected with non-commutative fuzzy logic systems (such that non-commutative residuared lattices, pseudo-BL/pseudo-MTL algebras).

BCK-algebras and BCI-algebras were introduced by Iséki [10] as algebras induced by Meredith's implicational logics BCK and BCI. The name of BCK-algebra and BCI-algebra originates from the combinatories B, C, K, I in combinatory logic. The notion of pseudo-BCK algebra was introduced by G. Georgescu and A. Iorgulescu in [11] as a non-commutative extension of BCK-algebras. Then, as common generalization of pseudo-BCK algebras and BCI-algebras, W.A. Dudek and Y.B. Jun introduced the concept of pseudo-BCI algebra in [12]. In fact, there are many other non-classical logic algebraic systems related to BCK- and BCI-algebras, such as BCC-algebra, BZ-algebra and so forth, some monographs and papers on these topics can be found in [7–9,13–18].

Pseudo-BCI-algebras are algebraic models of some extension of a noncommutative version of the BCI-logic, the corresponding logic is called pseudo-BCI logic [19]. P. Emanovský and J. Kühr studied some properties of pseudo-BCI algebras, X.L. Xin et al. [20] investigated monadic pseudo BCI-algebras and corresponding logics and some authors discussed the filter (ideal) theory of pseudo-BCI algebras [21–28]. Moreover, some notions of period, state and soft set are applied to pseudo-BCI algebras [29–31].

In this paper, we further study the structure characterizations of pseudo-BCI algebras. By using the notions of quasi-maximal element, quasi-left unit element, KG-union and direct product, we give the structure theorem of the class of pseudo-BCI algebras in which every element is quasi-maximal (call they QM-pseudo-BCI algebras). Moreover, we introduce weak associative property in pseudo-BCI algebras, discuss basic properties of weak associative pseudo-BCI algebra (WA-pseudo-BCI algebra) and establish the structure theorem of WA-pseudo-BCI algebra.

It should be noted that the original definition of pseudo-BCI/BCK algebra is different from the definition used in this paper. They are dual. We think that the logical semantics of this algebraic structure can be better represented by using the present definition.

2. Preliminaries

Definition 1 ([10,16]). *An algebra $(A; \rightarrow, 1)$ of type (2,0) is called a BCI-algebra if the following conditions are satisfied for all x, y, z from A:*

(1) $x \rightarrow y \leq (y \rightarrow z) \rightarrow (x \rightarrow z)$,
(2) $x \leq (x \rightarrow y) \rightarrow y$,
(3) $x \leq x$,
(4) $x \leq y$, $y \leq x$ *imply* $x = y$, *where* $x \leq y$ *means* $x \rightarrow y = 1$. *An algebra $(A; \rightarrow, 1)$ of type (2,0) is called a BCK-algebra if it is a BCI-algebra and satisfies:*
(5) $x \rightarrow 1 = 1, \ \forall x \in A$.

Definition 2 ([10,16]). *A BCK-algebra $(A; \rightarrow, 1)$ is called bounded if there exists unique element 0 such that $0 \rightarrow x = 1$ for any $x \in A$.*

Definition 3 ([13,14]). *A BCK-algebra $(A; \rightarrow, 1)$ is called quasi-alternating BCK-algebra if it satisfies the following axiom: $\forall\, x, y \in X$, $x \neq y$ implies $x \rightarrow y = y$.*

Definition 4 ([9,11]). *A pseudo-BCK algebra is a structure $(A; \leq, \rightarrow, \rightsquigarrow, 1)$, where "$\leq$" is a binary relation on A, "$\rightarrow$" and "$\rightsquigarrow$" are binary operations on A and "1" is an element of A, verifying the axioms: for all $x, y, z \in A$,*

(1) $x \rightarrow y \leq (y \rightarrow z) \rightsquigarrow (x \rightarrow z)$, $x \rightsquigarrow y \leq (y \rightsquigarrow z) \rightarrow (x \rightsquigarrow z)$,
(2) $x \leq (x \rightarrow y) \rightsquigarrow y$, $x \leq (x \rightsquigarrow y) \rightarrow y$
(3) $x \leq x$,
(4) $x \leq 1$,
(5) $x \leq y$, $y \leq x \Rightarrow x = y$,
(6) $x \leq y \rightsquigarrow x \rightarrow y = 1 \Leftrightarrow x \rightsquigarrow y = 1$.

If $(A; \leq, \rightarrow, \rightsquigarrow, 1)$ is a pseudo-BCK algebra satisfying $x \rightarrow y = x \rightsquigarrow y$ for all $x, y \in A$, then $(A; \rightarrow, 1)$ is a BCK-algebra.

Proposition 1 ([9,11]). *Let $(A; \leq, \rightarrow, \rightsquigarrow, 1)$ be a pseudo-BCK algebra, then A satisfy the following properties $(\forall x, y, z \in A)$:*

(1) $x \leq y \Rightarrow y \rightarrow z \leq x \rightarrow z$, $y \rightsquigarrow z \leq x \rightsquigarrow z$
(2) $x \leq y, y \leq z \Rightarrow x \leq z$,
(3) $x \rightsquigarrow (y \rightarrow z) = y \rightarrow (x \rightsquigarrow z)$,
(4) $x \leq y \rightarrow z \Leftrightarrow y \leq x \rightsquigarrow z$,
(5) $x \rightarrow y \leq (z \rightarrow x) \rightarrow (z \rightarrow y)$, $x \rightsquigarrow y \leq (z \rightsquigarrow x) \rightsquigarrow (z \rightsquigarrow y)$,
(6) $x \leq y \rightarrow x$, $x \leq y \rightsquigarrow x$,
(7) $1 \rightarrow x = x$, $1 \rightsquigarrow x = x$,
(8) $x \leq y \Rightarrow z \rightarrow x \leq z \rightarrow y$, $z \rightsquigarrow x \leq z \rightsquigarrow y$,
(9) $((y \rightarrow x) \rightsquigarrow x) \rightarrow x = y \rightarrow x$, $((y \rightsquigarrow x) \rightarrow x) \rightsquigarrow x = y \rightsquigarrow x$.

Definition 5 ([[12]). *A pseudo-BCI algebra is a structure $(A; \leq, \rightarrow, \rightsquigarrow, 1)$, where "$\leq$" is a binary relation on A, "$\rightarrow$" and "$\rightsquigarrow$" are binary operations on A and "1" is an element of A, verifying the axioms: for all $x, y, z \in A$,*

(1) $x \rightarrow y \leq (y \rightarrow z) \rightsquigarrow (x \rightarrow z), \; x \rightsquigarrow y \leq (y \rightsquigarrow z) \rightarrow (x \rightsquigarrow z)$,
(2) $x \leq (x \rightarrow y) \rightsquigarrow y, \, x \leq (x \rightsquigarrow y) \rightarrow y$,
(3) $x \leq x$,
(4) *if* $x \leq y$ *and* $y \leq x$, *then* $x = y$,
(5) $x \leq y$ *iff* $x \rightarrow y = 1$ *iff* $x \rightsquigarrow y = 1$.

Note that, every pseudo-BCI algebra satisfying $x \rightarrow y = x \rightsquigarrow y$ for all $x, y \in A$ is a BCI-algebra.

Proposition 2 ([12,22,24]). *Let $(A; \leq, \rightarrow, \rightsquigarrow, 1)$ be a pseudo-BCI algebra, then A satisfy the following properties $(\forall x, y, z \in A)$:*

(1) *if* $1 \leq x$, *then* $x = 1$,
(2) *if* $x \leq y$, *then* $y \rightarrow z \leq x \rightarrow z$ *and* $y \rightsquigarrow z \leq x \rightsquigarrow z$,
(3) *if* $x \leq y$ *and* $y \leq z$, *then* $x \leq z$,
(4) $x \rightsquigarrow (y \rightarrow z) = y \rightarrow (x \rightsquigarrow z)$,
(5) $x \leq y \rightarrow z$, *iff* $y \leq x \rightsquigarrow z$
(6) $x \rightarrow y \leq (z \rightarrow x) \rightarrow (z \rightarrow y), \; x \rightsquigarrow y \leq (z \rightsquigarrow x) \rightsquigarrow (z \rightsquigarrow y)$,
(7) *if* $x \leq y$, *then* $z \rightarrow x \leq z \rightarrow y$ *and* $z \rightsquigarrow x \leq z \rightsquigarrow y$,
(8) $1 \rightarrow x = x, \; 1 \rightsquigarrow x = x$,
(9) $((y \rightarrow x) \rightsquigarrow x) \rightarrow x = y \rightarrow x, \; ((y \rightsquigarrow x) \rightarrow x) \rightsquigarrow x = y \rightsquigarrow x$,
(10) $x \rightarrow y \leq (y \rightarrow x) \rightsquigarrow 1, \; x \rightsquigarrow y \leq (y \rightsquigarrow x) \rightarrow 1$,
(11) $(x \rightarrow y) \rightarrow 1 = (x \rightarrow 1) \rightsquigarrow (y \rightarrow 1), (x \rightsquigarrow y) \rightsquigarrow 1 = (x \rightsquigarrow 1) \rightarrow (y \rightarrow 1)$
(12) $x \rightarrow 1 = x \rightsquigarrow 1$.

Definition 6 ([10,24]). *A pseudo-BCI algebra A is said to be an anti-grouped pseudo-BCI algebra if it satisfies the following identities:*

$$for\ any\ x \in A,\ (x \rightarrow 1) \rightarrow 1 = x\ or\ (x \rightsquigarrow 1) \rightsquigarrow 1 = x.$$

Proposition 3 ([24]). *A pseudo-BCI algebra A is anti-grouped if and only if it satisfies:*

(G1) *for all* $x, y, z \in A$, $(x \rightarrow y) \rightarrow (x \rightarrow z) = y \rightarrow z$ *and*
(G2) *for all* $x, y, z \in A$, $(x \rightsquigarrow y) \rightsquigarrow (x \rightsquigarrow z) = y \rightsquigarrow z$.

Proposition 4 ([24]). *Let* $\mathbf{A} = (A; \leq, \rightarrow, \rightsquigarrow, 1)$ *be an anti-grouped pseudo-BCI algebra. Define* $\boldsymbol{\Phi}(\mathbf{A}) = (A; +, -, 1)$ *by*

$$x + y = (x \rightarrow 1) \rightarrow y \;=\; (y \rightsquigarrow 1) \rightsquigarrow x, \; \forall x, \; y \in A;$$

$$-x = x \rightarrow 1 = x \rightsquigarrow 1, \; \forall x \in A.$$

Then $\boldsymbol{\Phi}(\mathbf{A})$ *is a group. Conversely, let* $\mathbf{G} = (G; +, -, 1)$ *be a group. Define* $\boldsymbol{\Psi}(\mathbf{G}) = (G; \leq, \rightarrow, \rightsquigarrow, 1)$, *where*

$$x \rightarrow y = (-x) + y, \; x \rightsquigarrow y = y + (-x), \; \forall x, \; y \in G;$$

$$x \leq y\ if\ and\ only\ if\ (-x) + y = 1\ (or\ y + (-x) = 1), \; \forall x, \; y \in G.$$

Then, $\boldsymbol{\Psi}(\mathbf{G})$ *is an anti-grouped pseudo-BCI algebra. Moreover, the mapping* $\boldsymbol{\Phi}$ *and* $\boldsymbol{\Psi}$ *are mutually inverse.*

Definition 7 ([27]). *Let $(A; \leq, \rightarrow, \rightsquigarrow, 1)$ be a pseudo-BCI algebra. Denote*

$$K(A) = \{x \in A | x \leq 1\};$$

$$AG(A) = \{x \in A | (x \rightarrow 1) \rightarrow 1 = x\}.$$

We say that K(A) is the pseudo-BCK part of A and $AG(A)$ is the anti-grouped part of A.

Definition 8 ([28]). *A pseudo-BCI algebra A is said to be a T-type if it satisfies the following identities:*
(T1) for all $x \in A$, $(x \rightarrow 1) \rightarrow 1 = x \rightarrow 1$, or $(x \rightsquigarrow 1) \rightsquigarrow 1 = x \rightsquigarrow 1$.

Proposition 5 ([28]). *A pseudo-BCI algebra A is T-type if and only if it satisfies:*
(T2) for all $x \in A$, $x \rightarrow (x \rightarrow 1) = 1$, or $x \rightsquigarrow (x \rightsquigarrow 1) = 1$.

3. Some New Concepts and Results

By the definition of pseudo-BCI/BCK algebra, we know that the direct product of two pseudo-BCI/BCK algebras is a pseudo-BCI/BCK algebra. That is, we have the following lemma.

Lemma 1 ([20]). *Let $(X; \rightarrow_X, \rightsquigarrow_X, 1_X)$ and $(Y; \rightarrow_Y, \rightsquigarrow_Y, 1_Y)$ be two pseudo-BCI algebras. Define two binary operators $\rightarrow, \rightsquigarrow$ on $X \times Y$ as follwos: for any $(x_1, y_1), (x_2, y_2) \in X \times Y$,*

$$(x_1, y_1) \rightarrow (x_2, y_2) = (x_1 \rightarrow_X x_2, y_1 \rightarrow_Y y_2);$$

$$(x_1, y_1) \rightsquigarrow (x_2, y_2) = (x_1 \rightsquigarrow_X x_2 y_1 \rightsquigarrow_Y y_2);$$

and denote $1 = (1_X, 1_Y)$. Then $(X \times Y; \rightarrow, \rightsquigarrow, 1)$ is a pseudo-BCI algebra.

By the results in [18,20], we can easy to verify that the following lemma (the proof is omitted).

Lemma 2. *Let $(K; \rightarrow, \rightsquigarrow, 1)$ be a pseudo-BCK algebra, $(G; \rightarrow, \rightsquigarrow, 1)$ an anti-grouped pseudo-BCI algebra and $K \cap G = \{1\}$. Denote A = KG b and define the operations $\rightarrow, \rightsquigarrow$ on A as follows:*

$$x \rightarrow y = \begin{cases} x \rightarrow y & if\ x, y \in K\ or\ x, y \in G \\ y & if\ x \in K, \quad y \in G \\ x \rightarrow 1 & if\ y \in K\{1\}, \quad x \in G \end{cases}$$

$$x \rightsquigarrow y = \begin{cases} x \rightsquigarrow y & if\ x, y \in K\ or\ x, y \in G \\ y & if\ x \in K, \quad y \in G \\ x \rightsquigarrow 1 & if\ y \in K\{1\}, \quad x \in G \end{cases}$$

Then $(A; \rightarrow, \rightsquigarrow, 1)$ is a pseudo-BCI algebra.

Definition 9. *Let K be a pseudo-BCK algebra and G be an anti-grouped pseudo-BCI algebra, $K \cap G = \{1\}$. If the operators $\rightarrow, \rightsquigarrow$ are defined on $A = K \cup G$ according to Lemma 2, then $(A; \rightarrow, \rightsquigarrow, 1)$ is a pseudo-BCI algebra, we call A to be a KG-union of K and G and denote by $A = K \oplus_{KG} G$.*

Definition 10. *Let $(X, \leq)$ is a partial ordered set with 1 as a constant element. For x in X, we call x a quasi-maximal element of X, if for any $a \in X$, $x \leq a \Rightarrow x = a$ or $a = 1$.*

Definition 11. *Let $(G, *)$ be a grouoid, $x \in G$. Then x is called a quasi-left unit element of G, if it satisfies:*

$$\forall y \in G,\ x * y = y \text{ when } x \neq y.$$

Theorem 1. *Let $(A; \leq, \to, \rightsquigarrow, 1)$ be pseudo-BCK algebra. Then the following conditions are equivalent:*

(a1) $\forall x \in A$, *x is a quasi-maximal element;*
(a2) $\forall x \in A, y \in A - \{1\}$, *$x \leq y$ implies $x = y$;*
(a3) $\forall x \in A$, *x is a quasi-left unit elemen w.r.t $\to, \rightsquigarrow$, that is, $x \neq y$ implies $x \to y = y$ and $x \rightsquigarrow y = y$;*
(a4) $\forall x, y \in A$, *$x \neq y$ implies $x \to y = y$;*
(a5) $\forall x, y \in A$, *$x \neq y$ implies $x \rightsquigarrow y = y$.*

Proof. (a1) $\Rightarrow$ (a2): Suppose that $x \in A, y \in A - \{1\}$ *and* $x \leq y$.

Case 1: If $x = 1$, it is follows that $1 = x \leq y \leq 1$, that is, $x = y = 1$.

Case 2: If $x \neq 1$, by (a1) and Definition 10, from $x \leq y$ and $y \neq 1$, we have $x = y$. Therefore, (a2) hold.

(a2) $\Rightarrow$ (a3): For any x, y in A, by Proposition 1 (6) and Definition 4 (2), we have $x \leq y \to x$, $y \leq x \to y, x \leq (x \to y) \rightsquigarrow y$. Assume $x \neq y$. If $y \to x = 1$, then $x \to y \neq 1$ (since, if $x \to y = 1$, then form $y \to x = 1$ and $x \to y = 1$ we get $x = y$, this is contradictory to the hypothesis $x \neq y$). Thus, from $y \leq x \to y$ and $x \to y \neq 1$, using (a2) we have $y = x \to y$.

If $y \to x \neq 1$, from this and $x \leq y \to x$ *and* applying (a2), we have $x = y \to x$. Thus,

(i) when $(x \to y) \rightsquigarrow y = 1$, we can get $x \to y \leq y \leq x \to y$, that is, $y = x \to y$;
(ii) when $(x \to y) \rightsquigarrow y \neq 1$, from this and $x \leq (x \to y) \rightsquigarrow y$, using (a2) we have $x = (x \to y) \rightsquigarrow y$. Combine the aforementioned conclusion $x = y \to x$, we can get

$$x = y \to x = y \to ((x \to y) \rightsquigarrow y) = (x \to y) \rightsquigarrow (y \to y) = (x \to y) \rightsquigarrow 1 = 1,$$

It follows that $y = 1 \to y = x \to y$.

Therefore, based on the above cases we know that $x \neq y$ implies $y = x \to y$.

Similarly, we can prove that $x \neq y$ implies $y = x \rightsquigarrow y$.

(a3) $\Rightarrow$ (a4): Obviously.

(a4) $\Rightarrow$ (a5): Suppose $x \neq y$. Applying (a4), $x \to y = y$. Also, by Definition 4 (2), $x \leq (x \rightsquigarrow y) \to y$, thus $x \to [(x \rightsquigarrow y) \to y] = 1$.

Case 1: If $x \neq (x \rightsquigarrow y) \to y$, using (a4), $x \to [(x \rightsquigarrow y) \to y] = [(x \rightsquigarrow y) \to y]$. Hence, $(x \rightsquigarrow y) \to y = 1$. Moreover,

$$y \to (x \rightsquigarrow y) = x \rightsquigarrow (y \to y) = x \rightsquigarrow 1 = 1.$$

Therefore, $y = x \rightsquigarrow y$. Case 2: If $x = (x \rightsquigarrow y) \to y$, then $x \rightsquigarrow y = y$. In fact, if $x \rightsquigarrow y \neq y$, using (a4), $(x \rightsquigarrow y) \to y = y$, it follows that $x = y$, this is a contradiction with $x \neq y$.

By above results we know that (a5) hold.

(a5) $\Rightarrow$ (a1): Assume that $x \in X, a \in X$ and $x \leq a$. Then $x \rightsquigarrow a = 1$. If $x \neq a$, by (a5), $x \rightsquigarrow a = a$, then $a = x \rightsquigarrow a = 1$. This means that $x \leq a$ implies $x = a$ or $a = 1$. □

By Theorem 1 and Definition 3 we get

Corollary 1. *Let $(A; \leq, \to, \rightsquigarrow, 1)$ be a pseudo-BCK algebra. Then every element of A is quasi-maximal if and only if A is a quasi-alternating BCK-algebra.*

4. The Class of Pseudo-BCI Algebras in Which Every Element is Quasi-Maximal

Example 1. *Let A = {a, b, c, d, e, f, g, 1}. Define operations $\to$ and $\rightsquigarrow$ on A as following Cayley Tables 1 and 2. Then A is pseudo-BCI algebra in which every element is quasi-maximal.*

Table 1. The Cayley table of operation →.

→	*a*	*b*	*c*	*d*	*e*	*f*	*g*	*1*
a	1	b	c	d	e	f	g	1
b	a	1	c	d	e	f	g	1
c	c	c	1	f	g	d	e	c
d	d	d	e	1	c	g	f	d
e	f	f	g	c	1	e	d	f
f	e	e	d	g	f	1	c	e
g	g	g	c	d	e	f	1	g
1	a	b	c	d	e	f	g	1

Table 2. The Cayley table of operation ⇝.

⇝	*a*	*b*	*c*	*d*	*e*	*f*	*g*	*1*
a	1	b	c	d	e	f	g	1
b	a	1	c	d	e	f	g	1
c	c	c	1	f	g	d	e	c
d	d	d	e	1	c	g	f	d
e	f	f	g	c	1	e	d	f
f	e	e	d	g	f	1	c	e
g	g	g	c	d	e	f	1	g
1	a	b	c	d	e	f	g	1

Definition 12. *A pseudo-BCI/BCK algebra A is said to be a QM-pseudo-BCI/BCK algebra if every element of A is quasi-maximal.*

Theorem 2. *Let (A; ≤, →, ⇝, 1) be a pseudo-BCI algebra. Then A is a QM-pseudo-BCI algebra if and only if it satisfies:*

$$for\ any\ x,\ y \in A - \{1\},\ x \leq y \Rightarrow x = y.$$

Proof. If A is a QM-pseudo-BCI algebra, by Definitions 10 and 12, the above condition is satisfied.

Conversely, assume that $x,\ y \in A$, $x \leq y$. If $x = 1$, then $1 = x \leq y$, it follows that $x = y = 1$, by Proposition 2 (1). If $x \neq 1, y \neq 1$, then $x = y$ by the condition. This means that x is a quasi- maximal element in A, hence, A is a QM-pseudo-BCI algebra. □

By Theorem 1 we know that a pseudo-BCK algebra is a QM-pseudo-BCK algebra if and only if it is a quasi-alternating BCK-algebra. It will be proved that any QM-pseudo-BCI algebra is constructed by the combination of a quasi-alternating BCK-algebra and an anti-grouped pseudo- BCI algebra (a group-like algebra).

Lemma 3 ([27]). *Let A be a pseudo-BCI algebra, K(A) the pseudo-BCK part of A. If AG(A) = (A − K(A))∪{1} is subalgebra of A, then (∀x, y ∈ A)*

(1) *If $x \in K(A)$ and $y \in A - K(A)$, then $x \to y = x \rightsquigarrow y = y$.*
(2) *If $x \in A - K(A)$ and $y \in K(A)$, then $x \to y = x \rightsquigarrow y = x \to 1$.*

Applying the results in [24,27] we can easy to verify that the following lemma is true (the proof is omitted).

Lemma 4. *Let A be an anti-grouped pseudo-BCI algebra. Then*

(1) *for any x, y in A, $x \leq y$ implies $x = y$;*

(2) *for any* x, y *in* A, $x = (x \to y) \rightsquigarrow y = (x \rightsquigarrow y) \to y$.

Theorem 3. *Let A be a pseudo-BCI algebra, K(A) the pseudo-BCK part of A and AG(A) the anti-grouped part of A. The following statements are equivalent:*

(1) A is a QM-pseudo-BCI algebra;
(2) *K(A) is quasi-alternating BCK-algebras and AG(A) = (A − K(A)) ∪ {1};*
(3) $\forall x, y \in A$, $x \neq y$ *implies* $(x \to y) \rightsquigarrow y = (x \to 1) \to 1$;
(4) $\forall x, y \in A$, $x \neq y$ *implies* $(x \rightsquigarrow y) \to y = (x \to 1) \to 1$;
(5) $\forall x, y \in A$, $x \neq y$ *implies* $(x \to y) \to y = (x \to 1) \to 1$;
(6) $\forall x, y \in A$, $x \neq y$ *implies* $(x \rightsquigarrow y) \rightsquigarrow y = (x \to 1) \to 1$.

Proof. (1) ⇒ (2): Suppose that A is a QM-pseudo-BCI algebra. Then, for all $x, y \in K(A)$, by Corollary 1, we can know $K(A)$ is quasi-alternating BCK-algebras. If $x \in A - K(A)$, then $x \to 1 \neq 1$ and $(x \to 1) \to 1 \neq 1$. Since $x \leq (x \to 1) \to 1$, by Definition 12 we have $x = (x \to 1) \to 1$. Thus, $(A - K(A)) \cup \{1\} \subseteq AG(A)$. On the other hand, obviously, $AG(A) \subseteq (A - K(A)) \cup \{1\}$. Hence $AG(A) = (A - K(A)) \cup \{1\}$.

(2) ⇒ (3): Assume that (2) hold. For any x, y in A, $x \neq y$,

Case 1: $x, y \in K(A)$. Then $x \to 1 = y \to 1 = 1$. Because $K(A)$ is quasi-alternating BCK-algebra, using Theorem 1, $x \to y = y$. Thus

$$(x \to y) \rightsquigarrow y = y \rightsquigarrow y = 1 = 1 \to 1 = (x \to 1) \to 1.$$

Case 2: $x, y \in AG(A)$. Since $AG(A)$ is an anti-grouped pseudo-BCI subalgebra of A, then by Lemma 4 we get

$$(x \to y) \rightsquigarrow y = x = (x \to 1) \to 1.$$

Case 3: $x \in K(A), y \in AG(A)$. Then $x \to 1 = 1$. Applying Lemma 3 (1), $x \to y = y$. Then

$$(x \to y) \rightsquigarrow y = y \rightsquigarrow y = 1 = 1 \to 1 = (x \to 1) \to 1.$$

Case 4: $x \in AG(A)$, $y \in K(A)$. Then $x = (x \to 1) \to 1$, $y \to 1 = 1$. Applying Lemma 3 (2), $x \to y = x \to 1$. When $x = 1$, then $(x \to y) \rightsquigarrow y = (x \to 1) \to 1$; when $x \neq 1$, then $x \to 1 \in A - K(A)$, using Lemma 3 (2),

$$(x \to 1) \rightsquigarrow y = (x \to 1) \to 1$$

Hence,

$$(x \to y) \rightsquigarrow y = (x \to 1) \rightsquigarrow y = (x \to 1) \to 1.$$

(3) ⇒ (1): Assume that $x \leq y$ and $x \neq y$. We will prove that $y = 1$. By (3), we have

$$y = 1 \rightsquigarrow y = (x \to y) \rightsquigarrow y = (x \to 1) \to 1.$$

Case 1: when $x \in K(A)$, then $x \to 1 = 1$, so $y = 1$. Case 2: when $x \in X - K(A)$, then $(x \to 1) \to 1 = x$, so $y = x$, this is a contradiction with $x \neq y$.

Therefore, for all $x \in A$, x is a quasi-maximal element of A.

(4) ⇒ (2): Suppose (4) hold. For any x, y in A.

If $x, y \in K(A)$, $x \neq y$, by (4),

$$(x \rightsquigarrow y) \to y = (x \to 1) \to 1 = 1.$$

Then, $x \rightsquigarrow y \leq y$. Since $K(A)$ is a pseudo-BCK subalgebra of A, using Proposition 1 (6), $y \leq x \rightsquigarrow y$. It follows that $y \leq x \rightsquigarrow y \leq y$, that is, $x \rightsquigarrow y = y$. Thus, applying Theorem 1, $K(A)$ is a quasi-alternating BCK-algebra.

If $x \in A - K(A)$, we prove that $(x \to 1) \to 1 = x$. Assume $(x \to 1) \to 1 \neq x$, by (4), we have

$$\{[(x \to 1) \to 1] \rightsquigarrow x\} \to x = \{[(x \to 1) \to 1] \to 1\} \to 1.$$

Using Proposition 2 (9) and (12),

$$\{[(x \to 1) \to 1] \to 1\} \to 1 = (x \to 1) \to 1.$$

Thus

$$\{[(x \to 1) \to 1] \rightsquigarrow x\} \to x = (x \to 1) \to 1.$$

Moreover, applying Proposition 2 (9), (11) and (12) we have

$$\begin{aligned} & \{[(x \to 1) \to 1] \rightsquigarrow x\} \to 1 \\ = & \{[(x \to 1) \to 1] \rightsquigarrow 1\} \to (x \rightsquigarrow 1) \\ = & \{[(x \to 1) \rightsquigarrow 1] \to 1\} \to (x \rightsquigarrow 1) \\ = & (x \to 1) \to (x \rightsquigarrow 1) \\ = & 1. \end{aligned}$$

This means that $((x \to 1) \to 1) \to x \in K(A)$. By Lemma 3 (1),

$$\{[(x \to 1) \to 1] \rightsquigarrow x\} \to x = x.$$

Hence, $(x \to 1) \to 1 = x$. This is contraction with $(x \to 1) \to 1 \neq x$. Therefore, $(x \to 1) \to 1 = x$ and $x \in AG(A)$. It follows that $(A - K(A)) \cup \{1\} \subseteq AG(A)$. Obviously, $AG(A) \subseteq (A - K(A)) \cup \{1\}$. So $AG(A) = (A - K(A)) \cup \{1\}$.

(2) $\Rightarrow$ (4): It is similar to (2) $\Rightarrow$ (3). It follows that (4) $\Leftrightarrow$ (2).

Similarly, we can prove (5) $\Leftrightarrow$ (2), (6) $\Leftrightarrow$ (2). □

Theorem 4. *Let $(A; \leq, \to, \rightsquigarrow, 1)$ be a pseudo-BCI algebra, $AG(A)$ the anti-grouped part of A, $K(A)$ the pseudo-BCK part of A. Then A is a QM-pseudo-BCI algebra if and only if $K(A)$ is a quasi-alternating BCK-algebra and $A = K(A) \oplus_{KG} AG(A)$.*

Proof. If A is a QM-pseudo-BCI algebra, then $K(A)$ is a quasi-alternating BCK-algebra and $A = K(A) \oplus_{KG} AG(A)$, by Lemma 3 and Theorem 3.

Conversely, if $K(A)$ is a quasi-alternating BCK-algebra, then every element in $K(A)$ is quasi-maximal; if $A = K(A) \oplus_{KG} AG(A)$, then $AG(A) = (A - K(A)) \cup \{1\}$, it follows that every element in $A - K(A)$ is quasi-maximal. By Definition 12, we know that A is a QM-pseudo-BCI algebra. □

5. Weak Associative Pseudo-BCI Algebras

Definition 13. *A pseudo-BCI/BCK algebra A is said to be weak associative, if it satisfies:*

$$for\ any,\ y, z \in A,\ (x \to y) \to z = x \to (y \to z)\ when\ (x \neq y,\ x \neq z).$$

Example 2 *Let $A = \{a, b, c, d, e, f, 1\}$. Define operation $\to$ on A as following Cayley Table 3. Then A is a weak associative pseudo-BCI algebra, where $\rightsquigarrow = \to$.*

Table 3. The Cayley table of the operation $\rightarrow$.

$\rightarrow$	***a***	***b***	***c***	***d***	***e***	***f***	***1***
a	1	*b*	*c*	*d*	*e*	*f*	1
b	*a*	1	*c*	*d*	*e*	*f*	1
c	*a*	*b*	1	*d*	*e*	*f*	1
d	*d*	*d*	*d*	1	*f*	*e*	*d*
e	*e*	*e*	*e*	*f*	1	*d*	*e*
f	*f*	*f*	*f*	*e*	*d*	1	*f*
1	*a*	*b*	*c*	*d*	*e*	*f*	1

Theorem 5. *Let $(A; \leq, \rightarrow, \rightsquigarrow, 1)$ be a weak associative pseudo-BCI algebra. Then A is a QM-pseudo-BCI algebra and a T-type pseudo-BCI algebra.*

Proof. For any x, y in A, $x \neq y$, then (by Definition 13)

$$(x \rightarrow y) \rightarrow y = x \rightarrow (y \rightarrow y) = x \rightarrow 1.$$

Thus, if $x \neq 1$, then $(x \rightarrow 1) \rightarrow 1 = x \rightarrow 1$. Obviously, when $x = 1$, $(x \rightarrow 1) \rightarrow 1 = x \rightarrow 1$. Hence, from Definition 13 we get that for any x, y in A, $x \neq y \Rightarrow (x \rightarrow y) \rightarrow y = (x \rightarrow 1) \rightarrow 1$. Applying Theorem 3 (5) we know that A is a QM-pseudo-BCI algebra.

Moreover, we already prove that $(x \rightarrow 1) \rightarrow 1 = x \rightarrow 1$ for any x in A, by Definition 8 we know that A is a T-type pseudo-BCI algebra. □

The inverse of Theorem 5 is not true. Since $(d \rightarrow c) \rightarrow c \neq d \rightarrow 1$, so the QM-pseudo-BCI algebra in Example 1 is not weak associative. The following example shows that a T-type pseudo-BCI algebra may be not a QM-pseudo-BCI algebra.

Example 3. *Let A = {a, b, c, d, 1}. Define operations $\rightarrow$ and $\rightsquigarrow$ on A as following Cayley Tables 4 and 5. Then A is a T-type pseudo-BCI algebra but it is not a QM-pseudo-BCI algebra, since*

$$(b \rightarrow c) \rightarrow a = a \neq 1 = b \rightarrow (c \rightarrow a).$$

Table 4. The operation $\rightarrow$ in the T-type pseudo-BCI algebra.

$\rightarrow$	***a***	***b***	***c***	***d***	***1***
a	1	1	1	*d*	1
b	*b*	1	1	*d*	1
c	*b*	*b*	1	*d*	1
d	*d*	*d*	*d*	1	*d*
1	*a*	*b*	*c*	*d*	1

Table 5. The operation $\rightsquigarrow$ in the T-type pseudo-BCI algebra.

$\rightsquigarrow$	***a***	***b***	***c***	***d***	***1***
a	1	1	1	*d*	1
b	*c*	1	1	*d*	1
c	*a*	*b*	1	*d*	1
d	*d*	*d*	*d*	1	*d*
1	*a*	*b*	*c*	*d*	1

Lemma 5 ([16,24]). *Let $(A; \rightarrow, 1)$ be a BCI-algebra. Then the following statements are equivalent:*

(1) *A is associative, that is,* $(x \to y) \to z = x \to (y \to z)$ *for any x, y, z in A;*
(2) *for any x in A,* $x \to 1 = x$*;*
(3) *for all x, y in A,* $x \to y = y \to x$.

Theorem 6. *Let* $(A; \leq, \to, \rightsquigarrow, 1)$ *be a weak associative pseudo-BCI algebra, AG(A) the anti-grouped part of A, K(A) the pseudo-BCK part of A. Then*

(1) *K(A) is quasi-alternating BCK-algebra and AG(A) = (A − K(A))∪{1};*
(2) *For any x in AG(A),* $x \to 1 = x \rightsquigarrow 1 = x$*;*
(3) *For any x, y in A,* $x \to y = x \rightsquigarrow y$, that is, A is a BCI-algebra;
(4) *AG(A) is an Abel group, that is, AG(A) is associative BCI-algebra.*

Proof. (1) It follows from Theorems 5 and 3.

(2) For any x in $AG(A)$, then $(x \to 1) \to 1 = x$. We will prove that $x \to 1 = x$.
If $x = 1$, obviously, $x \to 1 = x$.
If $x \neq 1$, then $(x \to 1) \to 1 = x \to 1$ by Definition 13. Thus,

$$x \to 1 = (x \to 1) \to 1 = x.$$

Applying Proposition 2 (12) we have

$$x \rightsquigarrow 1 = x \to 1 = x.$$

(3) For any x, y in A,

(i) when x, y in $K(A)$, by (1), $K(A)$ is a BCK-algebra, so $x \to y = x \rightsquigarrow y$;
(ii) when x, y in $(A - K(A))$, by (1) and (2), applying Proposition 2 (11),

$$x \to y = (x \to y) \to 1 = (x \to 1) \rightsquigarrow (y \to 1) = x \rightsquigarrow y;$$

(iii) when x in $K(A)$, y in $(A - K(A))$, using Lemma 3 (1), $x \to y = x \rightsquigarrow y$;
(iv) when y in $K(A)$, x in $(A - K(A))$, using Lemma 3 (2), $x \to y = x \rightsquigarrow y$;

Therefore, for all x, y in A, $x \to y = x \rightsquigarrow y$. It follows that A is a BCI-algebra.

(4) Applying (2), by Lemma 5 we know that $AG(A)$ is an Abel group, that is, AG(A) is associative BCI-algebra. □

From Theorems 6 and 4 we immediately get

Theorem 7. *Let* $(A; \leq, \to, \rightsquigarrow, 1)$ *be a pseudo-BCI algebra, AG(A) the anti-grouped part of A, K(A) the pseudo-BCK part of A. Then A is a weak associative pseudo-BCI algebra if and only if K(A) is a quasi-alternating BCK-algebra, AG(A) is an Abelian group and* $A = K(A) \oplus_{KG} AG(A)$.

Theorem 8. *Let* $(A; \leq, \to, \rightsquigarrow, 1)$ *be a pseudo-BCI algebra. Then the following conditions are equivalent:*

(1) *for any* $x, y, z \in A$, $(x \to y) \to z = x \to (y \to z)$ *when* $(x \neq y, x \neq z)$*;*
(2) *for any* $x, y, z \in A$, $(x \rightsquigarrow y) \rightsquigarrow z = x \rightsquigarrow (y \rightsquigarrow z)$ *when* $(x \neq y, x \neq z)$*;*
(3) *for any* $x, y, z \in A$, $(x \to y) \rightsquigarrow z = x \to (y \rightsquigarrow z)$ *when* $(x \neq y, x \neq z)$*;*
(4) *for any* $x, y, z \in A$, $(x \rightsquigarrow y) \to z = x \rightsquigarrow (y \to z)$ *when* $(x \neq y, x \neq z)$.

Proof. $(1) \Rightarrow (2)$: It follows from Definition 13 and Theorem 6.

$(2) \Rightarrow (1)$: Similar to the discussion process from Definition 13 to Theorem 6, we can obtain a result similar to Theorem 6. That is, from (2) we can get that A is a BCI-algebra. Hence, (2) implies (1).

Similarly, (3) ⇔ (1) and (4) ⇔ (1). □

Finally, we discuss the relationships among general pseudo-BCI algebras, QM-pseudo-BCI algebras and weak associative pseudo-BCI algebras (WA-pseudo-BCI algebras).

In fact, in every T-type pseudo-BCI algebra, there is a maximal WA-pseudo-BCI subalgebra. That is, if $(A; \leq, \rightarrow, \rightsquigarrow, 1)$ is a T-type pseudo-BCI algebra, $AG(A)$ the anti-grouped part of A, $K(A)$ the pseudo-BCK part of A, then $K_{qm}(A) \cup AG(A)$ is a WA-pseudo-BCI subalgebra of A, where $K_{qm}(A)$ is the set of all quasi-maximal element in $K(A)$. For example, $\{c, d, 1\}$ is a WA-pseudo-BCI subalgebra of the pseudo-BCI algebra A in Example 3.

In general, in every pseudo-BCI algebra, there is a maximal QM-pseudo-BCI subalgebra. That is, if $(A; \leq, \rightarrow, \rightsquigarrow, 1)$ is a pseudo-BCI algebra, $AG(A)$ the anti-grouped part of A, $K(A)$ the pseudo-BCK part of A, then $K_{qm}(A) \cup AG(A)$ is a QM-pseudo-BCI subalgebra of A, where $K_{qm}(A)$ is the set of all quasi-maximal element in $K(A)$.

6. Conclusions

In the study of pseudo-BCI algebras, the structures of various special pseudo-BCI algebras are naturally an important problem. At present, the structures of several subclasses such as quasi-alternating pseudo-BCI algebras and anti-grouped pseudo-BCI algebras are clear. In this paper, we have studied an important subclass of pseudo-BCI algebras, that is, QM-pseudo-BCI algebras in which every element is quasi-maximal. We obtain a very clear structure theorem of this subclass. At the same time, we have studied a class of more special pseudo-BCI algebras, that is, weak associative (WA) pseudo-BCI algebras in which every element is weak associative and obtained the structure theorem of this subclass. These results enrich the research content of pseudo-BCI algebras and clearly presented the relationships between various subclasses, which can be illustrated as Figure 1. Finally, we show that the two types of pseudo-BCI algebras are very important, since (1) every pseudo-BCI algebra contains a subalgebra which is QM-pseudo-BCI algebra, (2) every T-type pseudo-BCI algebra contains a subalgebra which is WA-pseudo-BCI algebra. As a further study direction, we will discuss the integration of related topics in the light of some new research findings in [32–34].

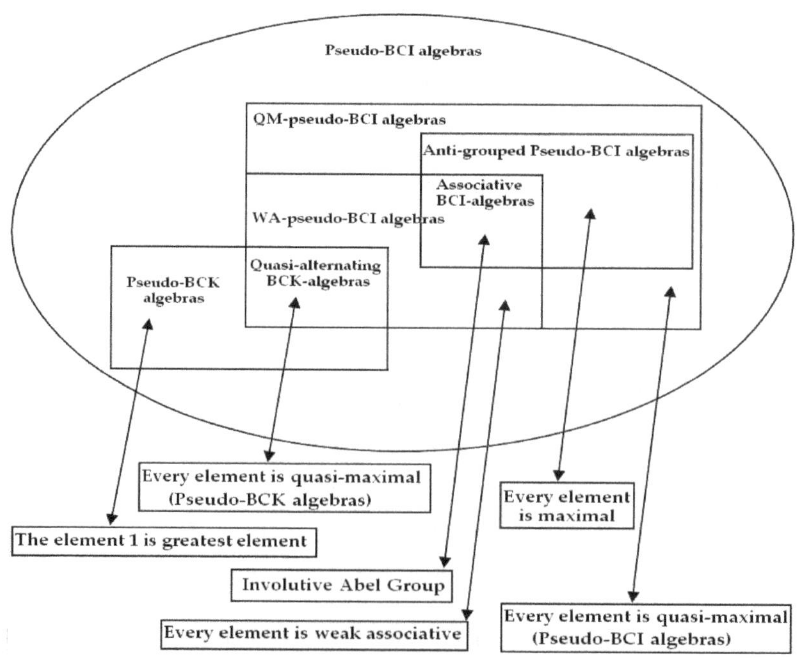

Figure 1. Main results in this paper.

Author Contributions: The contributions of the authors are roughly equal. X.Z. and X.W. initiated the research; X.W. wrote the draft and X.Z. completed final version.

References

1. Klement, E.P.; Mesiar, R.; Pap, E. *Triangular Norms*; Springer: Dordrecht, The Netherlands, 2000.
2. Flaminio, T. Strong non-standard completeness for fuzzy logics. *Soft Comput.* **2008**, *12*, 321–333. [CrossRef]
3. Aguzzoli, S.; Flaminio, T.; Ugolini, S. Equivalences between subcategories of MTL-algebras via Boolean algebras and prelinear semihoops. *J. Logic Comput.* **2017**, *27*, 2525–2549. [CrossRef]
4. Hájek, P. *Metamathematics of Fuzzy Logic*; Springer: Dordrecht, The Netherlands, 1998.
5. Esteva, F.; Godo, L. Monoidal t-norm based logic: Towards a logic for left-continuous t-norms. *Fuzzy Sets Syst.* **2001**, *124*, 271–288. [CrossRef]
6. Wang, G.J.; Zhou, H.J. *Introduction to Mathematical Logic and Resolution Principle*; Alpha International Science Ltd.: Oxford, UK, 2009.
7. Zhang, X.H. *Fuzzy Logics and Algebraic Analysis*; Science Press: Beijing, China, 2008.
8. Zhang, X.H.; Dudek, W.A. BIK+-logic and non-commutative fuzzy logics. *Fuzzy Syst. Math.* **2009**, *23*, 8–20.
9. Iorgulescu, A. *Implicative-Groups vs. Groups and Generalizations*; Bucharest Academy of Economic Studies: Bucharest, Romania, 2018.
10. Iséki, K. An algebra related with a propositional calculus. *Proc. Jpn. Acad.* **1966**, *42*, 26–29. [CrossRef]
11. Georgescu, G.; Iorgulescu, A. Pseudo-BCK algebras: An extension of BCK algebras. In *Combinatorics, Computability and Logic*; Springer: London, UK, 2001; pp. 97–114.
12. Dudek, W.A.; Jun, Y.B. Pseudo-BCI algebras. *East Asian Math. J.* **2008**, *24*, 187–190.
13. Pu, Y.S. Quasi-alternating BCI-algebra and others. *J. Hanzhong Teach. Coll.* **1985**, *3*, 1–8.
14. Zhang, X.H.; Yang, Y.B. A kind of BCI-algebra and its structure. *J. Northwest Norm. Univ. (Nat. Sci.)* **1992**, *28*, 27–31.
15. Dudek, W.A.; Zhang, X.H. On atoms in BCC-algebras, Discuss. *Math. Algebra Stoch. Methods* **1995**, *15*, 81–85.
16. Huang, Y.S. *BCI-Algebra*; Science Press: Beijing, China, 2006.
17. Dudek, W.A.; Zhang, X.H.; Wang, Y.Q. Ideals and atoms of BZ-algebras. *Math. Slov.* **2009**, *59*, 387–404.
18. Zhang, X.H. BCC-algebras and residuated partially-ordered groupoid. *Math. Slov.* **2013**, *63*, 397–410.
19. Dymek, G.; Kozanecka-Dymek, A. Pseudo-BCI-logic. *Bull. Sect. Logic* **2013**, *42*, 33–41.
20. Emanovský, P.; Kühr, J. Some properties of pseudo-BCK- and pseudo-BCI-algebras. *Fuzzy Sets Syst.* **2018**, *339*, 1–16. [CrossRef]
21. Xin, X.L.; Fu, Y.L.; Lai, Y.Y.; Wang, J.T. Monadic pseudo BCI-algebras and corresponding logics. *Soft Comput.* **2018**, 1–12. [CrossRef]
22. Jun, Y.B.; Kim, H. S.; Neggers, J. On pseudo-BCI ideals of pseudo-BCI algebras. *Mat. Vesnik* **2006**, *58*, 39–46.
23. Dymek, G. Atoms and ideals of pseudo-BCI-algebras. *Comment. Math.* **2012**, *52*, 73–90.
24. Zhang, X.H.; Jun, Y.B. Anti-grouped pseudo-BCI algebras and anti-grouped filters. *Fuzzy Syst. Math.* **2014**, *28*, 21–33.
25. Zhang, X.H. Fuzzy anti-grouped filters and fuzzy normal filters in pseudo-BCI algebras. *J. Intell. Fuzzy Syst.* **2017**, *33*, 1767–1774. [CrossRef]
26. Zhang, X.H.; Gong, H.J. Implicative pseudo-BCK algebras and implicative pseudo-filters of pseudo-BCK algebras. In Proceedings of the 2010 IEEE International Conference on Granular Computing, San Jose, CA, USA, 14–16 August 2010; Volume 62, pp. 615–619.
27. Zhang, X.H. Pseudo-BCK part and anti-grouped part of pseudo-BCI algebras. In Proceedings of the IEEE International Conference on Progress in Informatics and Computing, Shanghai, China, 10–12 December 2010; Volume 1, pp. 127–131.
28. Zhang, X.H.; Lu, Y.F.; Mao, X.Y. T-type pseudo-BCI algebras and T-type pseudo-BCI filters. In Proceedings of the 2010 IEEE International Conference on Granular Computing, San Jose, CA, USA, 14–16 August 2010; pp. 839–844.
29. Dymek, G. On a period of elements of pseudo-BCI algebras. *Discuss. Math. Gener. Algebra Appl.* **2015**, *35*, 21–31. [CrossRef]
30. Xin, X.L.; Li, Y.J.; Fu, Y.L. States on pseudo-BCI algebras. *Eur. J. Pure Appl. Math.* **2017**, *10*, 455–472.
31. Zhang, X.H.; Park, C.; Wu, S.P. Soft set theoretical approach to pseudo-BCI algebras. *J. Intell. Fuzzy Syst.* **2018**, *34*, 559–568.

32. Zhang, X.H.; Smarandache, F.; Liang, X.L. Neutrosophic duplet semi-group and cancellable neutrosophic triplet groups. *Symmetry* **2017**, *9*, 275. [CrossRef]
33. Zhang, X.H.; Bo, C.X.; Smarandache, F.; Park, C. New operations of totally dependent- neutrosophic sets and totally dependent-neutrosophic soft sets. *Symmetry* **2018**, *10*, 187. [CrossRef]
34. Zhang, X.H.; Bo, C.X.; Smarandache, F.; Dai, J.H. New inclusion relation of neutrosophic sets with applications and related lattice structure. *Int. J. Mach. Learn. Cyber.* **2018**, *9*, 1753–1763. [CrossRef]

13

Maximum Detour–Harary Index for Some Graph Classes

Wei Fang [1], Wei-Hua Liu [2,*], Jia-Bao Liu [3], Fu-Yuan Chen [4], Zhen-Mu Hong [5] and Zheng-Jiang Xia [5]

1 College of Information & Network Engineering, Anhui Science and Technology University, Fengyang 233100, China; fangw@ahstu.edu.cn
2 College of Information and Management Science, Henan Agricultural University, Zhengzhou 450002, China
3 School of Mathematics and Physics, Anhui Jianzhu University, Hefei 230601, China; liujiabaoad@163.com
4 Institute of Statistics and Applied Mathematics, Anhui University of Finance and Economics, Bengbu 233030, China; accfy2016@163.com
5 School of Finance, Anhui University of Finance and Economics, Bengbu 233030, China; zmhong@mail.ustc.edu.cn (Z.-M.H.); 120150025@aufe.edu.cn (Z.-J.X.)
* Correspondence: liuwhnuc@sina.com

Abstract: The definition of a Detour–Harary index is $\omega H(G) = \frac{1}{2}\sum_{u,v\in V(G)}\frac{1}{l(u,v|G)}$, where G is a simple and connected graph, and $l(u,v|G)$ is equal to the length of the longest path between vertices u and v. In this paper, we obtained the maximum Detour–Harary index about unicyclic graphs, bicyclic graphs, and cacti, respectively.

Keywords: Detour–Harary index; maximum; unicyclic; bicyclic; cacti

1. Introduction

In recent years, chemical graph theory (CGT) has been fast-growing. It helps researchers to understand the structural properties of a molecular graph, for example, References [1–3].

A simple graph is an undirected graph without multiple edges and loops. Let G be a simple and connected graph, and $V(G)$ and $E(G)$ be the vertex set and edge set of G, respectively. For vertices u, v of G, $d_G(v_1, v_2)$ (or $d(v_1, v_2)$ for short) is the distance between v_1 and v_2, which equals to the length of the shortest path between v_1 and v_2 in G; $l(v_1, v_2|G)$ (or $l(v_1, v_2)$ for short) is the detour distance between v_1 and v_2, which equals to the longest path of a shortest path between v_1 and v_2 in G.

$G[S]$ is an induced subgraph of G, the vertex set is S, and the edge set is the set of edges of G and both ends in S. $G - S$ is the induced subgraph $G[V(G) \setminus S]$; when $S = \{w\}$, we write $G - w$ for short.

In 1947, Wiener introduced the first molecular topological index–Wiener index. The Wiener index has applications in many fields, such as chemistry, communication, and cryptology [4–7]. Moreover, the Wiener index was studied from a purely graph-theoretical point of view [8–10]. In Reference [11], Wiener gave the definition of the Wiener index:

$$W(G) = \frac{1}{2}\sum_{u,v\in V(G)} d(u,v).$$

The Harary index was independently introduced by Plavšić et al. [12] and by Ivanciuc et al. [13] in 1993. In References [12,13], they gave the definition of the Harary index:

$$H(G) = \frac{1}{2}\sum_{u,v\in V(G)} \frac{1}{d(u,v)}.$$

In Reference [13], Ivanciuc gave the definition of the Detour index:

$$\omega(G) = \frac{1}{2} \sum_{u,v \in V(G)} l(u,v|G).$$

Lukovits [14] investigated the use of the Detour index in quantitative structure–activity relationship (QSAR) studies. Trinajstić and his collaborators [15] analyzed the use of the Detour index, and compared its application with Wiener index. They found that the Detour index in combination with the Wiener index is very efficient in the structure-boiling point modeling of acyclic and cyclic saturated hydrocarbons.

In this paper, we introduce a new graph invariant reciprocal to the Detour index, namely, the Detour–Harary index, as

$$\omega H(G) = \frac{1}{2} \sum_{u,v \in V(G)} \frac{1}{l(u,v|G)}.$$

Let G be a simple and connected graph, $V(G) = n$ and $E(G) = m$. If $m = n - 1$, then G is a tree; if $m = n$, then G is a unicyclic graph; if $m = n + 1$, then G is a bicyclic graph.

Suppose $\mathcal{U}_n(\mathcal{B}_n$, respectively) is the set of unicyclic (bicyclic, respectively) graphs set with n vertices. Any bicyclic graph G can be obtained from $\theta(p,q,l)$-graph or $\theta(p,q,l)$-graph G_0 by attaching trees to the vertices, where $p, q, l \geq 1$, and at most one of them is equal to 1. We denote G_0 be the kernel of G (Figure 1).

If each block of G is either a cycle or an edge, then we called graph G a *cactus* graph. Suppose $\mathcal{C}_n^k$ be the set of all cacti with n-vertices and k cycles. Obviously, $\mathcal{C}_n^0$ are trees, $\mathcal{C}_n^1$ are unicyclic graphs, and $\mathcal{C}_n^2$ are bicyclic graphs with exactly two cycles.

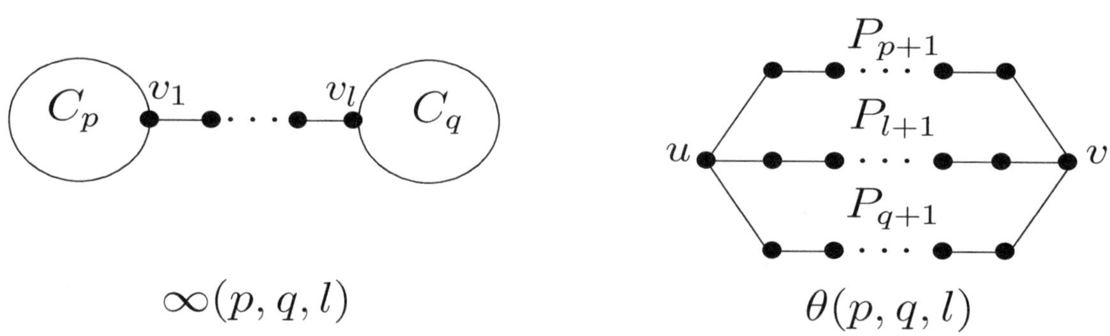

Figure 1. ∞-graph and θ-graph.

There are more results about cacti and bicyclic graphs [16–25]. More results about Harary index can be found in References [26–34], and more results about Detour index can be found in References [14,35–39].

Note that the Detour–Harary index is the same as Harary index for a tree graph; we study the Detour–Harary index of topological structures containing cycles. In this paper, we gave the maximum Detour–Harary index among $\mathcal{U}_n$,$\mathcal{B}_n$ and $\mathcal{C}_n^k$ $(k \geq 3)$, respectively.

2. Preliminaries

In this section, we introduce useful lemmas and graph transformations.

Lemma 1. *[40] Let G be a connected graph, x be a cut-vertex of G, and u and v be vertices occurring in different components that arise upon the deletion of vertex x. Then*

$$l(u,v|G) = l(u,x|G) + l(x,v|G).$$

2.1. Edge-Lifting Transformation

The edge-lifting transformation [41]. Let G_1 and G_2 be two graphs with $n_1 \geq 2$ and $n_2 \geq 2$ vertices. $u_0 \in V(G_1)$ and $v_0 \in V(G_2)$, G is the graph obtained from G_1 and G_2 by adding an edge between u_0 and v_0. G' is the graph obtained by identifying u_0 to v_0 and adding a pendent edge to $u_0(v_0)$. We called graph G' the edge-lifting transformation of graph G (see Figure 2).

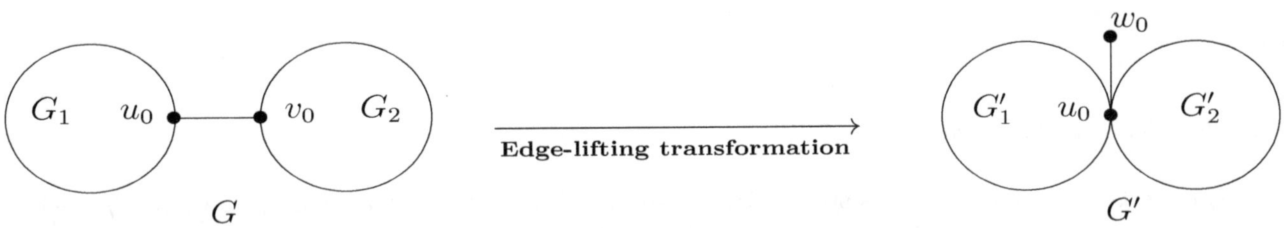

Figure 2. Edge-lifting transformation.

Lemma 2. *Let graph G' be the edge-lifting transformation of graph G. Then $\omega H(G) < \omega H(G')$.*

Proof. By the definition of $\omega H(G)$ and Lemma 1,

$$\begin{aligned}
\omega H(G) =\ & \omega H(G_1) + \omega H(G_2) + \sum_{x \in V(G_1)\backslash\{u_0\}} \frac{1}{l(v_0, x|G)} + \sum_{y \in V(G_2)\backslash\{v_0\}} \frac{1}{l(u_0, y|G)} \\
& + \frac{1}{l(u_0, v_0|G)} + \sum_{\substack{x \in V(G_1)\backslash\{u_0\} \\ y \in V(G_2)\backslash\{v_0\}}} \frac{1}{l(x, y|G)} \\
=\ & \omega H(G_1) + \omega H(G_2) + \sum_{x \in V(G_1)\backslash\{u_0\}} \frac{1}{1 + l(u_0, x|G)} + \sum_{y \in V(G_2)\backslash\{v_0\}} \frac{1}{1 + l(v_0, y|G)} \\
& + 1 + \sum_{\substack{x \in V(G_1)\backslash\{u_0\} \\ y \in V(G_2)\backslash\{v_0\}}} \frac{1}{l(u_0, x|G) + 1 + l(v_0, y|G)}, \\
\omega H(G') =\ & \omega H(G_1') + \omega H(G_2') + \sum_{x' \in V(G_1')\backslash\{u_0\}} \frac{1}{l(w_0, x'|G')} + \sum_{y' \in V(G_2')\backslash\{u_0\}} \frac{1}{l(w_0, y'|G')} \\
& + \frac{1}{l(u_0, w_0|G')} + \sum_{\substack{x' \in V(G_1')\backslash\{u_0\} \\ y' \in V(G_2')\backslash\{u_0\}}} \frac{1}{l(x', y'|G')} \\
=\ & \omega H(G_1') + \omega H(G_2') + \sum_{x' \in V(G_1')\backslash\{u_0\}} \frac{1}{1 + l(u_0, x'|G')} + \sum_{y' \in V(G_2')\backslash\{u_0\}} \frac{1}{1 + l(u_0, y'|G')} \\
& + 1 + \sum_{\substack{x' \in V(G_1')\backslash\{u_0\} \\ y' \in V(G_2')\backslash\{u_0\}}} \frac{1}{l(u_0, x'|G') + l(u_0, y'|G')}.
\end{aligned}$$

Obviously,

$$\begin{aligned}
\omega H(G_1) &= \omega H(G_1'); \\
\omega H(G_2) &= \omega H(G_2'); \\
l(u_0, x|G) &= l(u_0, x'|G'), \text{ where } x \in V(G_1) \setminus \{u_0\} \text{ and } x' \in V(G_1') \setminus \{u_0\}; \\
l(v_0, y|G) &= l(u_0, y'|G'), \text{ where } y \in V(G_2) \setminus \{v_0\} \text{ and } y' \in V(G_2') \setminus \{u_0\}.
\end{aligned}$$

Then

$$\omega H(G) - \omega H(G') = \sum_{\substack{x \in V(G_1)\setminus\{u_0\} \\ y \in V(G_2)\setminus\{v_0\}}} \frac{1}{l(x, u_0|G) + 1 + l(v_0, y|G)} - \sum_{\substack{x' \in V(G_1')\setminus\{u_0\} \\ y' \in V(G_2')\setminus\{u_0\}}} \frac{1}{l(x', u_0|G') + l(u_0, y'|G')} < 0.$$

□

2.2. Cycle-Edge Transformation

Suppose $G \in \mathcal{C}_n^l$ is a cactus as shown in Figure 3. $C_p = v_1 v_2 \cdots v_p v_1$ is a cycle of G; G_i is a cactus, and $v_i \in V(G_i)$, $1 \le i \le p$; $W_{v_i} = N_G(v_i) \cap V(G_i)$, $1 \le i \le p$. G' is the graph obtained from G by deleting the edges from v_i to $W_{v_i} (2 \le i \le p)$, while adding the edges from v_1 to $W_{v_i} (2 \le i \le p)$.

We called graph G' the cycle-edge transformation of graph G (see Figure 3).

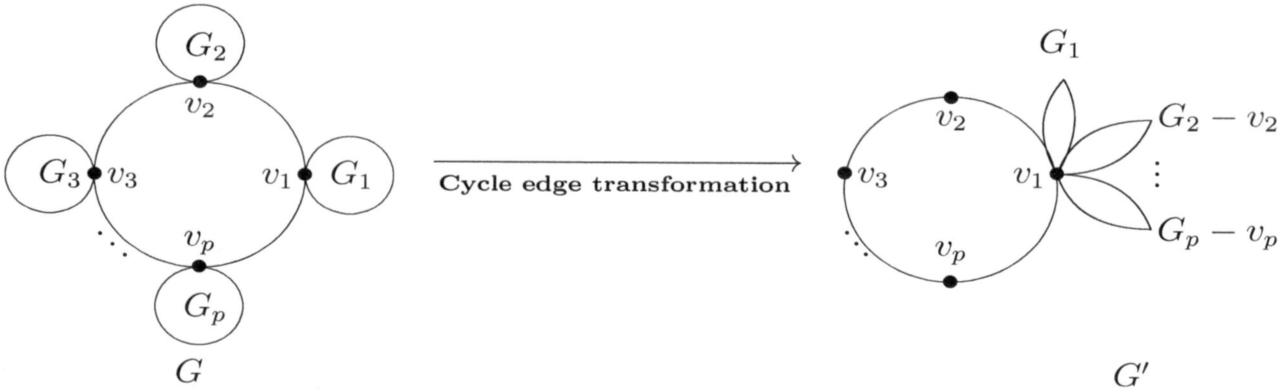

Figure 3. Cycle-edge transformation.

Lemma 3. *Suppose $G \in \mathcal{C}_n^l$ is a cactus, $p \ge 3$, and G' is the cycle-edge transformation of G (see Figure 3). Then, $\omega H(G) \le \omega H(G')$, and the equality holds if and only if $G \cong G'$.*

Proof. Let $V_i = V(G_i - v_i)$, $1 \le i \le p$. By the definition of $\omega H(G)$ and Lemma 1,

$$\begin{aligned}\omega H(G) &= \omega H(C_p) + \frac{1}{2}\sum_{i=1}^{p}\sum_{x,y \in V_i} \frac{1}{l(x,y|G)} + \frac{1}{2}\sum_{i=1}^{p}\sum_{j=1}^{p}\sum_{\substack{x \in V_i \\ y \in V_j \\ i \neq j}} \frac{1}{l(x,y|G)} + \sum_{i=1}^{p}\sum_{\substack{x \in V_i \\ y \in V(C_p)}} \frac{1}{l(x,y|G)} \\ &= \omega H(C_p) + \frac{1}{2}\sum_{i=1}^{p}\sum_{x,y \in V_i} \frac{1}{l(x,y|G)} + \frac{1}{2}\sum_{i=1}^{p}\sum_{j=1}^{p}\sum_{\substack{x \in V_i \\ y \in V_j \\ i \neq j}} \frac{1}{l(x,v_i|G) + l(v_i,v_j|G) + l(v_j,y|G)} \\ &\quad + \sum_{i=1}^{p}\sum_{\substack{x \in V_i \\ y \in V(C_p)}} \frac{1}{l(x,v_i|G) + l(v_i,y|G)},\end{aligned}$$

$$\begin{aligned}\omega H(G') &= \omega H(C_p) + \frac{1}{2}\sum_{i=1}^{p}\sum_{x,y\in V_i}\frac{1}{l(x,y|G')} + \frac{1}{2}\sum_{i=1}^{p}\sum_{j=1}^{p}\sum_{\substack{x\in V_i\\ y\in V_j\\ i\neq j}}\frac{1}{l(x,y|G')} + \sum_{i=1}^{p}\sum_{\substack{x\in V_i\\ y\in V(C_p)}}\frac{1}{l(x,y|G')}\\ &= \omega H(C_p) + \frac{1}{2}\sum_{i=1}^{p}\sum_{x,y\in V_i}\frac{1}{l(x,y|G')} + \frac{1}{2}\sum_{i=1}^{p}\sum_{j=1}^{p}\sum_{\substack{x\in V_i\\ y\in V_j\\ i\neq j}}\frac{1}{l(x,v_1|G')+l(v_1,y|G')}\\ &\quad + \sum_{i=1}^{p}\sum_{\substack{x\in V_i\\ y\in V(C_p)}}\frac{1}{l(x,v_1|G')+l(v_1,y|G')}.\end{aligned}$$

Obviously,

$$\begin{aligned}\sum_{i=1}^{p}\sum_{x,y\in V_i}\frac{1}{l(x,y|G)} &= \sum_{i=1}^{p}\sum_{x,y\in V_i}\frac{1}{l(x,y|G')};\\ l(x,v_i|G) &= l(x,v_1|G'), \text{where } x\in V_i;\\ l(v_j,y|G) &= l(v_1,y|G'), \text{where } y\in V_j;\\ \sum_{i=1}^{p}\sum_{\substack{x\in V_i\\ y\in V(C_p)}}\frac{1}{l(x,v_i|G)+l(v_i,y|G)} &= \sum_{i=1}^{p}\sum_{\substack{x\in V_i\\ y\in V(C_p)}}\frac{1}{l(x,v_1|G')+l(v_1,y|G')}.\end{aligned}$$

Then

$$\begin{aligned}\omega H(G)-\omega H(G') &= \frac{1}{2}\sum_{i=1}^{p}\sum_{j=1}^{p}\sum_{\substack{x\in V_i\\ y\in V_j\\ i\neq j}}\frac{1}{l(x,v_i|G)+l(v_i,v_j|G)+l(v_j,y|G)}\\ &\quad -\frac{1}{2}\sum_{i=1}^{p}\sum_{j=1}^{p}\sum_{\substack{x\in V_i\\ y\in V_j\\ i\neq j}}\frac{1}{l(x,v_1|G')+l(v_1,y|G')} < 0.\end{aligned}$$

The proof is completed. □

2.3. Cycle Transformation

Suppose $G \in \mathcal{C}_n^l$ is a cactus, as shown in Figure 4. $C_p = v_1v_2\cdots v_pv_1$ is a cycle of G, and G_1 is a simple and connected graph, $v_1 \in V(G_1)$. G' is the graph obtained from G by deleting the edges from v_i to $v_{i+1}(2 \le i \le p-1)$, meanwhile, adding the edges from v_1 to $v_i(3 \le i \le p-1)$.

We called graph G' is the *cycle transformation* of G (see Figure 4).

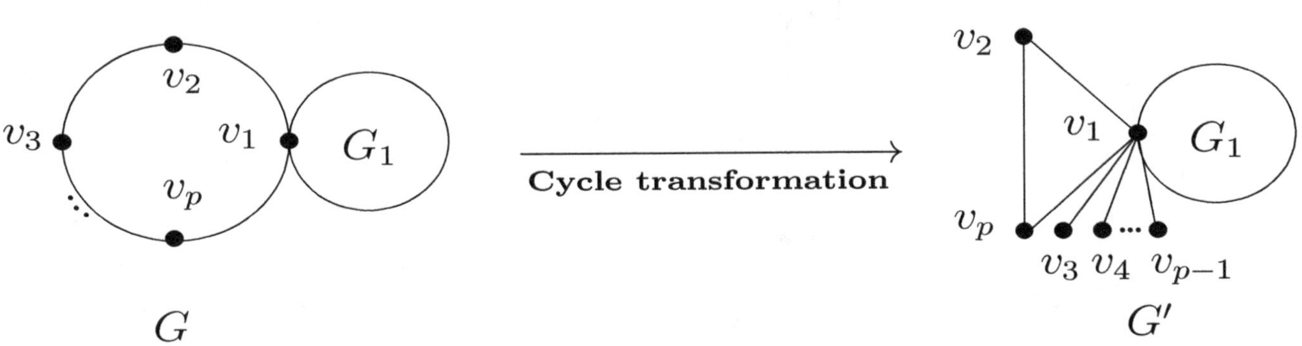

Figure 4. Cycle transformation.

Lemma 4. *Suppose graph G is a simple and connected graph with $p \geq 4$, and G' is the cycle transformation of G(see Figure 4). Then, $\omega H(G) < \omega H(G')$.*

Proof. Let $V(C_p) = \{v_1, v_2, \cdots, v_p\}$, $V_1 = V(C_p - v_1)$, $V_2 = V(G_1 - v_1)$. By the definition of $\omega H(G)$,

$$\begin{aligned}
\omega H(G) &= \omega H(G_1) + \sum_{x,y \in V(C_p)} \frac{1}{l(x,y|G)} + \sum_{\substack{x \in V_1 \\ y \in V_2}} \frac{1}{l(x,y|G)} \\
&= \omega H(G_1) + \sum_{x,y \in V(C_p)} \frac{1}{l(x,y|G)} + \sum_{\substack{x \in V_1, \\ y \in V_2}} \frac{1}{l(x,v_1|G) + l(v_1,y|G)}, \\
\omega H(G') &= \omega H(G_1) + \sum_{x,y \in V(C_p)} \frac{1}{l(x,y|G')} + \sum_{\substack{x \in V_1, \\ y \in V_2}} \frac{1}{l(x,y|G')} \\
&= \omega H(G_1) + \sum_{x,y \in V(C_p)} \frac{1}{l(x,y|G')} + \sum_{\substack{x \in V_1, \\ y \in V_2}} \frac{1}{l(x,v_1|G') + l(v_1,y|G')},
\end{aligned}$$

Obviously,

$$\begin{aligned}
l(x,y|G) &\geq l(x,y|G'), \text{where } x, y \in V_1; \\
l(x,v_1|G) &> 2 \geq l(x,v_1|G'), \text{where } x \in V_1; \\
l(v_1,y|G) &= l(v_1,y|G'), \text{where } y \in V_2.
\end{aligned}$$

Then

$$\begin{aligned}
\omega H(G) - \omega H(G') = &\left(\sum_{x,y \in V(C_p)} \frac{1}{l(x,y|G)} - \sum_{x,y \in V(C_p)} \frac{1}{l(x,y|G')} \right) \\
&+ \left(\sum_{\substack{x \in V_1, \\ y \in V_2}} \frac{1}{l(x,v_1|G) + l(v_1,y|G)} - \sum_{\substack{x \in V_1, \\ y \in V_2}} \frac{1}{l(x,v_1|G') + l(v_1,y|G')} \right) < 0.
\end{aligned}$$

□

3. Maximum Detour–Harary Index of Unicyclic Graphs

For any unicyclic graph $G \in \mathcal{U}_n$, by repeating edge-lifting transformations, cycle-edge transformations, cycle transformations, or any combination of these on G, we get U_1 from G, where graph U_1 is defined in Figure 5.

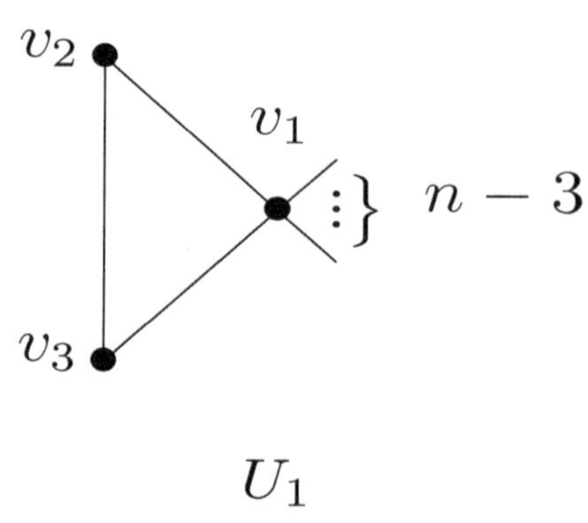

Figure 5. Unicyclic graph U_1.

Theorem 1. *Let U_1 be defined as Figure 5. Then, U_1 is the unique graph that attains the maximum Detour–Harary index among all graphs in $\mathcal{U}_n (n \geq 3)$, and $\omega H(U_1) = \frac{3n^2-n-6}{12}$.*

Proof. By Lemmas 2–4, U_1 is the unique graph which attains the maximum Detour–Harary index of all graphs in $\mathcal{U}_n$. We then calculate the value $\omega H(U_1)$.

Let $V(U_1) = \{v_1, v_2, \cdots, v_n\}$. It can be checked directly that

$$\sum_{i=2}^{n} \frac{1}{l(v_1, v_i|U_1)} = n-2;$$

$$\sum_{1\leq i\leq n, i\neq 2} \frac{1}{l(v_2, v_i|U_1)} = \sum_{1\leq j\leq n, j\neq 3} \frac{1}{l(v_3, v_j|U_1)} = \frac{1}{2} + \frac{1}{2} + \frac{n-3}{3} = \frac{n}{3};$$

$$\sum_{1\leq i\leq n, i\neq 4} \frac{1}{l(v_4, v_i|U_1)} = 1 + \frac{n-4}{2} + \frac{2}{3} = \frac{3n-2}{6}.$$

Then

$$\begin{aligned}\omega H(U_1) &= \frac{1}{2}[\sum_{i=2}^{n} \frac{1}{l(v_1, v_i|U_1)} + 2\sum_{1\leq i\leq n, i\neq 2} \frac{1}{l(v_2, v_i|U_1)} + (n-3)\sum_{i=1}^{n} \frac{1}{l(v_4, v_i|U_1)}] \\ &= \frac{3n^2-n-6}{12}.\end{aligned}$$

The proof is completed. □

4. Maximum Detour–Harary Index of Bicyclic Graphs

For any bicyclic graph $G \in \infty(p,q,l)$ with exactly two cycles, by repeating edge-lifting transformations, cycle-edge transformations, cycle transformations, or any combination of these on G, we get B_1 from G, where graph B_1 is defined in Figure 6.

For any bicyclic graph $G \in \theta(p,q,l)$ with n vertices, by repeating edge-lifting transformations on G, we get B_2 from G, where graph B_2 is defined in Figure 7.

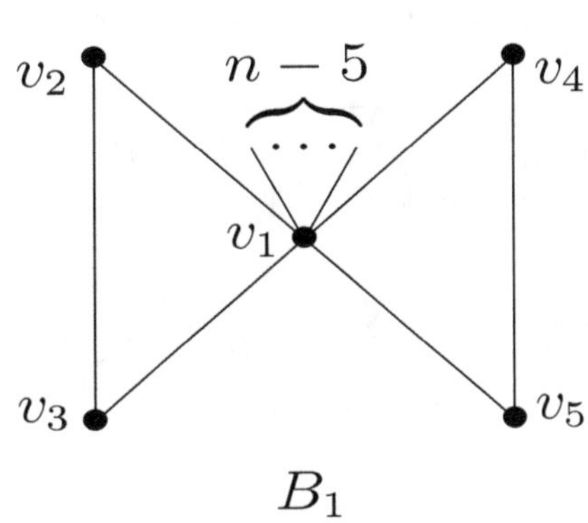

Figure 6. Bicyclic graph B_1.

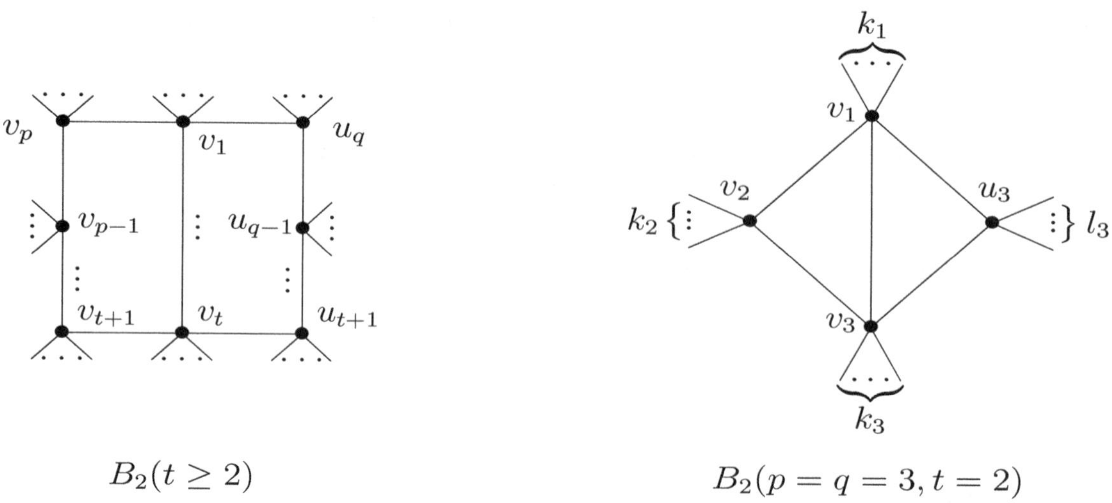

Figure 7. Bicyclic graph $B_2(t \geq 2)$.

Theorem 2. *Let B_2, B_3 be defined as Figures 7 and 8. Then, $\omega H(B_2) \leq \omega H(B_3)$, and the equality holds if and only if $B_2 \cong B_3$.*

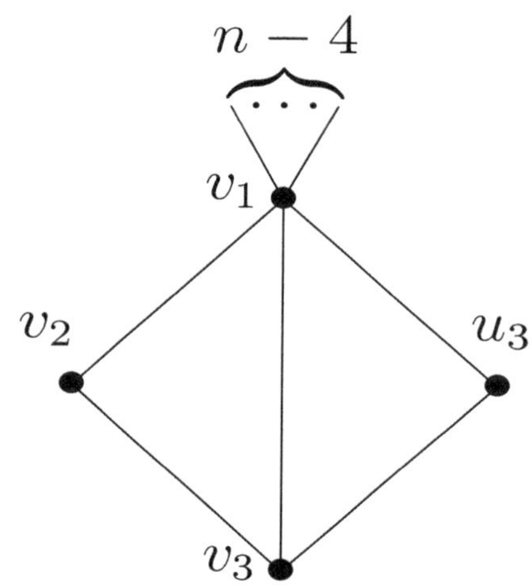

Figure 8. Bicyclic graph $B_2(t \geq 2)$.

Proof. Case 1. $B_2 = B_3$. Obviously, $\omega H(B_2) = \omega H(B_3)$.

Case 2. $B_2 \neq B_3$ and $p = q = 3, t = 2$(see Figures 7 and 8).

Let $V_1 = \{v_1, v_2, v_3, u_3\}$, $W_{v_i} = \{w \mid wv_i \in E(B_2) \text{ and } d_{B_2}(w) = 1\}$ and $|W_{v_i}| = k_i$, $W_{u_3} = \{w \mid wu_3 \in E(B_2) \text{ and } d_{B_2}(w) = 1\}$ and $|W_{u_3}| = l_3, k_i + l_3 = n - 4$ for $1 \leq i \leq 3$.

$$\omega H(B_2) = \sum_{x,y \in V_1} \frac{1}{l(x,y|B_2)} + \sum_{\substack{x \in V_1, \\ y \in V(B_2) - V_1}} \frac{1}{l(x,y|B_2)} + \sum_{x,y \in V(B_2) - V_1} \frac{1}{l(x,y|B_2)},$$

$$\omega H(B_3) = \sum_{x,y \in V_1} \frac{1}{l(x,y|B_3)} + \sum_{\substack{x \in V_1, \\ y \in V(B_3) - V_1}} \frac{1}{l(x,y|B_3)} + \sum_{x,y \in V(B_3) - V_1} \frac{1}{l(x,y|B_3)}.$$

Easily,

$$\sum_{x,y \in V_1} \frac{1}{l(x,y|B_2)} = \sum_{x,y \in V_1} \frac{1}{l(x,y|B_3)} \tag{1}$$

$$\begin{aligned}\sum_{\substack{x\in V_1,\\ y\in V(B_2)-V_1}} \frac{1}{l(x,y|B_2)} &= \sum_{w\in V(B_2)-V_1} \frac{1}{l(v_1,w|B_2)} + \sum_{w\in V(B_2)-V_1} \frac{1}{l(v_2,w|B_2)} \\ &+ \sum_{w\in V(B_2)-V_1} \frac{1}{l(v_3,w|B_2)} + \sum_{w\in V(B_2)-V_1} \frac{1}{l(u_3,w|B_2)} \\ &= (1\cdot k_1 + \frac{1}{4}\cdot k_2 + \frac{1}{3}\cdot k_3 + \frac{1}{4}\cdot l_3) + (\frac{1}{4}\cdot k_1 + 1\cdot k_2 + \frac{1}{4}\cdot k_3 + \frac{1}{4}\cdot l_3) \\ &+ (\frac{1}{3}\cdot k_1 + \frac{1}{4}\cdot k_2 + 1\cdot k_3 + \frac{1}{4}\cdot l_3) + (\frac{1}{4}\cdot k_1 + \frac{1}{4}\cdot k_2 + \frac{1}{4}\cdot k_3 + 1\cdot l_3) \\ &= \frac{11(k_1+k_3)}{6} + \frac{7(k_2+l_3)}{4},\end{aligned}$$

$$\begin{aligned}\sum_{\substack{x\in V_1,\\ y\in V(B_3)-V_1}} \frac{1}{l(x,y|B_3)} &= \sum_{w\in V(B_3)-V_1} \frac{1}{l(v_1,w|B_3)} + \sum_{w\in V(B_3)-V_1} \frac{1}{l(v_2,w|B_3)} \\ &+ \sum_{w\in V(B_3)-V_1} \frac{1}{l(v_3,w|B_3)} + \sum_{w\in V(B_3)-V_1} \frac{1}{l(u_3,w|B_3)} \\ &= 1\cdot(n-4) + \frac{1}{4}\cdot(n-4) + \frac{1}{3}\cdot(n-4) + \frac{1}{4}\cdot(n-4) \\ &= \frac{11(n-4)}{6} \\ &= \frac{11(k_1+k_2+k_3+l_3)}{6}, \qquad (\text{since } k_i + l_3 = n-4 \text{ for } 1\le i\le 3)\end{aligned}$$

Then,

$$\sum_{\substack{x\in V_1,\\ y\in V(B_2)-V_1}} \frac{1}{l(x,y|B_2)} - \sum_{\substack{x\in V_1,\\ y\in V(B_3)-V_1}} \frac{1}{l(x,y|B_3)} = \frac{1}{12}(k_2+l_3) \ge 0, \tag{2}$$

the equality holds if and only if $k_2 = l_3 = 0$.

On the other hand $\frac{1}{l(x,y|B_2)} \le \frac{1}{l(x,y|B_3)} = \frac{1}{2}$, where $x, y \in V(B_2) - V_1$, then

$$\sum_{x,y\in V(B_3)-V_1} \frac{1}{l(x,y|B_2)} \le \sum_{x,y\in V(B_3)-V_1} \frac{1}{l(x,y|B_3)}, \tag{3}$$

the equality holds if $k_1 = n-4$ or $k_2 = n-4$ or $k_3 = n-4$ or $l_3 = n-4$.

By (1)–(3) and $B_2 \ne B_3$, we have $\omega H(B_2) < \omega H(B_3)$.

Case 3. $B_2 \ne B_3$ and $p+q-t > 4$.

It can be checked directly that

$$\omega H(B_2) \le \underbrace{(1+1+\cdots+1)}_{n-p-q+t} + \tfrac{1}{2}\binom{n-p-q+t}{2} + \tfrac{1}{4}\left[\binom{n}{2} - (n-p-q+t) - \binom{n-p-q+t}{2}\right],$$

$$\omega H(B_3) = \underbrace{(1+1+\cdots+1)}_{n-4} + \tfrac{1}{2}\left[1+\binom{n-4}{2}\right] + \tfrac{1}{3}[5+(n-4)] + \tfrac{1}{4}[2(n-4)].$$

B_2, B_3 are bicyclic graphs and $|V(B_2)| = |V(B_3)| = n$. Since $p+q-t > 4$, then $n-p-q+t \le n-5$ and $\binom{n-p-q+t}{2} < \binom{n-4}{2}$, we have $\omega H(B_2) < \omega H(B_3)$.

The proof is completed. □

Theorem 3. *Let B_1, B_3 be defined as Figures 6 and 8. Then,*

$$\max\{\omega H(\mathcal{B}_n)\} = \begin{cases} \omega H(B_3) = \frac{13}{6}, & \text{if } n = 4, \\ \omega H(B_1) = \omega H(B_3) = \frac{3n^2-5n-2}{12}, & \text{if } n \ge 5. \end{cases}$$

Proof. Let $G \in \infty(p,q,l)$, by Lemmas 2–4, we have $\omega H(G) \le \omega H(B_1)$, and the equality holds if and only if $G \cong B_1$.

For any bicyclic graph with $G \in \theta(p,q,l)$, by Lemmas 2–4 and Theorem 2, we have $\omega H(G) \le \omega H(B_3)$, and the equality holds if and only if $G \cong B_3$. Thus, $\max\{\omega H(\mathcal{B}_n)\} = \max\{\omega H(B_1), \omega H(B_3)\}$.

It can be checked directly that

$$\omega H(B_1) = (n-5) + \frac{1}{2}[\binom{n-5}{2} + 6] + \frac{1}{3}[4(n-5)] + \frac{1}{4}\cdot 4 = \frac{3n^2-5n-2}{12}, n \ge 5;$$

$$\omega H(B_3) = (n-4) + \frac{1}{2}\binom{n-4}{2} + \frac{1}{3}(n-4) + \frac{1}{4}[2(n-4)] = \frac{3n^2-5n-2}{12}, n \ge 4.$$

Therefore

$$\max\{\omega H(\mathcal{B}_n)\} = \begin{cases} \omega H(B_3) = \frac{13}{6}, & \text{if } n = 4, \\ \omega H(B_1) = \omega H(B_3) = \frac{3n^2-5n-2}{12}, & \text{if } n \ge 5. \end{cases}$$

The proof is completed. □

5. Maximum Detour–Harary Index of Cacti

For any cactus graph $G \in \mathcal{C}_n^k (k \ge 3)$, by repeating edge-lifting transformations, cycle-edge transformations, cycle transformations, or any combination of these on G, we get $\mathcal{C}_1$ from G, where graph $\mathcal{C}_1$ is defined in Figure 9.

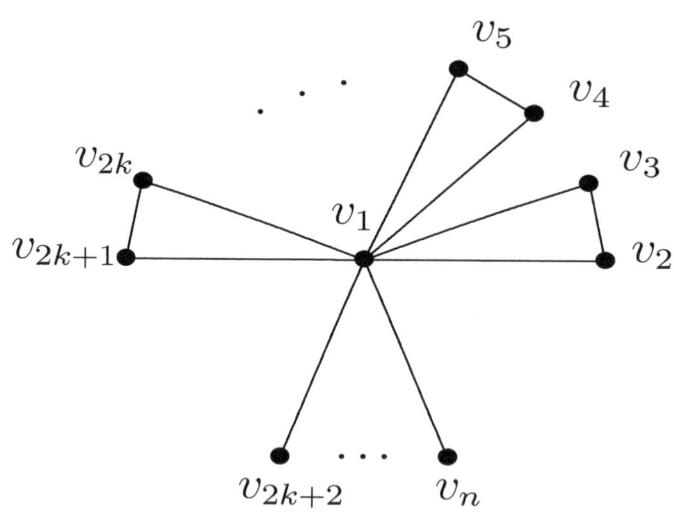

Figure 9. Cactus graph $\mathcal{C}_1 (k \ge 3)$.

Theorem 4. *Let $\mathcal{C}_1$ be defined as Figure 9. Then, $\mathcal{C}_1$ is the unique cactus graph in $\mathcal{C}_n^k (k \ge 3)$ that attains the maximum Detour–Harary index, and $\omega H(\mathcal{C}_1) = \frac{3n^2+2k^2-4nk+3n-2k-6}{12}$.*

Proof. By Lemmas 2–4, $\mathcal{C}_1$ is the unique graph that attains the maximum Detour–Harary index of all graphs in $\mathcal{C}_n^k (k \ge 3)$.

Let $V(\mathcal{C}_1) = \{v_1, v_2, \cdots, v_n\}$, and it can be checked directly that

$$\sum_{i=2}^{n} \frac{1}{l(v_1, v_i|\mathcal{C}_1)} = 1 \cdot (n - 2k - 1) + \frac{1}{2} \cdot 2k = n - k - 1;$$

$$\sum_{1 \leq i \leq n, i \neq 2} \frac{1}{l(v_2, v_i|\mathcal{C}_1)} = \frac{1}{2} \cdot 2 + \frac{1}{3} \cdot (n - 2k - 1) + \frac{1}{4} \cdot (2k - 2) = \frac{1}{3}n - \frac{1}{6}k + \frac{1}{6};$$

$$\sum_{j=1}^{n-1} \frac{1}{l(v_n, v_j|\mathcal{C}_1)} = 1 + \frac{1}{2} \cdot (n - 2k - 2) + \frac{1}{3} \cdot 2k = \frac{1}{2}n - \frac{1}{3}k.$$

Then,

$$\begin{aligned} \omega H(\mathcal{C}_1) &= \frac{1}{2}[(n - k - 1) + 2k \cdot (\frac{1}{3}n - \frac{1}{6}k + \frac{1}{6}) + (n - 2k - 1) \cdot (\frac{1}{2}n - \frac{1}{3}k)] \\ &= \frac{3n^2 + 2k^2 - 4nk + 3n - 2k - 6}{12}. \end{aligned}$$

The proof is completed. □

Author Contributions: Conceptualization, W.F. and W.-H.L.; methodology, F.-Y.C.; Z.-J.X. and J.-B.L.; writing—original draft preparation, W.F. and Z.-M.H; writing—review and editing, W.-H.L.

References

1. Imran, M.; Ali, M.A.; Ahmad, S.; Siddiqui, M.K.; Baig, A.Q. Topological sharacterization of the symmetrical structure of bismuth tri-iodide. *Symmetry* **2018**, *10*, 201. [CrossRef]
2. Liu, J.; Siddiqui, M.K.; Zahid, M.A.; Naeem, M.; Baig, A.Q. Topological Properties of Crystallographic Structure of Molecules. *Symmetry* **2018**, *10*, 265. [CrossRef]
3. Shao, Z.; Siddiqui, M.K.; Muhammad, M.H. Computing zagreb indices and zagreb polynomials for symmetrical nanotubes. *Symmetry* **2018**, *10*, 244. [CrossRef]
4. Dobrynin, A.; Entringer, R.; Gutman, I. Wiener Index of Trees: Theory and Applications. *Acta Appl. Math.* **2001**, *66*, 211–249. [CrossRef]
5. Alizadeh, Y.; Andova, V.; Zar, S.K.; Skrekovski, R.V. Wiener dimension: Fundamental properties and (5,0)-nanotubical fullerenes. *MATCH Commun. Math. Comput. Chem.* **2014**, *72*, 279–294.
6. Needham, D.E.; Wei, I.C.; Seybold, P.G. Molecular modeling of the physical properties of alkanes. *J. Am. Chem. Soc.* **1988**, *110*, 4186–4194. [CrossRef]
7. Vijayabarathi, A.; Anjaneyulu, G.S.G.N. Wiener index of a graph and chemical applications. *Int. J. ChemTech Res.* **2013**, *5*, 1847–1853.
8. Gutman, I.; Cruz, R.; Rada, J. Wiener index of Eulerian graphs. *Discret. Appl. Math.* **2014**, *162*, 247–250. [CrossRef]
9. Lin, H. Extremal Wiener index of trees with given number of vertices of even degree. *MATCH Commun. Math. Comput. Chem.* **2014**, *72*, 311–320.
10. Lin, H. Note on the maximum Wiener index of trees with given number of vertices of maximum degree. *MATCH Commun. Math. Comput. Chem.* **2014**, *72*, 783–790.
11. Wiener, H. Structural determination of paraffin boiling points. *Am. Chem. Soc.* **1947**, *69*, 17–20. [CrossRef]
12. Plavšić, D.; Nikolić, S.; Trinajstić, N.; Mihalić, Z. On the Harary index for the characterization of chemical graphs. *J. Math. Chem.* **1993**, *12*, 235–250. [CrossRef]
13. Ivanciuc, O.; Balaban, T.S.; Balaban, A.T. Reciprocal distance matrix, related local vertex invariants and topological indices. *J. Math. Chem.* **1993**, *12*, 309–318. [CrossRef]
14. Lukovits, I. The Detour index. *Croat. Chem. Acta* **1996**, *69*, 873–882.
15. Trinajstić, N.; Nikolić, S.; Lučić, B.; Amić, D.; Mihalić, Z. The Detour matrix in chemistry. *J. Chem. Inf. Comput. Sci.* **1997**, *37*, 631–638. [CrossRef]

16. Chen, S. Cacti with the smallest, second smallest, and third smallest Gutman index. *J. Comb. Optim.* **2016**, *31*, 327–332. [CrossRef]
17. Chen, Z.; Dehmer, M.; Shi, Y.; Yang, H. Sharp upper bounds for the Balaban index of bicyclic graphs. *MATCH Commun. Math. Comput. Chem.* **2016**, *75*, 105–128.
18. Fang, W.; Gao, Y.; Shao, Y.; Gao, W.; Jing, G.; Li, Z. Maximum Balaban index and sum-Balaban index of bicyclic graphs. *MATCH Commun. Math. Comput. Chem.* **2016**, *75*, 129–156.
19. Fang, W.; Wang, Y.; Liu, J.-B.; Jing, G. Maximum Resistance-Harary index of cacti. *Discret. Appl. Math.* **2018**. [CrossRef]
20. Gutman, I.; Li, S.; Wei, W. Cacti with n-vertices and t-cycles having extremal Wiener index. *Discret. Appl. Math.* **2017**, *232*, 189–200. [CrossRef]
21. Ji, S.; Li, X.; Shi, Y. Extremal matching energy of bicyclic graphs. *MATCH Commun. Math. Comput. Chem.* **2013**, *70*, 697–706.
22. Liu, J.; Pan, X.; Yu, L.; Li, D. Complete characterization of bicyclic graphs with minimal Kirchhoff index. *Discret. Appl. Math.* **2016**, *200*, 95–107. [CrossRef]
23. Wang, H.; Hua, H.; Wang, D. Cacti with minimum, second-minimum, and third-minimum Kirchhoff indices. *Math. Commun.* **2010**, *15*, 347–358.
24. Wang, L.; Fan, Y.; Wang, Y. Maximum Estrada index of bicyclic graphs. *Discret. Appl. Math.* **2015**, *180*, 194–199. [CrossRef]
25. Lu, Y.; Wang, L.; Xiao, P. Complex Unit Gain Bicyclic Graphs with Rank 2,3 or 4. *Linear Algebra Appl.* **2017**, *523*, 169–186. [CrossRef]
26. Furtula, B.; Gutman, I.; Katanić, V. Three-center Harary index and its applications. *Iran. J. Math. Chem.* **2016**, *7*, 61–68.
27. Feng, L.; Lan, Y.; Liu, W.; Wang, X. Minimal Harary index of graphs with small parameters. *MATCH Commun. Math. Comput. Chem.* **2016**, *76*, 23–42.
28. Hua, H.; Ning, B. Wiener index, Harary index and hamiltonicity of graphs. *MATCH Commun. Math. Comput. Chem.* **2017**, *78*, 153–162.
29. Li, X.; Fan, Y. The connectivity and the Harary index of a graph. *Discret. Appl. Math.* **2015**, *181*, 167–173. [CrossRef]
30. Xu, K.; Das, K.C. On Harary index of graphs. *Discret. Appl. Math.* **2011**, *159*, 1631–1640. [CrossRef]
31. Xu, K. Trees with the seven smallest and eight greatest Harary indices. *Discret. Appl. Math.* **2012**, *160*, 321–331. [CrossRef]
32. Xu, K.; Das, K.C. Extremal unicyclic and bicyclic graphs with respect to Harary Index. *Bull. Malaysian Math. Sci. Soc.* **2013**, *36*, 373–383.
33. Xu, K.; Wang, J.; Das, K.C.; Klavžar, S. Weighted Harary indices of apex trees and k-apex trees. *Discret. Appl. Math.* **2015**, *189*, 30–40. [CrossRef]
34. Zhou, B.; Cai, X.; Trinajstić, N. On Harary index. *J. Math. Chem.* **2008**, *44*, 611–618. [CrossRef]
35. Fang, W.; Yu, H.; Gao, Y.; Jing, G.; Li, Z.; Li, X. Minimum Detour index of cactus graphs. *Ars Comb.* **2019**, in press.
36. Qi, X.; Zhou, B. Detour index of a class of unicyclic graphs. *Filomat* **2010**, *24*, 29–40.
37. Qi, X.; Zhou, B. Hyper-Detour index of unicyclic graphs. *MATCH Commun. Math. Comput. Chem.* **2011**, *66*, 329–342.
38. Rücker, G.; Rücker, C. Symmetry-aided computation of the Detour matrix and the Detour index. *J. Chem. Inf. Comput. Sci.* **1998**, *38*, 710–714. [CrossRef]
39. Zhou, B.; Cai, X. On Detour index. *MATCH Commun. Math. Comput. Chem.* **2010**, *44*, 199–210.
40. Qi, X. Detour index of bicyclic graphs. *Util. Math.* **2013**, *90*, 101–113.
41. Deng, H. On the Balaban index of trees. *MATCH Commun. Math. Comput. Chem.* **2011**, *66*, 253–260.

Fixed Points Results in Algebras of Split Quaternion and Octonion

Mobeen Munir [1,*], Asim Naseem [2], Akhtar Rasool [1], Muhammad Shoaib Saleem [3] and Shin Min Kang [4,5,*]

1 Division of Science and Technology, University of Education, Lahore 54000, Pakistan; akhtar.ue@gmail.com
2 Department of Mathematics, Government College University, Lahore 54000, Pakistan; asimroz@gmail.com
3 Department of Mathematics, University of Okara, Okara 56300, Pakistan; shaby45@yahoo.com
4 Department of Mathematics and RINS, Gyeongsang National University, Jinju 52828, Korea
5 Center for General Education, China Medical University, Taichung 40402, Taiwan
* Correspondence: mmunir@ue.edu.pk (M.M.); smkang@gnu.ac.kr (S.M.K.)

Abstract: Fixed points of functions have applications in game theory, mathematics, physics, economics and computer science. The purpose of this article is to compute fixed points of a general quadratic polynomial in finite algebras of split quaternion and octonion over prime fields $\mathbb{Z}_p$. Some characterizations of fixed points in terms of the coefficients of these polynomials are also given. Particularly, cardinalities of these fixed points have been determined depending upon the characteristics of the underlying field.

Keywords: fixed point; split-quaternion; quadratic polynomial; split-octonion

Subject Classification (2010): 30C35; 05C31

1. Introduction

Geometry of space-time can be understood by the choice of convenient algebra which reveals hidden properties of the physical system. These properties are best describable by the reflections of symmetries of physical signals that we receive and of the algebra using in the measurement process [1–3]. Thus, we need normed division algebras with a unit element for the better understanding of these systems. For these reasons, higher dimension algebras have been an immense source of inspiration for mathematicians and physicists as their representations pave the way towards easy understanding of universal phenomenons. These algebras present nice understandings towards general rotations and describe some easy ways to consider geometric problems in mechanics and dynamical systems [4,5].

Quaternion algebra have been playing a central role in many fields of sciences such as differential geometry, human imaging, control theory, quantum physics, theory of relativity, simulation of particle motion, 3D geo-phones, multi-spectral images, signal processing including seismic velocity analysis, seismic waveform de-convolution, statistical signal processing and probability distributions (see [6–8] and references therein). It is known that rotations of 3D-Minkowski spaces can be represented by the algebra of split quaternions [5]. Applications of these algebras can be traced in the study of Graphenes, Black holes, quantum gravity and Gauge theory. A classical application of split quaternion is given in [1] where Pavsic discussed spin gauge theory. Quantum gravity of $2+1$ dimension has been described by Carlip in [2] using split quaternions. A great deal of research is in progress where authors are focused on considering matrices of quaternions and split-quaternions [9–12]. The authors in [13] gave a fast structure-preserving method to compute singular value decomposition of quaternion matrices. Split quaternions play a vital role in geometry and physical models in four-dimensional spaces as the elements of split quaternion are used to express Lorentzian rotations [14]. Particularly, the geometric and physical applications of split quaternions require solving split quaternionic equations [15,16]. Similarly, octonion and split octonion algebras play important role in mathematical physics. In [8], authors discussed ten dimensional space-time with help of these eight dimensional algebras. In [16], authors gave comprehensive applications of split octonions in geometry. Anastasiou developed M-theory algebra with the help of octonions [3].

This article mainly covers finite algebras of split quaternion and split octonion over prime fields $\mathbb{Z}_p$. Split quaternion algebra over $\mathbb{R}$ was in fact introduced by James Cockle in 1849 on already established quaternions by Hamilton in 1843. Both of these algebras are actually associative, but non-commutative, non-division ring generated by four basic elements. Like quaternion, it also forms a four dimensional real vector space equipped with a multiplicative operation. However, unlike the quaternion algebra, the split quaternion algebra contains zero divisors, nilpotent and nontrivial idempotents. For a detailed description of quaternion and its generalization (octonions), please follow [15–18]. As mathematical structures, both are algebras over the real numbers which are isomorphic to the algebra of 2×2 real matrices. The name split quaternion is used due to the division into positive and negative terms in the modulus function. The set $(1, \hat{\imath}, \hat{\jmath}, \hat{k})$ forms a basis. The product of these elements are $\hat{\imath}^2 = -1$, $\hat{\jmath}^2 = 1 = \hat{k}^2$, $\hat{\imath}\hat{\jmath} = \hat{k} = -\hat{\jmath}\hat{\imath}$, $\hat{\jmath}\hat{k} = -\hat{\imath} = -\hat{k}\hat{\jmath}$, $\hat{k}\hat{\imath} = \hat{\jmath} = -\hat{\imath}\hat{k}$, $\hat{\imath}\hat{\jmath}\hat{k} = 1$. It follows from the defining relations that the set $(\pm 1, \pm i, \pm j, \pm k)$ is a group under split quaternion multiplication which is isomorphic to the dihedral group of a square. Following Table 1 encodes the multiplication of basis split quaternions.

Table 1. Split quaternion multiplication table.

$\cdot$	$\mathbf{1}$	$\hat{\imath}$	$\hat{\jmath}$	$\hat{k}$
1	1	$\hat{\imath}$	$\hat{\jmath}$	$\hat{k}$
$\hat{\imath}$	$\hat{\imath}$	-1	$\hat{k}$	$-\hat{\jmath}$
$\hat{\jmath}$	$\hat{\jmath}$	$-\hat{k}$	1	$-\hat{\imath}$
$\hat{k}$	$\hat{k}$	$\hat{\jmath}$	$\hat{\imath}$	1

The split octonion is an eight-dimensional algebraic structure, which is non-associative algebra over some field with basis 1, $\acute{t}_1$, $\acute{t}_2$, $\acute{t}_3$, $\acute{t}_4$, $\acute{t}_5$, $\acute{t}_6$ and $\acute{t}_7$. The subtraction and addition in split octonions is computed by subtracting and adding corresponding terms and their coefficients. Their multiplication is given in this table. The product of each term can be given by multiplication of the coefficients and a multiplication table of the unit split octonions is given following Table 2.

Table 2. Split octonions' multiplication table.

$\cdot$	$\acute{t}_1$	$\acute{t}_2$	$\acute{t}_3$	$\acute{t}_4$	$\acute{t}_5$	$\acute{t}_6$	$\acute{t}_7$
$\acute{t}_1$	-1	$\acute{t}_3$	$-\acute{t}_2$	$-\acute{t}_7$	$\acute{t}_6$	$-\acute{t}_5$	$\acute{t}_4$
$\acute{t}_2$	$-\acute{t}_3$	-1	$\acute{t}_1$	$-\acute{t}_6$	$-\acute{t}_7$	$\acute{t}_4$	$\acute{t}_5$
$\acute{t}_3$	$\acute{t}_2$	$-\acute{t}_1$	-1	$\acute{t}_5$	$-\acute{t}_4$	$-\acute{t}_7$	$\acute{t}_6$
$\acute{t}_4$	$\acute{t}_7$	$\acute{t}_6$	$-\acute{t}_5$	1	$-\acute{t}_3$	$\acute{t}_2$	$\acute{t}_1$
$\acute{t}_5$	$-\acute{t}_6$	$\acute{t}_7$	$\acute{t}_4$	$\acute{t}_3$	1	$-\acute{t}_1$	$\acute{t}_2$
$\acute{t}_6$	$\acute{t}_5$	$-\acute{t}_4$	$\acute{t}_7$	$-\acute{t}_2$	$\acute{t}_1$	1	$\acute{t}_3$
$\acute{t}_7$	$-\acute{t}_4$	$-\acute{t}_5$	$-\acute{t}_6$	$-\acute{t}_1$	$-\acute{t}_2$	$-\acute{t}_3$	1

From the table, we get very useful results:

$$\acute{t}_i^2 = -1, \forall i = 1, ..., 3,$$

$$\acute{t}_i^2 = 1, \forall i = 4, ..., 7,$$

$$\acute{t}_i \acute{t}_j = -\acute{t}_j \acute{t}_i, \forall i \neq j.$$

Brand in [19] computed the roots of a quaternion over $\mathbb{R}$. Strictly speaking, he proved mainly De Moivres theorem and then used it to find nth roots of a quaternion. His approach paved way for finding roots of a quaternion in an efficient and intelligent way. Ozdemir in [20] computed the roots of a split quaternion. In [21], authors discussed Euler's formula and De Moivres formula for quaternions. In [15], authors gave some geometrical applications of the split quaternion. It is important to mention that these two algebras can also be constructed for $\mathbb{Z}_p$ over prime finite fields of characteristic P. In this way, we obtain finite algebras with entirely different properties. Recently, the ring of quaternion over $\mathbb{Z}_p$ was studied by Michael Aristidou in [22,23], where they computed the idempotents and nilpotents in $\mathbb{H}/\mathbb{Z}_p$. In [18], authors computed the roots of a general quadratic polynomial in algebra of split quaternion over $\mathbb{R}$. They also computed fixed points of general quadratic polynomials in the same sittings. A natural question arises as to what happens with the same situations over $\mathbb{Z}_p$. Authors in [24] discussed split-quaternion over Z_p in algebraic settings.

In the present article, we first obtain the roots of a general quadratic polynomial in the algebra of split quaternion over $\mathbb{Z}_p$. Some characterizations of fixed points in terms of the coefficients of these polynomials are also given. As a consequence, we give some computations about algebraic properties of particular classes of elements in this settings. We also give examples as well as the codes that create these examples with ease. For a computer program, we refer to Appendix A at the end of the article. We hope that our results will be helpful in understanding the communication in machine language and cryptography.

Definition 1. *Let $x \in \mathbb{H}_s$, $x = a_0 + a_1\hat{i} + a_2\hat{j} + a_3\hat{k}$ where $a_i \in \mathbb{R}$. The* conjugate *of x is defined as $\bar{x} = a_0 - a_1\hat{i} - a_2\hat{j} - a_3\hat{k}$. The square of pseudo-norm of x is given by*

$$N(x) = x\bar{x} = a_0^2 + a_1^2 - a_3^2 - a_4^2. \tag{1}$$

Definition 2. *Let $x = a_0 + \sum_{i=1}^{7} a_i f_i \in \mathbb{O}_s/\mathbb{Z}_p$. The* conjugate *of x is defined as*

$$\begin{aligned} \overline{x} &= \overline{a_0 + \sum_{i=1}^{7} a_i f_i} \\ &= a_0 + \sum_{i=1}^{7} a_i \overline{f_i} \\ &= a_0 - \sum_{i=1}^{7} a_i f_i \\ &= a_0 + \sum_{i=1}^{7} a_i' f_i, \end{aligned}$$

where $a_i' = -a_i$ where $i = 1, 2, ..., 7$. The square of pseudo-norm of x is given by

$$N(x) = x\overline{x} = \sum_{i=0}^{3} a_i^2 - \sum_{i=4}^{7} a_i^2.$$

2. Main Results

In this section, we formulate our main results. At first, we give these results for split quaternions and then we move towards split octonions.

2.1. *Some Fixed Points Results of Quadratic Functions in Split Quaternions over the Prime Field*

We first solve a general quadratic polynomial in algebra of split quaternion. As a consequence, we find fixed points of associated functions in this algebra.

Theorem 1. *The quadratic equation $ax^2 + bx + c = 0$ $a, b, c \in \mathbb{Z}_p$, where p is an odd prime and $p \nmid a$, has root $x = a_0 + a_1\hat{i} + a_2\hat{j} + a_3\hat{k} \in \mathbb{H}_s/\mathbb{Z}_p$ if and only if $a_0 = \frac{p-b}{2a}$ and $a_1^2 - a_2^2 - a_3^2 = (\frac{p^2-b^2}{4a^2}) + \frac{c}{a}$.*

Proof.

$$x = a_0 + a_1\hat{i} + a_2\hat{j} + a_3\hat{k}, \tag{2}$$

$$x^2 = (a_0 + a_1\hat{i} + a_2\hat{j} + a_3\hat{k})^2, \tag{3}$$

$$\begin{aligned} &= a_0^2 - a_1^2 + a_2^2 + a_3^2 + 2a_0a_1\hat{i} + 2a_0a_2\hat{j} + 2a_0a_3\hat{k} \\ &= a_0^2 + a_0^2 - \|x\| + 2a_0a_1\hat{i} + 2a_0a_2\hat{j} + 2a_0a_3\hat{k} \\ &= 2a_0^2 - \|x\| + 2a_0a_1\hat{i} + 2a_0a_2\hat{j} + 2a_0a_3\hat{k} \\ &= 2a_0(a_0 + a_1\hat{i} + a_2\hat{j} + a_3\hat{k}) - \|x\| \\ &= 2a_0x - \|x\|. \end{aligned}$$

Putting x and x^2 into $ax^2 + bx + c = 0$, we have

$$\begin{aligned} 2aa_0x - a\|x\| + bx + c &= 0, \\ (2aa_0 + b)x - a\|x\| + c &= 0, \\ (2aa_0 + b)(a_0 + a_1\hat{i} + a_2\hat{j} + a_3\hat{k}) - a(a_0^2 + a_1^2 - a_2^2 - a_3^2) + c &= 0, \\ (2aa_0 + b)a_0 + (2aa_0 + b)(a_1\hat{i} + a_2\hat{j} + a_3\hat{k}) - a(a_0^2 + a_1^2 - a_2^2 - a_3^2) + c &= 0. \end{aligned}$$

Comparing vector terms in the above equation, we get

$$2aa_0 + b = 0, \tag{4}$$

$$a_0 = \frac{-b}{2a} = \frac{p-b}{2a}. \tag{5}$$

Comparing constant terms, we get

$$(2aa_0 + b)a_0 - a(a_0^2 + a_1^2 - a_2^2 - a_3^2) + c = 0, \tag{6}$$

$$(2aa_0 + b)a_0 - aa_0^2 + c = a(a_1^2 - a_2^2 - a_3^2), \tag{7}$$

$$aa_0^2 + ba_0 + c = a(a_1^2 - a_2^2 - a_3^2), \tag{8}$$

$$(aa_0 + b)a_0 + c = a(a_1^2 - a_2^2 - a_3^2), \tag{9}$$

$$(a(\frac{p-b}{2a}) + b)\frac{p-b}{2a} + c = a(a_1^2 - a_2^2 - a_3^2), \tag{10}$$

$$\frac{p^2-b^2}{4a^2} + \frac{c}{a} = a_1^2 - a_2^2 - a_3^2. \tag{11}$$

□

On the basis of the above results 2.1, we arrive at a new result given as

Theorem 2. *The fixed point of function $f(x) = x^2 + (b+1)x + c$ where $a, b, c \in \mathbb{Z}_p$, p is an odd prime and $p \nmid a$ is $x = a_0 + a_1\hat{i} + a_2\hat{j} + a_3\hat{k} \in \mathbb{H}_s/\mathbb{Z}_p$ if and only if $a_0 = \frac{p-b}{2a}$ and $a_1^2 - a_2^2 - a_3^2 = (\frac{p^2-b^2}{4a^2}) + \frac{c}{a}$.*

Proof. It is enough to give a new relation $f(x) = g(x) + x$, where $g(x) = x^2 + bx + c$. Then, existence of fixed points for $f(x)$ is equivalent to the solutions of $g(x)$. Then, the required result is immediate from the above theorem. □

Theorem 3. *Let p be an odd prime, $p \nmid a$, if $x = a_0 + a_1\hat{i} + a_2\hat{j} + a_3\hat{k} \in \mathbb{H}_s/\mathbb{Z}_p$ is a root of quadratic equation $x^2 + bx + c = 0$, where $a, b, c \in \mathbb{Z}_p$. Then, conjugate of x i.e., $\bar{x} = a_0 - a_1\hat{i} - a_2\hat{j} - a_3\hat{k} \in \mathbb{H}_s/\mathbb{Z}_p$ is also the root of quadratic equation $x^2 + bx + c = 0$.*

Proof. The proof follows simply by using condition of Theorem 1 applied on the conjugate of x. □

Theorem 4. *Let p be an odd prime, $p \nmid a$, if $x = a_0 + a_1\hat{i} + a_2\hat{j} + a_3\hat{k} \in \mathbb{H}_s/\mathbb{Z}_p$ be the fixed point of function $f(x) = x^2 + (b+1)x + c$, where $a, b, c \in Z_p$. Then, the conjugate of x i.e., $\bar{x} = a_0 - a_1\hat{i} - a_2\hat{j} - a_3\hat{k} \in \mathbb{H}_s/\mathbb{Z}_p$ also be the fix point of function $f(x) = x^2 + (b+1)x + c$.*

Proof. Again, it is enough to use relation $f(x) = g(x) + x$ where $g(x) = x^2 + bx + c$. Then, the existence of fixed points for $f(x)$ is equivalent to the solutions of $g(x)$. Then, the required result is immediate from the above theorem. □

The following two theorems are new results about the number of fixed points of $f(x) = x^2 + (b+1)x + c$.

Theorem 5. $|Fix(f)| = \begin{cases} p^2, & b = 0, c = 0, \\ p^2 + p + 2, & c = 0, b \neq 0. \end{cases}$

Proof. We split the proof in cases.

Case 1: For $c = 0$ and $b = 0$, we obtain two $\mathbb{H}_s/\mathbb{Z}_p \cong \mathbb{M}_2(\mathbb{Z}_p)$,where p is prime. It is easy to see that $\mathbb{H}_s/\mathbb{Z}_p$ and $\mathbb{M}_2(\mathbb{Z}_p)$ are isomorphic as algebras, the map $\varphi : \mathbb{H}_s/\mathbb{Z}_p \longmapsto \mathbb{M}_2(\mathbb{Z}_p)$ is defined as $\varphi(a_0 + a_1\hat{i} + a_2\hat{j} + a_3\hat{k}) = a_0\binom{10}{01} + a_1\binom{0p-1}{10} + a_2\binom{0p-1}{p-10} + a_3\binom{p-10}{01}$. As $\mathbb{H}_s/\mathbb{Z}_p \cong \mathbb{M}_2(\mathbb{Z}_p)$, so we find the number of nilpotent elements in $\mathbb{M}_2(\mathbb{Z}_p)$. It is well-known by Fine and Herstein that the probability that $n \times n$ matrix over a Galois field having p^{α} elements have $p^{\alpha.n}$ nilpotent elements. As in our case, $\alpha = 1$ and $n = 2$, thus the probability that the 2×2 matrix over $\mathbb{Z}_p$ has p^{-2} nilpotent elements:

$$\frac{|nil(\mathbb{M}_2(\mathbb{Z}_p))|}{|(\mathbb{M}_2(\mathbb{Z}_p|))} = p^{-2}, \tag{12}$$

$$\frac{|nil(\mathbb{M}_2(\mathbb{Z}_p))|}{p^4} = p^{-2}, \tag{13}$$

$$|nil(\mathbb{M}_2(\mathbb{Z}_p))| = p^2. \tag{14}$$

Case 2: For $c = 0$ and $b \neq 0$, we obtain as many points as there are matrices $M_2(\mathbb{Z}_p)$ because of the above isomorphism, and, using the argument given in 2, we arrive at the result. □

Theorem 6. *Let $b \neq 0$ and $c \neq 0$. Then,* $|Fix(f)| = \begin{cases} p^2 - p, & p \equiv 1(mod3), \\ p^2 + p, & p \equiv 2(mod3), \\ 3, & p = 3. \end{cases}$

Proof. Case 1: For $p = 3$, there is nothing to prove.
Case 2: For $p \equiv 1(mod3)$, we have two further cases:

case I: If $p \equiv 3(mod4)$,
$x^2 + y^2 = z$ has a unique solution for z = 0.
$x^2 + y^2 = z$ has $(p+1)$ options for $z \neq 0$, thus $(p+1)(p-1)$ options in all.
Thus, we get that $x^2 + y^2 = z$ has total number of solutions $(p+1)(p-1)+1 = p^2 - 1 + 1 = p^2$.
Now, when $z = 0$, we get no solution for a_1:

$$\begin{aligned} \sharp &= 1(p+1) + 2(\frac{p-1}{2})(p+1) && (15) \\ &= p+1+p^2-1 && (16) \\ &= p^2+p. && (17) \end{aligned}$$

case II: If $p \equiv 1(mod4)$,
$x^2 + y^2 = z$ has $(2p-1)$ solutions for z = 0.
$x^2 + y^2 = z$ has $(p-1)$ options for $z \neq 0$, thus $(p-1)(p-1)$ options in all.
Thus, we get that $x^2 + y^2 = z$ has total number of solutions
$(p-1)(p-1) + (2p-1) = p^2 - p - p + 1 + 2p - 1 = p^2$.
Now, when $z = 0$, we get two solutions for a_1:

$$\begin{aligned} \sharp &= 2(2p-1) + 2(\frac{p-3}{2})(p-1) + 1(p-1) && (18) \\ &= 4p-2+p^2-p-3p+3+p-1 && (19) \\ &= p^2+p. && (20) \end{aligned}$$

Case 3: For $p \equiv 2(mod3)$, we have two further cases:
case I: If $p \equiv 3(mod4)$
$x^2 + y^2 = z$ has a unique solution for z = 0.
$x^2 + y^2 = z$ has $(p+1)$ options for $z \neq 0$. So $(p+1)(p-1)$ options in all.
Thus we get, $x^2 + y^2 = z$ has total number of solutions $(p+1)(p-1)+1 = p^2 - 1 + 1 = p^2$
Now, when $z = 0$, we get no solution for a_1:

$$\begin{aligned} \sharp &= 1(2) + 2(\frac{p-3}{2})(p+1) + 1(p+1) && (21) \\ &= 2+p^2+p-3p-3+p+1 && (22) \\ &= p^2-p. && (23) \end{aligned}$$

case II: If $p \equiv 1(mod4)$,
$x^2 + y^2 = z$ has $(2p-1)$ solutions for z = 0.
$x^2 + y^2 = z$ has $(p-1)$ options for $z \neq 0$. So $(p-1)(p-1)$ options in all.
Thus we get, $x^2 + y^2 = z$ has total number of solutions
$(p-1)(p-1) + (2p-1) = p^2 - p - p + 1 + 2p - 1 = p^2$.
Now, when $z = 0$, we get two solutions for a_1.

$$\begin{aligned} \sharp &= 1(p-1) + 2(\frac{p-1}{2})(p-1) && (24) \\ &= p-1+p^2-p-p+1 && (25) \\ &= p^2-p. && (26) \end{aligned}$$

□

2.2. *Some Algebraic Consequences about $\mathbb{H}_s/\mathbb{Z}_p$*

We can understand the algebraic structure of $\mathbb{H}_s/\mathbb{Z}_p$ with ease. The following results are simple facts obtained from the previous section.

Corollary 1. *Let p be an odd prime, an element*

$$x = a_0 + a_1\hat{i} + a_2\hat{j} + a_3\hat{k} \in \mathbb{H}_s/\mathbb{Z}_p \tag{27}$$

is idempotent $\Leftrightarrow a_0 = \frac{p+1}{2}$ and $a_1^2 - a_2^2 - a_3^2 = \frac{p^2-1}{4}$.

Proof. Taking $a = 1$, $b = p - 1$ and $c = p$ in the above theorem, we have

$$\begin{aligned} x^2 + (p-1)x + p &= 0, \\ x^2 - x &= 0, \\ x^2 &= x \end{aligned}$$

has root

$$x = a_0 + a_1\hat{i} + a_2\hat{j} + a_3\hat{k},$$

where

$$\begin{aligned} a_0 &= \frac{p-b}{2a} \\ &= \frac{p+1}{2}, \end{aligned}$$

and

$$\begin{aligned} a_1^2 - a_2^2 - a_3^2 &= \frac{p^2 - b^2}{4a^2} + \frac{c}{a} = \frac{p^2 - (-1)^2}{4(1)^2} + \frac{0}{1} \\ &= \frac{p^2 - 1}{4}. \end{aligned}$$

In other words, we can say x is idempotent. □

We also present similar results but without proof as they can be derived similarly.

Corollary 2. *Let p be an odd prime an element and*

$$x = a_0 + a_1\hat{i} + a_2\hat{j} + a_3\hat{k} \in \mathbb{H}_s/\mathbb{Z}_p \tag{28}$$

is idempotent if and only if $a_0 = \frac{p+1}{2}$ and $\|x\| = 0$.

Corollary 3. *Let p be an odd prime and $x \in \mathbb{H}_s/\mathbb{Z}_p$. If x is an idempotent, then $\|x\| = 0$.*

Corollary 4. *Let p be an odd prime. If $x \in \mathbb{H}_s/\mathbb{Z}_p$ is idempotent, then $\bar{x}$ is also an idempotent.*

Corollary 5. *Let p be an odd prime. If $x \in \mathbb{H}_s/\mathbb{Z}_p$ and x is of the form $x = a_0$. If x is idempotent, then it is either 0 or 1.*

Corollary 6. *Let p be an odd prime and $x \in \mathbb{H}_s/\mathbb{Z}_p$ of the form*

$$x = a_0 + a_1\hat{\imath} + a_2\hat{\jmath} + a_3\hat{k}, \tag{29}$$

where at least one $a_i \neq 0$. Then, x is not an idempotent.

Corollary 7. *Let p be an odd prime, and the quadratic equation $x^2 = 0$ has root $x = a_0 + a_1\hat{\imath} + a_2\hat{\jmath} + a_3\hat{k} \in \mathbb{H}_s/\mathbb{Z}_p$, where $a_0 = \frac{p}{2}$ and $a_1^2 - a_2^2 - a_3^2 = \frac{p^2}{4}$.*

Proof. Taking $a = 1, b = 0$ and $c = 0$ in the above theorem, we have that

$$\begin{aligned} x^2 + (p)x + o &= 0, \\ x^2 &= 0 \end{aligned}$$

has root

$$x = a_0 + a_1\hat{\imath} + a_2\hat{\jmath} + a_3\hat{k},$$

where

$$\begin{aligned} a_0 &= \frac{p-b}{2a} = \frac{p-0}{2} \\ &= \frac{p}{2}, \end{aligned}$$

and

$$\begin{aligned} a_1^2 - a_2^2 - a_3^2 &= \frac{p^2 - b^2}{4a^2} + \frac{c}{a} = \frac{p^2 - (0)^2}{4(1)^2} + \frac{0}{1} \\ a_1^2 - a_2^2 - a_3^2 &= \frac{p^2}{4}. \end{aligned}$$

In other words, we can say x is nilpotent. □

2.3. *Some Fixed Points Results of Quadratic Functions in Split Octonions over the Prime Field*

Theorem 7. *The quadratic equation $ax^2 + bx + c = 0$ where $a, b, c \in \mathbb{Z}_p$, p is an odd prime and $p \nmid a$ has root $x = a_0 + \sum_{i=1}^{7} a_i t_i \in \mathbb{O}_s/\mathbb{Z}_p$ if and only if $a_0 = \frac{p-b}{2a}$ and $\sum_{i=1}^{3} a_i^2 - \sum_{i=4}^{7} a_i^2 = (\frac{p^2-b^2}{4a^2}) + \frac{c}{a}$.*

Proof.

$$ax^2 + bx + c = 0.$$

Take $x = a_0 + \sum_{i=1}^{7} a_i f_i$, we have

$$
\begin{aligned}
x^2 &= (a_0 + \sum_{i=1}^{7} a_i f_i)^2 \\
&= (a_0)^2 + (\sum_{i=1}^{7} a_i f_i)^2 + 2a_0 \sum_{i=1}^{7} a_i f_i \\
&= (a_0)^2 - \sum_{i=1}^{7} a_i^2 + 2a_0 \sum_{i=1}^{7} a_i f_i where (\sum_{i=1}^{7} a_i f_i)^2 = -\sum_{i=1}^{7} a_i^2 \\
&= (a_0)^2 + a_0^2 - ||x|| + 2a_0 \sum_{i=1}^{7} a_i f_i where ||x|| = a_0 + \sum_{i=1}^{7} a_i^2 \\
&= 2(a_0)^2 - ||x|| + 2a_0 \sum_{i=1}^{7} a_i f_i \\
&= 2(a_0)^2 + 2a_0 \sum_{i=1}^{7} a_i f_i - ||x||, \\
&= 2a_0 x - ||x||.
\end{aligned}
$$

Putting it in the above equation, we get

$$
\begin{aligned}
a(2a_0 x - ||x||) + bx + c &= 0, && (30) \\
2aa_0 x - a||x|| + bx + c &= 0, && (31) \\
(2aa_0 + b)x - a||x|| + c &= 0. && (32)
\end{aligned}
$$

Here, $x = a_0 + \sum_{i=1}^{7} a_i f_i$ and $||x|| = a_0^2 + \sum_{i=1}^{3} a_i^2 - \sum_{i=3}^{7} a_i^2$, we have

$$
\begin{aligned}
(2aa_0 + b)(a_0 + \sum_{i=1}^{7} a_i f_i) - a[a_0^2 + \sum_{i=1}^{3} a_i^2 - \sum_{i=3}^{7} a_i^2] + c &= o, \\
(2aa_0 + b)a_0 + (2aa_0 + b)\sum_{i=1}^{7} a_i f_i - aa_0^2 - a\sum_{i=1}^{3} a_i^2 + a\sum_{i=3}^{7} a_i^2 + c &= o,
\end{aligned}
$$

Comparing vector terms on both sides, we have

$$
\begin{aligned}
(2aa_0 + b)a_i &= 0, \\
2aa_0 + b &= 0, \\
a_0 &= \frac{-b}{2a}, \\
a_0 &= \frac{p - b}{2a}.
\end{aligned}
$$

Comparing constant terms on both sides, we have

$$
\begin{aligned}
(2aa_0 + b)a_0 - aa_0^2 - a\sum_{i=1}^{3} a_i^2 + a\sum_{i=4}^{7} a_i^2 + c &= o, \\
2aa_0^2 + ba_0 - aa_0^2 + c &= a\sum_{i=1}^{3} a_i^2 - a\sum_{i=4}^{7} a_i^2, \\
a_0(aa_0 + b) + c &= a[\sum_{i=1}^{3} a_i^2 - \sum_{i=4}^{7} a_i^2],
\end{aligned}
$$

where $a_0 = \frac{p-b}{2a}$.

$$\begin{aligned}
(\frac{p-b}{2a})(a(\frac{p-b}{2a})+b)+c &= a[\sum_{i=1}^{3} a_i^2 - \sum_{i=4}^{7} a_i^2], \\
(\frac{p-b}{2a})(\frac{p+b}{2})+c &= a[\sum_{i=1}^{3} a_i^2 - \sum_{i=4}^{7} a_i^2], \\
(\frac{p^2-b^2}{4a})+c &= a[\sum_{i=1}^{3} a_i^2 - \sum_{i=4}^{7} a_i^2], \\
(\frac{p^2-b^2}{4a^2})+\frac{c}{a} &= \sum_{i=1}^{3} a_i^2 - \sum_{i=4}^{7} a_i^2.
\end{aligned}$$

□

Theorem 8. *The fixed points of function $f(x) = ax^2 + (b+1)x + c$ are $x = a_0 + \sum_{i=1}^{7} a_i t_i \in \mathbb{O}_s/\mathbb{Z}_p$, where $a_0 = \frac{p-b}{2a}$ and $\sum_{i=1}^{3} a_i^2 - \sum_{i=4}^{7} a_i^2 = (\frac{p^2-b^2}{4a^2}) + \frac{c}{a}$.*

Proof. It is enough to use relation $f(x) = g(x) + x$ where $g(x) = ax^2 + bx + c$. Then, the existence of fixed points for $f(x)$ is equivalent to the solutions of $g(x)$. Then, the required result is immediate from the above theorem. □

Corollary 8. *The fixed point of function $f(x) = x^2 + x$ are $x = a_0 + \sum_{i=1}^{7} a_i t_i \in \mathbb{O}_s/\mathbb{Z}_p$ where $a_0 = \frac{p}{2}$ and $\sum_{i=1}^{3} a_i^2 - \sum_{i=4}^{7} a_i^2 = \frac{p^2}{4}$.*

Proof. It is obvious from the above theorem, only by taking $a = 1$, $b = 0$ and $c = 0$. □

Theorem 9. *Let p be an odd prime. If $x = a_0 + \sum_{i=1}^{7} a_i t_i \in \mathbb{O}_s/\mathbb{Z}_p$ is the root of the quadratic equation $ax^2 + bx + c = 0$ $a, b, c \in Z_p$, then $\overline{x} = a_0 + \sum_{i=1}^{7} \acute{a}_i t_i \in \mathbb{O}_s/\mathbb{Z}_p$ is also the root of the quadratic equation $ax^2 + bx + c = 0$ $a, b, c \in Z_p$.*

Proof.

$$\begin{aligned}
\overline{x} &= \overline{a_0 + \sum_{i=1}^{7} a_i t_i} = a_0 + \sum_{i=1}^{7} a_i \overline{t_i} = a_0 - \sum_{i=1}^{7} a_i t_i && (33) \\
&= a_0 + \sum_{i=1}^{7} \acute{a}_i t_i, && (34)
\end{aligned}$$

where $\acute{a}_i = -a_i$ where $i = 1, 2, ..., 7$ as

$$a_0 = \frac{p-b}{2a}$$

and

$$\begin{aligned}
\sum_{i=1}^{3} \acute{a}_i^2 - \sum_{i=4}^{7} \acute{a}_i^2 &= \sum_{i=1}^{3} (-a_i)^2 - \sum_{i=4}^{7} (-a_i)^2 && (35) \\
&= \sum_{i=1}^{3} (a_i)^2 - \sum_{i=4}^{7} (a_i)^2 && (36) \\
&= \frac{p^2-b^2}{4a^2} + \frac{c}{a}. && (37)
\end{aligned}$$

It implies that $\overline{x}$ is the root of the quadratic equation $ax^2 + bx + c = 0$ $a, b, c \in Z_p$. □

Theorem 10. *If the function $f(x) = ax^2 + (b+1)x + c$ has fixed point $x = a_0 + \sum_{i=1}^{7} a_i f_i \in \mathbb{O}_s/\mathbb{Z}_p$, then $\overline{x} = a_0 + \sum_{i=1}^{7} \acute{a}_i f_i \in \mathbb{O}_s/\mathbb{Z}_p$ also is the fixed point of function $f(x) = ax^2 + (b+1)x + c$.*

Proof. It is enough to use relation $f(x) = g(x) + x$, where $g(x) = ax^2 + bx + c$. Then, the existence of fixed points for $f(x)$ is equivalent to the solutions of $g(x)$. Then, the required result is immediate from the above theorem. □

3. Some Algebraic Consequences about $\mathbb{O}_s/\mathbb{Z}_p$

Proposition 1. *Let p be an odd prime and an element*

$$x = a_0 + \sum_{i=1}^{7} a_i f_i \in \mathbb{O}_s/\mathbb{Z}_p$$

is idempotent $\Leftrightarrow a_0 = \frac{p+1}{2}$ *and*

$$\sum_{i=1}^{3} a_i^2 - \sum_{i=4}^{7} a_i^2 = \frac{p^2-1}{4}.$$

Proof. Taking $a = 1$, $b = p - 1$ and $c = p$ in the above theorem, we have

$$x^2 + (p-1)x + p = 0, \tag{38}$$

$$x^2 - x = 0, \tag{39}$$

$$x^2 = x \tag{40}$$

has root

$$x = a_0 + \sum_{i=1}^{7} a_i f_i, \tag{41}$$

where

$$a_0 = \frac{p-b}{2a} = \frac{p-p+1}{2} \tag{42}$$

$$= \frac{1}{2}, \tag{43}$$

and

$$\sum_{i=1}^{3} a_i^2 - \sum_{i=4}^{7} a_i^2 = \frac{p^2-b^2}{4a^2} + \frac{c}{a} = \frac{p^2-(-1)^2}{4(1)^2} + \frac{0}{1} \tag{44}$$

$$= \frac{p^2-1}{4}. \tag{45}$$

In other words, we can say that x is idempotent. □

Proposition 2. *Let p be an odd prime and element*

$$x = a_0 + \sum_{i=1}^{7} a_i f_i \in \mathbb{O}_s/\mathbb{Z}_p \tag{46}$$

is idempotent if and only if $a_0 = \frac{p+1}{2}$ *and* $\|x\| = 0$.

Proposition 3. *Let p be an odd prime and $x \in \mathbb{O}_s/\mathbb{Z}_p$. If x is an idempotent, then $\|x\| = 0$.*

Proposition 4. *Let p be an odd prime. If $x \in \mathbb{O}_s/\mathbb{Z}_p$ is idempotent, then $\bar{x}$ is also an idempotent.*

Proposition 5. *Let p be an odd prime. If $x = a_0 \in \mathbb{O}_s/\mathbb{Z}_p$ is idempotent, then it is either 0 or 1.*

Proposition 6. *Let p be an odd prime and $x \in \mathbb{O}_s/\mathbb{Z}_p$ be of the form*

$$x = \sum_{i=1}^{7} a_i f_i, \tag{47}$$

where at least one $a_i \neq 0$. Then, x is not an idempotent.

Proposition 7. *Let p be an odd prime and the quadratic equation $x^2 = 0$ has root $x = a_0 + \sum_{i=1}^{7} a_i f_i \in \mathbb{O}_s/\mathbb{Z}_p$, where $a_0 = \frac{p}{2}$ and $\sum_{i=1}^{3} a_i^2 - \sum_{i=4}^{7} a_i^2 = \frac{p^2}{4}$.*

Proof. Taking $a = 1, b = 0$ and $c = 0$ in the above theorem, we have

$$x^2 + (p)x + o = 0, \tag{48}$$
$$x^2 = 0 \tag{49}$$

has root

$$x = a_0 + \sum_{i=1}^{7} a_i f_i, \tag{50}$$

where

$$a_0 = \frac{p-b}{2a} = \frac{p-0}{2} \tag{51}$$
$$= \frac{p}{2} \tag{52}$$

and

$$\sum_{i=1}^{3} a_i^2 - \sum_{i=4}^{7} a_i^2 = \frac{p^2 - b^2}{4a^2} + \frac{c}{a} = \frac{p^2 - (0)^2}{4(1)^2} + \frac{0}{1} \tag{53}$$
$$= \frac{p^2}{4}. \tag{54}$$

In other words, we can say that x is nilpotent. □

Using results of the previous section and programs mentioned in the Appendix A, we can give many examples.

4. Examples

In this section, we add examples relating to the previous section. These results are generated by the codes given in Appendix A. These along with other examples can be created using codes, and results can be applied to crypto systems and communication channel systems.

Example 1. *We find all solutions of* $x^2 - x = 0$ *over* $\mathbb{H}_s/\mathbb{Z}_7$*. As above, we see that, if* $x \in \mathbb{H}_s/\mathbb{Z}_7$*, then* $a_0 = \frac{7-(-1)}{2} = 4$ *and following are the values of* a_1*,* a_2 *and* a_3*, respectively, satisfying the equation* $-a_1^2 + a_2^2 + a_3^2 = -5$ *or* $a_1^2 - a_2^2 - a_3^2 = 5$*.*

$$\begin{array}{cccccccc}
(0,1,1) & (0,1,6) & (0,3,0) & (0,4,0) & (0,1,1) & (0,6,1) & (0,6,6) & (1,1,3) \\
(1,1,4) & (1,3,1) & (1,3,6) & (1,4,1) & (1,4,6) & (1,6,4) & (2,2,3) & (2,2,4) \\
(2,3,2) & (2,3,3) & (2,3,5) & (2,4,2) & (2,4,3) & (2,4,5) & (2,5,3) & (2,5,4) \\
(3,0,2) & (3,0,5) & (3,2,0) & (3,3,3) & (3,3,4) & (3,4,3) & (3,4,4) & (3,5,0) \\
(4,0,2) & (4,0,5) & (4,2,0) & (4,3,3) & (4,3,4) & (4,4,3) & (4,4,4) & (4,5,0) \\
(5,0,6) & (5,2,3) & (5,2,4) & (5,3,2) & (5,3,5) & (5,4,2) & (5,4,5) & (5,5,3) \\
(5,5,4) & (5,6,0) & (6,1,4) & (6,3,6) & (6,4,1) & (6,4,6) & (6,6,3) & (6,6,4)
\end{array}.$$

Example 2. *We compute all solutions of* $2x^2 + x = 0$ *over* $\mathbb{H}_s/\mathbb{Z}_5$*. As above, we see that, if* $x \in \mathbb{H}_s/\mathbb{Z}_5$*, then* $a_0 = 1$ *and following are the values of* a_1*,* a_2 *and* a_3*, respectively, satisfying the equation* $-a_1^2 + a_2^2 + a_3^2 = -4$ *or* $a_1^2 - a_2^2 - a_3^2 = 4$*.*

$$\begin{array}{ccccc}
(0,0,1) & (1,4,1) & (2,2,4) & (3,1,2) & (3,4,2) \\
(0,0,4) & (1,4,4) & (2,3,1) & (3,1,3) & (3,4,3) \\
(0,1,0) & (2,0,0) & (2,3,4) & (3,2,1) & (4,1,1) \\
(0,4,0) & (2,1,2) & (2,4,2) & (3,2,4) & (4,1,4) \\
(1,1,1) & (2,1,3) & (2,4,3) & (3,3,1) & (4,4,1) \\
(1,1,4) & (2,2,1) & (3,0,0) & (3,3,4) & (4,4,4)
\end{array}.$$

Example 3. *We compute all solutions of* $x^2 + x + 1 = 0$ *over* $\mathbb{H}_s/\mathbb{Z}_7$*. As above, we see that, if* $x \in \mathbb{H}_s/\mathbb{Z}_7$*, then* $a_0 = 3$ *and following are the values of* a_1*,* a_2 *and* a_3*, respectively satisfying the equation* $-a_1^2 + a_2^2 + a_3^2 = -6$ *or* $a_1^2 - a_2^2 - a_3^2 = 6$*.*

$$\begin{array}{ccccccc}
(0,0,1) & (1,0,3) & (6,0,3) & (2,1,2) & (5,1,2) & (3,1,3) & (4,1,3) \\
(0,0,6) & (1,0,4) & (6,0,4) & (2,1,5) & (5,1,5) & (3,1,4) & (4,1,4) \\
(0,1,0) & (1,1,6) & (6,1,6) & (2,2,1) & (5,2,1) & (3,3,1) & (4,3,1) \\
(0,2,2) & (1,3,0) & (6,3,0) & (2,2,6) & (5,2,6) & (3,3,6) & (4,3,6) \\
(0,2,5) & (1,4,0) & (6,4,0) & (2,5,1) & (5,5,1) & (3,4,1) & (4,4,1) \\
(0,5,2) & (1,6,1) & (6,6,1) & (2,5,6) & (5,5,6) & (3,4,6) & (4,4,6) \\
(0,5,5) & (1,1,1) & (6,1,1) & (2,6,2) & (5,6,2) & (3,6,3) & (4,6,2) \\
(0,6,0) & (1,6,6) & (6,6,6) & (2,6,5) & (5,6,5) & (3,6,4) & (4,6,5)
\end{array}.$$

Example 4. *We compute all solutions of* $x^2 = 0$ *over* $\mathbb{H}_s/\mathbb{Z}_5$*. As above, we see that, if* $x \in \mathbb{H}_s/\mathbb{Z}_5$*, then* $a_0 = 0$ *and following are the values of* a_1*,* a_2 *and* a_3*, respectively, satisfying the equation* $-a_1^2 + a_2^2 + a_3^2 = 0$ *or* $a_1^2 - a_2^2 - a_3^2 = 0$*.*

$$\begin{array}{ccccc}
(0,0,0) & (0,3,4) & (1,4,0) & (2,0,3) & (3,3,0) \\
(0,1,2) & (0,4,2) & (4,0,1) & (2,2,0) & (0,3,1) \\
(0,1,3) & (0,4,3) & (4,0,4) & (2,3,0) & (1,1,0) \\
(0,2,1) & (1,0,1) & (4,1,0) & (3,0,2) & (2,0,2) \\
(0,2,4) & (1,0,4) & (4,4,0) & (3,0,3) & (3,2,0)
\end{array}.$$

Example 5. *We compute all solutions of* $x^2 - x = 0$ *over* $\mathbb{O}_s/\mathbb{Z}_3$ *(idempotents in the split octonion algebra). As above we see that* $x = a_0 + \sum_{i=1}^{7} a_i f_i \in \mathbb{O}_s/\mathbb{Z}_3$ *where* $a_0 = \frac{3-(-1)}{2} = 2$ *and following is the values of* a_1*,* a_2*,* a_3*,* a_4*,* a_5*,* a_6 *and* a_7 *respectively satisfying the equation* $\sum_{i=1}^{3} a_i^2 - \sum_{i=4}^{7} a_i^2 = (\frac{p^2-b^2}{4a^2}) + \frac{c}{a} = 2$ *.* *We do so by putting values for* $p = 3$*,* $a = 1$*,* $b = -1$*,* $c = 0$ *in above given code.*

$(2,1,0,1,1,0,2); (2,1,0,1,1,1,0); (2,1,0,1,1,2,0); (2,1,0,1,2,0,1);$
$(2,1,0,1,2,0,2); (2,1,0,1,2,1,0); (2,1,0,1,2,2,0); (2,1,0,2,0,1,1);$
$(2,1,0,2,0,1,2); (2,1,0,2,0,2,1); (2,1,0,2,0,2,2); (2,1,0,2,1,0,1);$
$(2,1,0,2,1,0,2); (2,1,0,2,1,1,0); (2,1,0,2,1,2,0); (2,1,0,2,2,0,1);$
$(2,1,0,2,2,0,2); (2,1,0,2,2,1,0); (2,1,0,2,2,2,0); (2,1,1,0,0,0,1);$
$(2,1,1,0,0,0,2); (2,1,1,0,0,1,0); (2,1,1,0,0,2,0); (2,1,1,0,1,0,0);$
$(2,1,1,0,2,0,0); (2,1,1,1,0,0,0); (2,1,1,1,1,1,1); (2,1,1,1,1,1,2);$
$(2,1,1,1,1,2,1); (2,1,1,1,1,1,1); (2,1,1,1,1,1,2); (2,1,1,1,1,2,1);$
$(2,1,1,1,1,2,2); (2,1,1,1,2,1,1); (2,1,1,1,2,1,2); (2,1,1,1,2,2,1);$
$(2,1,1,1,2,2,2); (2,1,1,2,0,0,0); (2,1,1,2,1,1,1); (2,1,1,2,1,1,2);$
$(2,1,1,2,1,2,1); (2,1,1,2,1,2,2); (2,1,1,2,2,1,1); (2,1,1,2,2,1,2);$
$(2,1,1,2,2,2,1); (2,1,1,2,2,2,2); (2,1,2,0,0,0,1); (2,1,2,0,0,0,2);$
$(2,1,2,0,0,1,0); (2,1,2,0,0,2,0); (2,1,2,0,1,0,0); (2,1,2,0,2,0,0);$
$(2,1,2,1,0,0,0); (2,1,2,1,1,1,1); (2,1,2,1,1,1,2); (2,1,2,1,1,2,1);$
$(2,1,2,1,1,2,2); (2,1,2,1,2,1,1); (2,1,2,1,2,1,2); (2,1,2,1,2,2,1);$
$(2,1,2,1,2,2,2); (2,1,2,2,0,0,0); (2,1,2,2,1,1,1); (2,1,2,2,1,1,2);$
$(2,1,2,2,1,2,1); (2,1,2,2,1,2,2); (2,1,2,2,2,1,1); (2,1,2,2,2,1,2);$
$(2,1,2,2,2,2,1); (2,1,2,2,2,2,2); (2,2,0,0,0,0,0); (2,2,0,0,1,1,1);$
$(2,2,0,0,1,1,2); (2,2,0,0,1,2,1); (2,2,0,0,1,2,2); (2,2,0,0,1,2,2);$
$(2,2,0,0,2,1,1); (2,2,0,0,2,1,2); (2,2,0,0,2,2,1); (2,2,0,0,2,2,2);$
$(2,2,0,1,0,1,1); (2,2,0,1,0,1,2); (2,2,0,1,0,2,1); (2,2,0,1,0,2,2);$
$(2,2,0,1,1,0,1); (2,2,0,1,1,0,2); (2,2,0,1,1,1,0); (2,2,0,1,1,2,0);$
$(2,2,0,1,2,0,1); (2,2,0,1,2,0,2); (2,2,0,1,2,1,0); (2,2,0,1,2,2,0);$
$(2,2,0,2,0,1,1); (2,2,0,2,0,1,1); (2,2,0,2,0,1,2); (2,2,0,2,0,2,1);$
$(2,2,0,2,0,2,2); (2,2,0,2,1,0,1); (2,2,0,2,1,0,2); (2,2,0,2,1,1,0);$
$(2,2,0,2,1,2,0); (2,2,0,2,2,0,1); (2,2,0,2,2,0,2); (2,2,0,2,2,1,0);$
$(2,2,0,2,2,2,0); (2,2,1,0,0,0,1); (2,2,1,0,0,0,2); (2,2,1,0,0,1,0);$
$(2,2,1,0,0,2,0); (2,2,1,0,1,0,0); (2,2,1,0,2,0,0); (2,2,1,1,0,0,0);$
$(2,2,1,1,1,1,1); (2,2,1,1,1,1,1); (2,2,1,1,1,1,2); (2,2,1,1,1,2,1);$
$(2,2,1,1,1,2,2); (2,2,1,1,2,1,1); (2,2,1,1,2,1,2); (2,2,1,1,2,2,1);$
$(2,2,1,1,2,2,2); (2,2,1,2,0,0,0); (2,2,1,2,1,1,1); (2,2,1,2,1,1,2);$
$(2,2,1,2,1,2,1); (2,2,1,2,1,2,2); (2,2,1,2,2,1,1); (2,2,1,2,2,1,2);$
$(2,2,1,2,2,2,1); (2,2,1,2,2,2,2); (2,2,2,0,0,0,1); (2,2,2,0,0,0,2);$
$(2,2,2,0,0,1,0); (2,2,2,0,0,2,0); (2,2,2,0,1,0,0); (2,2,2,0,2,0,0);$
$(2,2,2,1,0,0,0); (2,2,2,1,1,1,1); (2,2,2,1,1,1,2); (2,2,2,1,1,2,1);$
$(2,2,2,1,1,2,2); (2,2,2,1,2,1,2); (2,2,2,1,2,2,1); (2,2,2,1,2,2,2);$
$(2,2,2,2,0,0,0); (2,2,2,2,1,1,1); (2,2,2,2,1,1,2); (2,2,2,2,1,2,1);$
$(2,2,2,2,1,2,2); (2,2,2,2,2,1,1); (2,2,2,2,2,1,2); (2,2,2,2,2,2,1);$
$(2,2,2,2,2,2,2).$

5. Conclusions and Further Directions

In this article, we produced some general results about fixed points of a general quadratic polynomial in algebras of split quaternion and octonion over $\mathbb{Z}_p$. We not only characterized these points in terms of the coefficients of these polynomials but also gave the cardinality of these points and also the programs that produced fixed points. We arrived at the following new results for a general quadratic function.

Theorem 11. $|Fix(f)| = \begin{cases} p^2, & b=0, c=0, \\ p^2+p+2, & c=0, b\neq 0. \end{cases}$

Theorem 12. *Let $b \neq 0$ and $c \neq 0$. Then,* $|Fix(f)| = \begin{cases} p^2-p, & p\equiv 1(mod3); \\ p^2+p, & p\equiv 2(mod3); \\ 3, & p=3. \end{cases}$

We also give the following two new results for the fixed points of a general quadratic quaternionic equation without proofs. Proofs are left as an open problem.

Theorem 13. $|Fix(f)| = \begin{cases} p^6, & b=0, c=0; \\ p^6+p^3, & c=0, b\neq 0. \end{cases}$

Theorem 14. *Let $b \neq 0$ and $c \neq 0$. Then,* $|Fix(f)| = \begin{cases} p^6+p^3, & p\equiv 1(mod3); \\ p^6-p^3, & p\equiv 2(mod3); \\ p^6, & p=3. \end{cases}$

We like to remark that new results can be obtained for a general cubic polynomials in these algebras.

Author Contributions: M.M. conceived the idea and drafted the manuscript. Computations have been done by A.R., A.N. and M.S.S., S.M.K. edited and made final corrections. All authors read and approved the final manuscript.

Acknowledgments: The authors are thankful to the reviewers for their valuable comments and suggestions. The authors are thankful to the University of Education Lahore for providing us a platform to present this article at ICE 2018.

Appendix A. Computer Codes

Here, we put together some programs to compute fixed points and roots easily.

Appendix A.1. Program for Finding Solutions of the Quadratic Equation in $\mathbb{H}_s/\mathbb{Z}_p$

Following codes, count and print the number of solutions of quadratic equation $ax^2+bx+c=0$ in $\mathbb{H}_s/\mathbb{Z}_p$. These codes print the string a_1, a_2, a_3 with the understanding that the co-efficient $a_0 = \frac{p-b}{2a}$ is fixed in $\mathbb{H}_s/\mathbb{Z}_p$ and satisfying the relation $a_1^2 - a_2^2 - a_3^2 = (\frac{p^2-b^2}{4a^2}) + \frac{c}{a}$ or $-a_1^2+a_2^2+a_3^2 = (\frac{-p^2+b^2}{4a^2}) - \frac{c}{a}$ for $\mathbb{H}_s/\mathbb{Z}_p$.

CODE: This code will give solutions of the quadratic equation only by putting values for p, a, b, c, where p is an odd prime and $a, b, c\varepsilon\mathbb{Z}_p$.

```
#include<iostream>
```

```
#include<conio.h>
using namespace std;

main()
{
int a1, a2, a3, p, n, a, b, c, count;
count=0;
cout<<"Enter value for p: ";
cin>>p;
cout<<"Enter value for a: ";
cin>>a;
cout<<"Enter value for b: ";
cin>>b;
cout<<"Enter value for c: ";
cin>>c;
n=((p*p-b*b)/(4*a*a))+(c/a);
while(n<0)
n=n+p;
for(int i=0; i<p; i++)
{
a3=i;
for(int j=0; j<p; j++)
{
a2=j;
for(int k=0; k<p; k++)
{
a1=k;
int sum=(a1*a1)-(a2*a2)-(a3*a3);
while(sum<0)
sum=sum+p;
if(sum%p==n)
{
count++;
cout<<a1<<" "<<a2<<" "<<a3<<endl;
}
}
}
}
cout<<"\nCount: "<<count;
getch();
}
```

Appendix A.2. Program for Finding Roots of the Quadratic Equation in $\mathbb{O}_s/\mathbb{Z}_p$

Following codes, count and print the number of solutions of quadratic equation $ax^2 + bx + c = 0$ in $\mathbb{O}_s/\mathbb{Z}_p$. These codes print the string $a_1, a_2, a_3, a_4, a_5, a_6, a_7$ with the understanding that the co-efficient $a_0 = \frac{p-b}{2a}$ is fixed in $\mathbb{O}_s/\mathbb{Z}_p$ and satisfying the relation $\sum_{i=1}^{3} a_i^2 - \sum_{i=4}^{7} a_i^2 = (\frac{p^2-b^2}{4a^2}) + \frac{c}{a}$ for $\mathbb{O}_s/\mathbb{Z}_p$.

CODE: This code will give solutions of the quadratic equation only by putting values for p, a, b, c, where p is an odd prime and $a, b, c \in \mathbb{Z}_p$.

```
#include <iostream>
```

```
#include <fstream>
using namespace std;
int main(){
  int a1,a2,a3,a4,a5,a6,a7;
  int sum=0;
  int p;
  int n=0;
  int count=2;
  int totalCount=0;
  cout<<"Enter value of p: ";
  cin>>p;
  n = ((p*p -1)/4)%p;
  for(int i=0;i<p;i++)
  {
   a1 = i;
   for(int j=0; j<p; j++)
   {
    a2 = j;
    for(int k=0; k<p; k++)
    {
     a3 = k;
      for(int l=0; l<p;l++)
      {
       a4 = l;
       for(int m=0; m<p; m++)
       {
        a5 = m;
        for(int q=0; q<p; q++)
        {
         a6 = q;
         for(int r=0; r<p;r++)
         {
          a7 = r;
          totalCount++;
          cout<<a1<<" "<<a2<<" "<<a3<<" "<<a4<<" "<<a5
              <<" "<<a6<<" "<<a7<<endl;
          //dataFile << a1 << endl;
          sum = a1*a1+a2*a2+a3*a3-a4*a4-a5*a5-a6*a6-a7*a7;
          if(sum%p == n)
           count++;
          }
         }
        }
       }
      }
     }
    }
    cout<<"Total Count is: "<<totalCount<<endl;
    cout<<"Count is: "<<count<<endl;
    system("pause");
```

```
return 0;
}
```

References

1. Pavsic, M. Spin gauge theory of gravity in clifford space: A realization of kaluza-klein theory in four-dimensional spacetime. *Int. J. Mod. Phys. A* **2006**, *21*, 5905–5956. [CrossRef]
2. Carlip, S. *Quantum Gravity in* $2+1$ *Dimensions*; Cambridge University Press: Cambridge, UK, 2003.
3. Anastasiou, A.; Borsten, L.; Duff, M.J. An octonionic formulation of the M-theory algebra. *J. High Energy Phys.* **2014**, *2014*, 22. [CrossRef]
4. Kula, L.; Yayli, Y. Split Quaternions and Rotations in Semi Euclidean Space. *J. Korean Math. Soc.* **2007**, *44*, 1313–1327. [CrossRef]
5. Ozdemir, M.; Ergin, A. Rotations with Timelike Quaternions in Minkowski 3-Space. *J. Geom. Phys.* **2006**, *56*, 322–336. [CrossRef]
6. Dzhunushaliev, V. Cosmological constant, supersymmetry, nonassociativity, and big numbers. *Eur. Phys. J. C* **2015**, *75*, 86. [CrossRef]
7. Dzhunushaliev, V. A non-associative quantum mechanics. *Found. Phys. Lett.* **2006**, *19*, 157–167. [CrossRef]
8. Pavsic, M. Kaluza-klein theory without extra dimensions, curved Clifford space. *Phys. Lett. B* **2005**, *614*, 85–95. [CrossRef]
9. Alagz, Y.; Oral, K.; Yce, S. Split Quaternion Matrices. *Misk. Mathem. Not.* **2012**, *13*, 223–232.
10. Erdogdu, M.; Ozedemir, M. On Eigenvalues of Split Quaternion Matrics. *Adv. Appl. Clifford Algebras* **2013**, *23*, 615–623. [CrossRef]
11. Zhang, F. Quaternions and matrices of quaternions. *Linear Algebra Appl.* **1997**, *251*, 21–57. [CrossRef]
12. Lin, Y.; Wang, Q. Completing a 2×2 bock matrix of real quaternions with a partial specified inverse. *J. Appl. Math.* **2013**, *2013*, 271–278. [CrossRef]
13. Li, Y.; Wei, M.; Zhang, F.; Zhao, J. A fast structure-preserving method for computing the singular value decomposition of quaternion matrices. *Appl. Math. Comput.* **2014**, *235*, 157–167. [CrossRef]
14. Chung, K.; Sudbery, A. Octonions and the Lorentz and conformal groups of ten-dimensional space-time. *Phys. Lett. B* **1987**, *198*, 161–164. [CrossRef]
15. Gogberashvili, M.; Sakhelashvili, O. Geometrical Applications of Split Octonions. *Adv. Math. Phys.* **2015**, *2015*, 196708. [CrossRef]
16. Conway, J.H.; Smith, D.A. On Quaternions and Octonions: Their Geometry, Arithematic, and Symmetry. *Bull. Am. Math. Soc.* **2005**, *42*, 229–243.
17. John, C.B. The octonions. *Bull. Am. Math. Soc.* **2002**, *39*, 145–205.
18. Kwun, Y.C.; Munir, M.; Nazeer, W.; Kang, S. Some Fixed points Results of Quadratic Functions in Split Quaternions. *J. Funct. Spaces* **2106**, *2016*, 3460257. [CrossRef]
19. Brand, L. The roots of a quaternion. *Am. Math. Mon.* **1942**, *49*, 519–520. [CrossRef]
20. Ozdemir, M. The Roots of a Split Quaternion. *Appl. Math. Lett.* **2009**, *22*, 258–263. [CrossRef]
21. Cho, E. Euler's formula and De Moivre's formula for quaternions. *Mo. J. Math. Sci.* **1999**, *11*, 80–83.
22. Aristiou, M.; Demetre, A. Idempotent Elements in Quaternion Rings over $\mathbb{Z}_p$. *Int. J. Algebraic* **2012**, *6*, 249–254.
23. Miguel, C.J.; Serodio, R. On the Structure of Quaternion Rings over $\mathbb{Z}_p$. *Int. J. Algebra* **2011**, *5*, 1313–1325.
24. Munir, M.; Nizami, A.; Nazeer, W.; Kang, S. On Elements of Split Quaternions over Z_p. *Glob. J. Pure Appl. Math.* **2016**, *12*, 4253–4271.

Permissions

All chapters in this book were first published in MDPI; hereby published with permission under the Creative Commons Attribution License or equivalent. Every chapter published in this book has been scrutinized by our experts. Their significance has been extensively debated. The topics covered herein carry significant findings which will fuel the growth of the discipline. They may even be implemented as practical applications or may be referred to as a beginning point for another development.

The contributors of this book come from diverse backgrounds, making this book a truly international effort. This book will bring forth new frontiers with its revolutionizing research information and detailed analysis of the nascent developments around the world.

We would like to thank all the contributing authors for lending their expertise to make the book truly unique. They have played a crucial role in the development of this book. Without their invaluable contributions this book wouldn't have been possible. They have made vital efforts to compile up to date information on the varied aspects of this subject to make this book a valuable addition to the collection of many professionals and students.

This book was conceptualized with the vision of imparting up-to-date information and advanced data in this field. To ensure the same, a matchless editorial board was set up. Every individual on the board went through rigorous rounds of assessment to prove their worth. After which they invested a large part of their time researching and compiling the most relevant data for our readers.

The editorial board has been involved in producing this book since its inception. They have spent rigorous hours researching and exploring the diverse topics which have resulted in the successful publishing of this book. They have passed on their knowledge of decades through this book. To expedite this challenging task, the publisher supported the team at every step. A small team of assistant editors was also appointed to further simplify the editing procedure and attain best results for the readers.

Apart from the editorial board, the designing team has also invested a significant amount of their time in understanding the subject and creating the most relevant covers. They scrutinized every image to scout for the most suitable representation of the subject and create an appropriate cover for the book.

The publishing team has been an ardent support to the editorial, designing and production team. Their endless efforts to recruit the best for this project, has resulted in the accomplishment of this book. They are a veteran in the field of academics and their pool of knowledge is as vast as their experience in printing. Their expertise and guidance has proved useful at every step. Their uncompromising quality standards have made this book an exceptional effort. Their encouragement from time to time has been an inspiration for everyone.

The publisher and the editorial board hope that this book will prove to be a valuable piece of knowledge for researchers, students, practitioners and scholars across the globe.

List of Contributors

Nafaa Chbili, Shamma Al Dhaheri and Amna A. E. Abunamous
Department of Mathematical Sciences, United Arab Emirates University, Al Ain 15551, UAE

Mei Y. Tahnon
College of Engineering, United Arab Emirates University, Al Ain 15551, UAE

Tyler Reese, Randy Paffenroth and Joseph D. Fehribach
Department of Mathematical Sciences, Worcester Polytechnic Institute, 100 Institute Road, Worcester, MA 01609-2280, USA

Yilun Shang
Department of Computer and Information Sciences, Faculty of Engineering and Environment, Northumbria University, Newcastle NE1 8ST, UK

Salama Nagy Daoud
Department of Mathematics, Faculty of Science, Taibah University, Al-Madinah 41411, Saudi Arabia
Department of Mathematics and Computer Science, Faculty of Science, Menoufia University, Shebin El Kom 32511, Egypt

Jia-Bao Liu
School of Mathematics and Physics, Anhui Jianzhu University, Hefei 230601, China

Jingqian Wang
College of Electrical & Information Engineering, Shaanxi University of Science & Technology, Xi'an 710021, China

Abdollah Alhevaz and Maryam Baghipur
Faculty of Mathematical Sciences, Shahrood University of Technology, Shahrood, Iran

Hilal Ahmad Ganie
Department of Mathematics, University of Kashmir, Srinagar 190006, India

Aykut Emniyet
Department of Mathematics, Osmaniye Korkut Ata University, 8000 Osmaniye, Turkey

Memet Şahin
Department of Mathematics, Gaziantep University, 27310 Gaziantep, Turkey

Lifeng Li
School of Science, Xi'an University of Posts and Telecommunications, Xi'an 710121, China
Shaanxi Key Laboratory of Network Data Analysis and Intelligent Processing, Xi'an University of Posts and Telecommunications, Xi'an 710121, China

Qinjun Luo
School of Statistics, Xi'an University of Finance and Economics, Xi'an 710100, China

Ahmed A. Elsonbaty
Department of Mathematics, Faculty of Science, Taibah University, Al-Madinah 41411, Saudi Arabia
Department of Mathematics, Faculty of Science, Ain Shams University, Cairo 11566, Egypt

Jihyun Choi and Jae-Hyouk Lee
Department of Mathematics, Ewha Womans University 52, Ewhayeodae-gil, Seodaemun-gu, Seoul 03760, Korea

Xiaoying Wu
Department of Mathematics, Shaanxi University of Science & Technology, Xi'an 710021, China

Xiaohong Zhang
Department of Mathematics, Shaanxi University of Science & Technology, Xi'an 710021, China
Department of Mathematics, Shanghai Maritime University, Shanghai 201306, China

Wei Fang
College of Information & Network Engineering, Anhui Science and Technology University, Fengyang 233100, China

Wei-Hua Liu
College of Information and Management Science, Henan Agricultural University, Zhengzhou 450002, China

Fu-Yuan Chen
Institute of Statistics and Applied Mathematics, Anhui University of Finance and Economics, Bengbu 233030, China

Zhen-Mu Hong and Zheng-Jiang Xia
School of Finance, Anhui University of Finance and Economics, Bengbu 233030, China

Mobeen Munir and Akhtar Rasool
Division of Science and Technology, University of Education, Lahore 54000, Pakistan

Asim Naseem
Department of Mathematics, Government College University, Lahore 54000, Pakistan

Muhammad Shoaib Saleem
Department of Mathematics, University of Okara, Okara 56300, Pakistan

Shin Min Kang
Department of Mathematics and RINS, Gyeongsang National University, Jinju 52828, Korea
Center for General Education, China Medical University, Taichung 40402, Taiwan

Index

www.ingramcontent.com/pod-product-compliance
Lightning Source LLC
LaVergne TN
LVHW060538210926
848485LV00005B/1220